OXFORD ENGINEERING S

GENERAL EDITC

J. M. BRADY   C. E. BRENNEN   W. ᴋ. ᴌᴀᴛᴏᴄᴋ TAYLOR
M. Y. HUSSAINI   T. V. JONES   J. VAN BLADEL

THE OXFORD ENGINEERING SCIENCE SERIES

Convective Boiling and Condensation

THIRD EDITION

JOHN G. COLLIER F.R.S. F.ENG

AND

JOHN R. THOME D.PHIL

CLARENDON PRESS · OXFORD

OXFORD

UNIVERSITY PRESS

Great Clarendon Street, Oxford OX2 6DP

Oxford University Press is a department of the University of Oxford.
It furthers the University's objective of excellence in research, scholarship,
and education by publishing worldwide in

Oxford New York

Athens Auckland Bangkok Bogotá Buenos Aires Calcutta
Cape Town Chennai Dar es Salaam Delhi Florence Hong Kong Istanbul
Karachi Kuala Lumpur Madrid Melbourne Mexico City Mumbai
Nairobi Paris São Paulo Singapore Taipei Tokyo Toronto Warsaw

with associated companies in Berlin Ibadan

Oxford is a registered trade mark of Oxford University Press
in the UK and in certain other countries

Published in the United States
by Oxford University Press Inc., New York

© John G. Collier and John R. Thome, 1994

The moral rights of the author have been asserted
Database right Oxford University Press (maker)

1st edition (McGraw-Hill) 1972
2nd edition (McGraw-Hill) 1982
3rd edition (Oxford University Press) 1994
3rd edition in paperback 1996
Reprinted 1999

A catalogue record for this book is available from the British Library

Library of Congress Cataloging in Publication Data
Collier, John G. (John Gordon), 1935–
Convective boiling and condensation / John G. Collier and John R. Thome—3rd ed.
(Oxford engineering science series, 38)
Includes bibliographical references and index.
1. Heat—Transmission. 2. Ebullition. 3. Condensation. 4. Fluid
dynamics. I. Thome, John R. II. Title. III. Series.
TP363.C643 1994 660'.28427—dc20 93-33482
ISBN 0 19 856296 9

Printed in Great Britain by Bookcraft Ltd, Midsomer Norton, Avon

In memory of John G. Collier

John Gordon Collier passed away on 18 November 1995.

He was presently Chairman of Nuclear Electric plc, England's state-owned nuclear power utility. Previously he held the posts of Chairman of the UKAEA (United Kingdom Atomic Energy Authority), Director-General of the Generating and Construction Development Division of the Central Electricity Generating Board, Head of the Atomic Energy Technical Unit at Harwell, Head of the Chemical Engineering Division, also at Harwell, and previously responsible positions with Atomic Energy of Canada Ltd and Atomic Power Constructions Ltd.

John Collier was born on 22 January 1935. He received a 1st Class Honors Degree BSc (Chemical Engineering) from University College, London in 1956. After graduating, he began his work on reactor heat transfer systems at AERE, Harwell, notably for steam and gas-cooled heavy water moderated reactors. From there, John joined the reactor heat transfer team designing the CANDU reactor system in Canada, but later returned to Harwell to lead the research on use of liquid metals as reactor coolants.

John Collier was one of the world's leading authorities on two-phase flow and boiling, as evidenced by his authorative book *Convective boiling and condensation* first published in 1972, that has been used widely as a textbook and reference in leading universities and nuclear research laboratories. In fact, he played a key role in establishing two-phase flow and heat transfer as a scientific discipline and in the development of the nuclear power industry in the United Kingdom.

John Collier was a Fellow of the Royal Society, a Fellow of the Royal Academy of Engineering, a Fellow of the Institutions of Chemical, Mechanical and Nuclear Engineering, and a Fellow of the Institute of Energy. He held an honorary Doctorate of Science from Cranfield Institute of Technology and an honorary Doctorate of Engineering from Bristol University. John was a Calvin Rice Lecturer and an honorary lifetime member of the American Society of Mechanical Engineers.

As co-author of the 3rd edition of *Convective boiling and condensation*, I have collaborated with John over the past few years and have had the pleasure of knowing him since I was at Oxford University as a student. On behalf of the international heat transfer community, I wish to extend this tribute to John G. Collier for his dedication to two-phase flow and heat transfer engineering and for his leadership in the safe use of nuclear power.

John R. Thome

To Ellen and to Carla

PREFACE TO THIRD EDITION

It is now over 20 years since the first edition of this book appeared. The original author (J.G.C.) is no longer active in research and development in the field of thermo-hydraulics and the preparation of this third edition has therefore been a collaborative effort. Much of the extensive new material included relates to the particular interests of the second author (J.R.T.) in petrochemical and refrigeration equipment design and operation.

Having stood the test of time the overall structure of the book has been left unchanged and the significant new developments in the field of two-phase flow and boiling have been set into the earlier framework. Again there has been fairly extensive revision and correction of the text. In Chapter 1 the unified flow pattern model of Taitel (1990) has been included whilst the equations of two-phase flow in Chapter 2 have been updated and generalized. The opportunity has also been taken to include in Chapter 3 the most recent extensions to the annular flow models which allow for simultaneous departures from hydrodynamic and thermodynamic equilibrium. Extensive new information relating to the design of industrial evaporators, reboilers, and condensers is described in Chapters 4 *et seq.* This includes the latest heat transfer and critical heat flux correlations for in-tube boiling of pure liquids and mixtures and, in addition, boiling and condensation outside tube bundles.

Finally, the problems set at the end of each chapter have been retained and expanded. Copies of the worked solutions to these problems are available via the authors.

J.G.C.
J.R.T.
1994

PREFACE TO FIRST EDITION

The purpose in writing this book is to present the current state of understanding in the fields of convective boiling and condensation. Convective boiling (or condensation) is the term used to describe the process of boiling (or condensation) in the presence of a forced flow and the text is mainly devoted to the problems of estimating heat transfer rates and pressure losses in conduits where boiling or condensation is occurring.

The text is aimed primarily at assisting the design engineer in the power plant and process industries. He is faced with the solution of difficult technical problems involving the boiling and condensation of various fluids and mixtures and often has to design equipment on the basis of very limited information. The literature on two-phase flow is voluminous and much of it—that in the form of laboratory reports and that published in the USSR—is relatively inaccessible. The writer has for some time felt the need to set down in a logical manner the basic principles, so that future research work may readily be put into the correct perspective. The fields of boiling and condensation are at present more of an art than a science and the subject remains, and will remain for some time to come, largely empirical. The writer is, therefore, under no delusions: some of the information presented will be out-dated before long. It is to be hoped that later editions may provide the opportunity to revise such material.

Much of the material in this book was first published in a series of course notes, produced for Summer School Courses on Two-Phase Flow and Boiling Heat Transfer given at the Thayer School of Engineering, Dartmouth College, New Hampshire (1965, 1966, 1968) and at Stanford University, California (1967). These courses were given jointly by Professor G. B. Wallis and the writer. Professor Wallis has already published much of his material in his book on One-Dimensional Two-Phase Flow. The material in the original notes has been rearranged and considerably revised.

After the introduction, which includes a section on definition of terms and a description of the flow patterns encountered in convective boiling and condensation, the next two chapters are devoted to a development of the basic models of the fluid flow and to the empirical methods used to estimate pressure loss and void fraction in pipes and other components such as bends, valves, orifices, etc. Chapter 4 introduces the subject of boiling heat transfer using the novel concept of a 'boiling map' which enables the various mechanisms of convective boiling to be clearly delineated. This concept is carried through all the following chapters on boiling. The topic of pool boiling has deliberately been restricted to a brief mention in Chapter 4. Those readers familiar with the field will appreciate the problems of adequately covering this topic which, despite intensive research work, has defied a general understanding because of the inherent obstacle of characterizing the nucleation properties of surfaces. Chapters 5 and 6 cover the heat transfer and fluid dynamic aspects respectively of subcooled boiling and the complications introduced by the departure from thermodynamic equilibrium. The prediction of heat transfer rates in saturated convective boiling is dealt with in Chapter 7. The most important boundary on the boiling map covering the sharp deterioration of heat transfer coefficient in convective boiling—the so-called 'critical heat flux' condition—is discussed in detail in Chapters 8 and 9. The subject of condensation is covered in Chapter 10. Chapter 11 discusses various methods which have been proposed to improve the performance of boiling and condensing heat transfer surfaces. The later parts of this chapter are devoted to the fouling of heat transfer surfaces under boiling conditions and the influence this has on the thermo-hydraulic performance of the unit. The subject of two-component two-phase heat and mass transfer is reviewed in Chapter 12 although this subject is not strictly related to boiling and condensation.

The units used throughout the book are SI (Système International d'Unités) with the exception that the bar is used as the unit of pressure (1 bar = 10^5 N/m^2). British Engineering Units are given in brackets. A number of examples have been included in the text to demonstrate the application of the various recommended methods. Useful tables of fluid thermodynamic and transport properties are included in an Appendix at the back of the book.

J.G.C.
1971

PREFACE TO SECOND EDITION

In almost a decade since the appearance of the first edition of this book, our knowledge of the subjects of two-phase flow, boiling and condensation has continued to increase.

Significant progress has been made in the formulation and solution of the equations of two-phase flow, in the modelling of annular flow and in the understanding of the flow and heat transfer properties of horizontal two-phase flows. Extensive research has been carried out in the region where the critical heat flux has been exceeded and the existence of a *transition boiling* region in convective as well as pool boiling has been established. Significant developments in instrumentation for studying two-phase flows, particularly associated with transient conditions, have been made over the period. Finally, there have been important improvements in our understanding of binary and multi-component boiling and condensation.

In preparing the second edition I have decided to retain the structure adopted for the first edition. There has, however, been a fairly extensive revision of the text. A number of corrections have been made but to a large extent these revisions have taken the form of bringing up to date the factual material in the light of advances made in our knowledge since 1972.

Chapter 12, which previously dealt with two-component two-phase heat and mass transfer has been completely rewritten and now covers both pool and convective boiling of binary and multi-component mixtures. Finally, since this text has been used to teach postgraduate courses on Two-Phase Flow and Heat Transfer the problems set out at the end of each chapter have been retained. Copies of the worked solutions to these problems are available from the author at Harwell, Oxfordshire, United Kingdom.

J.G.C.
1980

CONTENTS

exceeded. Boiling on tube bundles. References and further reading. Example. Problems.

NOTATION

Symbol	Description	SI units	British eng. units	Introduced in Chapter
A	heat transfer area of a tube bundle	m^2	ft^2	9
A	flow area	m^2	ft^2	1
A	total wetted surface area of finned tube	m^2	ft^2	11
A	parameter in eq. (8.15)	W/m^2	$Btu/h\ ft^2$	8
A	amplitude of vibration	mm	N/A†	11
A	dimensionless group—eq. (10.67)	—	—	10
A_c	flow area at vena contracta	m^2	ft^2	3
A_f	surface area on fins	m^2	ft^2	11
A_f	flow area occupied by liquid phase	m^2	ft^2	1
A_{fc}	flow area occupied by phase at vena contracta	m^2	ft^2	3
A_{fo}	flow area occupied by liquid phase at orifice	m^2	ft^2	3
A_g	flow area occupied by gaseous phase	m^2	ft^2	1
A_{gc}	flow area occupied by gaseous phase at vena contracta	m^2	ft^2	3
A_{go}	flow area occupied by gaseous phase at orifice	m^2	ft^2	3
A_i	internal tube surface area	m^2	ft^2	11
A_{min}	minimum cross flow area between adjacent tubes— eq. (7.95)	m^2	ft^2	7
A_o	external tube surface area	m^2	ft^2	11
A_o	flow area at orifice	m^2	ft^2	3
A_p	cross-sectional area of a fin	m^2	ft^2	11
A_r	base area between fins of low finned tube	m^2	ft^2	11

† Refers to 'not applicable' or 'not appropriate'—equation in dimensional form.

(*Continued*)

Symbol	Description	SI units	British eng. units	Introduced in Chapter
A_w	log mean tube area	m^2	ft^2	11
A'	parameter used in eq. (8.22) and the Bowring CHF correlation	W/m^2	$Btu/h\ ft^2$	8
A^1	parameter used in Macbeth CHF correlation in eq. (9.20)	N/A	in.	9
A_1–A_3	parameters in eq. (2.76)	—	—	2
A^*	constant in Mostinski correlation in eq. (4.49)	—	—	4
a	parameter defined by eq. (12.76)	—	—	12
a	velocity ratio given by eq. (10.23)	—	—	10
a	parameter given by eq. (10.42)	—	—	10
a'	empirical factors in eqs. (6.50), (6.51)	—	—	6
a_1–a_5	coefficients given by eq. (10.97)	—	—	10
B	parameter in eq. (8.15)	$kg/m^2\ s$	$lb/h\ ft^2$	8
B	parameter defined by eq. (3.86)	—	—	3
B	bubble growth parameter	$m/s^{1/2}$	$ft/s^{1/2}$	4
B	parameter used in eq. (4.63)	K^{-1}	$°R^{-1}$	4
B	parameter defined by eq. (5.27)	$m\ K$	$ft\ °R$	5
B	dimensionless group—eq. (10.67)	—	—	10
B	constant in eq. (12.36)	—	—	12
B_0	scaling factor $(=1.0)$	—	—	12
b	ratio of height of superheated layer to bubble height	—	—	4
b	bubble growth parameter eq. (4.31)	—	—	4
C	constant used in eq. (10.112)	—	—	10
C	unspecified constant in eq. (10.151)	—	—	10

Symbol	Description	SI units	British eng. units	Introduced in Chapter
C	homogeneous mean droplet concentration in vapour core	kg/m^3	lb/ft^3	3
C	parameter used in Chisholm correlation (eq. 2.71)	—	—	2
C	coefficient in eq. (6.7)	—	—	6
C	constant used in eq. (7.91)	—	—	7
C	constant used in Katto correlation	—	—	8
$\bar{C}$	parameter defined by eq. (2.74)	—	—	2
C_A	parameter defined by eq. (3.11)	—	—	3
C_D	coefficient of discharge for orifice	—	—	3
C_E	equilibrium homogeneous droplet concentration in vapour core	kg/m^3	lb/ft^3	3
C_B, C_F, C_S	coefficients in eq. (6.6)	—	—	6
C_c	contraction coefficient (A_c/A_2)	—	—	3
C_f	contraction coefficient for liquid phase ⎫*	—	—	3
C_g	contraction coefficient for gas phase ⎭	—	—	3
C_g, C_f	concentrations of salt in steam and water phases respectively	kg/m^3	lb/ft^3	11
C_o	flow distribution parameter	—	—	3
C_{sf}	constant in Rohsenow correlation	—	—	4
C'	coefficient in eq. (6.7)	—	—	6
C'	parameter used in eq. (8.22)	m^{-1}	ft^{-1}	8
C^+	dimensionless concentration	—	—	12
C^1	parameters used in Macbeth CHF correlation in eq. (9.21)	N/A	in.	9
C_1	constant in eq. (10.46)	—	—	10

* corresponding to two-phase Δp

(*Continued*)

Symbol	Description	SI units	British eng. units	Introduced in Chapter
C_2	parameter used in Chisholm correlation (eq. 2.71)	—	—	2
c	empirical constant in eqs. (7.83) + (7.91)	—	—	7
$\tilde{c}$	molar concentration	kmol/m^3	lbmol/ft^3	10
c_b	condensate flooded fraction of finned tube circumference	—	—	11
c_p	specific heat	J/kg K	Btu/lb°F	4
$\tilde{c}_p$	molar specific heat	J/kmol K	Btu/lbmol°F	12
c_{pf}	specific heat of liquid phase	J/kg K	Btu/lb°F	4
c_{pg}	specific heat of gaseous phase	J/kg K	Btu/lb°F	4
D	pipe diameter characteristic dimension	m	ft	1
D	molar diffusion coefficient	m^2/s	ft^2/s	10
D	mass of magnetite deposited per unit area of tube	kg/m^2	N/A	11
D_B	tube bundle diameter	m	ft	9
D_R	diameter of rod	N/A	in.	9
D_d	bubble departure diameter	m	ft	4
D_d	rate of deposition of droplets	kg/m^{2}s	lb/ft^{2}s	3
D_e	hydraulic equivalent diameter	m	ft	1, 6
D_f	hydraulic diameter of liquid flow	m	ft	2
D_g	hydraulic diameter of gas flow	m	ft	2
D_h	heated equivalent diameter	m	ft	6
D_i	internal diameter of annulus or tube	m	ft	1, 9
D_o	external diameter of annulus or tube	m	ft	1, 9
D_o	standard tube ref. diameter (Steiner–Taborek)	m	ft	7
D_0	fin tip dia. of low finned tube	m	ft	11

Symbol	Description	SI units	British eng. units	Introduced in Chapter
D_r	root dia. of finned tube	m	ft	11
d	orifice diameter	m	ft	3
d	any diameter within annular liquid film	m	ft	3
d	wire diameter	m	ft	11
d	coil diameter	m	ft	3
d	droplet diameter	m	ft	7
d_c	dispersed bubble diameter given by eq. (1.22)	m	ft	1
d_{CR}	droplet diameter at dryout point	m	ft	7
d_{CRIT}	critical bubble diameter given by eq. (1.24)/(1.25)	m	ft	1
E	flow weighted mixture internal energy per unit mass	J/kg (N m/kg)	Btu/lb (ft pdl/lb)	2
E	rate of entrainment of droplets	kg/m² s	lb/ft² s	3
E	convection enhancement factor (eq. 7.36)	—	—	7
E_2	stratified flow correction factor (eq. 7.38)	—	—	7
e	fin height	m	ft	11
e	fraction of liquid entrained	—	—	3
F	parameter defined by eq. (1.19)	—	—	1
F	force due to pressure variation over upstream face of orifice (eq. 3.64)	N	pdl	3
F	time averaged percentage area of surface covered by bubbles	—	—	5
F	factor appearing in Chen correlation in eq. (7.19)	—	—	7
F	factor defined by eq. (9.7)	—	—	9
F	factor in eq. (10.68)	—	—	10
F	factor in eq. (10.102)	—	—	10
F	frequency of vibration	Hz	Hz	11

(Continued)

Symbol	Description	SI units	British eng. units	Introduced in Chapter
F	constant used in eq. (7.34)	—	—	7
$\bar{F}$	net force exerted by homogeneous fluid in overcoming friction	N	pdl	2
F_E	entrance factor given by eq. (8.45)	—	—	8
$F(M)$	residual correction factor eq. (7.53)	—	—	7
F_{NcB}	nucleate boiling correction factor given by eq. (7.49)	—	—	7
F_{PF}	pressure correction factor eq. (4.52)/(4.54)	—	—	4
F_{TP}	two-phase convection multiplier eq. (7.47)/(7.48)	—	—	7
F_b	bundle boiling factor eq. (7.84)	—	—	7
F_c	mixture boiling correction factor eq. (7.84)	—	—	7
F_f	force exerted by liquid phase in overcoming friction	N	pdl	2
F_g	force exerted by vapour phase in overcoming friction	N	pdl	2
F_k	force exerted by phase k in overcoming friction	N	pdl	2
F_{tb}	critical heat flux bundle factor	—	—	9
F_x	boiling number ratio	—	—	8
F^1	factor in eq. (10.103)	—	—	10
F_1, F_2, F_3, F_4	parameters in Bowring CHF correlation in eq. (8.24) and Table 8.3	—	—	8
f	friction parameter in eq. (3.64)	—	—	3
f	friction factor	—	—	3
f	frequency of bubble departure	Hz	Hz	4
f_{TP}	two-phase friction factor	—	—	2

Symbol	Description	SI units	British eng. units	Introduced in Chapter
f_{SCB}	friction factor in subcooled boiling	—	—	6
f_f	friction factor based on liquid alone flow	—	—	2
f_{fo}	friction factor based on total flow assumed liquid	—	—	2
f_g	friction factor based on gas alone flow	—	—	2
f_{go}	friction factor based on total flow assumed as gas	—	—	2
f_i	interfacial friction factor	—	—	3
f_i^1	apparent interfacial friction factor	—	—	10
$f(p)$	function of pressure in Biasi CHF correlation eq. (8.26)	N/A	N/A	8
f_s	friction factor for straight pipe (eq. 3.93)	—	—	2
$f(\ )$ fn$(\ )$	function of$(\ )$	N/A	N/A	
G	mass velocity	kg/m^2 s	lb/ft^2 s	1
G_c	mass velocity at vena contracta	kg/m^2 s	lb/ft^2 s	3
G_d	vaporized mass diffusion fraction	—	—	12
G_e	equivalent mass velocity defined in eq. (10.113)	kg/m^2 s	lb/ft^2 s	10
G_f	mass velocity of liquid phase alone	kg/m^2 s	lb/ft^2 s	1
G_{fE}	entrained liquid mass velocity	kg/m^2 s	lb/ft^2 s	3
G_{fF}	liquid film mass velocity	kg/m^2 s	lb/ft^2 s	3
G_g	mass velocity of gaseous phase alone	kg/m^2 s	lb/ft^2 s	1
$\Delta G(r)$	free energy of formation of nucleus, radius r	N m	pdl ft	4
G^*	reference mass velocity in Chisholm correlation	kg/m^2 s	lb/ft^2 s	2
g	acceleration due to gravity	m/s^2	ft/s^2	1
g	Gibbs free energy	J/kmol	Btu/lb mol	12

(Continued)

Symbol	Description	SI units	British eng. units	Introduced in Chapter
H	height of ribbon	m	ft	4
H'	dimensionless characteristic height defined by eq. (4.60)	—	—	4
h	heat transfer coefficient	W/m² °C	Btu/h ft² °F	1
h	Planck's constant	kg m²/s	lb ft²/s	4
h_D	dropwise condensation coefficient	W/m² °C	Btu/h ft² °F	10
h_I	'ideal' heat transfer coefficient defined by eq. (12.34)	W/m² °C	Btu/h ft² °F	12
h_{NcB}	nucleate boiling heat transfer coefficient	W/m² °C	Btu/h ft² °F	7
h_{SCB}	heat transfer coefficient in subcooled boiling	W/m² °C	Btu/h ft² °F	6
h_{TP}	two-phase heat transfer coefficient	W/m² °C	Btu/h ft² °F	7
h_b	heat transfer coefficient on flooded portion of finned tube	W/m² °C	Btu/h ft² °F	11
h_c	convective heat transfer coefficient	W/m² °C	Btu/h ft² °F	4
h_f	heat transfer coefficient on fins	W/m² °C	Btu/h ft² °F	11
h_f	level of liquid in stratified flow	m	ft	1
h_f	local condensing heat transfer coefficient	W/m² °C	Btu/h ft² °F	10
h_f	heat transfer coefficient for liquid alone flow	W/m² °C	Btu/h ft² °F	7
h_f	heat transfer coefficient for liquid film	W/m² °C	Btu/h ft² °F	7
h_{fN}	heat transfer coefficient for the N th tube row	W/m² °C	Btu/h ft² °F	10
$(h_f)_{Nu}$	heat transfer coefficient from Nusselt theory	W/m² °C	Btu/h ft² °F	10
h_{fo}	heat transfer coefficient for total flow assumed liquid	W/m² °C	Btu/h ft² °F	5
h_{f1}	heat transfer coefficient for the first tube row	W/m² °C	Btu/h ft² °F	10
$\bar{h}_f^*$	dimensionless heat transfer coefficient in eq. (10.87)	—	—	10

Symbol	Description	SI units	British eng. units	Introduced in Chapter
h_g	sensible heat transfer coefficient at interface	W/m² °C	Btu/h ft² °F	10
h'_g	modified sensible heat transfer coefficient at interface in eq. (10.42)	W/m² °C	Btu/h ft² °F	10
h_{go}	heat transfer coefficient for total flow assumed vapour	W/m² °C	Btu/h ft² °F	7
h_h	heat transfer coefficient on base area between fins	W/m² °C	Btu/h ft² °F	11
h_i	interfacial heat transfer coefficient	W/m² °C	Btu/h ft² °F	10
h_o	reference heat transfer coefficient used by Gorenflo	W/m² °C	Btu/h ft² °F	4
h_o, h_i	outside and inside film heat transfer coefficients respectively	W/m² °C	Btu/h ft² °F	11
$h(p)$	function in Biasi CHF correlation eq. (8.27)	N/A	N/A	8
h_r	radiation heat transfer coefficient	W/m² °C	Btu/h ft² °F	4
$\bar{i}$	molar enthalpy	J/kmol	Btu/lb mol	12
i	enthalpy of fluid	J/kg	Btu/lb	1
i'_B	fluid enthalpy at point B	J/kg	Btu/lb	6
i_f	enthalpy of saturated liquid	J/kg	Btu/lb	1
i_{fg}	latent heat of vaporization	J/kg	Btu/lb	1
i'_{fg}	modified latent heat of vaporization in eq. (4.66) or eq. (10.66)	J/kg	Btu/lb	4, 10
i_g	enthalpy of saturated vapour	J/kg	Btu/lb	1
i_k	enthalpy of phase k per unit mass	J/kg	Btu/lb	2
$(\Delta i_{SUB})_i$	enthalpy inlet subcooling	J/kg	Btu/lb	8
J	mechanical equivalent of heat	1	2.5×10^4 ft pdl/Btu	1
J	drift flux of bulk fluid towards interface	m/s	ft/s	10
J	diffusive molar flux	kmol/m² s	lb mol/ft² s	10

(Continued)

Symbol	Description	SI units	British eng. units	Introduced in Chapter
J_a	molar flux of non-condensable gas	kmol/m² s	lb mol/ft² s	10
J_g	molar flux of vapour	kmol/m² s	lb mol/ft² s	10
j	volumetric flux (superficial velocity)	m/s	ft/s	1
$\|j\|$	molecular flux	kg/m² s	lb/ft² s	10
j_D	dropwise condensation mass flux	kg/m² s	lb/ft² s	10
j_F	filmwise condensation mass flux	kg/m² s	lb/ft² s	10
j_f	superficial velocity of liquid phase	m/s	ft/s	1
j_f^*	dimensionless liquid volumetric flux	—	—	1
j_g	superficial velocity of gas phase	m/s	ft/s	1
j_g^*	dimensionless vapour volumetric flux	—	—	1
j_{gf}, j_{fg}	drift flux	m/s	ft/s	3
j_{gMAX}	limiting volumetric vapour flux at surface eq. (4.56)	m/s	ft/s	4
K	proportionality constant eq. (11.21)	—	—	11
K	rate of energy dissipation in bubbly flow eq. (1.22)	J/kg s	ft pdl/lb s	1
K	constant in eq. (4.58)	—	—	4
K	Dean Number $[\mathrm{Re}(D/d)^{0.5}]$	—	—	3
K	spreading coefficient in eq. (9.16)	m	ft	9
K	constant in eq. (9.12) see Table 9.3	—	—	9
K	equilibrium constant	—	—	12
K	parameter used in eq. (8.29)	—	—	8
K	parameter defined by eq. (1.21)	—	—	1
K	constant in equation for bubble rise velocity	—	—	1
K, k	mass transfer coefficient	m/s	ft/s	10
K_D	partition coefficient eq. (11.20)	—	—	11

Symbol	Description	SI units	British eng. units	Introduced in Chapter
K_f	Blasius constant in eq. (2.60)	—	—	2
K_g	self diffusion coefficient in vapour	m^2/s	ft^2/s	7
K_p	pressure correction factor eq. (7.88)	—	—	7
K_1	constant given by eq. (7.80)	N/A	N/A	7
K_1–K_3	parameters in Katto correlation	—	—	8
k	exponent used in eq. (8.14)	—	—	8
k	Boltzmann constant	$kg\ m^2/s^2\ K$	$lb\ ft^2/s^2\ °R$	4
k	droplet mass transfer coefficient	m/s	ft/s	3
k_{TP}	effective two-phase thermal conductivity	$W/m\ °C$	$Btu/h\ ft\ °F$	7
k_w	thermal conductivity of the wall	$W/m\ °C$	$Btu/h\ ft\ °F$	11
k_{eff}	effective conductivity of deposit	$W/m\ °C$	$Btu/h\ ft\ °F$	11
k_f	thermal conductivity of liquid phase	$W/m\ °C$	$Btu/h\ ft\ °F$	4
k_{fo}	loss coefficient in Chisholm correlation see eq. (3.92)	—	—	3
k_g	thermal conductivity of gas phase	$W/m\ °C$	$Btu/h\ ft\ °F$	4
k_p	thermal conductivity of particulate phase	$W/m\ °C$	$Btu/h\ ft\ °F$	11
L	length	m	ft	4
L	length of tube bundle	m	ft	9
L	length of channel (or equivalent length of channel)	m	ft	2
L_b	length of vapour bubble	m	ft	3
L_f	fin characteristic length	m	ft	11
L'	dimensionless characteristic dimension (eq. 4.60)	—	—	4
M	parameter used in eq. (8.18)	W/m^2	$Btu/h\ ft^2$	8
M	molecular weight	$kg/kmol$	$lb/lb\ mol$	4

(*Continued*)

Symbol	Description	SI units	British eng. units	Introduced in Chapter
M	parameter in eq. (6.31)	—	—	6
m	index	—	—	3
m	mass of one molecule	kg	lb	4
m	index in eq. (4.44)	—	—	4
m	empirical exponent eq. (7.83)/(7.91)	—	—	7
N	number of molecules per unit volume	m^{-3}	ft^{-3}	4
N	number of new bubbles appearing on surface	$m^{-2} s^{-1}$	$ft^{-2} s^{-1}$	5
N	dimensionless parameter eq. (7.31)	—	—	7
N	number of tubes in crossflow eq. (7.100)	—	—	7
N	number of tube rows in flow direction	—	—	7
N	number of droplets passing through unit area per unit time	$m^{-2} s^{-1}$	$ft^{-2} s^{-1}$	7
N	parameter used in eq. (8.18)	W/m^2	$Btu/h\,ft^2$	8
$N(r)$	number of nuclei of radius r per unit volume	m^{-3}	ft^{-3}	4
n	tube number from top tube row	—	—	10
n	Blasius index used in eq. (2.60)	—	—	2
n	number of wires	—	—	11
n	index	—	—	3
n	index in eq. $pv^n = $ constant	—	—	3
n	index used in eq. (4.40)	—	—	4
n	index used in eq. (4.45)	—	—	4
n	figure reflecting distortion of temperature profile in boundary layer by bubble nucleus	—	—	4, 5
n	empirical exponent in eq. (7.83)/(7.91)	—	—	7
n	exponent used in eq. (8.41)	—	—	8

Symbol	Description	SI units	British eng. units	Introduced in Chapter
n	exponent used in eq. (10.112)	—	—	10
nf	heat flux correction term eq. (4.53)/(4.55)	—	—	4
P	wetted perimeter	m	ft	2
P'	dimensionless characteristic perimeter defined by eq. (4.60)	—	—	4
P_f	fin pitch along axis of tube	m	ft	11
P_h	heated perimeter	m	ft	6
P_r	reduced pressure p/p_{CRIT} eq. (3.18)	—	—	3
P_{ro}	standard reduced pressure	—	—	4
p	static pressure	N/m^2(bar)	pdl/ft^2(psia)	1
p_{CRIT}	critical pressure	N/m^2(bar)	pdl/ft^2(psia)	3
p_{SAT}	saturation pressure	N/m^2(bar)	pdl/ft^2(psia)	1
p_a	partial pressure of non-condensable gas	N/m^2	pdl/ft^2	4
p_{ai}	partial pressure of non-condensable gas at interface	N/m^2	pdl/ft^2	10
p_{am}	log mean partial pressure of non-condensable gas	N/m^2	pdl/ft^2	10
p_{ao}	partial pressure of non-condensable gas in bulk stream	N/m^2	pdl/ft^2	10
p_f	pressure in liquid space	N/m^2	pdl/ft^2	4
p_g	pressure in vapour space	N/m^2	pdl/ft^2	4
p_{gi}	partial pressure of vapour at interface	N/m^2	pdl/ft^2	10
p_∞	vapour pressure over planar surface	N/m^2	pdl/ft^2	4
Q	volumetric rate of flow	m^3/s	ft^3/s	1
Q_f	volumetric rate of flow of liquid phase	m^3/s	ft^3/s	1
Q_g	volumetric rate of flow of gas phase	m^3/s	ft^3/s	1
Q_{wf}	heat transferred to fluid across channel wall	W/m	Btu/h ft	2

(Continued)

Symbol	Description	SI units	British eng. units	Introduced in Chapter
q	heat absorbed from surroundings	J/kg	Btu/lb	2
R	radius of curvature of bend	m	ft	1
R	universal gas constant	$\dfrac{Nm}{kmol\ K}$	$\dfrac{pdl\ ft}{lbmol\ °R}$	4
R	radius of bubble cavity	m	ft	4
R	radius of cylinder or sphere	m	ft	4
$\dot{R}$	velocity of bubble cavity wall	m/s	ft/s	4
$\ddot{R}$	acceleration of bubble cavity wall	m/s^2	ft/s^2	4
R_p	surface roughness	μm	N/A	4
R_{po}	standard surface roughness	μm	N/A	4
R^+	dimensionless bubble radius in eq. (4.28)	—	—	4
R'	dimensionless characteristic radius defined by eq. (4.60)	—	—	4
R^1	density ratio in Katto correlation	—	—	8
R_0	initial radius of bubble cavity	m	ft	4
$\dot{r}$	velocity of spherical liquid element	m/s	ft/s	4
r	dimension of a cluster of molecules	m	ft	4
r	ratio of upstream to downstream pressure eq. (3.85)	—	—	3
r	radial distance from axis of pipe	m	ft	1
r	pipe or rod radius	m	ft	3
r	bubble or droplet radius	m	ft	6
r_B	bubble radius at detachment	m	ft	6
r_c	cavity mouth radius	m	ft	4
r_1	acceleration multiplier for homogeneous flow	—	—	2

Symbol	Description	SI units	British eng. units	Introduced in Chapter
r_2	acceleration multiplier for separated flow	—	—	2
r_1, r_2	radii of curvature of interface	m	ft	4
r^*	radius of spherical vapour or liquid nucleus	m	ft	4
S	boiling suppression factor used in Gungor–Winterton correlation eq. (7.37)	— —	— —	7 7
S	bundle boiling suppression factor eq. (7.90)	—	—	7
S	spacing between bubbles	m	ft	6
S	suppression factor used in Chen correlation	—	—	7
S	spacing between heated sides of rectangular duct	m	in	9
S	parameter used in Barnett cluster CHF correlation	N/A	—	9
S	breadth of rectangular duct	m	ft	1
S_R	shear force ratio in eq. (3.69)	—	—	3
S_m	length of profile of one side of a fin	m	ft	11
S_1, S_2	force exerted by phase k on other phases at interface	N	pdl	2
S_2	stratified flow correction factor eq. (7.39)	—	—	7
s	length along profile of fin	m	ft	11
s_1, s_2	transverse and longitudinal tube pitch respectively	m	ft	7
T	parameter defined by eq. (1.27)	—	—	1
T	parameter used in Chisholm correlation in eq. (2.75)	—	—	2
T	temperature	°C(K)	°F(°R)	1
T_B	liquid temperature at bubble tip	°C	°F	6

(*Continued*)

Symbol	Description	SI units	British eng. units	Introduced in Chapter
T_B^+	dimensionless temperature at bubble tip given by eq. (6.9)	—	—	6
T_{FDB}	wall temperature corresponding to fully developed boiling	°C	°F	5
T_{SAT}	saturation temperature	°C	°F	1
T_c	critical temperature	°C	°F	4
T_{fi}	inlet bulk liquid temperature	°C	°F	4
$T_f(z)$	bulk liquid temperature at axial position (z)	°C	°F	4
$T_{f\infty}$	liquid temperature well away from wall	°C	°F	4
T_g	vapour temperature	°C	°F	4
T_{gi}	interfacial temperature	°C	°F	10
T_{go}	bulk temperature of condensing fluid	°C	°F	10
T_i	interfacial temperature	°C	°F	7
T_w	wall temperature	°C	°F	4
$(T_w)_{onb}$	wall temperature at onset of boiling	°C	°F	4
$(T_w)_{SCB}$	wall temperature under subcooled boiling	°C	°F	5
$(T_w)_{SPL}$	wall temperature under single-phase liquid conditions	°C	°F	5
T^+	dimensionless temperature given by eq. (7.12)	—	—	7
T^*	dimensionless temperature defined by eq. (6.34)	—	—	6
T_0	local primary fluid temperature	°C	°F	11
T_1	triple point temperature	°C	°F	4
ΔT	temperature difference	°C	°F	1
$\Delta T_{SAT}(z)$	superheat at axial position (z)	°C	°F	1
$(\Delta T_{SAT})_{ONB}$	wall superheat necessary to cause nucleation	°C	°F	5
$(\Delta T_{SUB})_d$	subcooling at point of bubble detachment	°C	°F	6

Symbol	Description	SI units	British eng. units	Introduced in Chapter
$(\Delta T_{\text{SUB}})_{\text{FDB}}$	subcooling at onset of fully developed subcooled boiling	°C	°F	6
$(\Delta T_{\text{SUB}})_i$	inlet subcooling	°C	°F	5
$\Delta T_{\text{SUB}}(z)$	subcooling at axial position (z)	°C	°F	1
ΔT_e	effective superheat into which bubble grows	°C	°F	7
ΔT_f	temperature difference between wall and bulk liquid	°C	°F	5
t	plate or tube thickness	m	ft	3
t	time	s	s	4
t_b	fin thickness at base	m	ft	11
t_g	gas phase residence time	s	s	1
t_g	bubble growth time	s	s	4
t_t	fin thickness at tip	m	ft	11
t^+	dimensionless bubble growth period—eq. (4.28)	—	—	4
U	overall heat transfer coefficient	W/m² °C	Btu/h ft² °F	11
u	velocity	m/s	ft/s	1
$\bar{u}$	average velocity of homogeneous fluid	m/s	ft/s	2
u_c	velocity at vena contracta eq. (3.63)	m/s	ft/s	3
u_d	velocity of droplet	m/s	ft/s	7
u_f	actual velocity of liquid phase	m/s	ft/s	1
u_{fc}	velocity of liquid phase at vena contracta eq. (3.65)	m/s	ft/s	3
u_{fo}	velocity in channel if total flow assumed liquid	m/s	ft/s	1
$u_{f\infty}$	free stream liquid velocity	m/s	ft/s	6
u_g	actual velocity of gaseous phase	m/s	ft/s	1
u_{gc}	velocity of gaseous phase at vena contracta eq. (3.66)	m/s	ft/s	3

(*Continued*)

Symbol	Description	SI units	British eng. units	Introduced in Chapter
u_{gf}	relative velocity between phases	m/s	ft/s	3
u_{gj}	weighted mean drift velocity	m/s	ft/s	3
$u_{g\infty}$	free stream vapour velocity	m/s	ft/s	10
u_i	interfacial liquid velocity	m/s	ft/s	3
u_m	maximum velocity	m/s	ft/s	3
u_n	vapour velocity at the nth tube row based on min. flow c.s.a.	m/s	ft/s	10
u_y	velocity at a distance y from surface	m/s	ft/s	10
u^*	friction velocity $(=\sqrt{(\tau/\rho)}$	m/s	ft/s	5
v	specific volume	m³/kg	ft³/lb	1
$\bar{v}$	average specific volume of homogeneous fluid	m³/kg	ft³/lb	2
v_f	specific volume of liquid	m³/kg	ft³/lb	2
v_{fg}	difference in specific volumes of saturated liquid and vapour	m³/kg	ft³/lb	2
v_g	specific volume of vapour or gas	m³/kg	ft³/lb	2
W	mass rate of flow	kg/s	lb/s	1
W	total power supplied to heated channel up to critical heat flux condition	W	Btu/h	9
W_f	mass rate of flow of liquid phase	kg/s	lb/s	1
W_g	mass rate of flow of vapour phase	kg/s	lb/s	1
W_{fc}	film flow-rate assuming constant shear stress	kg/s	lb/s	3
W_{fi}	condensate inundation rate from above tubes on tube i	kg/s	lb/s	10
W_{fN}	condensate rate on the Nth tube row	kg/s	lb/s	10
W_{fv}	film flow-rate allowing for gravitational forces	kg/s	lb/s	3

Symbol	Description	SI units	British eng. units	Introduced in Chapter
W'	dimensionless group in Katto correlation	—	—	8
w	rectangular duct width	m	ft	1
X	Martinelli parameter defined by eq. (2.67)	—	—	1, 2
X	parameter defined by eq. (5.35)	$\dfrac{\mathrm{K}}{[\mathrm{W/m^2}]^{0.5}}$	$\dfrac{\mathrm{R}}{[\mathrm{Btu/h\,ft^2}]^{0.5}}$	5
X	parameter used in eq. (8.29)	—	—	8
X_1–X_5	parameters in Katto correlation	—	—	8
x	mass fraction of more volatile component of binary mixture in liquid phase	—	—	12
x	mass vapour quality	—	—	1
x'	pseudo-mass vapour quality in subcooled region	—	—	6
x_{CRIT}	critical mass quality	—	—	8
x_{d}	negative mass quality at bubble detachment	—	—	6
x_{i}	inlet vapour quality	—	—	8
x_{ieq}	effective inlet vapour quality	—	—	8
x_{m}	modified mass vapour quality in eq. (3.60)	—	—	3
Y	parameter in Taitel's unified flow pattern model eq. (1.20)	—	—	1
Y	parameter used in Groeneveld post dryout correlation	—	—	7
Y	dimensionless variable used in eq. (7.90)	—	—	7
Y	parameter used in Shah correlation	—	—	8
Y_{B}	distance from wall to bubble tip	m	ft	6

(*Continued*)

Symbol	Description	SI units	British eng. units	Introduced in Chapter
Y_B^+	dimensionless distance from wall to bubble tip	—	—	6
Y_1–Y_6	constants in Kirby CHF correlation	—	—	9
y	distance measured from boundary	m	ft	1
y	parameter defined by eq. (3.84)	—	—	3
y	mass fraction in vapour phase	—	—	12
y	twist ratio of helical insert	—	—	11
y^+	dimensionless distance from wall eq. (7.14)	—	—	7
y_0–y_5	constants in Macbeth CHF correlation	—	—	9
Z	dimensionless length given by eq. (6.55)	—	—	6
Z	parameter defined by eq. (3.71)	—	—	3
Z	parameter given by eq. (3.15)	—	—	3
Z^+	dimensionless length defined by eq. (6.37)	—	—	6
Z'	dimensionless length in Katto correlation	—	—	8
z	axial co-ordinate	m	ft	1
z_{CR}, z_{CRIT}	length of tube to point of critical heat flux	m	ft	7
z_{FDB}	length of tube to start of fully developed boiling	m	ft	5, 6
z_{NB}	length of tube under non-boiling conditions	m	ft	5
z_{SAT}	boiling length (measured from $x = 0$)	m	ft	8
z_{SC}	length of tube under subcooled conditions	m	ft	4
z_d	length of tube to bubble detachment point	m	ft	6
z_{eq}	equivalent length of a tube	m	ft	8
z^*	length of tube to point of thermal equilibrium	m	ft	5, 6

Symbol	Description	SI units	British eng. units	Introduced in Chapter
z^*	dimensionless distance in eq. (10.84)	—	—	10
z_0	arbitrarily chosen length associated with start of void formation (eq. 6.34) *et seq.*	m m	ft ft	6 6
Greek				
α	void fraction	—	—	1
α	angle between any radius and vertical	deg	deg	10
α	relative volatility	—	—	12
α	parameter used in eq. (9.16)	—	—	9
α	half angle of condensate retained on a finned tube	radians	radians	11
α_B	void fraction in free stream	—	—	6
α_f	thermal diffusivity of liquid	m^2/s	ft^2/s	4
α_m	maximum value of void fraction	—	—	3
α_w	void fraction at wall	—	—	6
α^*	void fraction at axial distance z^*	—	—	6
β	volumetric quality	—	—	1
β	included angle of surface cavity	deg	deg	4
β	coefficient of expansion	K^{-1}	$°R^{-1}$	4
β	ratio of mean liquid film to interfacial velocity	—	—	10
β_f	liquid mass transfer coefficient	—	—	12
Γ	film mass flow-rate per unit width	kg/m s	lb/ft s	10
Γ_g, Γ_f	mass transfer rate to gas, liquid phase	kg/m s	lb/ft s	2
Γ_k	mass transfer rate to phase k from various interphase mass transfers	kg/m s	lb/ft s	2
$\Gamma()$	correction factor allowing for net motion of vapour	—	—	10

(Continued)

Symbol	Description	SI units	British eng. units	Introduced in Chapter
	in vicinity of interface in eq. (10.22)			
Γ'_α	local mass flow-rate per unit length of tube at radius angle α to vertical	kg/m s	lb/ft s	10
Γ_N	local mass flow-rate of condensate per unit length from Nth tube	kg/m s	lb/ft s	10
γ	parameter defined by eq. (2.58)	—	—	2
γ	parameter defined by eq. (9.30)	m^2/N	ft^2/pdl	9
γ	exponent used in eq. (10.119)	—	—	10
δ	thickness of diffusion layer	m	ft	10
δ	film thickness	m	ft	1
δ	parameter defined by eq. (2.59)	—	—	2
δ	thermal boundary layer thickness	m	ft	4
δ	dimensionless axial heat flux gradient taken at point of average heat flux	—	—	9
δ^*	dimensionless film thickness in eq. (10.83)	—	—	10
ε	internal energy	J/kg	ft pdl/lb	2
ε	pipe roughness	m	ft	2
ε	emissivity of heating surface	—	—	4
ε	ratio of ϕ_a/ϕ_c in eqn. (6.18)	—	—	6
ε	eddy diffusivity of momentum	m^2/s	ft^2/s	7
ε_H	eddy diffusivity of heat	m^2/s	ft^2/s	7
ε_{go}	mass flux to interface	kg/m^2 s	lb/ft^2 s	10
ε_k	internal energy per unit mass of phase k	J/kg	ft pdl/lb	2
ζ	aspect ratio of fin	—	—	11

Symbol	Description	SI units	British eng. units	Introduced in Chapter
η	empirical factor used in eq. (6.4)	$m^3\,°C/J$	$ft^3\,°F/Btu$	6
η	heat flux efficiency	—	—	11
η	surface efficiency of finned tube	—	—	11
η_f	fin efficiency	—	—	11
Θ	constant in eq. (12.30)	—	—	12
θ	lifetime of average bubble	s	s	5
θ	angle to horizontal plane	deg.	deg.	1
θ	contact angle measured through liquid	deg.	deg.	4
θ'	effective contact angle in eq. (4.18)	deg.	deg.	4
θ_m	angle for average value of heat transfer coefficient	deg.	deg.	10
θ_m	rotational angle	radians	radians	11
$\Delta\theta$	temperature difference	$°C$	$°F$	12
$\Delta\theta_E$	excess superheat	$°C$	$°F$	12
$\Delta\theta_1$	'ideal' superheat defined by eq. (12.31)	$°C$	$°F$	12
$\Delta\theta_{bp}$	boiling point range of mixture	$°C$	$°F$	12
Λ	constant in eq. (12.32)	—	—	12
λ	factor used in Baker flow pattern chart in eq. (1.17)	—	—	1
λ	parameter used in Chisholm correlation	—	—	2
λ	collision frequency	s^{-1}	s^{-1}	4, 10, 12
λ_c	most critical wavelength	m	ft	4
λ_d	Taylor instability wavelength (eq. 4.61)	m	ft	4
μ	viscosity	$N\,s/m^2$	$lb/ft\,s$	1
$\bar{\mu}$	mean viscosity of homogeneous fluid	$N\,s/m^2$	$lb/ft\,s$	2
μ_b	viscosity of bulk liquid	$N\,s/m^2$	$lb/ft\,s$	11
μ_f	liquid viscosity	$N\,s/m^2$	$lb/ft\,s$	1
μ_{fg}	difference in viscosity between liquid and gas phases	$N\,s/m^2$	$lb/ft\,s$	2

(*Continued*)

Symbol	Description	SI units	British eng. units	Introduced in Chapter
μ_g	vapour viscosity	$N\,s/m^2$	$lb/ft\,s$	2
μ_w	viscosity of water at 1 atm. and 20°C	$N\,s/m^2$	$lb/ft\,s$	1
v	kinematic viscosity	m^2/s	ft^2/h	1
ζ	parameter in eq. (5.20)	—	—	5
ζ	ratio of heat flux on a rod to maximum heat flux in cluster	—	—	9
ζ	parameter defined by eq. (10.145)	—	—	10
ρ	density	kg/m^3	lb/ft^3	1
$\bar{\rho}$	average density of homogeneous fluid	kg/m^3	lb/ft^3	2
ρ_A	density of air at 1 atm. and 20°C	kg/m^3	lb/ft^3	1
ρ_c	average density of vapour core in eq. (3.38)	kg/m^3	lb/ft^3	3
ρ_f	liquid density	kg/m^3	lb/ft^3	1
ρ_g	gas density	kg/m^3	lb/ft^3	1
ρ_g^*	fictitious vapour density in eq. (10.78)	kg/m^3	lb/ft^3	10
ρ_w	density of water at 1 atm. and 20°C	kg/m^3	lb/ft^3	1
σ	surface tension	N/m	pdl/ft	1
σ	ratio A_1/A_2 eq. (3.58)	—	—	3
σ	Stefan–Boltzmann constant	$W/m^2\,K^4$	$Btu/h\,ft^2\,°R^4$	4
σ_e	evaporation coefficient	—	—	10
$\sigma_{(r)}$	surface tension of drop radius r	N/m	pdl/ft	10
σ_w	surface tension of water at 1 atm. and 20°C	N/m	pdl/ft	1
σ_∞	surface tension for planar interface	N/m	pdl/ft	10
$\sigma(\sigma_c)$	condensation coefficient	—	—	10
τ	shear stress	N/m^2	pdl/ft^2	3

Symbol	Description	SI units	British eng. units	Introduced in Chapter
τ	effective temperature rise of liquid pulled in and pushed out by bubble	°C	°F	6
τ_f	wall shear stress for liquid phase flowing alone	N/m^2	pdl/ft^2	3
τ_g	wall shear stress for gas phase flowing alone	N/m^2	pdl/ft^2	3
$(\tau_i)_o$	shear stress in absence of mass transfer	N/m^2	pdl/ft^2	10
τ_i	interfacial shear stress	N/m^2	pdl/ft^2	3
τ_i^*	dimensionless interfacial shear stress in eq. (10.85)	—	—	10
τ_{kn}	interfacial shear stress between phase k and n	N/m^2	pdl/ft^2	2
τ_{knz}	z component of interfacial shear stress between phase k and n	N/m^2	pdl/ft^2	2
τ_{kw}	wall shear stress between phase k and channel wall	N/m^2	pdl/ft^2	2
τ_w	wall shear stress	N/m^2	pdl/ft^2	2
ϕ	factor given by eq. (4.17) or eq. (10.8)	—	—	4, 10
ϕ	surface heat flux	W/m^2	$Btu/h\,ft^2$	1
$\bar{\phi}$	average heat flux around perimeter of tube	W/m^2	$Btu/h\,ft^2$	9
ϕ_{CRIT}	critical heat flux	W/m^2	$Btu/h\,ft^2$	4
ϕ_{FDB}	fully developed boiling heat flux	W/m^2	$Btu/h\,ft^2$	6
ϕ_{MAX}	peak heat flux around non-uniformly heated perimeter of tube	W/m^2	$Btu/h\,ft^2$	9
ϕ_{ONB}	surface heat flux to cause boiling	W/m^2	$Btu/h\,ft^2$	5
ϕ_{SAT}	heat flux required to produce just saturated liquid at tube exit	W/m^2	$Btu/h\,ft^2$	6
ϕ_{SCB}	surface heat flux transferred by bubble nucleation	W/m^2	$Btu/h\,ft^2$	5

(Continued)

Symbol	Description	SI units	British eng. units	Introduced in Chapter
ϕ_{SPL}	surface heat flux transferred by single phase convection	W/m^2	Btu/h ft^2	5
ϕ'_{SPL}	maximum heat flux which can be sustained without subcooled boiling in eq. (6.53)	W/m^2	Btu/h ft^2	6
ϕ_a	heat flux removed by bubble agitation	W/m^2	Btu/h ft^2	6
ϕ_e	heat flux removed as latent heat	W/m^2	Btu/h ft^2	6
ϕ_{fo}^2	two-phase frictional multiplier based on pressure gradient for total flow assumed liquid	—	—	2
ϕ_{f}^2	two-phase frictional multiplier based on pressure gradient for liquid alone flow	—	—	2
ϕ_{g}^2	two-phase frictional multiplier based on pressure gradient for gas alone flow	—	—	2
ϕ_{go}^2	two-phase frictional multiplier based on pressure gradient for total flow assumed gas	—	—	3
ϕ_i	interfacial heat flux	W/m^2	Btu/h ft^2	7, 12
ϕ_i	heat flux on the ith tube in Honda correlation	W/m^2	Btu/h ft^2	10
$\dot\phi_k$	specific rate of internal heat generation within phase k	W/m^3	Btu/h ft^3	2
ϕ_n	heat flux on the nth tube in Honda correlation	W/m^2	Btu/h ft^2	10
ϕ_0	standard heat flux	W/m^2	Btu/h ft^2	4
ϕ_r	radiative heat flux	W/m^2	Btu/h ft^2	7
χ	constant in eq. (12.63)	—	—	12
ψ	factor used in Baker flow pattern chart eq. (1.18)	—	—	1
ψ	half apex angle of a fin	deg	deg	11

Symbol	Description	SI units	British eng. units	Introduced in Chapter
ψ	parameter used in Chisholm correlation eq. (2.73)	—	—	2
ψ	parameter used in eq. (4.40)	N/A	N/A	4
ψ_{CHF}	Ahmed's critical heat flux parameter	—	—	9
Ω	mass velocity correction factor in Baroczy correlation	—	—	2
Ω	parameter appearing in eq. (9.8)	N/A	in^{-1}	9
ω	angular rotation	rad/s	rad/s	11

Gradients and differences

Symbol	Description	SI units	British eng. units	Introduced in Chapter
(dp/dz)	pressure gradient	$N/m^2\ m$	$pdl/ft^2\ ft$	1
$(dp/dz\ F)$	pressure gradient due to friction	$N/m^2\ m$	$pdl/ft^2\ ft$	2
$(dp/dz\ F)_f$	frictional pressure gradient assuming liquid alone flow	$N/m^2\ m$	$pdl/ft^2\ ft$	1, 2
$(dp/dz\ F)_{fo}$	frictional pressure gradient assuming total flow to be liquid	$N/m^2\ m$	$pdl/ft^2\ ft$	2
$(dp/dz\ F)_g$	frictional pressure gradient assuming gas alone flow	$N/m^2\ m$	$pdl/ft^2\ ft$	1, 2
$(dp/dz\ F)_{go}$	frictional pressure gradient assuming total flow to be gas	$N/m^2\ m$	$pdl/ft^2\ ft$	3
$(dp/dz)_{SCB}$	pressure gradient in subcooled boiling	$N/m^2\ m$	$pdl/ft^2\ ft$	6
$(dp/dz\ a)$	pressure gradient due to acceleration	$N/m^2\ m$	$pdl/ft^2\ ft$	2
$(dp/dz\ fF)$	pressure gradient due to friction in liquid phase	$N/m^2\ m$	$pdl/ft^2\ ft$	2
$(dp/dz)'_{fo}$	pressure gradient for total flow assumed liquid at same heat flux, velocity and temperature	$N/m^2\ m$	$pdl/ft^2\ ft$	6
$(dp/dz\ gF)$	pressure gradient due to friction in gas phase	$N/m^2\ m$	$pdl/ft^2\ ft$	2

(Continued)

Symbol	Description	SI units	British eng. units	Introduced in Chapter
$(dp/dz\ z)$	pressure gradient due to static head	$N/m^2\ m$	$pdl/ft^2\ ft$	2
Δp	pressure loss or difference	N/m^2	pdl/ft^2	1
Δp_F	frictional pressure drop	N/m^2	pdl/ft^2	3
Δp_{SAT}	change in vapour pressure corresponding to temperature change ΔT_{SAT}	N/m^2	pdl/ft^2	7
Δp_{SCB}	pressure drop in subcooled boiling	N/m^2	pdl/ft^2	6
Δp_{SPL}	pressure drop in unheated tube at same temperature and velocity	N/m^2	pdl/ft^2	6
Δp_{TP}	pressure drop for two-phase flow	N/m^2	pdl/ft^2	3
Δp_a	accelerational pressure drop	N/m^2	pdl/ft^2	3
Δp_e	change in vapour pressure corresponding to temperature change ΔT_e	N/m^2	pdl/ft^2	7
Δp_f	pressure drop for liquid alone flow	N/m^2	pdl/ft^2	3
Δp_g	pressure drop for gas alone flow	N/m^2	pdl/ft^2	3
Δp_z	static head pressure drop	N/m^2	pdl/ft^2	3

Dimensionless numbers

Symbol	Description	SI units	British eng. units	Introduced in Chapter
Bo	Boiling number	—	—	7
C_0	parameter used in Shah correlation	—	—	7
Fr	Froude number $[u_{fo}^2/gD]$	—	—	3
$(Fr)_\theta$	Froude number in eq. (1.36)	—	—	1
Ga	Galileo number $\left[\dfrac{D^3\rho_f(\rho_f - \rho_g)g}{\mu_f^2}\right]$	—	—	10
Gr	Grashof number $\left[\dfrac{\beta g\ \Delta T D^3\rho^2}{\mu^2}\right]$	—	—	4, 10

Symbol	Description	SI units	British eng. units	Introduced in Chapter
Gr*	modified Grashof number $\left[\dfrac{z^3 g \rho_g (\rho_f - \rho_g)}{\mu_g^2}\right]$	—	—	7
Ja	Jakob number $\left[\dfrac{\rho_f c_{\rho f} \, \Delta T_{\text{SAT}}}{\rho_g i_{fg}}\right]$	—	—	4
Nu	Nusselt number $[hD/k]$	—	—	4
Nu_f	Nusselt number $[h_{\text{TP}} \delta / k_f]$	—	—	7
$\overline{\mathrm{Nu}}_z$	Nusselt number $[\bar{h}(z) z / k_g]$	—	—	7
Pe	Peclet number $[GD c_p / k]$	—	—	6
Pr	Prandtl number $[c_P \mu / k]$	—	—	4
Pr_f	Prandtl number at bulk liquid temperature	—	—	5
Pr_w	Prandtl number at wall temperature	—	—	5
Pr_{TP}	effective two-phase Prandtl number	—	—	7
Pr*	modified Prandtl number $[i_{fg} \mu_g / k_g \, \Delta T]$	—	—	7
Re	Reynolds number $[GD/\mu]$	—	—	2
Re	film Reynolds number $[4\Gamma_z / \mu_f]$	—	—	10
Re_e	equivalent Reynolds number $[G_e D / \mu_f]$	—	—	10
$\mathrm{Re}_{\text{trans}}$	Reynolds number for transition from laminar to turbulent flow	—	—	3
Re_{TP}	effective two-phase Reynolds number	—	—	7
Sc	Schmidt number $[\mu / \rho D]$	—	—	10
Sh	Sherwood number $[K \delta / D]$	—	—	12
Sn	Scriven number defined in eq. (12.25)	—	—	12
St	Stanton number $[\phi / G c_p \, \Delta T]$	—	—	6
We	Weber number $[G^2 D / \bar{\rho} \sigma]$	—	—	2

Subscripts

Symbol	Description
1, 2 A, B (i), (ii)	referring to stations 1, 2 or stages, 1, 2 or components A, B, or (i), (ii) etc.
B	tube bundle
em	evaporation microlayer
f	liquid or film
fg, gf	change of phase—liquid to vapour or vice versa or interface between gas and liquid
fo	liquid only
fw	fraction of wall relating to liquid phase; interface between wall and liquid phase
g	gas or vapour
go	gas only
gw	fraction of wall relating to gaseous phase; interface between wall and gas phase
k	phase k
kn	contact or interface between phase k and phase n
kw, wk	fraction of wall relating to phase k; interface between wall and phase k
NB, NcB	nucleate boiling
nc	natural convection
rm	relaxation microlayer
TB	transition boiling
tf	thin film
tt	liquid turbulent/gas turbulent
tv	liquid turbulent/gas viscous
vt	liquid viscous/gas turbulent
vv	liquid viscous/gas viscous
w	wall
CRIT, cr	critical
EXP	experimental
F + S	friction + static head
FDB	fully developed boiling
FILM	evaluated at some mean temperature between wall and bulk
GRAV	gravity
i	interface, ideal
MAX	maximum
MIN	minimum
NOM	nominal
nu	non-uniform
o	away from interface
ONB	onset of nucleate boiling
r	reduced property, radiation
REF	reference

Symbol	Description
T	total
u	uniform
vib	vibration present

Superscripts

$^{-}$	denotes average property
$\sim$	denotes molar quantity

1

INTRODUCTION

1.1 Objectives

The subject of boiling and condensation under conditions of natural or
forced convection is an extremely important one. The design of water-
tube and fire-tube boilers, pipe stills, water-cooled nuclear reactors, refrigeration and air-conditioning equipment, heat pumps, petrochemical plant
reboilers and condensers, surface condensers and many other major items
of chemical and power plant is dependent upon a knowledge of the fluid
dynamic and heat transfer processes occurring during convective boiling and
condensation.

Convective boiling will be defined as being the addition of heat to a flowing
liquid in such a way that generation of vapour occurs. This definition
therefore excludes the process of flashing where vapour generation occurs
solely as a result of a reduction in system pressure. In many applications,
however, the two processes do occur simultaneously and therefore cannot
be clearly separated. Condensation is conversely defined as the removal of
heat from the system in such a way that vapour is converted into liquid.

This text will primarily be concerned with single-component systems, i.e.,
a pure liquid and its vapour. Much of the information presented will be
devoted to one such system, namely the water/steam system. Many other
fluid systems are, however, of industrial importance and these include
refrigerants, organic liquids, cryogenic liquids (such as oxygen, nitrogen, etc.),
and liquid metals. Important applications, particularly in the process
chemical industries, involve the use of multi-component systems. Rather less
information has been published on the processes of boiling and condensation
with multi-component systems. The available information is reviewed in
Chapter 12.

It is the prime purpose of this text to present in a logical and coherent
manner the current state of the art in estimating the important engineering
parameters in convective boiling and condensation. The usual design
variables of interest are the rates of heat (or mass) transfer and the pressure
losses in a particular context. It is often important to know those phenomena
which might impose some limitation upon the performance of a specific item
of equipment such as a change in hydrodynamic or heat transfer process or
regime. In boiling and condensation the hydrodynamic and the heat transfer
processes are very closely linked. This coupling is much closer than exists

in single-phase flows. The addition or removal of heat from a two-phase flow causes variations in the amount and distribution of each phase and the flow pattern or topology of the flow. These changes in turn induce variations in the local heat transfer processes. Because of the continuous change of all the thermal and hydraulic properties of the flow, the situation at any axial point in the channel can never be fully developed either thermally or hydrodynamically. This facet of the problem will be reiterated on many occasions. In addition, because the local situation is not an equilibrium situation, it is necessary to know the variation of the flow properties upstream of the point being considered. This information is required to define the magnitude and direction of the departure from equilibrium. Add to these difficulties the fact that certain situations involve time varying properties and others involve departures from thermodynamic equilibrium and one begins to realize the magnitude of the problem being tackled.

Because of these difficulties the subject to date is largely empirical and the design engineer is confronted with a vast amount of published literature which must be digested and placed into context. It is hoped that this book will provide a basis for any such exercise. In addition, considerable skill and knowledge of the subject is required to choose the particular published treatment best suited to the design problem in hand. Many such treatments have only a limited range of application and must not be applied outside this range. Here again guidance will be given on the basis of both authors direct experience.

Certain aspects of boiling will not be treated in this book. The prime omissions are the subjects of flashing and critical flow and the stability and transient response of boiling and condensing systems. These areas have been omitted, because they are not directly relevant to the development of the subject as treated in this book.

1.2 Sources of information

Examination of the problems of convective boiling and condensation has been in progress since the nineteenth century but research in this field has been greatly stimulated since the Second World War by the advent of the nuclear and rocket technologies. An often reproduced graph, first prepared by Gouse (1964), shows the cumulative number of papers and reports available in the published literature on boiling and two-phase flow against the year. The graph indicates an exponential rise in the number of publications available with time. The rate of increase in the number of publications appears to double around the year 1940 and the total number of published papers in the literature by 1970 was around 10,000. Yet despite the efforts of the many investigators in most of the highly developed countries of the world, the mechanics of boiling and condensation remain a very poorly understood phenomenon. Even so, empirical methods for designing two-phase heat transfer equipment have improved significantly in the last decade.

The engineer requires some assistance in gaining access to this vast amount of information.

Apart from the present text, a number of books dealing with the subject of two-phase heat transfer have appeared over the past 20 years. They include the books by Tong (1965), by Wallis (1969), by Hsu and Graham (1976), by Chisholm (1983) and by Whalley (1987). In addition a considerable number of lecture courses have been given and the lectures subsequently published, e.g., Butterworth and Hewitt (1977) and Ginoux (1978). Summer schools have been held under the sponsorship of the NATO Scientific Affairs Division and various international short courses presented (Kakac and Ishii 1983; Kakac *et al.* 1988). Texts have also appeared which concentrate more specifically on a particular aspect, such as annular flow (Hewitt and Hall-Taylor 1970) instrumentation (Hewitt 1978) and enhanced boiling (Thome 1990). Handbooks have also been published such as those edited by Hetsroni (1982) and by Schlünder (1983). Finally, books dealing with the thermal design of boilers, evaporators and condensers have been produced (Smith 1986; Kakac 1991; Cumo and Naviglio 1991).

Also of value to the engineer entering this field of endeavour are critical reviews of the entire field or part of it. Particularly authorative and therefore recommended are the volumes entitled *Multiphase science and technology* edited by Hewitt *et al.* (1990), the series *Advances in heat transfer* edited by Irvine and Hartnett (1964 *et seq.*) and also keynote lectures given at the International Heat Transfer Conferences. In addition a journal specifically devoted to two-phase flow problems, i.e., *International Journal of Multiphase Flow*, started publication in 1974.

Much of the information on boiling and two-phase flow is contained in the reports of government and private research laboratories rather than as published papers in the journals of learned societies. The distribution and availability of these reports may be limited but this does not invalidate their usefulness and the authors have drawn freely on such sources of information.

1.3 Units

It has been decided to use the International (SI) System of units in this book. In this system the basic units are:

Length	1 metre	L
Mass	1 kilogramme	M
Time	1 second	T

The unit of force is the newton and is that force required to give a mass of 1 kilogramme an acceleration of 1 metre per second per second.

The equations in the book are written so that any consistent set of units based on length, mass, and time may be used. Thus the reader may wish to work in the foot-pound-second (FPS) system. In this system the basic units are:

Length	1 foot	L
Mass	1 pound	M
Time	1 second	T

The unit of force is the poundal and is that force required to give a mass of 1 pound an acceleration of 1 foot per second per second.

In the field of engineering in the United States an additional unit is introduced, namely the pound force. This force is that required to give an acceleration of g (32.2 ft/s²) to a mass of 1 pound. Thus the pound force is related to the poundal; viz.

$$1 \text{ pound force} = 32.2 \text{ poundals} = 32.2 \left[\frac{(\text{pound mass})(\text{foot})}{(\text{second})^2} \right]$$

The constant (32.2) in this relationship is often referred to as g_c or g_0. In the present book only the units of mass, length, and time will be employed and the acceleration due to gravity will only be involved when work is done in the earth's gravitational field. Throughout the book quantities in the FPS system will be shown in parentheses. Some conversion factors are shown in Table 1.1.

The unit of temperature in the SI system is the degree centigrade, known in that system as the kelvin, and in the FPS system is the degree Fahrenheit. The unit of heat is the SI system is the watt-second (or joule). Normally it is necessary to relate thermal and mechanical energy via the mechanical

Table 1.1   Units

Quantity	Dimension	SI	FPS	Conversion
Mass	M	1 kg	1 lb	1 lb = 0.45359 kg
Length	L	1 m	1 ft	1 ft = 0.3048 m
Time	T	1 s	1 s	
Force	$\dfrac{ML}{T^2}$	1 N	1 pdl	1 pdl = 0.1382 N
Temperature	θ	1°C	1°F	1°F = 0.555°C
Heat	$M\theta$	1 J	1 Btu	1 Btu = 1055.1 J

equivalent of heat (J). Thus,

$$J \times \text{thermal energy} = \text{mechanical energy}$$

Happily, in the SI system the units of heat and work are identical and J is unity. The unit of heat in the FPS system is the British thermal unit (Btu) and the value of J is 2.50×10^4 foot poundals/Btu.

In some cases the hour is used as the unit of time in the British system of units in place of the second (FPH system). In this case,

$$1 \text{ pound force} = 4.173 \times 10^8 \left[\frac{(\text{pound mass})(\text{foot})}{(\text{hour})^2} \right]$$

The unit of pressure in the SI system of units is N/m^2. This is a rather small quantity and throughout this text the bar will be used as an alternative,

$$1 \text{ bar} = 10^5 \text{ N/m}^2$$

1.4 Notation

A full notation is given at the beginning of the book. This section introduces the primary variables used throughout the work and derives some simple relationships between them for the case of one-dimensional flow.

To distinguish between gas and liquid phases the subscripts 'g' for gas and 'f' for liquid will be used. The subscript 'g' is conventionally used to denote conditions in the saturated vapour phase. The choice of 'f' rather than 'l' was made to keep the convention in line with the steam and other thermodynamic tables. The subscript 'fg' refers to the difference between gas (or vapour) and liquid properties at some specified condition (usually the saturation temperature).

Consider a channel in which two phases are flowing concurrently. The flow is steady and one-dimensional, i.e., there are no changes in system properties in directions normal to the direction of flow. At any instant of time, a given point within the channel may be occupied by the liquid phase while an instant later it will be occupied by the gas phase. A *local time-average gas fraction* may be defined by measuring the cumulative residence time of the gas phase Σt_g over a total time interval, t. Such a measurement may be made using electrical or optical probes (Hewitt 1978). Alternatively, at any instant of time a line passing through the channel normal to the channel axis will lie partly in the liquid phase and partly in the gas phase. *An instantaneous line-average gas fraction* may be defined as the length of the line submerged in the gas phase over the total length of the line within the channel. Such a measurement may be made using the attenuation of an X-ray or γ-ray beam (Hewitt 1978). Similarly, *an instantaneous area (or volume) average gas fraction* may be defined as the area (or volume) of the channel occupied by the gas phase divided by the cross sectional area (or total volume) of the channel. The *volume average gas fraction* may be measured using quick-closing valves at the entrance and exit of the channel (Hewitt 1978). The space- and time-average gas fractions will be equal only in special circumstances when the flow is steady and one-dimensional. In this text the area-average gas fraction will be referred to as the *void-fraction* and will be denoted by α. If the cross-sectional area of the channel is A and the cross-sectional areas occupied by the gas and liquid phases are A_g and A_f respectively then the *void fraction* is given by

$$\alpha = \frac{A_g}{A}, \qquad (1 - \alpha) = \frac{A_f}{A} \qquad (1.1)$$

A fuller discussion of the various space- and time-averaging processes and the resulting equations of two-phase flow has been presented by Delhaye *et al.* (1979). The mass rate of flow will be represented by the symbol W and will be the sum of the individual phase flow rates W_f and W_g. The mean velocity of an individual phase is denoted by the symbol u. The volumetric rate of flow is represented by the symbol Q and will be the sum of the individual phase volumetric flow rates Q_f and Q_g. In boiling and condensation it is often convenient to use the fraction of the total mass flow which is composed of vapour or liquid. Thus, the mass quality, x, is defined as

$$x = \frac{W_g}{W_g + W_f}, \qquad (1 - x) = \frac{W_f}{W_g + W_f} \qquad (1.2)$$

The rate of mass flow divided by the flow area is given the name 'mass velocity' and the symbol G. Thus,

$$G = \frac{W}{A} = \rho u = \frac{u}{v} \qquad (1.3)$$

$$W_g = GAx, \qquad W_f = GA(1 - x) \qquad (1.4)$$

$$u_g = \frac{W_g}{\rho_g A_g}, \qquad u_f = \frac{W_f}{\rho_f A_f} \qquad (1.5)$$

$$u_g = \frac{Q_g}{A_g}, \qquad u_f = \frac{Q_f}{A_f} \qquad (1.6)$$

$$u_g = \frac{Gx}{\rho_g \alpha}, \qquad u_f = \frac{G(1 - x)}{\rho_f(1 - \alpha)} \qquad (1.7)$$

Sometimes it is necessary to use the fraction of the total volumetric flow which is composed of vapour or liquid. Thus the volumetric quality, β, is defined as

$$\beta = \frac{Q_g}{Q_g + Q_f}, \qquad (1 - \beta) = \frac{Q_f}{Q_g + Q_f} \qquad (1.8)$$

The rate of volumetric flow divided by the flow area is given the name 'volumetric flux' (or sometimes 'superficial velocity') and the symbol j. Thus,

$$j = \frac{Q}{A}, \qquad j_g = \frac{Q_g}{A}, \qquad j_f = \frac{Q_f}{A} \qquad (1.9)$$

Or using eqs. (1.1) and (1.6),

$$j_g = u_g \alpha = j\beta = \frac{Gx}{\rho_g}, \qquad j_f = u_f(1 - \alpha) = j(1 - \beta) = \frac{G(1 - x)}{\rho_f} \qquad (1.10)$$

$$G_g = j_g \rho_g = Gx, \qquad G_f = j_f \rho_f = G(1 - x), \qquad G = G_g + G_f \qquad (1.11)$$

One relationship which will be useful in later sections expresses the ratio of the gas phase velocity to the liquid phase velocity (known as the 'slip ratio') in terms of the mass quality and the void fraction. Thus,

$$\frac{u_g}{u_f} = \frac{W_g \rho_f A_f}{W_f \rho_g A_g} = \left(\frac{x}{1 - x}\right)\left(\frac{\rho_f}{\rho_g}\right)\left(\frac{1 - \alpha}{\alpha}\right) \qquad (1.12)$$

The choice of symbols for co-ordinates is limited by the use of x for mass quality. z will be used as the co-ordinate in the direction of flow along the axis of a conduit and the radial distance from the axis of a pipe will be denoted by r. The use of y will be restricted to the distance measured from a boundary such as a surface or interface.

The rate of change of static pressure in the direction of flow will be represented by (dp/dz). Positive values for this quantity denote a rise of pressure with respect to axial distance. Engineers are mainly concerned with the *loss* of pressure from one position to another and this will be denoted by the sumbol Δp. Thus,

$$\Delta p = -\int_1^2 (dp/dz)\, dz \qquad (1.13)$$

In the case of heat transfer the symbol ϕ will be used to denote 'heat flux' or the amount of heat flowing through unit surface area in unit time. The symbol T will be used for temperature with appropriate subscripts, 'w' for wall, 'f' for liquid, 'g' for vapour, etc. The normal convention in most thermodynamic tables is to use h for enthalpy. In the present text i will be used for enthalpy to avoid confusion with heat transfer coefficient which will be denoted by the symbol h.

The thermodynamic equilibrium temperature at which vapour and liquid

exist together at a particular pressure will be called the saturation temperature (T_{SAT}). The corresponding pressure will be the saturation pressure
(p_{SAT}). The enthalpy of liquid at the saturation temperature will be denoted
by i_f and of vapour i_g. The difference between these two quantities is the
latent heat of vaporization and using the normal convention this will be
denoted by the symbol i_{fg}. Provided both liquid and vapour phases are in
thermodynamic equilibrium, i.e., exist at the saturation temperature, then an
alternative definition of mass quality x can be given on the basis of
thermodynamic properties. Thus,

$$x = \frac{i - i_f}{i_{fg}} \tag{1.14}$$

Only when thermodynamic equilibrium exists are the values of x obtained
from eqs. (1.2) and (1.14) identical.

In boiling and condensing systems it is often convenient to use the
saturation temperature as a datum and refer other system temperatures to
this datum. Material at a temperature higher than the saturation temperature
is said to be 'superheated' with respect to the saturation temperature and
the difference in temperature from the datum is denoted by the symbol ΔT_{SAT}.
Material at a temperature below the saturation temperature is said to be
'subcooled' with respect to the saturation temperature and the difference in
temperature from the datum is denoted by the symbol ΔT_{SUB}. Thus,

$$T - T_{SAT} = \Delta T_{SAT} \tag{1.15}$$

$$T_{SAT} - T = \Delta T_{SUB} \tag{1.16}$$

Inevitably it is not possible to avoid duplication of some symbols and where
this occurs it is hoped that the meaning will be clear. Each new symbol used
will be defined as it is introduced and a reference to this definition is given
in the Notation.

1.5   Methods of analysis

The methods used to analyse a two-phase flow are extensions of those already
well tried for single-phase flows. The procedure invariably is to write down
the basic equations governing the conservation of mass, momentum, and
energy, often in a one-dimensional form and to seek to solve these equations
by the use of various simplifying assumptions. Three main types of assumption have been made, viz.,

(a) *The 'homogeneous' flow model.* In this, the simplest approach to the
problem, the two-phase flow is assumed to be a single-phase flow having

pseudo-properties arrived at by suitably weighting the properties of the individual phases.

(b) *The 'separated' flow model.* In this approach the two phases of the flow are considered to be artificially segregated. Two sets of basic equations can now be written, one for each phase. Alternatively, the equations can be combined. In either case information must be forthcoming about the area of the channel occupied by each phase (or alternatively, about the velocities of each phase) and about the frictional interactions with the channel wall. In the former case additional information concerning the frictional interaction between the phases is also required. This information is inserted into the basic equations, either from separate empirical relationships in which the void fraction and the wall shear stress are related to the primary variables, or on the basis of simplified models of the flow.

(c) *The 'flow pattern' models.* In this more sophisticated approach the two phases are considered to be arranged in one of three or four definite prescribed geometries. These geometries are based on the various configurations or flow patterns found when a gas and a liquid flow together in a channel. The basic equations are solved within the framework of each of these idealized representations. In order to apply these models it is necessary to know when each should be used and to be able to predict the transition from one pattern to another.

This latter approach (Wallis 1969) is still under development and much of the information presented in the following chapters is based on the first two approaches only. However, considerable strides have been made in the development of flow pattern models in the past two decades and are included in several sophisticated two-phase heat exchanger and pipe flow design methods now. The use of the digital computer is needed to ease the considerable complication of design calculations this inevitably brings to the engineer. In addition, finite difference methods widely used for single-phase flows are now starting to be used to evaluate two-phase processes.

1.6 Flow patterns and transition models

The analysis of single-phase flow is made easier if it can be established that the flow is either laminar or turbulent and whether any separation or secondary flow effect occurs. This information is equally useful in the study of gas–liquid flow. However, perhaps of greater importance in the latter case is the topology or geometry of the flow. When a liquid is vaporized in a heated channel the liquid and the vapour generated take up a variety of configurations known as flow patterns. The particular flow pattern depends on the conditions of pressure, flow, heat flux, and channel geometry. Each has a descriptive name and in the design of a heat exchanger it is desirable to know what the flow pattern or successive flow patterns are so that a hydrodynamic or heat transfer theory appropriate to that pattern can be chosen. Often, successful *troubleshooting* of malfunctioning heat exchangers is dependent on knowledge of the flow pattern in identifying the source and solution of the problem.

Various techniques (Hewitt 1978) are available for the study of two-phase flow patterns in heated and unheated channels. In transparent channels at low velocities it is possible to distinguish the flow pattern by direct visual observation. At higher velocities, where the pattern becomes indistinct, flash and cine photography can be used to slow the flow down and extend the range. However, reflection and refraction at multiple interfaces often means that the interpretation of visual observations and photographs is open to considerable uncertainty. This is particularly so at high flow rates. The development of X-radiography (Hewitt 1978; Derbyshire *et al.* 1964) has helped to reduce this uncertainty and has also allowed investigation in opaque channels with heated walls. Another technique for examining heated channels involves the coating of a transparent tube with a very thin metallic conducting layer in which heat is generated but which does not greatly reduce the transparency. Various other ingenious techniques have been used, for example, to look down the axis of a tube containing a two-phase flow (Arnold and Hewitt 1967) and to examine flow patterns in boiling systems at high pressures. Various types of probe; electrical, hot wire, pressure (Chaudry *et al.* 1965), and optical have been developed (Hewitt 1978) to study the structure of the flow and the signals from these probes can be used to provide indirect information from which it is possible to deduce the flow pattern.

Because the name given to a flow pattern is to a large extent subjective, there exists in the literature a multitude of terms purporting to describe the various possible phase distributions. In the present text only those patterns which are clearly distinguishable and generally recognized are listed. Subdivisions do exist but the transitions between one pattern and another are seldom well defined. The indeterminate region between patterns is sometimes referred to by hyphenating the names of the patterns bounding the region, e.g., the region between the bubbly and slug regions might be referred to as the bubbly-slug region.

1.6.1   *Flow patterns in vertical co-current flow*

The flow patterns encountered in vertical upwards co-current flow are shown schematically in Fig. 1.1 together with actual photographs of each pattern.

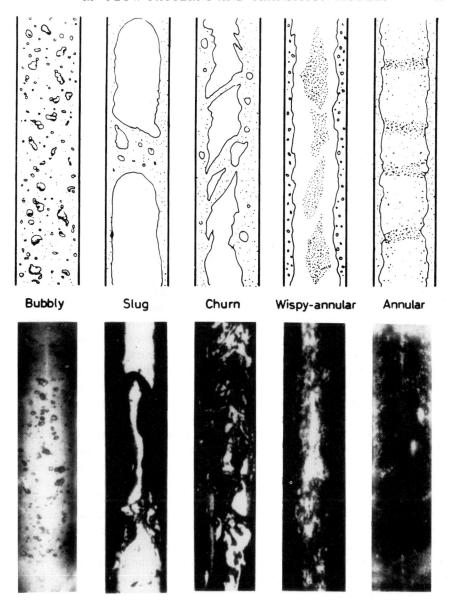

Fig. 1.1. Flow patterns in vertical co-current flow.

(a) *Bubbly flow*. In bubbly flow the gas or vapour phase is distributed as discrete bubbles in a continuous liquid phase. At one extreme the bubbles may be small and spherical and at the other extreme the bubbles may be large with a spherical cap and a flat tail. In this latter state although the size of bubbles does not approach the diameter of the pipe, there may be some confusion with slug flow.

(b) *Slug flow*. In slug flow the gas or vapour bubbles are approximately the diameter of the pipe. The nose of the bubble has a characteristic spherical cap and the gas in the bubble is separated from the pipe wall by a slowly descending film of liquid. The liquid flow is contained in liquid slugs which separate successive gas bubbles. These slugs may or may not contain smaller entrained gas bubbles carried in the wake of the large bubble. The length of the main gas bubble can vary considerably.

(c) *Churn flow*. Churn flow is formed by the breakdown of the large vapour bubbles in the slug flow. The gas or vapour flows in a more or less chaotic manner through the liquid which is mainly displaced to the channel wall. The flow has an oscillatory or time varying character; hence, the descriptive name 'churn' flow. This region is also sometimes referred to as semi-annular or slug-annular flow.

(d) *Wispy-annular flow*. Wispy-annular flow has been identified as a distinct flow pattern primarily as a result of the work of Hewitt and Hall-Taylor (1970). The flow in this region takes the form of a relatively thick liquid film on the walls of the pipe together with a considerable amount of liquid entrained in a central gas or vapour core. The liquid in the film is aerated by small gas bubbles and the entrained liquid phase appears as large droplets which have agglomerated into long irregular filaments or wisps. This region occurs at high mass velocities and because of the aerated nature of the liquid film could be confused with high velocity bubbly flow.

(e) *Annular flow*. In annular flow a liquid film forms at the pipe wall with a continuous central gas or vapour core. Large amplitude coherent waves are usually present on the surface of the film and the continuous break up of these waves forms a source for droplet entrainment which occurs in varying amounts in the central gas core. In this case, as distinct from the wispy-annular pattern, the droplets are separate rather than agglomerated.

No satisfactory general method has yet been developed to allow the correct flow pattern to be designated for a specified local flow condition. There are a variety of reasons for this deficiency. One reason, already touched on, is that the flow pattern is more a subjective judgement than an objective measurement. A second primary reason is that although the flow pattern is a strong function of the local parameters such as the volumetric quality β, other less easily defined variables such as the method of forming the two-phase flow, the amount of the departure from local hydrodynamic equilibrium, and the presence of trace contaminants in the system all considerably influence the particular pattern. One example is the formation of bubbly flow by the injection of air through a porous wall into water. Close to the injector bubbly flow will exist to relatively high void fractions. Downstream of the injector in relatively pure tap water, coalescence of bubbles will result in bubbly-slug or slug flow. In water containing trace amounts of a surface-active agent this coalescence will be inhibited.

1.6.2 *Flow patterns in vertical heated channels*

The formation of a two-phase mixture by vapour generation in a vertical heated tubular channel represents an important special case. The presence of a heat flux through the channel wall alters the flow pattern from that which would have occurred in a long unheated channel at the same local flow conditions. These changes occur due to two main reasons; firstly, the departure from thermodynamic equilibrium coupled with the presence of radial temperature profiles in the channel and secondly, the departure from local hydrodynamic equilibrium throughout the channel. Figure 1.2 shows a schematic representation of a vertical tubular channel heated by a uniform low heat flux and fed at its base with liquid just below the saturation temperature.

In the initial single-phase region the liquid is being heated to the saturation temperature. A thermal boundary layer forms at the wall and a radial temperature profile is set up. At some position up the tube the wall temperature will exceed the saturation temperature and the conditions for the formation of vapour (nucleation) at the wall are satisfied. These conditions are discussed in Chapters 4 and 5. Vapour is formed at preferred positions or sites on the surface of the tube. Vapour bubbles grow from these sites finally detaching to form a bubbly flow. With the production of more vapour the bubble population increases with length and coalescence takes place to form slug flow which in turn gives way to annular flow further along the channel. Close to this point the formation of vapour at sites on the wall may cease and further vapour formation will be as a result of evaporation at the liquid film–vapour core interface. Increasing velocities in the vapour core will cause entrainment of liquid in the form of droplets. The depletion of the liquid from the film by this entrainment and by evaporation finally causes the film to dry out completely. Droplets continue to exist and are slowly evaporated until only single-phase vapour is present.

There are a number of features concerning the above description which require further comment. Representative locations where the liquid reaches the saturation temperature ($x = 0$) and where the liquid is totally evaporated

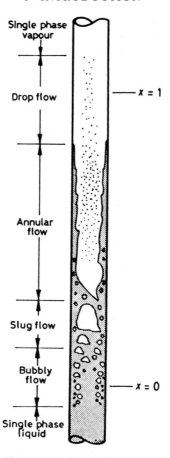

Single phase
vapour

Drop flow

— x = 1

Annular
flow

Slug flow

Bubbly
flow

— x = 0

Single phase
liquid

Fig. 1.2. Flow patterns in a vertical evaporator tube.

($x = 1$) (found on the basis of simple heat balance calculations) are shown
on the diagram. In this example vapour is seen to form before the liquid
mixed mean (mixing cup) temperature reaches the saturation temperature.
This effect occurs as a result of the radial temperature profile in the liquid
which allows the conditions for vapour formation at the wall to be satisfied
before the mean liquid temperature reaches the saturation temperature.
Under certain circumstances, however, the conditions for vapour formation
at the wall may not be satisfied until after the mean liquid temperature has
exceeded the saturation temperature. In this instance the growth rate of the
initial vapour bubble formed at the wall may be so great that slug flow is
entered directly without any bubbly flow region. A second feature shown in
Fig. 1.2 is the presence of a flow pattern not previously mentioned; that of

drop flow. In unheated channels, liquid invariably exists on the channel surface either as a film or, in extreme cases, as rivulets. Only in the case of a heated wall where any liquid at the wall is evaporated does true drop flow occur. The presence of radial temperature profiles and lack of thermodynamic equilibrium mean that close to the point of theoretical complete evaporation ($x = 1$) liquid droplets often exist in the presence of superheated vapour.

Physical processes which result in flow pattern changes such as bubble coalescence, droplet formation and break-up, require finite distances and times to occur. If the rate of change of the local flow conditions (expressed as the vapour quality, x), i.e., dx/dz, is large then these processes are not completed and the departure from hydrodynamic equilibrium conditions is such that the regions over which certain flow patterns occur are expanded or compressed or, in some cases, a particular flow pattern may disappear. One example is that of slug flow in a heated channel which normally occurs over only a short interval in a long channel but which may be completely absent in a short channel.

1.6.3 *Flow patterns in horizontal co-current flow*

The flow patterns observed in co-current two-phase flow in horizontal and inclined tubular channels are complicated by asymmetry of the phases resulting from the influence of gravity. The generally accepted flow patterns as given by Alves (1954) are shown diagrammatically in Fig. 1.3.

(a) *Bubbly flow.* This flow pattern is similar to that in vertical flow except that the vapour bubbles tend to travel in the upper half of the pipe. At moderate velocities of both vapour and liquid phases the entire pipe cross-section contains bubbles whilst at still higher velocities a flow pattern equivalent to the wispy-annular pattern is entered. This pattern is sometimes referred to as froth flow.

(b) *Plug flow.* This is similar to slug flow in the vertical direction. Again the gas bubbles tend to travel in the upper half of the pipe.

(c) *Stratified flow.* This pattern only occurs at very low liquid and vapour velocities. The two phases flow separately with a relatively smooth interface.

(d) *Wavy flow.* As the vapour velocity is increased the interface becomes disturbed by waves travelling in the direction of flow.

(e) *Slug flow.* A further increase in vapour velocity causes the waves at the interface to be picked up to form a frothy slug which is propagated along the channel at a high velocity. The upper surface of the tube behind the wave is wetted by a residual film which drains into the bulk of the liquid.

(f) *Annular flow.* A still higher vapour velocity will result in the formation

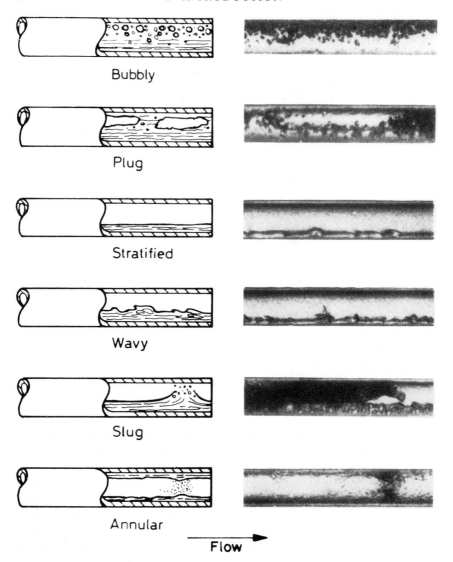

Fig. 1.3. Flow patterns in horizontal flow.

of a gas core with a liquid film around the periphery of the pipe. The film may not be continuous around the entire circumference but it will, of course, be thicker at the base of the pipe. Alves also delineated a spray or drop flow region where the majority of the flow was entrained in the gas core and dispersed as droplets.

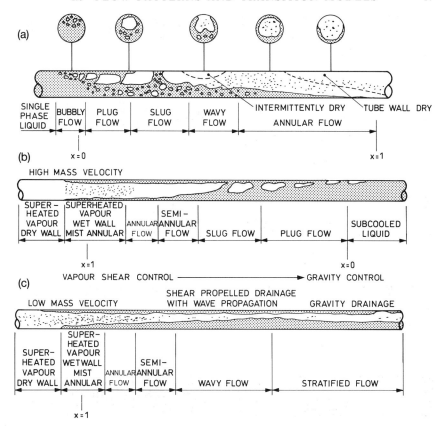

Fig. 1.4. Two-phase flow patterns in horizontal tubes: (a) evaporation; (b) condensation with high liquid loading; (c) condensation with low liquid loading.

Flow patterns formed during the generation of vapour in horizontal tubular channels are influenced by departures from thermodynamic and hydrodynamic equilibrium in the same way as for vertical flow. Asymmetric phase distributions and stratification introduce additional complications. Figure 1.4(a) shows a schematic representation of a horizontal tubular channel heated by a uniform low heat flux and fed with liquid just below the saturation temperature. The sequence of flow patterns shown corresponds to a relatively low inlet velocity (< 1 m/s). Important points to note from a heat transfer viewpoint are the possibility of intermittent drying and rewetting of the upper surfaces of the tube in slug and wavy flow and the progressive drying out over long tube lengths of the upper circumference of the tube wall in annular flow. At higher inlet liquid velocities the influence of gravity is less obvious, the phase distribution becomes more symmetrical and the flow patterns become closer to those seen in vertical flow.

Figures 1.4(b) and (c) illustrate the flow patterns existing in condensation inside horizontal tubes (Schlünder 1983). At the inlet film condensation around the circumference of the tube produces an annular flow with some droplets entrained in the central high velocity vapour core. As condensation continues, the vapour velocity falls and reduces the influence of vapour shear on the condensate and the influence of gravity forces increases. At high flow rates slug and bubble flows eventually are reached while at low flow rates large magnitude waves and then stratified flow are formed.

The design of many heat exchangers using horizontal tubular elements necessitates that these elements be interconnected using 180° return bends to form a serpentine arrangement. In this case the influence of the return bend on the flow pattern is considerable. This topic is discussed in section 1.6.5. The effect of a return bend on the flow pattern may be seen up to 50 pipe diameters downstream of the bend.

1.6.4   *Flow pattern maps and transitions*

Despite the present deficiencies in the understanding of the various flow patterns and the transitions from one pattern to another, there is a widely felt need for simple methods to give some idea of the particular pattern likely to occur for a given set of local flow parameters. One method of representing the various transitions is in the form of a flow pattern map. The respective patterns may be represented as areas on a graph, the co-ordinates of which are the actual superficial phase velocities (j_f or j_g) or generalized parameters containing these velocities. The flow pattern is also influenced by a number of secondary variables but it is impossible to represent their influence using only a two-dimensional plot. The use of the actual superficial phase velocities for the axes of the map restricts its application to one particular situation but, whereas the choice of a more generalized parameter may be adequate to represent one particular transition, it is unlikely that this same parameter will also be suitable for a different transition governed by a different balance of forces. An alternative and more flexible method which overcomes this difficulty is to examine each transition individually and derive a criterion valid for that particular transition. This latter approach is, however, still being developed.

1.6.4.1   *Vertical flow*   Figure 1.5 shows a flow pattern map obtained (Hewitt and Roberts 1969) from observations on low-pressure air–water and high-pressure steam–water flow in small diameter (1–3 cm) vertical tubes. The axes represent the superficial momentum fluxes of the liquid ($\rho_f j_f^2$) and vapour ($\rho_g j_g^2$) phases respectively. These superficial momentum fluxes can also be expressed in terms of the mass velocity (G) and the vapour

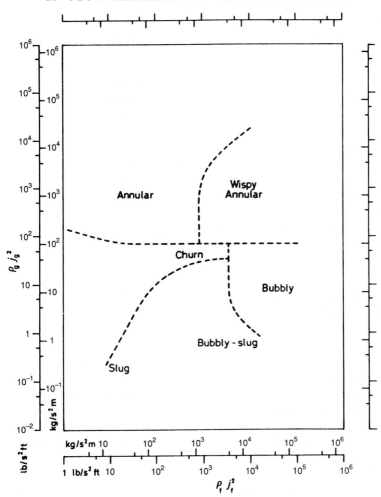

Fig. 1.5. Flow pattern map for vertical flow (Hewitt and Roberts 1969).

quality (x), viz,

$$\rho_f j_f^2 = \frac{[G(1-x)]^2}{\rho_f}; \qquad \rho_g j_g^2 = \frac{[Gx]^2}{\rho_g}$$

This map must be regarded as no more than a rough guide. It is clear that the momentum fluxes alone are not adequate to represent the influence of fluid physical properties or channel diameter.

1.6.4.2   *Horizontal flow*   One flow pattern map, that of Baker (1954) is widely used in the petrochemical industry. The original diagram given by

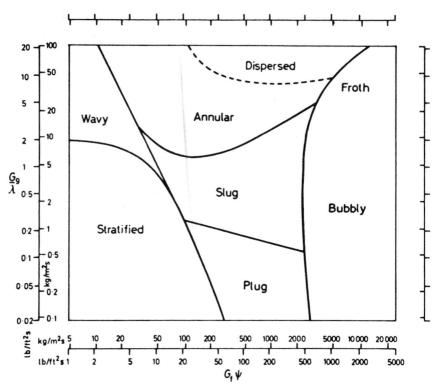

Fig. 1.6. Flow pattern map for horizontal flow (Baker 1954).

Baker has been transposed in Fig. 1.6. G_g and G_f are the superficial mass velocities of the vapour and liquid phases respectively given by eq. (1.11). The factors λ and ψ are given by the relationships

$$\lambda = \left[\left(\frac{\rho_g}{\rho_A}\right)\left(\frac{\rho_f}{\rho_W}\right)\right]^{1/2} \tag{1.17}$$

and

$$\psi = \left(\frac{\sigma_W}{\sigma}\right)\left[\left(\frac{\mu_f}{\mu_W}\right)\left(\frac{\rho_W}{\rho_f}\right)^2\right]^{1/3} \tag{1.18}$$

The subscripts A and W refer to the values of the physical properties for air and water respectively at atmospheric pressure and temperature. Figure 1.6 may be compared with the flow map for vertical channels (Fig. 1.5). Both diagrams indicate that for an air–water flow at atmospheric pressure ($\lambda = \psi = 1$) bubbly flow is entered for superficial liquid velocities (j_f) above 2.5 m/s and annular flow occurs for superficial gas velocities above 10 m/s.

1.6.4.3 *Taitel unified flow pattern model* A comprehensive method of predicting flow pattern transitions over the complete range of pipe inclinations has been put forward by Taitel (1990). This model relies on previous work by Taitel and Dukler (1976, 1977), Dukler (1978) and Barnes and Taitel (1986). Taitel use the term *intermittent flow* to cover elongated bubble, slug and churn flows. A considerable number of transition borders are considered.

(a) *Stratified to non-stratified* (Taitel and Dukler 1976) In horizontal flows it is assumed that stratified flow will prevail unless the interface becomes unstable as a result of Kelvin–Helmholz instability. This instability arises at the interface of two fluid layers of different density flowing horizontally with different velocities. For this transition Fig. 1.7 is used with the co-ordinates F and X, where X is the Lockhart–Martinelli parameter defined later by eq. (2.67) and F is given by

$$F = \left(\frac{\rho_g}{\rho_f - \rho_g}\right)^{0.5} \frac{j_g}{(Dg \cos \theta)^{0.5}} \tag{1.19}$$

where D is the pipe diameter and θ is the angle of inclination of the pipe to the horizontal (positive for downflow).

For inclined pipes a further parameter Y has to be introduced:

$$Y = \frac{(\rho_f - \rho_g)g \sin \theta}{(dp/dz_F)_g} \tag{1.20}$$

Figure 1.7 is valid only for horizontal flow ($Y = 0$). Each pipe inclination requires a separate map. First the equilibrium level in stratified flow (h_f/D) is calculated as a function of X and Y using Fig. 1.8. Then using this same diagram a modified 'X' for the horizontal case ($Y = 0$) is established which has the same value of equilibrium liquid level. Then the previous procedure using Fig. 1.7 is used to determine the flow pattern on the basis of F and the modified value of 'X'.

For the transition from stratified to wavy flow in horizontal tubes Fig. 1.7 can be used with the co-ordinates K and X, where K is given by

$$K = \left[\frac{\rho_g j_g^2 j_f}{(\rho_f - \rho_g)gv_f \cos \theta}\right]^{0.5} \tag{1.21}$$

where v_f is the liquid kinematic viscosity.

(b) *Bubbly flow transition* A number of mechanisms exist which determine if bubbly flow will be present:

(i) Individual bubbles will collide and coalesce to form larger bubbles. Radovcich and Moissis (1962) examined the behaviour of an idealized model

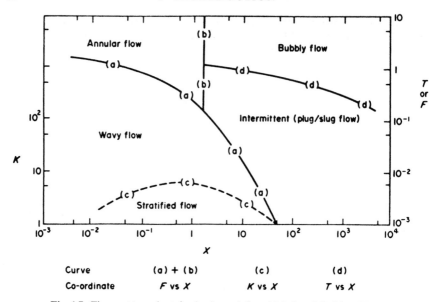

Curve	(a) + (b)	(c)	(d)
Co-ordinate	F vs X	K vs X	T vs X

Fig. 1.7. Flow pattern chart for horizontal flow (Taitel and Dukler 1976).

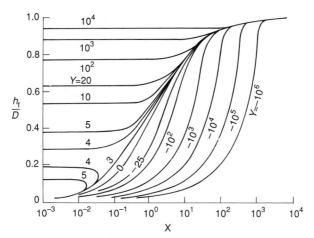

Fig. 1.8. Equilibrium liquid level in stratified flow.

of bubbly flow and came to the conclusion that for void fractions (α) below about 0.1 the collision frequency would be relatively low. Above this value the frequency rose steeply until at $\alpha = 0.3$ the number of collisions was such that a rapid transition to slug flow was to be expected. Taitel recommends a transition condition of $\alpha > 0.25$.

(ii) Turbulent fluctuations in the flow will tend to break up and disperse bubbles. Dispersed bubble flow will exist if the stable maximum diameter of the dispersed bubbles d_c is less than the critical diameter d_{CRIT}. The value of d_c is given by

$$d_c = (0.725 + 4.25\beta^{0.5})(\sigma/\rho_f)^{0.6}K^{-0.4} \qquad (1.22)$$

where β is the volumetric quality given by eq. (1.9) and K is the rate of energy dissipation per unit mass

$$K = \frac{2f_{TP}}{D}j^3 \qquad (1.23)$$

where j is the volumetric flux given by eq. (1.9) and f_{TP} is the friction factor based on this volumetric flux given by eq. (2.41). The critical diameter is given by

$$d_{CRIT} = 2\left[\frac{0.4\sigma}{(\rho_f - \rho_g)g}\right]^{0.5} \qquad (1.24)$$

(iii) Gravity forces in a horizontal or inclined pipe cause bubbles to rise and coalesce at the top of the pipe to form elongated bubbles. The critical bubble size for this to occur is

$$d_{CRIT} = \frac{3}{8}\left(\frac{\rho_f}{\rho_f - \rho_g}\right)\left(\frac{f_{TP}j^2}{g\cos\theta}\right) \qquad (1.25)$$

Finally at high void fractions, greater than 50 per cent, dispersed bubbles tend to coalesce, the flow pattern is not stable and intermittent (slug or churn) flow occurs.

Bubbly flow also does not take place in small pipes. The pipe diameter must be greater than

$$D > \frac{2.34\sin^2\theta}{(0.35\sin\theta + 0.54\cos\theta)^2}\left[\frac{(\rho_f - \rho_g)\sigma}{g\rho_f^2}\right]^{0.5} \qquad (1.26)$$

Bubbly flow is more characteristic of vertical and steeply inclined pipes, typically $\theta > 55$–$70°$. Taitel gives an equation to calculate the critical angle below which bubbly flow will not take place because buoyancy forces dominate.

For the transition from bubbly flow to intermittent flow, Fig. 1.7 can be used with the co-ordinates T and X, where T is given by

$$T = \left[\left(\frac{dp}{dz}F\right)_f \bigg/ (\rho_f - \rho_g)g\cos\theta\right]^{0.5} \qquad (1.27)$$

where $(dp/dzF)_f$ is the frictional pressure gradient assuming the liquid to flow alone in the pipe (see Chapter 2).

(c) *Transition to annular flow*   For transition from bubbly to annular flow in horizontal pipes (Taitel and Dukler 1976) it will be seen from Fig. 1.7 that X is constant at a value of 1.6. In vertical flow the transition between intermittent (churn) flow and annular flow is most easily understood by considering a vertical tube having a section of porous wall at its midpoint and fed at its base with an upwards gas flow. Liquid fed through the porous wall section will form a film which will travel *downwards* if the gas flow is low or *upwards* if the gas flow is high. Clearly, the transition between these two regions, the 'flow reversal' point can be associated with the lower limit of annular flow. Wallis (1969) has found that the critical gas velocity and the 'flow reversal' point can be characterized by the criterion

$$j_g^* = 0.9 \tag{1.28}$$

where

$$j_g^* = j_g \rho_g^{0.5} [gD(\rho_f - \rho_g)]^{-0.5} \tag{1.29}$$

This criterion may be used to give an approximate prediction of the churn flow–annular flow transition. Jones and Zuber (1978) have examined this transition for various geometries. They recommend

$$j_g^* = 4\left(\frac{\rho_g}{\rho_f}\right)^{0.5} [j_f^* + K] \tag{1.30}$$

where K is the constant in the equation for the bubble rise velocity (cf. 0.35 for round tubes) and j_f^* is given by

$$j_f^* = j_f \rho_f^{0.5} [gD(\rho_f - \rho_g)]^{-0.5} \tag{1.31}$$

Values for K are given in Table 1.2 for various geometries.

Taitel and Dukler (1977) propose the following equations for the transition between churn (or slug) flow and annular flow

$$\frac{j_g \rho_g^{0.5}}{[g(\rho_f - \rho_g)\sigma]^{0.25}} = 3.09 \frac{[1 + 20X + X^2]^{0.5} - X}{[1 + 20X + X^2]^{0.5}} \tag{1.32}$$

This equation simplifies to

$$j_g = 3.09 \rho_g^{-0.5} [g(\rho_f - \rho_g)\sigma]^{0.25} \qquad \text{for } X \ll 1 \tag{1.33}$$

and to

$$j_g = 30.9 \frac{[g(\rho_f - \rho_g)\sigma]^{0.25}}{\rho_g^{0.5} X} \qquad \text{for } X \gg 1 \tag{1.34}$$

X is the Lockhart–Martinelli parameter defined by eq. (2.67).

Table 1.2 Values of the coefficient K and characteristic dimension D for use in eq. (1.30)

Geometry	K	Characteristic dimension D
Tube	0.35	Tube diameter D
Rectangular duct	$0.23 + 0.13 \dfrac{S}{w}$	Duct width w
Annulus	$0.35 + 0.058 \dfrac{D_i}{D_e}$	External diameter D_0
Rod bundle	$0.35 + 0.8 \left[1 - \dfrac{D_e}{D_0} \right]^3$	Hydraulic equivalent diameter D_e

(d) *Annular flow–wispy annular flow transition*   This transition is difficult to distinguish visually and the information upon which to base a transition criterion is sparse. Bennett *et al.* (1965) observed that for high-pressure steam–water flow at pressures of 34.5 bar (500 psia) and 69 bar (1000 psia) respectively this transition could be represented by approximately constant superficial liquid velocities (j_f) of about 1.05 m/s (3.5 ft/s) and 1.35 m/s (4.5 ft/s). Filaments of agglomerated entrainment may be detected by an electrical probe (Hewitt 1978) and on the basis of such measurements Wallis (1969) has suggested that at high liquid flow rates this transition is given by

$$j_g = \left[7 + 0.06 \frac{\rho_f}{\rho_g} \right] j_f \tag{1.35}$$

for the case where $j_f^* > 1.5$.

The wispy annular region usually corresponds to the region where j_g^* is greater than unity and j_f^* is greater than 2.5 to 3.

Taitel's unified method is somewhat complex and it does require the use of a logical sequence to determine the dominant flow pattern. The recommended sequence is:

1. *Bubbly flow*   Check if the flow pattern is bubbly flow. Dispersed bubble flow occurs if

(a) d_c given by eq. (1.22) is less than d_{CRIT} given by eq. (1.24) or (1.25) and
(b) the void fraction is less than 0.5.

If these criteria are not satisfied, check for bubbly flow. Bubbly flow occurs if

(a) the pipe diameter is large (eq. (1.26))
(b) the inclination of the pipe is steep ($\theta > 55$–$70°$)
(c) the void fraction is low ($\alpha < 0.25$)
(d) the flow is not annular.

2. *Stratified flow*   If the flow is not bubbly, check if it is stratified via the Kelvin–Holmholz analysis (Fig. 1.7 and Fig. 1.8).

3. *Annular flow*   If the flow is not bubbly or stratified flow, next check for annular flow (Fig. 1.7) and eqns. (1.28)–(1.34).

4. *Intermittent flow*   Finally, if the flow is not bubbly, stratified, or annular, it must be one of the following intermittent flows:

elongated bubble flow	$\alpha \simeq 0$
slug flow	$0 < \alpha \leqslant 0.50$
churn flow	$0.5 \leqslant \alpha$

Taitel states that this sequence of tests, best done via a computer program, provides a unique flow pattern map for a wide range of parameters.

1.6.5   *Flow patterns in other applications*

A limited amount of information is available in the published literature on the flow patterns likely to be encountered in a variety of specific applications. The source of this information is listed below under the particular application.

(a) *Vertical downward flow.* Golan and Stenning (1969) and also Oshinowo and Charles (1974) have given tentative maps for co-current downwards flow. Annular flow occupies the major area since at low liquid flow rates a falling liquid film can occur with no gas flow. Slug flow and bubbly flow also occur but only at liquid velocities greater than the bubble rise velocity.

(b) *Inclined channels.* Flow patterns in inclined pipes are discussed by Kosterin (1949) and by Brigham et al. (1957). Serious stratification of the phases is observed only for very low superficial velocities and inclinations close to the horizontal.

(c) *Rectangular channels.* A comprehensive study and review of high-pressure steam–water flow patterns in rectangular channels has been reported by Hosler (1967). The flow patterns and transitions encountered were similar to those observed in circular channels.

(d) *Internal grooves, helical inserts, obstructions, expansions and contractions.* Papers by Nishikawa et al. (1968) and by Zarnett and Charles (1968) have discussed the flow patterns which occur when the inside surface of the tube is grooved or a helical ribbon is inserted into a smooth tube. Gardner and Neller (1969) have commented on the flow past ring-type obstructions and Richardson (1958) has considered the changes in flow pattern at contractions and expansions.

(e) *Bends and coils.* The profound influence of bends on two-phase flow patterns has already been mentioned. In general a bend or system of bends acts to separate the phases. For example, a bend will induce the coalescence of bubbles to form slug flow and will separate entrained droplets in annular flow. The presence of a curvature in the pipe induces a secondary flow in the fluid in the form of a double spiral; outwards across a diameter of the tube and towards the centre of curvature along the pipe walls. At low superficial velocities the action of gravitational forces and the fact that vapour phase tends to flow faster than the liquid phase greatly complicates the picture.

Gardner and Neller (1969) have considered the case of a vertical pipe joined to a horizontal pipe via a 90° bend. In this case the momentum of the upflowing liquid is attempting to carry it to the outside of the bend whilst gravitational forces are tending to make it fall to the inside of the bend. This balance between counteracting forces was expressed by a Froude number

$$(Fr)_\theta = \frac{j^2}{gR \sin \theta} \tag{1.36}$$

where j is the total volumetric flux, R is the radius of curvature, and θ the angle of the radius to the horizontal. Gardner and Neller confirmed that values of $(Fr)_\theta$ less than unity resulted in the vapour phase moving to the outside of the bend while values greater than unity caused it to hug the inside radius. This simple approach is probably only valid for the bubbly and slug flow patterns. These ideas can, however, be extended to the case of vertical tubes connected by 180° return bends considered by Golan and Stenning (1969).

Zahn (1964) has visually examined the important case of evaporation in horizontal tubes connected by 180° vertical return bends. Liquid initially thrown to the outside of the bend rapidly drained to the base of the horizontal tube immediately after the bend causing a dry patch to form on the upper part of the tube. A considerable distance may be needed before this dry area is rewetted intermittently by developing liquid slugs or permanently by deposition of droplets entrained from the lower film. The effect on heat transfer rates will be discussed in a later chapter.

The flow patterns observed for two-phase flow in horizontal helical coils (Boyce *et al.* 1968) are similar to those seen in horizontal tubes with the notably absence of bubbly flow. The most significant effects of coiling are seen in annular flow. At the high velocities necessary for annular flow it might be expected that the liquid phase would be forced

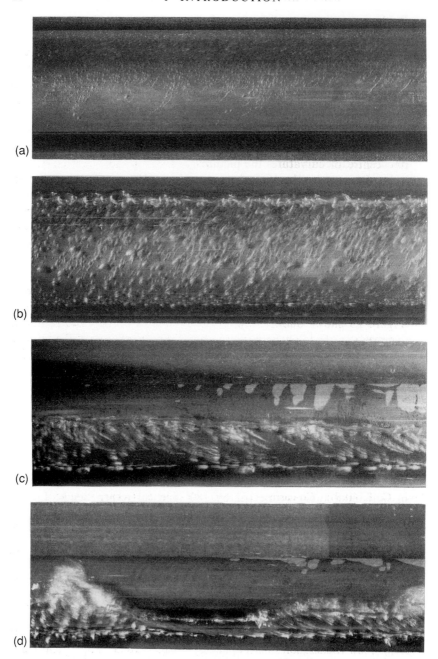

Fig. 1.9. Flow patterns in horizontal annulus. (a) subcooled boiling, (b) saturated boiling, (c) stratified wavy flow with evaporating film, (d) large amplitude wave (Kattan *et al.* 1992).

to the outside of the coil. In fact, however, it has been observed that the liquid travels as a film on the inside surfaces of the coil at low flows. Banerjee *et al.* (1967) have explained this phenomenon on the basis that the momentum of the gas phase is higher than that of the liquid phase due to the considerably higher velocity outweighing the much lower density. High-speed cine photography looking down the axis of a coiled tube has been employed by Hewitt (1978). In the annular flow pattern, secondary flow in the gas core induces a corresponding circulation in the liquid film. Liquid is entrained over an arc of channel wall close to the centre of curvature and is carried with the gas flow across the tube diameter to be deposited on the further wall some distance downstream. The liquid is then returned to the inside surface of the coil still further downstream via the circulation in the film. This intense peripheral circulation of the liquid film and the efficient separation of the entrained droplets ensures that the likelihood of dry patches is considerably reduced in coiled tubes. Similar mechanisms occur in bends when annular flow is present at the entry to the bend.

(f) *Annuli.* Flow patterns in horizontal evaporating flows of refrigerant R-123 in an annulus have been photographed by Kattan *et al.* (1992). Figure 1.9 shows bubbly flow in the subcooled boiling and saturated regimes, stratified flow with intermittent thin-film wetting of the exposed tube (in the light areas the film has dried out) and passage of a large-amplitude wave.

References and further reading

Alves, G. E. (1954). 'Co-current liquid–gas flow in a pipeline contactor'. *Chem. Process. Engng*, **50**(9), 449–456.

Arnold, C. R. and Hewitt, G. F. (1967). 'Further developments in the photography of two-phase gas–liquid flow'. *J. Photo. Sci.*, **15**, 97.

Baker, O. (1954). 'Design of pipe lines for simultaneous flow of oil and gas'. *Oil and Gas J.*, July, 26.

Banerjee, S., Rhodes, E., and Scott, D. S. (1967). 'Film inversion of co-current two-phase flow in helical coils'. *AIChE J.*, **13**(1), 189–191.

Barnea, D. and Taitel, Y. (1986). 'Flow pattern transition in two-phase gas–liquid flows'. *Encyclopedia of fluid mechanics*, Vol. 3, Gulf Publishing, pp. 403–474.

Bennett, A. W., Hewitt, G. F., Kearsey, H. A., Keeys, R. K. F., and Lacey, P. M. C. (1965). 'Flow visualization studies of boiling at high pressure'. Paper presented at Symposium on Boiling Heat Transfer in Steam Generating Units and Heat Exchangers, Manchester 15–16 September.

Boyce, B. E., Collier, J. G., and Levy, J. (1968). 'Hold up and pressure drop measurements in the two-phase flow of air–water mixtures in helical coils'. Paper presented at International Symposium on Research in Co-current Gas–Liquid flow, University of Waterloo, September.

Brigham, W. E., Holstein, E. D., and Huntington, R. L. (1957). 'Two-phase concurrent flow of liquids and air through inclined pipe'. *Oil and Gas J.* 11 November, 145.

Butterworth, D. and Hewitt, G. F. (1977). *Two-phase flow and heat transfer.* Oxford University Press.

Chaudry, A. B., Emerton, A. C., and Jackson, R. (1965). 'Flow regimes in the co-current upwards flow of water and air'. Paper B2 presented at Symposium on Two-phase Flow, Exeter, June.

Chisholm, D. (1983) *Two-phase flow in pipelines and heat exchangers,* George Godwin, London and New York, in association with The Institution of Chemical Engineers.

Cumo, M. and Naviglio, A. (1991). *Thermal hydraulic design of components for steam generation plants,* CRC Press, Boca Raton, FL.

Delhaye, J. M., Giot, M., and Riethmuller, M. L. (1979). *Two-phase flows in nuclear reactors.* Hemisphere Pub. Corp./McGraw-Hill.

Derbyshire, R. T. P., Hewitt, G. F., and Nicholls, B. (1964). 'X-radiography of two-phase gas–liquid flow'. AERE-M 1321.

Dukler, A. E. (1978). 'Modelling two-phase flow and heat transfer'. Keynote paper KS-11 presented at 6th International Heat Transfer Conference, Toronto, Canada, August.

Gardner, G. C. and Neller, P. H. (1969). 'Phase distributions in flow of an air–water mixture round bends and past obstructions at the wall of a 76 mm bore tube'. Paper 12 presented at Symposium on Fluid Mechanics and Measurements in Two-phase Flow Systems, Leeds University, 24–25 September.

Ginoux, J. J. (ed.) (1978). *Two-phase flows and heat transfer (with application to reactor design problems).* Hemisphere Pub. Corp./McGraw-Hill.

Golan, L. P. and Stenning, A. H. (1969). 'Two-phase vertical flow maps'. Paper No. 14 presented at Symposium on Fluid Mechanics and Measurements in Two-phase Flow Systems, Leeds University, 24–25 September.

Gouse, S. W. Jr. (1964). 'An introduction to two-phase gas–liquid flow'. Report 8734-3 Engineering Projects Lab, MIT.

Hetsroni, G. (1982). *Handbook of multiphase systems,* Hemisphere Pub. Corp./McGraw-Hill, Washington, DC.

Hewitt, G. F. (1978). *Measurement of two-phase flow parameters.* Academic Press, London.

Hewitt, G. F. and Hall-Taylor, N. S. (1970). *Annular two-phase flow.* Pergamon Press, Oxford.

Hewitt, G. F. and Roberts, D. N. (1969). 'Studies of two-phase flow patterns by simultaneous X-ray and flash photography'. AERE-M 2159, HMSO.

Hewitt, G. F., Delhaye, J. M., and Zuber, N. (eds.) (1982–90). *Multiphase Science and Technology,* Vols. 1–5. Hemisphere, McGraw-Hill.

Hosler, E. R. (1967). 'Flow patterns in high pressure two-phase (steam–water) flow with heat addition'. AIChE preprint 22 presented at 9th National U.S. Heat Transfer Conference, Seattle, August.

Hsu, Y. Y. and Graham, R. W. (1976). *Transport processes in boiling and two-phase systems.* Hemisphere Pub. Corp./McGraw-Hill.

Irvine, T. J., Jr. and Hartnett, J. P. (eds.) (1964 *et seq*). *Advances in heat transfer*. (eds.), Academic Press, Orlando, FL.

Jones, O. C. and Zuber, N. (1978). 'Slug-annular transition with particular reference to narrow rectangular ducts'. International Seminar, Momentum, Heat and Mass Transfer in Two-phase Energy and Chemical Systems, Dubrovnik, Yugoslavia, September.

Kakac, S. (ed.) (1991). *Boilers, evaporators and condensers*. Wiley.

Kakac, S. and Ishii, M. (eds.) (1983). *Advances in two-phase flow and heat transfer*. Martinus Nijhoff, in cooperation with NATO Scientific Affairs Division.

Kakac, S., Bergles, A. E., and Fernandes, E. O. (eds.) (1988). *Two-phase flow heat exchangers—thermal-hydraulic fundamentals and design*. Kluwer Academic Publishers, in cooperation with NATO Scientific Affairs Division.

Kattan, N., Thome, J. R., and Favrat, D. (1992). 'Convective boiling and two-phase flow patterns in an annulus'. Presented at 10th National Heat Transfer Congress held at Genoa, Italy 25–27 June, 309–320.

Kosterin, S. I. (1949). 'An investigation of the influence of the diameter and inclination of a tube on the hydraulic resistance and flow structure of gas–liquid mixtures'. *Izvest. Akad. Nauk. S.S.S.R. Otdel Tekh. Nauk*, **12**, 1824.

Nishikawa, K. *et al.* (1968). 'Two-phase annular flow in a smooth tube and grooved tubes'. Paper presented at International Symposium on Research in Co-current Gas–Liquid Flow. University of Waterloo, September.

Oshinowo, T. and Charles, M. E. (1974). 'Vertical two-phase flow. Part I. Flow pattern correlations'. *The Canadian J. Chem. Engng.*, **52**, 25–35.

Radovcich, N. A. and Moissis, R. (1962). 'The transition from two-phase bubble flow to slug flow'. Report No. 7-7673-22, Dept. of Mech. Engng., MIT.

Richardson, B. L. (1958). 'Some problems in horizontal two-phase two-component flow'. ANL-5949.

Schlünder, E. U. (editor in chief) (1983). *Heat exchanger design handbook*. Hemisphere Pub. Corp., Washington, DC.

Smith, R. A. (1986). *Vaporizers—selection, design and operation*. Longmans Scientific and Technical.

Taitel, Y. (1990). 'Flow pattern transition in two-phase flow'. Paper presented at 9th Int. Heat Transfer Conference, Jerusalem.

Taitel, Y. and Dukler, A. E. (1976). 'A model for predicting flow regime transitions in horizontal and near horizontal gas–liquid flow'. *AIChE J.* **22**, 47–55.

Taitel, Y. and Dukler, A. E. (1977). 'Flow regime transitions for vertical upward gas–liquid flow: a preliminary approach through physical modelling. Paper presented at AIChE 70th Annual Meeting, New York, Session on Fundamental Research in Fluid Mechanics.

Thome, J. R. (1990). *Enhanced boiling heat transfer*. Hemisphere Pub. Corp., Washington, DC.

Tong, L. S. (1965). *Boiling heat transfer and two-phase flow*. Wiley, New York.

Wallis, G. B. (1969). *One dimensional two-phase flow*. McGraw-Hill, New York.

Whalley, P. E. (1987). *Boiling, condensation and gas–liquid flows*. Clarendon Press, Oxford.

Zahn, W. R. (1964). 'A visual study of two-phase flow while evaporating in horizontal tubes'. *J. Heat Transfer*. August, 417–429.

Zarnett, G. D. and Charles, M. E. (1968). 'Co-current gas–liquid flow in horizontal tubes with internal spiral ribs'. Paper presented at International Symposium on Research in Co-current Gas–Liquid Flow. University of Waterloo, September.

Problems

1. In a vertical annular flow pattern the thickness of the liquid film on the tube wall is δ. The tube diameter is D. If $\delta \ll D$ what is the value of the void fraction, α?

   Answer: $(1 - \alpha) = \dfrac{4\delta}{D}$

2. Derive an expression for β, the volumetric quality, in terms of the mass quality, x, and the specific volumes of the phases.

   Answer: $\beta = \dfrac{1}{1 + [(1 - x)v_f/xv_g]}$

3. Derive an expression for the ratio u_g/u_f (the slip ratio) in terms of α and β only.

   Answer: $\dfrac{u_g}{u_f} = \left(\dfrac{\beta}{1 - \beta}\right)\left(\dfrac{1 - \alpha}{\alpha}\right)$

4. Derive an expression for j in terms of the inlet velocity to an evaporating tube, u_{fo}, the mass quality, x, and the phase specific volumes, v_g and v_f.

   Answer: $j = u_{fo}\left[1 + x\left(\dfrac{v_g}{v_f} - 1\right)\right]$

5. Using Fig. 1.5 evaluate the most likely flow pattern occurring in a 2.54 cm i.d. vertical boiler tube when the system pressure is 30 bar, 70 bar, and 170 bar, the mass quality, x, is 1 per cent, 10 per cent, and 50 per cent, and the mass velocity, G, is 500 kg/m^2 s and 2000 kg/m^2 s respectively.

6. Compare the location of the churn flow–annular flow pattern boundary as predicted by Fig. 1.5 and eq. (1.28) respectively.

7. Repeat the exercise given in Problem 5 but in this case assume the tube is horizontal and use Fig. 1.6.

8. Repeat the exercise given in Problem 5 but only for the 30 bar case using Taitel's unified method. Compare the results with those from Fig. 1.5.

9. Determine the flow patterns for vertical and horizontal flow of refrigerant R-12 (using Figs. 1.5 and 1.6) inside a 10 mm internal diameter tube at 0°C (3.09 bar) for a flow rate of 0.04 kg/s and vapour quality of 20%. (Properties: liquid density = 1397 kg/m³, vapour density = 18.1 kg/m³, liquid viscosity = 0.267 kN s/m² and vapour viscosity = 0.01183 kN s/m²).

2

THE BASIC MODELS

2.1 Introduction

In this chapter the two most important models of two-phase gas–liquid flow—the 'homogeneous' model and the 'separated flow' model—will be discussed. These treatments have been used in engineering design calculations in a number of diverse applications. Such applications often involve high pressures and temperatures and specialized treatments as applied to particular flow patterns may not be possible since the flow pattern or, more likely, the several successive flow patterns, may be unknown. The treatments discussed have therefore been applied to the calculation of pressure drops and density without particular reference to the flow pattern.

2.2 Basic equations of two-phase flow

The general equations for two-phase flows have received considerable attention and have been derived in a number of forms; see, in particular, Bouré (1978), Bouré and Reocreux (1972), Ishii (1990) and Delhaye (1990). The equations may be written in terms of the local instantaneous conditions or in terms of some space-or-time-average conditions. The averaging processes make the equations more tractable but, at the same time, useful information about the flow is lost at each simplifying step. In the present context it is appropriate to consider only very simple one-dimensional steady state relationships.

A simplified one-dimensional analysis of a multi-phase flow can be made by considering the system shown in Fig. 2.1. This shows a stratified multi-phase flow in an inclined channel under conditions where there is mass transfer between the phases. A stratified flow is chosen to allow the equations to be derived for the general case where each phase is in contact with the channel wall as well as having a common interface. Mean values of velocity and density of each phase are assumed to exist across any phase normal to the flow. It is further assumed that the pressure across any phase normal to the channel is uniform (which is strictly not true for a stratified flow) and that the sum of the areas occupied by the phases in any phase normal to the channel axis equals the channel cross-sectional area, A.

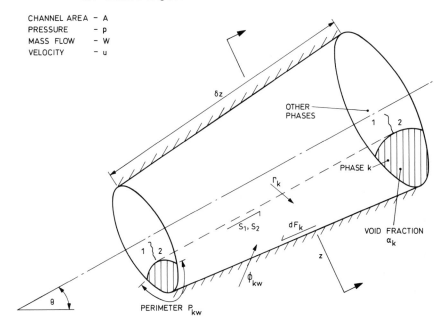

CHANNEL AREA – A
PRESSURE   – p
MASS FLOW   – W
VELOCITY   – u

Fig. 2.1. Simplified model for multi-phase flow in an element of channel.

2.2.1 *Conservation of mass*

The equations expressing the conservation of mass in the channel in the absence of any removal or addition of mass through the channel walls are

$$\frac{\partial}{\partial t}(A\alpha_k\rho_k) + \frac{\partial}{\partial z}(A\alpha_k\rho_k u_k) = \Gamma_k \qquad (2.1)$$

where α_k is the time-averaged void fraction of phase k, ρ_k is the density of phase k and u_k is the mass-weighted mean velocity of phase k. The term denoted by Γ_k represents the mass transfer (mass generation rate per unit length) to phase k from the various interphase mass transfers:

$$\sum_k \Gamma_k = 0$$

For steady state two-phase gas(g)/liquid(f) flow in a constant-area channel this reduces to

$$\left.\begin{aligned}\frac{d}{dz}(A_g\rho_g u_g) &= \Gamma_g \\[2mm] \frac{d}{dz}(A_f\rho_f u_f) &= \Gamma_f\end{aligned}\right\} \qquad (2.2)$$

and

$$\Gamma_g = -\Gamma_f = \frac{dW_g}{dz} = -\frac{dW_f}{dz} \qquad (2.3)$$

These equations are formed and may also be transposed to various other forms using the interrelationships derived in Section 1.4.

2.2.2 Conservation of momentum

The rate of creation of momentum of phase k within the control element plus the rate of inflow of momentum is balanced against the sum of the forces acting on that phase in the control element. The rate of creation plus the rate of inflow of phase k momentum directed along the duct axis is given by

$$\frac{\partial}{\partial t}(W_k\, \delta z) + \left(W_k u_k + \delta z\, \frac{\partial}{\partial z}(W_k u_k) \right) - W_k u_k$$

which simplifies to

$$\frac{\partial}{\partial t}(W_k\, \delta z) + \delta z\, \frac{\partial}{\partial z}(W_k u_k)$$

Other forms are possible since $W_k = A\alpha_k \rho_k u_k$.

This is balanced against the sum of the forces acting in the z direction on phase k in the control element plus the momentum generation due to mass transfer, given by

$$\left[A\alpha_k p - \left(A\alpha_k p + \delta z\, \frac{\partial}{\partial z}(A\alpha_k p) \right) - \left\{ p\left(-\delta z\, \frac{\partial}{\partial z}(A\alpha_k) \right) \right\} \right]$$

$$- A\alpha_k \rho_k\, \delta z\, g \sin\theta - \tau_{kw} P_{kw}\, \delta z + \sum_{1}^{n} \tau_{knz} P_{kn}\, \delta z + u_k \Gamma_k$$

The term in the square brackets represents the pressure forces on the element and the second term the gravitational forces. The third term represents the wall shear force (dF_k), where τ_{kw} is the wall shear stress between the phase k and the channel wall and P_{kw} is the contact perimeter between the wall and phase k. The fourth term is the sum of the interfacial shear forces (S_1, S_2, etc.) where τ_{knz} is the z component of the interfacial shear stress between phase k and phase n and P_{kn} is the contact perimeter between phase k and phase n. The final term is the rate of generation of momentum of phase k due to mass transfer assuming the mass transferred across the interface is accelerated to the mean velocity of the receiving phase. Of course, either P_{kw} or P_{kn} may be zero depending on the disposition of the phases. Furthermore, the interfacial shear stress may be influenced by interfacial mass transfer (see Section 10.7). If the sum of the forces on phase k is now equated to the rate of creation of momentum for that phase, we get

$$-A\alpha_k \frac{\partial p}{\partial z} \delta z - \tau_{kw} P_{kw} \delta z + \sum_1^n \tau_{knz} P_{kn} \delta z - A\alpha_k \rho_k \delta z g \sin \theta + u_k \Gamma_k$$

$$= \frac{\partial}{\partial t}(W_k \delta z) + \delta z \frac{\partial}{\partial z}(W_k u_k) \quad (2.4)$$

Thus for a steady-state two-phase gas(g)/liquid(f) flow in a constant area channel we have

$$-A_g \, dp - \tau_{gw} P_{gw} \, dz + \tau_{gf} P_{gf} \, dz - A_g \rho_g \, dz \, g \sin \theta + u_g \Gamma_g = W_g \, du_g \quad (2.5)$$

and

$$-A_f \, dp - \tau_{fw} P_{fw} \, dz + \tau_{fg} P_{fg} \, dz - A_f \rho_f \, dz \, g \sin \theta + u_f \Gamma_f = W_f \, du_f \quad (2.6)$$

Adding eqs. (2.5) and (2.6) and using the fact that conservation of momentum across the interface requires that

$$\tau_{gf} P_{gf} \, dz + u_g \Gamma_g = \tau_{fg} P_{fg} \, dz + u_f \Gamma_f$$

we get

$$-A \, dp - \tau_{gw} P_{gw} \, dz - \tau_{fw} P_{fw} \, dz - g \sin \theta [A_f \rho_f + A_g \rho_g]$$

$$= d(W_f u_f + W_g u_g) \quad (2.7)$$

This equation represents the basic differential momentum equation for this simplified one-dimensional approach. Note that the interfacial shear terms and the terms due to momentum exchange from mass transfer do not appear since they sum to zero. The net frictional force acting on each phase may be expressed in terms of the ones occupied by each phase as

$$(dF_g + S) = -\tau_{gw} P_{gw} \, dz - \tau_{gf} P_{gf} \, dz = -A_g \left(\frac{dp}{dz} gF\right) dz$$

$$(dF_f - S) = -\tau_{fw} P_{fw} \, dz + \tau_{gf} P_{gf} \, dz = -A_f \left(\frac{dp}{dz} fF\right) dz \qquad (2.8)$$

$$(dF_g + dF_f) = -\tau_{gw} P_{gw} \, dz - \tau_{fw} P_{fw} \, dz = -A \left(\frac{dp}{dz} F\right) dz$$

The term $(dp/dz \, F)$ represents that part of the overall static pressure gradient required to overcome friction. Substitution of eq. (2.8) into eq. (2.7) and rearrangement yields

$$\left(\frac{dp}{dz}\right) = \left(\frac{dp}{dz}F\right) + \left(\frac{dp}{dz}a\right) + \left(\frac{dp}{dz}z\right) \tag{2.9}$$

where again using of the relationships derived in Section 1.4:

$$-\left(\frac{dp}{dz}a\right) = \frac{1}{A}\frac{d}{dz}\left(W_g u_g + W_f u_f\right) = G^2 \frac{d}{dz}\left[\frac{x^2 v_g}{\alpha} + \frac{(1-x)^2 v_f}{(1-\alpha)}\right] \tag{2.10}$$

and

$$-\left(\frac{dp}{dz}z\right) = g \sin\theta\left[\frac{A_g}{A}\rho_g + \frac{A_f}{A}\rho_f\right] = g \sin\theta[\alpha\rho_g + (1-\alpha)\rho_f] \tag{2.11}$$

The above derivation illustrates the use of the momentum equation to relate the total static pressure gradient in terms of the separate components of friction, acceleration and static head. It should be explained at this point that the frictional component has been derived in terms of the total wall shear force $(dF_g + dF_f)$.

2.2.3 *Conservation of energy*

The differential energy balance is obtained by equating the rate of increase of total energy for phase k (internal energy plus kinetic energy) within the control element together with the rate at which total energy is convected into the control element to the rate at which heat is added to phase k together with the rate at which work is done on phase k plus the rate at which energy is transferred across the interface to the control element.

The rate of increase of total energy within the control element plus the rate at which energy enters the element in the absence of the addition or subtraction of mass through the channel walls is

$$\frac{\partial}{\partial t}\left[\alpha_k \rho_k\left(\varepsilon_k + \frac{u_k^2}{2}\right)A\,\delta z\right] + W_k\left(\varepsilon_k + \frac{u_k^2}{2}\right)\delta z$$

$$-\left[W_k\left(\varepsilon_k + \frac{u_k^2}{2}\right) - \delta z\frac{\partial}{\partial z}W_k\left(\varepsilon_k + \frac{u_k^2}{2}\right)\right] \tag{2.12}$$

where ε_k is the internal energy per unit mass of phase k.

The rate at which heat enters phase k within the control volume is

$$\phi_{kw}P_{kw}\,\delta z + \sum_1^n \phi_{kn}P_{kn}\,\delta z + \dot{\phi}_k A\alpha_k\,\delta z \tag{2.13}$$

where the three terms are the heat flow via the channel wall over the perimeter P_{kw}, the heat flow via the various interfaces with the other n phases and the internal heat generation for phase k within the control element itself. The rate at which work is done on phase k within the control element is

$$\left[\frac{W_k p}{\rho_k} - \left(\frac{W_k p}{\rho_k} + \delta z \frac{\partial}{\partial z}\left(\frac{W_k p}{\rho_k}\right)\right)\right] - W_k g \sin\theta \, \delta z - pA \, \delta z \frac{\partial \alpha_k}{\partial t}$$

$$+ \Gamma_k \frac{\delta z \, p}{\rho_k} + u_k \sum_1^n \tau_{kn} P_{kn} \, \delta z \qquad (2.14)$$

The term in the square bracket is the work done by pressure forces; the second term is the work done by body forces whilst the remaining terms represent the work done by pressure and shear forces at the interface with the other phases. Additional terms are contributed by the work done by shear forces at the control element surface ('frictional heating') but these are usually neglected except for very high velocity flows.

Finally we have the rate at which energy is added to phase k by virtue of the mass transfer across the interface:

$$\Gamma_k \, \delta z \left(\varepsilon_k + \frac{u_k^2}{2}\right) \qquad (2.15)$$

Equating eq. (2.12) to the sum of eqs. (2.13)–(2.15) we obtain

$$\frac{\partial}{\partial t} A\alpha_k \rho_k \left(\varepsilon_k + \frac{u_k^2}{2}\right) + \frac{\partial}{\partial z} W_k\left(i_k + \frac{u_k^2}{2}\right)$$

$$= -W_k g \sin\theta + \phi_{wk} P_{wk} + \sum_1^n \phi_{kn} P_{kn} + \dot{\phi}_k A\alpha_k$$

$$- pA \frac{\partial \alpha_k}{\partial t} + \Gamma_k\left(i_k + \frac{u_k^2}{2}\right) + u_k \sum_1^n \tau_{kn} P_{kn} \qquad (2.16)$$

where i_k is the enthalpy of phase k per unit mass

$$i_k = u_k + \frac{p}{\rho_k}$$

For a steady two-phase gas(g)/liquid(f) flow in a constant area channel with no internal heat generation ($\dot{\phi}_k = 0$), eq. (2.16) reduces to

$$d\left[W_g\left(i_g + \frac{u_g^2}{2}\right)\right] + W_g g \sin\theta \, \delta z = \phi_{wg} P_{wg} \, \delta z + \phi_{gf} P_{gf} \, \delta z + u_g \tau_{gf} P_{gf} \, \delta z$$

$$+ \Gamma_g \, \delta z\left(i_g + \frac{u_g^2}{2}\right) \qquad (2.17a)$$

and

$$d\left[W_f\left(i_f + \frac{u_f^2}{2}\right)\right] + W_f g \sin\theta\,\delta z = \phi_{wf}P_{wf}\,\delta z + \phi_{fg}P_{fg}\,\delta z + u_f\tau_{fg}P_{fg}\,\delta z$$

$$+ \Gamma_f\,\delta z\left(i_f + \frac{u_f^2}{2}\right) \qquad (2.17b)$$

Adding eqs. (2.17), noting that conservation of energy across the gas–liquid interface requires that

$$\Gamma_g\left(i_g + \frac{u_g^2}{2}\right) + \phi_{gf}P_{gf} + u_g\tau_{gf}P_{gf} = \Gamma_f\left(i_f + \frac{u_f^2}{2}\right) + \phi_{fg}P_{fg} + u_f\tau_{fg}P_{fg}$$

we arrive at the equation

$$\frac{d}{dz}\left[W_g i_g + W_f i_f\right] + \frac{d}{dz}\left[\frac{W_g u_g^2}{2} + \frac{W_f u_f^2}{2}\right] + (W_g + W_f)g\sin\theta = Q_{wl} \quad (2.18)$$

where $Q_{wl}\,(=\phi_{wf}P_{wf} + \phi_{wg}P_{wg})$ is the heat transferred to the fluid across the channel wall per unit channel length. Using the relationships derived from Section 1.4, eq. (2.18) may be written

$$-\frac{dp}{dz}\left[xv_g + (1-x)v_f\right] = \left\{\frac{dE}{dz} - \frac{Q_{wf}}{W}\right\}$$

$$+ \left\{p\,\frac{d}{dz}\left[xv_g + (1-x)v_f\right] + \frac{G^2}{2}\frac{d}{dz}\left[\frac{x^3v_g^2}{\alpha^2} + \frac{(1-x)^3v_f^2}{(1-\alpha)^2}\right]\right\} + g\sin\theta \quad (2.19)$$

where $E = x\varepsilon_g + (1-x)\varepsilon_f$ is the flow-weighted mixture internal energy per unit mass; W is the total mass flow rate.

Once again it will be seen that the total pressure gradient can be expressed in terms of a frictional dissipation term (first bracketed term), an accelerational head term (second bracketed term) and a static (potential) head term (final term). In the case of the energy equation the frictional dissipation term $(dE/dz - Q_{wl}/W)$ includes the dissipation of mechanical energy not only within the fluid due to friction at the channel walls but also at the interface due to the relative motion of the phases. For many applications the internal heat generation and dissipation, the kinetic energy and the potential energy are negligible compared with the enthalpy and the external heat inputs. This simplifies the energy relationships when it comes to calculating changes in quality along a heated channel, as we shall see in Chapter 4.

2.2.4 *Use of the momentum or energy equation to evaluate the pressure gradient*

It is possible to use either the momentum equation or the energy equation as a starting point for the evaluation of the pressure gradient in two-phase flow. When experimental pressure drop data are to be correlated the procedure is as follows:

Using the momentum equation
(a) Use the measured void fraction (α) or that calculated from some arbitrary model (i.e., 'homogeneous' or 'separated flow') to calculate the acceleration (eq. (2.10)) and the static head (eq. (2.11)) components respectively.
(b) Calculate the frictional pressure gradient by difference and correlate against the independent variables.

Using the energy equation
(a) Calculate the pressure drop due to the change in potential energy of the fluid. Void fraction is not required.
(b) Calculate the pressure drop due to the change in kinetic energy of the fluid using some arbitrary model (i.e., 'homogeneous' or 'separated flow').
(c) Calculate the frictional (or 'irreversible' or 'viscous dissipation') pressure loss by difference and correlate against the independent variables.

Both methods require information about the void fraction though its importance in the two cases may differ. The energy balance approach has some advantages in that it avoids the difficulty of a negative friction component (wall shear stress) which is sometimes found in slug or annular flow when there is a reverse flow of liquid at the channel wall.

The majority of workers have, however, adopted the momentum balance as the basic equation and have defined the frictional component in terms of the wall shear force (eq. (2.8)).

The benefits to be derived from adopting the energy balance approach have only been examined briefly (Hughmark and Pressburg 1961). What follows is therefore based almost entirely on the momentum balance approach. The object is to show how the basic models have been developed, to compare them with experimental data and to show how they may be corrected to take into account second order effects.

2.3 The homogeneous model

The 'homogeneous' model, also known as the 'friction factor' or 'fog flow' model, considers the two phases to flow as a single phase possessing mean fluid properties. The model has been in use in various forms in the steam generation, petroleum, and refrigeration industries for a considerable time.

2.3.1 Derivation of the model and assumptions

The basic premises upon which the model is based are the following assumptions:

(a) equal vapour and liquid velocities,
(b) the attainment of thermodynamic equilibrium between the phases, and
(c) the use of a suitably defined single-phase friction factor for two-phase flow.

From the consideration of flow patterns in Chapter 1 it might be expected that this model would be valid for the bubbly and wispy-annular flow patterns, particularly at high linear velocities and pressures. However, the model has often been applied indiscriminately to problems in which other flow patterns would be expected. For steady homogeneous flow the basic equations reduce to the following form:

Continuity $\qquad W = A\bar{\rho}\bar{u}$ $\qquad\qquad\qquad\qquad\qquad\qquad$ (2.20)

Momentum $\qquad -A\,dp - d\bar{F} - A\bar{\rho}g\sin\theta\,dz = W\,d\bar{u}$ $\qquad\quad$ (2.21)

Energy $\qquad \delta q - \delta w = di + d\left(\dfrac{\bar{u}^2}{2}\right) + g\sin\theta\,dz$ $\qquad$ (2.22)

where $di = \delta q + dE + \bar{v}\,dp$.

In the above equations $\bar{\rho}$, $\bar{v}$, and $\bar{u}$ represent the average density, specific volume, and velocity of the homogeneous fluid. $d\bar{F}$ represents the net total wall shear force. The homogeneous fluid specific volume $\bar{v}$ is defined as the total volumetric flow rate Q divided by the total mass flow rate W. Using the relationships presented in Section 1.4,

$$\bar{v} = \frac{Q}{W} = [xv_g + (1-x)v_f] = [v_f + xv_{fg}] = \frac{j}{G} = \frac{1}{\bar{\rho}} \qquad (2.23)$$

where $\bar{\rho}$ is the homogeneous fluid density.

From the premise (a) above,

$$u_f = u_g = \bar{u} \qquad\qquad\qquad\qquad (2.24)$$

so that

$$\bar{u} = G\bar{v} = j \qquad\qquad\qquad\qquad (2.25)$$

and

$$\alpha = \frac{xv_g}{\bar{v}} = \beta, \qquad (1-\alpha) = \frac{(1-x)v_f}{\bar{v}} = (1-\beta) \qquad (2.26)$$

The total wall shear force $d\bar{F}$ can be expressed in terms of a wall shear stress τ_w acting over the inside area of the channel:

$$d\bar{F} = \tau_w P \, dz \tag{2.27}$$

where τ_w can be expressed in terms of a friction factor f_{TP}:

$$\tau_w = f_{TP}\left(\frac{\bar{\rho}\bar{u}^2}{2}\right) \tag{2.28}$$

Remembering eq. (2.8),

$$-\left(\frac{dp}{dz}F\right) = \frac{1}{A}\frac{d\bar{F}}{dz} = \frac{\tau_w P}{A} = \frac{f_{TP}P}{A}\left(\frac{\bar{\rho}\bar{u}^2}{2}\right) \tag{2.29}$$

This is the familiar Fanning equation which for the case of a circular channel $(P/A = 4/D)$ becomes

$$-\left(\frac{dp}{dz}F\right) = \frac{2f_{TP}G^2\bar{v}}{D} = \frac{2f_{TP}Gj}{D} \tag{2.30}$$

From eq. (2.10)

$$-\left(\frac{dp}{dz}a\right) = G\frac{d(\bar{u})}{dz} = G^2\frac{d(\bar{v})}{dz} \tag{2.31}$$

Evaluating $d(\bar{v})/dz$ neglecting the compressibility of the liquid phase

$$\frac{d(\bar{v})}{dz} = v_{fg}\frac{dx}{dz} + x\frac{dv_g}{dp}\left(\frac{dp}{dz}\right) \tag{2.32}$$

From eq. (2.11) and eq. (2.26)

$$-\left(\frac{dp}{dz}z\right) = \bar{\rho}g\sin\theta = \frac{g\sin\theta}{\bar{v}} \tag{2.33}$$

The total static pressure gradient as evaluated from the homogeneous model can be represented by substitution of eqs. (2.30), (2.31), and (2.33) into eq. (2.9) and rearranging:

$$-\left(\frac{dp}{dz}\right) = \frac{\dfrac{2f_{TP}G^2v_f}{D}\left[1 + x\left(\dfrac{v_{fg}}{v_f}\right)\right] + G^2v_f\left(\dfrac{v_{fg}}{v_f}\right)\dfrac{dx}{dz} + \dfrac{g\sin\theta}{v_f[1 + x(v_{fg}/v_f)]}}{1 + G^2x\left(\dfrac{dv_g}{dp}\right)} \tag{2.34}$$

2.3.2 *The two-phase friction factor*

All the terms in eq. (2.34) are definable except one, the two-phase friction factor f_{TP}. To use the homogeneous model it is necessary to apply a suitably defined single-phase friction factor to two-phase flow. A number of different approaches have been made to the definition of this two-phase friction factor.

(a) The friction factor f_{TP} has been assumed equal to that which would have occurred had the total flow been assumed to be all liquid. This friction factor, denoted by the symbol f_{fo}, will be a function of the all-liquid Reynolds number (GD/μ_f) and the pipe relative roughness (ε/D). Equation (2.30) becomes

$$-\left(\frac{dp}{dz}F\right) = \frac{2f_{fo}G^2 v_f}{D}\left[1 + x\left(\frac{v_{fg}}{v_f}\right)\right] = -\left(\frac{dp}{dz}F\right)_{fo}\left[1 + x\left(\frac{v_{fg}}{v_f}\right)\right] \quad (2.35)$$

where

$$-\left(\frac{dp}{dz}F\right)_{fo}$$

is the frictional pressure gradient calculated from the Fanning equation for the total flow (liquid plus vapour) assumed to flow as liquid, i.e.,

$$-\left(\frac{dp}{dz}F\right)_{fo} = \frac{2f_{fo}G^2 v_f}{D} \quad (2.36)$$

The use of f_{fo} in the evaluation of the two-phase frictional pressure gradient does not allow extrapolation to the correct value when $x = 1$, i.e., with single-phase vapour flowing through the conduit. A second method overcomes this difficulty.

(b) The friction factor f_{TP} has been evaluated using a mean two-phase viscosity $\bar{\mu}$ in the normal friction factor relationships. The form of the relationship between $\bar{\mu}$ and the quality x must be chosen to satisfy the following limiting conditions:

$$x = 0, \bar{\mu} = \mu_f; \qquad x = 1, \bar{\mu} = \mu_g \quad (2.37)$$

Possible forms of the relationship are

$$\text{(i)} \quad \frac{1}{\bar{\mu}} = \frac{x}{\mu_g} + \frac{(1-x)}{\mu_f} \quad (2.38)$$

(an equation proposed by McAdams *et al.* (1942))

$$\text{(ii)} \quad \bar{\mu} = x\mu_g + (1-x)\mu_f \quad (2.39)$$

(the definition chosen by Cicchitti *et al.* 1960)

$$\text{(iii)} \quad \bar{\mu} = \bar{\rho}[xv_g\mu_g + (1 - x)v_f\mu_f] \tag{2.40}$$

(suggested by Dukler *et al.* 1964).

Assuming that the friction factor may be expressed in terms of the Reynolds number by the Blasius equation

$$f_{TP} = 0.079[GD/\bar{\mu}]^{-1/4} \tag{2.41}$$

it can be shown that for equation (2.38)

$$-\left(\frac{dp}{dz}F\right) = -\left(\frac{dp}{dz}F\right)_{fo}\left[1 + x\left(\frac{v_{fg}}{v_f}\right)\right]\left[1 + x\left(\frac{\mu_{fg}}{\mu_g}\right)\right]^{-1/4} \tag{2.42}$$

In eqs. (2.35) and (2.42) the two-phase frictional pressure gradient is expressed in terms of the single-phase pressure gradient for the total flow considered as liquid. This method of representation is often useful for comparison and calculation needs. In general

$$-\left(\frac{dp}{dz}F\right) = -\left(\frac{dp}{dz}F\right)_{fo}\phi_{fo}^2 \tag{2.43}$$

where ϕ_{fo}^2 is known as the two-phase frictional multiplier. Equation (2.38) is the most common definition of $\bar{\mu}$ and therefore values of the term

$$\phi_{fo}^2 = \left[1 + x\left(\frac{v_{fg}}{v_f}\right)\right]\left[1 + x\left(\frac{\mu_{fg}}{\mu_g}\right)\right]^{-1/4}$$

are given in Table 2.1 for the steam–water system as a function of quality and pressure.

2.3.3 Use of model to evaluate pressure loss

In order to be of use in the evaluation of pressure drop, eq. (2.34) must be integrated with respect to axial length. In general this must be a stepwise procedure over the interval for which the evaluation is required. Equation (2.34) may be integrated analytically in some cases, provided a number of simplifying assumptions are made. These could be, for example,

(a) that $|G^2x(dv_g/dp)| \ll 1$, i.e., the compressibility of the gaseous phase may be neglected. In many cases this approximation is valid. For instance with a steam–water flow at 82.75 bar (1200 psia), a mass velocity of 4900 kg/m² s (1000 lb/ft² s) and a value of $x = 0.10$ the term $G^2x(dv_g/dp)$ becomes -0.008.

(b) that the term (v_{fg}/v_f) and the friction factor f_{TP} remain constant over the length considered.

Table 2.1 Values of the two-phase frictional multiplier ϕ_{fo}^2 for the homogeneous model steam–water system

$$\phi_{fo}^2 = \left[1 + x\left(\frac{v_{fg}}{v_f}\right)\right]\left[1 + x\left(\frac{\mu_{fg}}{\mu_g}\right)\right]^{-1/4}$$

Steam quality % by wt.	Pressure, bar (psia)								
	1.01 (14.7)	6.89 (100)	34.4 (500)	68.9 (1000)	103 (1500)	138 (2000)	172 (2500)	207 (3000)	221.2 (3206)
1	16.21	3.40	1.44	1.19	1.10	1.05	1.04	1.01	1.0
5	67.6	12.18	3.12	1.89	1.49	1.28	1.16	1.06	1.0
10	121.2	21.8	5.06	2.73	1.95	1.56	1.30	1.13	1.0
20	212.2	38.7	7.8	4.27	2.81	2.08	1.60	1.25	1.0
30	292.8	53.5	11.74	5.71	3.60	2.57	1.87	1.36	1.0
40	366	67.3	14.7	7.03	4.36	3.04	2.14	1.48	1.0
50	435	80.2	17.45	8.30	5.08	3.48	2.41	1.60	1.0
60	500	92.4	20.14	9.50	5.76	3.91	2.67	1.71	1.0
70	563	104.2	22.7	10.70	6.44	4.33	2.89	1.82	1.0
80	623	115.7	25.1	11.81	7.08	4.74	3.14	1.93	1.0
90	682	127	27.5	12.90	7.75	5.21	3.37	2.04	1.0
100	738	137.4	29.8	13.98	8.32	5.52	3.60	2.14	1.0

Of particular interest is the case where liquid is evaporated from an inlet condition at the saturation temperature ($x = 0$) to a vapour–liquid mixture having a mass quality x. For a linear change of x over a length $L(dx/dz =$ constant) it follows that

$$\Delta p = \frac{2f_{TP}LG^2v_f}{D}\left[1 + \frac{x}{2}\left(\frac{v_{fg}}{v_f}\right)\right] + G^2v_f\left(\frac{v_{fg}}{v_f}\right)x + \frac{g\sin\theta L}{v_{fg}x}\ln\left[1 + x\left(\frac{v_{fg}}{v_f}\right)\right]$$

(2.44)

2.3.4 *The application of the homogeneous theory to experimental observations*

The friction factor for use in the homogeneous flow model can be calculated from single-phase flow correlations as discussed in Section 2.3.2 or, alternatively, estimated directly from measured two-phase pressure drops using the procedure outlined in Section 2.2.4. For instance, values of f_{TP} in the range 0.0029–0.0033 have been suggested for low-pressure flashing steam–water flow (Benjamin and Miller 1942; Bottomley 1936–37; Allen 1951) and values

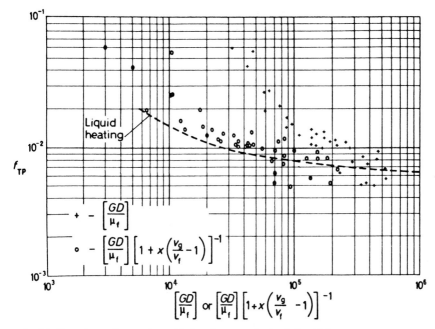

$$\left[\frac{GD}{\mu_f}\right] \text{ or } \left[\frac{GD}{\mu_f}\right]\left[1+x\left(\frac{v_g}{v_f}-1\right)\right]^{-1}$$

Fig. 2.2. Experimental two-phase friction factors, obtained for high pressure steam–water mixtures by Davidson *et al.* (1943).

of about 0.005 for analyses of circulation in high-pressure boilers (Lewis and Robertson 1940; Markson *et al.* 1942) and petroleum pipe stills (Dittus and Hildebrand 1942).

Davidson *et al.* (1943) reduced their high-pressure steam–water pressure drop data by the method outlined in Section 2.2.4. The experimental two-phase friction factor f_{TP} was plotted against the Reynolds number for all-liquid flow $[GD/\mu_f]$. Large discrepancies from the single-phase friction factor were observed at Reynolds numbers less than 2×10^5 (Fig. 2.2). It was found that considerably better agreement with the normal single-phase flow relationship (eq. (2.41)) was obtained if the experimental friction factor was plotted against the all-liquid Reynolds number multiplied by the ratio of the inlet to outlet mean specific volumes. This is equivalent to defining a two-phase viscosity $\bar{\mu}$ as

$$\bar{\mu} = \mu_f\left[1 + x\left(\frac{v_{fg}}{v_f}\right)\right] \qquad (2.45)$$

This equation is a further definition of two-phase viscosity; it does not extrapolate to the vapour phase viscosity as the mass quality x approaches unity. Five possible definitions of mean two-phase viscosity have been given.

One reason for the failure, so far, to establish an accepted definition is that the dependence of the friction factor on viscosity is small.

2.4 The separated flow model

The separated flow model considers the phases to be artificially segregated into two streams; one of liquid and one of vapour. In the model's simplest form each stream is assumed to travel at a mean velocity. For the case where the mean velocities of the two phases are equal the equations reduce to those of the homogeneous model. The separated flow model has been continuously developed since 1944 when Lockhart and Martinelli published their classic papers on two-phase gas–liquid flow.

2.4.1 *Derivation of the model and assumptions*

The basic premises upon which the separated flow model is based are the assumptions of:

(a) constant but not necessarily equal velocities for the vapour and liquid phases,
(b) the attainment of thermodynamic equilibrium between the phases,
(c) the use of empirical correlations or simplified concepts to relate the two-phase friction multiplier (ϕ^2) and the void fraction (α) to the independent variables of the flow.

Again from a consideration of the various flow patterns it would be expected that this model would be most valid for the annular flow pattern. The basic equations for a steady separated flow have been given in Section 2.2. The momentum equation may be rearranged to give

$$-\left(\frac{dp}{dz}\right) = -\left(\frac{dp}{dz}F\right) + G^2\frac{d}{dz}\left[\frac{x^2 v_g}{\alpha} + \frac{(1-x)^2 v_f}{(1-\alpha)}\right] + g\sin\theta[\alpha\rho_g + (1-\alpha)\rho_f]$$

$$(2.46)$$

As stated previously the frictional pressure gradient can be expressed in terms of the single-phase pressure gradient for the *total* flow considered as liquid. Thus

$$-\left(\frac{dp}{dz}F\right) = -\left(\frac{dp}{dz}F\right)_{fo}\phi_{fo}^2 = \left[\frac{2f_{fo}G^2 v_f}{D}\right]\phi_{fo}^2 \qquad (2.47)$$

Alternatively it may be expressed in terms of the single-phase pressure gradient for the *liquid* phase considered to flow alone in the channel

$$-\left(\frac{dp}{dz}F\right) = -\left(\frac{dp}{dz}F\right)_f \phi_f^2 = \left[\frac{2f_f G^2(1-x)^2 v_f}{D}\right]\phi_f^2 \qquad (2.48)$$

Now using the Blasius equation (cf. eq. (2.41))

$$\frac{f_f}{f_{fo}} = \left[\frac{1}{(1-x)}\right]^{1/4} \qquad (2.49)$$

and from eqs. (2.47) and (2.48)

$$\phi_{fo}^2 = \phi_f^2(1-x)^2\frac{f_f}{f_{fo}} = \phi_f^2(1-x)^{1.75} \qquad (2.50)$$

$$-\left(\frac{dp}{dz}a\right) = G^2\frac{d}{dz}\left[\frac{x^2 v_g}{\alpha} + \frac{(1-x)^2 v_f}{(1-\alpha)}\right] \qquad (2.51)$$

Upon expansion, neglecting the compressibility of the liquid phase

$$\frac{d}{dz}\left[\frac{x^2 v_g}{\alpha} + \frac{(1-x)^2 v_f}{(1-\alpha)}\right]$$

$$= \frac{dx}{dz}\left[\left\{\frac{2x v_g}{\alpha} - \frac{2(1-x)v_f}{(1-\alpha)}\right\} + \left(\frac{\partial\alpha}{\partial x}\right)_p\left\{\frac{(1-x)^2 v_f}{(1-\alpha)^2} - \frac{x^2 v_g}{\alpha^2}\right\}\right]$$

$$+ \frac{dp}{dz}\left[\frac{x^2}{\alpha}\frac{dv_g}{dp} + \left(\frac{\partial\alpha}{\partial p}\right)_x\left\{\frac{(1-x)^2 v_f}{(1-\alpha)^2} - \frac{x^2 v_g}{\alpha^2}\right\}\right] \qquad (2.52)$$

The total static pressure gradient as evaluated from the separated flow model can be represented by substitution of eqs. (2.47) and (2.52) into eq. (2.46) and rearranging,

$$-\left(\frac{dp}{dz}\right) = \left[\frac{2f_{fo}G^2 v_f}{D}\phi_{fo}^2 + G^2\frac{dx}{dz}\left[\left\{\frac{2x v_g}{\alpha} - \frac{2(1-x)v_f}{(1-\alpha)}\right\}\right.\right.$$

$$\left.+ \left(\frac{\partial\alpha}{\partial x}\right)_p\left\{\frac{(1-x)^2 v_f}{(1-\alpha)^2} - \frac{x^2 v_g}{\alpha^2}\right\}\right] + g\sin\theta[\rho_g\alpha + \rho_f(1-\alpha)]\right]$$

$$\div\left(1 + G^2\left[\frac{x^2}{\alpha}\left(\frac{dv_g}{dp}\right) + \left(\frac{\partial\alpha}{\partial p}\right)_x\left\{\frac{(1-x)^2 v_f}{(1-\alpha)^2} - \frac{x^2 v_g}{\alpha^2}\right\}\right]\right)$$

$$(2.53)$$

2.4.2   Use of model to evaluate pressure loss

In common with eq. (2.34) (homogeneous model) eq. (2.53) requires, in general, a stepwise integration. Analytical integration is, however, possible provided certain simplifying assumptions are made. These could be, for example,

(a) that

$$\left| G^2 \left[\frac{x^2}{\alpha} \left(\frac{dv_g}{dp} \right) - \left(\frac{\partial \alpha}{\partial p} \right)_x \left\{ \frac{(1-x)^2 v_f}{(1-\alpha)^2} - \frac{x^2 v_g}{\alpha^2} \right\} \right] \right| \ll 1$$

i.e., that the compressibility of the gaseous phase may be neglected;
(b) that the specific volumes (v_g and v_f) and the friction factor f_{fo} remain constant over the length considered.

For the particular case where a fluid is evaporated from liquid at the saturation temperature ($x = 0$) to a vapour–liquid mixture containing a mass quality x, with a linear change of x over a length L ($dx/dz = $ constant)

$$\Delta p = \frac{2 f_{fo} G^2 v_f L}{D} \left[\frac{1}{x} \int_0^x \phi_{fo}^2 \, dx \right] + G^2 v_f \left[\frac{x^2}{\alpha} \left(\frac{v_g}{v_f} \right) + \frac{(1-x)^2}{(1-\alpha)} - 1 \right]$$

$$+ \frac{Lg \sin \theta}{x} \int_0^x \left[\rho_g \alpha + \rho_f (1-\alpha) \right] dx \tag{2.54}$$

2.4.3   The evaluation of the two-phase multiplier (ϕ_{fo}^2) and the void fraction (α)

In order to apply eqs. (2.53) and (2.54) it is necessary to develop expressions for the two-phase multiplier (ϕ_{fo}^2) and the void fraction (α) in terms of the independent flow variables. This was first achieved by Martinelli and co-workers. The Martinelli model was successively developed in the period 1944–49 from a series of studies of isothermal two-phase two-component flow in horizontal tubes. These investigations were thus confined initially to the frictional component of eq. (2.46). These studies culminated in a paper by Lockhart and Martinelli (1949), which proposed a generalized method for calculating the frictional pressure gradient for isothermal two-component flow. A later extension of the work covered the estimation of the accelerative component and resulted in the well-known Martinelli–Nelson (1948) method for the prediction of pressure drop during forced circulation boiling and condensation.

2.4.3.1   The Lockhart–Martinelli (1949) correlation   In this treatment, a definite portion of the flow area is assigned to each phase and it is assumed that conventional friction pressure-drop equations can be applied to the flow

path of each phase. Interaction between the phases is thus ignored and this leads to an inconsistent result for the relationship between pressure gradient and void fraction as will be shown below.

Certain basic postulates were used to establish the correlation:

(a) Four flow regimes were defined on the basis of the behaviour of the flow (viscous or turbulent) when the respective phases were considered to pass alone through the channel.

(b) The liquid and gas phase pressure drops were considered equal irrespective of the details of the particular flow pattern. Also, since the initial concern was with the frictional component of the pressure drop, $(dp/dz\ F)$, only—the acceleration and static head components being assumed negligibly small—then the *frictional* pressure drop in the gas phase must equal the *frictional* drop in the liquid phase irrespective of the flow pattern details:

$$\left(\frac{dp}{dz}\,gF\right) = \left(\frac{dp}{dz}\,fF\right) = \left(\frac{dp}{dz}\,F\right) \tag{2.55}$$

The frictional pressure drop for the liquid flow may be written as

$$-\left(\frac{dp}{dz}\,fF\right) = \frac{2f_f \rho_f u_f^2}{D_f} \tag{2.56}$$

where D_f is the unknown 'hydraulic diameter' of the liquid flow. A similar expression can be written for the gas phase:

$$-\left(\frac{dp}{dz}\,gF\right) = \frac{2f_g \rho_g u_g^2}{D_g} \tag{2.57}$$

The hydraulic diameter of each phase is related to the cross-sectional area $(A_f$ or $A_g)$ through which the phase is flowing at any instant as follows:

$$A_f = \gamma\left(\frac{\pi}{4}\,D_f^2\right) \tag{2.58}$$

$$A_g = \delta\left(\frac{\pi}{4}\,D_g^2\right) \tag{2.59}$$

The friction factor f_f (or f_g) may be expressed in a Blasius type equation

$$f_f = K_f\left[\frac{\rho_f u_f D_f}{\mu_f}\right]^{-n} \tag{2.60}$$

By using eqs. (2.48), (2.56), (2.58), and (2.60) it may be shown that

$$\phi_f^2 = \gamma^{n-2}\left(\frac{D}{D_f}\right)^{5-n} \tag{2.61}$$

In other words, the ratio of the two-phase frictional pressure gradient to that which would exist if the liquid phase were to be flowing alone in the pipe is a function only of the fraction and shape of the flow area occupied by the liquid phase. Similarly for the gas phase,

$$\phi_g^2 = \left(\frac{dp}{dz}F\right)\Bigg/\left(\frac{dp}{dz}F\right)_g = \delta^{n-2}\left(\frac{D}{D_g}\right)^{5-n} \tag{2.62}$$

It is instructive to apply the Lockhart–Martinelli assumptions to the case of annular flow. Dividing eq. (2.58) by $A \ (=\pi D^2/4)$

$$\gamma = (1 - \alpha)(D/D_f)^2 \tag{2.63}$$

Substituting eq. (2.63) into (2.61),

$$\phi_f^2 = (1 - \alpha)^{n-2}\left(\frac{D}{D_f}\right)^{n+1} \tag{2.64}$$

For annular flow with a liquid film of thickness δ,

$$D_f = \frac{4\pi D\delta}{\pi D} = 4\delta; \qquad (1 - \alpha) = \frac{4\pi D\delta}{\pi D^2} = \frac{4\delta}{D}$$

Thus

$$\frac{D}{D_f} = \frac{D}{4\delta} = \frac{1}{(1 - \alpha)}$$

Then substituting into eq. (2.64),

$$\phi_f^2 = (1 - \alpha)^{n-2}(1 - \alpha)^{-(n+1)} = (1 - \alpha)^{-3} \tag{2.65}$$

This result is *incorrect*. The correct relationship for annular flow will be derived in Section 3.5.1. This is:

$$\phi_f^2 = (1 - \alpha)^{-2} \tag{2.66}$$

The discrepancy between eqs. (2.65) and (2.66) highlights the neglect of the interaction between the phases. Chisholm (1967) has corrected the above treatment by allowing for the interfacial shear force (S) between the phases.

Because of the above criticism it is prudent to consider the Lockhart–Martinelli correlation as purely empirical. Martinelli and his co-workers argued that the two-phase friction multipliers ϕ_f^2 and ϕ_g^2 could be correlated uniquely as a function of a parameter X, where

$$X^2 = \left(\frac{dp}{dz}F\right)_f\Bigg/\left(\frac{dp}{dz}F\right)_g \tag{2.67}$$

This was verified using their experimental data. The resulting graphical correlation is shown in Fig. 2.3 where ϕ (note: not ϕ^2) is plotted against X. All four flow regimes were correlated in this manner. In Chapter 3 it is

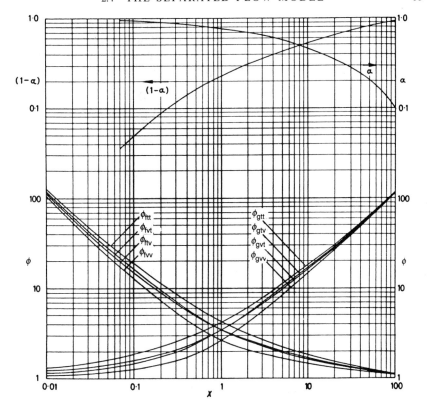

Fig. 2.3. Lockhart–Martinelli (1949) correlation.

shown that the parameter ϕ_f^2 and ϕ_g^2 can be related to the parameter X^2 by relationships of the form

$$\phi_f^2 = 1 + \frac{C}{X} + \frac{1}{X^2} \tag{2.68}$$

and

$$\phi_g^2 = 1 + CX + X^2 \tag{2.69}$$

The curves in Fig. 2.3 are well represented by eqs. (2.68) and (2.69) when C has the following values:

liquid	gas	C
turbulent — turbulent (tt)		20
viscous — turbulent (vt)		12
turbulent — viscous (tv)		10
viscous — viscous (vv)		5

To use the Lockhart–Martinelli correlation to calculate the two-phase friction pressure gradient, it is only necessary to calculate the friction pressure gradients for each phase flowing alone in the channel and then use Fig. 2.3 or alternatively eqs. (2.68) or (2.69). The correlation was developed for horizontal two-phase flow of two-component systems at low pressures (close to atmospheric) and its application to situations outside this range of conditions is not recommended.

It follows that if the parameter ϕ_f is a function of the parameter X then the void fraction α must also be a function of X. The correlation between α and X as derived by Lockhart and Martinelli is independent of the flow regime and is given in Fig. 2.3. The actual values tabulated in the original paper follow closely the relationship given in eq. (2.66) with $\phi_f^2 = \phi_{ftt}^2$. This expression may be combined with eq. (2.68) (with $C = 20$) to give an explicit equation for void fraction in terms of the parameter X.

2.4.3.2 *The Martinelli–Nelson (1948) correlation* The Lockhart–Martinelli correlation related to the adiabatic flow of low pressure air–liquid mixtures, but the information was purposely presented in a generalized manner to enable the application of the model to single component systems and to steam–water mixtures in particular.

For the prediction of pressure drops during forced circulation boiling Martinelli and Nelson (1948) assumed the flow regime would always be 'turbulent–turbulent'. The correlation of frictional pressure gradient is worked in terms of the parameter ϕ_{fo}^2 which is more convenient for boiling and condensation problems than ϕ_f^2 (see eqs. (2.47) and (2.48)). Thermodynamic equilibrium was assumed to exist at all points in the flow and the curve correlating ϕ_{ftt} in Fig. 2.3 arbitrarily applied to atmospheric pressure steam–water flow. A relationship between ϕ_f and X_{tt} was established for the critical pressure level by noting that as the pressure is increased towards the critical point, the densities and viscosities of the phases become similar. The relationship may be represented by eq. (2.68) with the value of $C = 1.36$ (Chisholm 1963). Knowing the curves for critical and atmospheric pressure, curves at intermediate pressures were established by trial and error using the data of Davidson *et al.* (1943) as a guide. Having established the curves of ϕ_f versus X_{tt} for a number of pressures, curves of ϕ_{fo}^2 as a function of mass quality x were plotted as shown in Fig. 2.4. Table 2.2 also gives values of ϕ_{fo}^2 as a function of mass quality and pressure. Table 2.1 gives values for the two-phase multiplier from the homogeneous model and the figures in the two tables can be directly compared. It will be observed that the deviation between the frictional multipliers for the homogeneous model and the Martinelli–Nelson correlation can be up to a factor of two. The term

$$\left[\frac{1}{x} \int_0^x \phi_{fo}^2 \, dx \right]$$

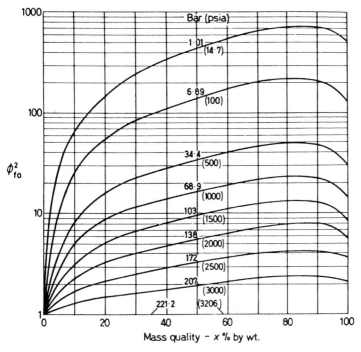

Fig. 2.4. Value of ϕ_{fo}^2 as a function of pressure and mass quality (Martinelli–Nelson 1948).

Table 2.2 Values of the two-phase frictional multiplier ϕ_{fo}^2 for the Martinelli–Nelson model steam–water system

Steam quality % by wt.	Pressure, bar (psia)								
	1.01	6.89	34.4	68.9	103	138	172	207	221.2
	(14.7)	(100)	(500)	(1000)	(1500)	(2000)	(2500)	(3000)	(3206)
1	5.6	3.5	1.8	1.6	1.35	1.2	1.1	1.05	1.00
5	30	15	5.3	3.6	2.4	1.75	1.43	1.17	1.00
10	69	28	8.9	5.4	3.4	2.45	1.75	1.30	1.00
20	150	56	16.2	8.6	5.1	3.25	2.19	1.51	1.00
30	245	83	23.0	11.6	6.8	4.04	2.62	1.68	1.00
40	350	115	29.2	14.4	8.4	4.82	3.02	1.83	1.00
50	450	145	34.9	17.0	9.9	5.59	3.38	1.97	1.00
60	545	174	40.0	19.4	11.1	6.34	3.70	2.10	1.00
70	625	199	44.6	21.4	12.1	7.05	3.96	2.23	1.00
80	685	216	48.6	22.9	12.8	7.70	4.15	2.35	1.00
90	720	210	48.0	22.3	13.0	7.95	4.20	2.38	1.00
100	525	130	30.0	15.0	8.6	5.90	3.70	2.15	1.00

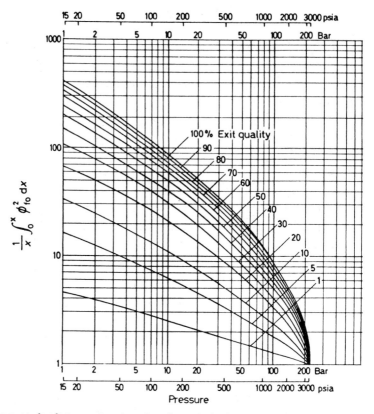

Fig. 2.5. $1/x \int_0^x \phi_{fo}^2 \, dx$ as a function of quality and absolute pressure steam–water (Martinelli–Nelson 1948).

is required in eq. (2.54). Martinelli and Nelson evaluated this integral from the graph shown in Fig. 2.4. This function is shown for the steam–water system in Fig. 2.5.

Values of the void fraction (α) are also required. Martinelli and Nelson used the curve shown in Fig. 2.3 for steam–water flow at atmospheric pressure and showed that, at the critical pressure $\alpha = \beta = x$. Knowing the curves of α versus X_{tt} for both atmospheric and critical pressures, curves at intermediate pressures were interpolated. These curves were then transposed to give values of α as a function of mass quality x with pressure as parameter (Fig. 2.6 (p. 57)). The term

$$\left[\frac{x^2}{\alpha} \left(\frac{v_g}{v_f} \right) + \frac{(1-x)^2}{(1-\alpha)} - 1 \right]$$

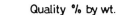

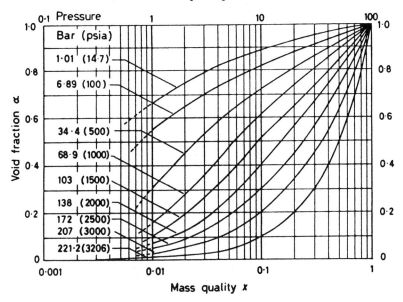

Fig. 2.6. Void fraction α as a function of quality and absolute pressure steam–water (Martinelli–Nelson 1948).

appears in eq. (2.54). To ease computation, values of this term, designated r_2, have been evaluated using the values of void fraction given in Fig. 2.6. The calculated values of r_2 are shown in Fig. 2.7 (p. 58) as a function of pressure with exit mass quality as parameter. The designation r_1 was given to the term $[(v_{fg}/v_f)x]$ which appears in eq. (2.44)—the homogeneous model.

2.4.3.3   *The Thom correlation*   An alternative set of consistent values for the terms

$$\phi_{fo}^2, \left[\frac{1}{x}\int_0^x \phi_{fo}^2 \, dx\right], \alpha, \text{ and } r_2$$

have been published by Thom (1964). These revised values were derived using an extensive set of experimental data for steam–water pressure drops obtained at Cambridge, England, on heated and unheated horizontal and vertical tubes. The alternative values of these quantities as suggested by Thom are given in Table 2.3.

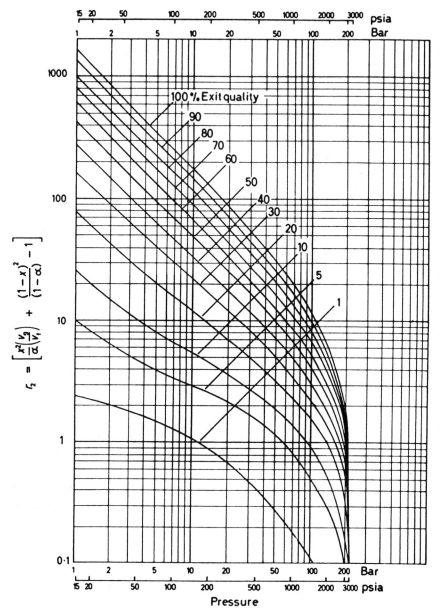

Fig. 2.7. Multiplier r_2 as a function of pressure of various exit qualities steam–water (Martinelli and Nelson 1948).

Pressure, bar (psia)

Steam quality % by wt.	17.2 (250) ϕ_{fo}^2	17.2 (250) $\frac{1}{x}\int_0^x \phi_{fo}^2\,dx$	41.3 (600) ϕ_{fo}^2	41.3 (600) $\frac{1}{x}\int_0^x \phi_{fo}^2\,dx$	86.1 (1250) ϕ_{fo}^2	86.1 (1250) $\frac{1}{x}\int_0^x \phi_{fo}^2\,dx$	145 (2100) ϕ_{fo}^2	145 (2100) $\frac{1}{x}\int_0^x \phi_{fo}^2\,dx$	207 (3000) ϕ_{fo}^2	207 (3000) $\frac{1}{x}\int_0^x \phi_{fo}^2\,dx$
1	2.12	1.49	1.46	1.11	1.10	1.03	—	—	—	—
5	6.29	3.71	2.86	2.09	1.62	1.31	1.21	1.10	1.02	—
10	11.1	6.30	4.78	3.11	2.39	1.71	1.48	1.21	1.08	1.06
20	20.6	11.4	8.42	5.08	3.77	2.47	2.02	1.46	1.24	1.12
30	30.2	16.2	12.1	7.00	5.17	3.20	2.57	1.72	1.40	1.18
40	39.8	21.0	15.8	8.80	6.59	3.89	3.12	2.01	1.57	1.26
50	49.4	25.9	19.5	10.6	8.03	4.55	3.69	2.32	1.73	1.33
60	59.1	30.5	23.2	12.4	9.49	5.25	4.27	2.62	1.88	1.41
70	68.8	35.2	26.9	14.2	10.19	6.00	4.86	2.93	2.03	1.50
80	78.7	40.1	30.7	16.0	12.4	6.75	5.45	3.23	2.18	1.58
90	88.6	45.0	34.5	17.8	13.8	7.50	6.05	3.53	2.33	1.66
100	98.86	49.93	38.30	19.65	15.33	8.165	6.664	3.832	2.480	1.740

Steam quality % by wt.	17.2 (250) α	17.2 (250) r_2	41.3 (600) α	41.3 (600) r_2	86.1 (1250) α	86.1 (1250) r_2	145 (2100) α	145 (2100) r_2	207 (3000) α	207 (3000) r_2
1	0.288	0.4125	0.168	0.2007	0.090	0.0955	0.0476	0.0431	0.0213	0.0132
5	0.678	2.169	0.512	1.040	0.340	0.4892	0.207	0.2182	0.102	0.0657
10	0.816	4.620	0.690	2.165	0.521	1.001	0.355	0.4431	0.193	0.1319
20	0.910	10.39	0.833	4.678	0.710	2.100	0.553	0.9139	0.350	0.2676
30	0.945	17.30	0.895	7.539	0.808	3.292	0.679	1.412	0.480	0.4067
40	0.964	25.37	0.930	10.75	0.866	4.584	0.767	1.937	0.589	0.5495
50	0.975	34.58	0.952	14.30	0.908	5.958	0.832	2.490	0.682	0.6957
60	0.984	44.93	0.967	18.21	0.936	7.448	0.881	3.070	0.763	0.8455
70	0.990	56.44	0.979	22.46	0.959	9.030	0.920	3.678	0.834	0.9988
80	0.994	69.09	0.988	27.06	0.976	10.79	0.952	4.512	0.895	1.156
90	0.997	82.90	0.995	32.01	0.989	12.48	0.978	5.067	0.951	1.316
100	1	98.10	1	37.30	1	14.34	1	5.664	1	1.480

2.4.4 *The application of the separated flow model to experimental observations*

The Lockhart–Martinelli–Nelson model has been used extensively for the correlation of experimental pressure gradients and void fraction measurements for both single- and two-component gas–liquid flow. Generally, it is found that the separated flow model is capable of more accurate predictions than the homogeneous model.

Two general observations can be made concerning the application of the Lockhart–Martinelli correlation.

(a) It has been widely recognized that the curves of experimental data plotted as ϕ_f (or ϕ_g) versus X (Fig. 2.3) are not smooth but show discontinuities of slope which may be associated, quite definitely, with changes of flow pattern (Sze-Foo Chen and Ibele 1962; Gazley and Bergelin 1949; Charvonia 1961; Kegel 1948; Dukler 1949).

(b) An effect of mass velocity upon the curves of ϕ_f versus X has also been widely reported (Sze-Foo Chen and Ibele 1962; Gazley and Bergelin 1949; Charvonia 1961; Kegel 1948; Dukler 1949). Data from other sources (Isbin 1959; Sher and Green 1959; Muscettola 1963) confirm, firstly, that there is a mass velocity effect with steam–water flow at high pressures; secondly, that the original Martinelli–Nelson correlation line corresponds to a mass velocity of 500–1000 kg/m² s and thirdly, that the homogeneous model yields values close to those obtained experimentally for mass velocities greater than 2000 kg/m² s.

This is perhaps to be expected since the Martinelli approach is based on a separated (annular) flow concept whilst the homogeneous model assumes a fully dispersed flow. Sufficient information is now available on the effect of mass velocity to add empirical corrections to the friction multipliers obtained from either the Martinelli–Nelson or the homogeneous models.

From the above discussion it might be expected that the void fraction (α) would also be a function of mass velocity. This is indeed the case for vertical up-flow at least. Quantitative data which illustrate the effect have been reported by Zuber *et al.* (1967) for the case of Freon at elevated pressures and by Hughmark and Pressburg (1961) for low-pressure air–liquid flow. Figure 2.8 shows data obtained by Zuber for the evaporation of Freon 22 in a vertical 1 cm i.d. heated tube. Effects associated with the absence of thermodynamic equilibrium are clearly visible at low void fractions. These effects are discussed in detail in Chapter 6. In the region of positive thermodynamic quality an increase in mass velocity increases the void fraction at a given value of x. The void fractions predicted from the Martinelli–Nelson correlation for the steam–water system at the same value of (ρ_f/ρ_g) are also shown and, with a little imagination, might be

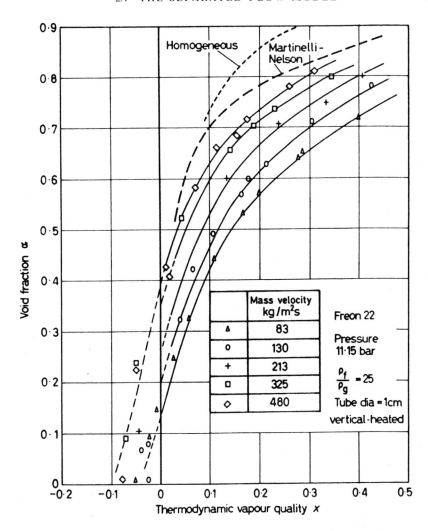

Fig. 2.8. Influence of mass velocity on void fraction (Zuber *et al.* (1967)).

expected to correspond to mass velocities near 1000 kg/m² s. The predictions from the homogeneous model yield even higher void fractions and might be expected to be valid at mass velocities greater than 2000 kg/m² s.

The reason for this influence of mass velocity will be more clearly seen in Chapter 3 where better correlations for void fractions will be developed.

2.5   Correlations for use with the homogeneous or separated flow models

Attempts to correct existing models for the influence of mass velocity on the frictional multiplier, ϕ_{fo}^2 have been published by Baroczy (1965), by Chisholm (1968), and by Friedel (1979).

2.5.1   The Baroczy correlation

The method of calculation proposed by Baroczy (1965) employs two separate sets of curves. The first of these is a plot of the two-phase frictional multiplier ϕ_{fo}^2 as a function of a physical property index $[(\mu_f/\mu_g)^{0.2}(v_f/v_g)]$ with mass quality x as parameter for a reference mass velocity of 1356 kg/m² s $(1 \times 10^6 \text{ lb/h ft}^2)$ (Fig. 2.9 and Table 2.4). The second is a plot of a correction

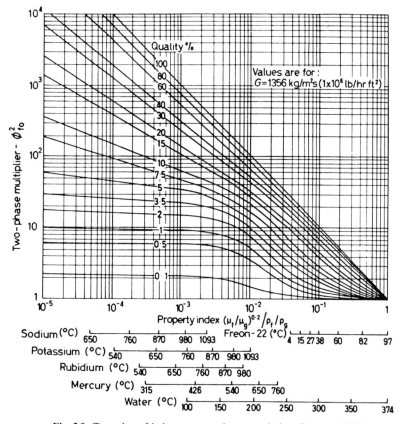

Fig. 2.9. Two-phase friction pressure drop correlation (Baroczy 1965).

Table 2.4 Co-ordinates of two-phase frictional multiplier ϕ_{fo}^2 for $G = 1356$ kg/m² s $(1 \times 10^6$ lb/h ft²) (Baroczy)

Physical property index $\left(\dfrac{\mu_f}{\mu_g}\right)^{0.2}$ $\left(\dfrac{\rho_f}{\rho_g}\right)$	\multicolumn{15}{c}{Vapour quality % by wt.}														
	0.1	0.5	1	2	3.5	5	7.5	10	15	20	30	40	60	80	100
0.0001	2.20	5.80	9.20	16.0	26.5	47.0	99.0	163	376	630	1300	2050	4300	6600	10,000
0.001	2.15	5.60	8.80	14.8	22.8	34.2	48.2	70.0	108	148	240	330	538	760	1,000
0.004	2.08	4.90	7.80	11.9	16.3	22.8	29.0	36.0	49.5	63.0	86.0	110	155	203	250
0.01	1.59	3.30	4.80	7.00	9.60	12.4	16.0	20.0	27.0	33.5	43.5	53.0	69.0	85.0	100
0.03	1.12	1.55	1.81	2.57	3.45	4.7	6.10	7.90	11.0	13.2	17.3	21.2	26.0	30.0	33.3
0.1	1.04	1.12	1.22	1.48	1.78	2.05	2.50	2.80	3.60	4.20	5.50	6.50	8.00	9.10	10.0
0.3	1.01	1.02	1.06	1.13	1.26	1.36	1.50	1.59	1.77	1.93	2.25	2.48	2.86	3.20	3.33
1	1	1	1	1	1	1	1	1	1	1	1	1	1	1	1

factor Ω expressed as a function of the same physical property index for mass velocites of 339, 678, 2712, and 4068 kg/m^2 s with mass quality as parameter (Fig. 2.10). This plot serves to correct the value of ϕ_{fo}^2 obtained from Fig. 2.9 to the appropriate value of mass velocity. Thus,

$$\left(\frac{dp}{dz}F\right) = \frac{2f_{fo}G^2v_f}{D}\phi_{fo(G=1356)}^2\Omega \tag{2.70}$$

The method proposed by Baroczy was tested against data from a wide range of systems including both liquid metals and refrigerants with satisfactory agreement between the measured and calculated values.

2.5.2   Chisholm's method

Chisholm (1968) found that eq. (2.68) provided a simple method of introducing the influence of mass velocity on ϕ_f^2. A most convenient general expression (Chisholm and Sutherland 1969) for the coefficient C is

$$C = [\lambda + (C_2 - \lambda)(v_{fg}/v_g)^{0.5}][(v_g/v_f)^{0.5} + (v_f/v_g)^{0.5}] \tag{2.71}$$

where $\lambda = 0.5(2^{(2-n)} - 2)$.

For rough tubes ($n = 0$ in Blasius equation (2.60)) $\lambda = 1$; for smooth tubes ($n = 0.25$) $\lambda = 0.68$. At the critical pressure ($v_f = v_g$; $v_{fg} = 0$) C takes on a value of 2 for rough tubes and 1.36 for smooth tubes. For steam–water flow in tubes at pressures above 30 bar (435 psia) Chisholm (1968) recommends the following procedure for the evaluation of ϕ_f^2:

(a) Mass velocity $G \leqslant G^*$
   For smooth tubes: $G^* = 2000$ kg/m^2 s (1.47×10^6 lb/h ft^2)
   Calculate the value of C from eq. (2.71) with $\lambda = 0.75$ (corresponding to $n = 0.2$) and $C_2 = (G^*/G)$
   For rough tubes: $G^* = 1500$ kg/m^2 s (1.1×10^6 lb/h ft^2)
   Calculate the value of C from eq. (2.71) with $\lambda = 1$ (corresponding to $n = 0$) and $C_2 = (G^*/G)$
(b) Mass velocity $G > G^*$
   For smooth tubes ($G > 2000$ kg/m^2 s) and rough tubes ($G > 1500$ kg/m^2 s)

$$\phi_f^2 = \left[1 + \frac{\bar{C}}{X} + \frac{1}{X^2}\right]\psi \tag{2.72}$$

where

$$\psi = \left[1 + \frac{C}{T} + \frac{1}{T^2}\right]\Big/\left[1 + \frac{\bar{C}}{T} + \frac{1}{T^2}\right] \tag{2.73}$$

$$\bar{C} = [(v_g/v_f)^{0.5} + (v_f/v_g)^{0.5}] \tag{2.74}$$

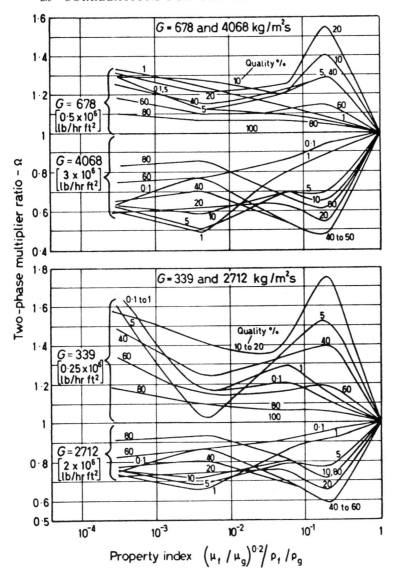

Fig. 2.10. Mass velocity correction vs property index (Baroczy 1965).

and

$$T = \left(\frac{x}{1-x}\right)^{(2-n)/2}\left(\frac{\mu_f}{\mu_g}\right)^{n/2}\left(\frac{v_f}{v_g}\right)^{1/2} \tag{2.75}$$

For smooth tubes the coefficient C in eq. (2.73) is evaluated from eq. (2.71) with $\lambda = 0.75$ (corresponding to $n = 0.2$) and $C_2 = (G^*/G)$

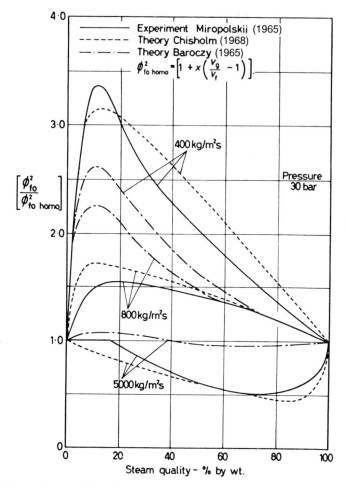

Fig. 2.11. Ratio of actual ϕ_{fo}^2 to ϕ_{fo}^2 from homogeneous theory. Comparison of experiment with theories of Chisholm (1968) and Baroczy (1965).

For rough rubes the coefficient C in eq. (2.74) is evaluated from eq. (2.71) with $\lambda = 1$ (corresponding to $n = 0$) and $C_2 = (G^*/G)$

The parameter ψ represents the ratio of ϕ_f^2 to that calculated from the homogeneous model. For values of mass velocity greater than G^* this parameter will be less than unity. Figure 2.11 shows the ratio of the actual value of ϕ_{fo}^2 to that calculated from the homogeneous model (eq. (2.42)) and compares experimental results from rough tubes from Miropolskii *et al.* (1965) for a pressure of 30 bar with the methods proposed by Baroczy (1965) and Chisholm (1968).

The method proposed by Baroczy can be used directly with systems other than steam–water. The Chisholm method may also be adapted by calculating the value of the property index $[(\mu_f/\mu_g)^{0.2}/(\rho_f/\rho_g)]$ for the system being considered, choosing the steam pressure with the same value of the property index and evaluating ϕ_f^2 as if it were for the steam–water system. This value of ϕ_f^2 may then be used in the remainder of the calculation. This adaptation of Chisholm's method is not recommended for values of the property index below 0.01. In this latter case the Baroczy method should be used. Chisholm (1983) has given alternative formulations of his method which may be more convenient, especially for steam–water flows in heated channels. The reader is referred to Chapter 7 of his book for further details.

2.5.3 The Friedel correlation

One of the most accurate two-phase pressure drop correlations is that of Friedel (1979). It was obtained by optimizing an equation for ϕ_{fo}^2 using a large data base of two-phase pressure drop measurements

$$\phi_{fo}^2 = A_1 + \frac{3.24 A_2 A_3}{\text{Fr}^{0.045}\, \text{We}^{0.035}} \tag{2.76}$$

where

$$A_1 = (1 - x)^2 + x^2 \left(\frac{\rho_f f_{go}}{\rho_g f_{fo}} \right)$$

$$A_2 = x^{0.78}(1 - x)^{0.224}$$

$$A_3 = \left(\frac{\rho_f}{\rho_g} \right)^{0.91} \left(\frac{\mu_g}{\mu_f} \right)^{0.19} \left(1 - \frac{\mu_g}{\mu_f} \right)^{0.7}$$

$$\text{Fr} = \frac{G^2}{g D \bar\rho^2}$$

$$\text{We} = \frac{G^2 D}{\bar\rho \sigma}$$

where f_{go} and f_{fo} are the friction factors defined by equations of the form eq. (2.36) for the total mass velocity G as all vapour and all liquid, respectively. D is the equivalent diameter, σ is the surface tension and $\bar\rho$ is the homogeneous density given by eq. (2.23). Standard deviations are about 40–50 per cent, which is large with respect to single-phase flows but quite good for two-phase flows. The correlation is valid for vertical upwards flow and for horizontal flow; a slightly different correlation was proposed by Friedel for vertical downwards flow.

Whalley (1980) has evaluated separated flow models against a large proprietary data bank and gives the following recommendation:

(a) For $(\mu_f/\mu_g) < 1000$: Utilize the Friedel (1979) correlation;
(b) For $(\mu_f/\mu_g) > 1000$ and $G > 100 \text{ kg/m}^2$ s: Utilize the most recent refinement of the Chisholm (1973) correlation;
(c) For $(\mu_f/\mu_g) > 1000$ and $G < 100 \text{ kg/m}^2$ s: Utilize the correlations of Lockhart and Martinelli (1949) and Martinelli and Nelson (1948).

For most fluids and operating conditions, (μ_f/μ_g) is less than 1000 and the Friedel correlation will be the preferred method.

2.6   Two-phase flow in inclined pipes

Lau et al. (1992) have made simultaneous measurements of pressure drop in both upflow and downflow of air–water mixtures in a 9.53 mm i.d. vertical tube. The correlations of Lockfort–Martinelli and of Chisholm satisfactorily represented the experimental data. However, it was observed that downflow frictional pressure drops were generally higher than for upflow. Negative frictional pressure drops associated with low liquid and gas flows in upflow are not encountered with downflow. Similarly, local maxima and minima in the pressure drop/flow curves characteristic of upflow do not occur in downflow. The downflow pressure drop is almost independent of gas flow in the falling film region ($Re_{f_o} < 1000$, $35 < Re_{g_o} < 6800$).

Equations (2.34) and (2.53) may be used as the basis for calculating the pressure gradient in an inclined pipe. The question arises as to whether the void fraction, α, is a function of the angle of inclination, θ. Beggs and Brill (1973) have made measurements of void fraction and pressure drop for air–water flow in pipes of 2.54 cm and 3.81 cm diameter. It was found that the void fraction for given air and water flow rates was indeed a function of the angle of inclination, θ. The liquid hold up at any value of θ, $[1 - \alpha(\theta)]$ may be expressed as the value for a horizontal pipe $[1 - \alpha(0)]$ for the given condition multiplied by a correction factor $F(\theta)$ which was found to be a function of both inclination angle θ and the volumetric quality β. Figure 2.12(a) shows this correction factor $F(\theta)$ as a function of θ and $(1 - \beta)$. The liquid hold up is seen to reach a maximum at an angle of approximately $+50°$ and a minimum at $-50°$ from the horizontal. The positive value of θ represents up flow and the negative value of θ, downflow. The two-phase frictional pressure gradient was examined on the basis of the homogeneous model using eq. (2.40), the Dukler definition of two-phase viscosity. The ratio between the observed homogeneous friction factor, f_{TP}, and that evaluated using the Dukler definition of two-phase viscosity, $\bar{f}$, was found to be a function of both the liquid hold-up and the volumetric quality as shown in

Fig. 2.12(b). The Beggs and Brill method correlated some 584 pressure gradient data points with a mean error of +1.11 per cent and an RMS error of 9.30 per cent. The method is recommended for use with inclined pipes.

2.7 The influence of imposed heat flux on the void fraction and pressure gradient

In addition to the effect of mass velocity on the friction multiplier Tarasova and Leont'ev (1965) and Tarasova *et al.* (1966) have found a complex

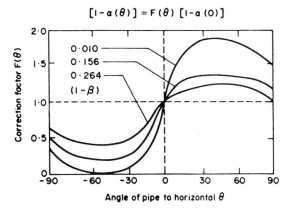

(a) Liquid hold-up as a function of angle of inclination

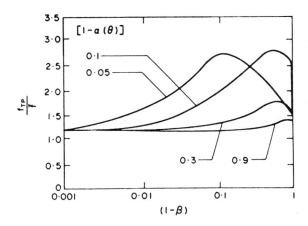

(b) Correction factor to homogeneous model based on Dukler definition of two-phase viscosity

Fig. 2.12. Two-phase flow in inclined pipes (Beggs and Brill 1973).

dependence on the heat flux imposed on the tube. A characteristic drop in the experimental value of ϕ_{fo}^2 was observed at the point along the tube at which dryout (see Chapters 8 and 9) occurred. This drop in ϕ_{fo}^2 might be expected to result from the decrease in effective roughness that the vapour core experiences as it passes the dryout transition. The disappearance of the slow moving liquid film also means an increased void fraction at this point. It is therefore recommended that, in the 'liquid deficient region' (see Chapter 4) corresponding to the 'drop flow' pattern, the homogeneous model be used to calculate ϕ_{fo}^2 and α.

At lower qualities the frictional pressure gradient is increased in the presence of a heat flux over that in an unheated tube. The estimation of pressure drop and void fraction in the absence of thermodynamic equilibrium under subcooled boiling and low quality conditions is discussed fully in Chapter 6. Tarasova and Leont'ev (1965), and Tarasova et al. (1966) have given an empirical equation relating the value of ϕ_{fo}^2 in heated and unheated tubes for the steam–water system as follows:

$$[\phi_{fo}^2]_{\text{heated tube}} = [\phi_{fo}^2]_{\text{unheated tube}} \left[1 + 4.4 \times 10^{-3} \left(\frac{\phi}{G} \right)^{0.7} \right] \quad (2.77)$$

where ϕ is the heat flux on the tube (watts/m^2) and G is the total mass velocity (kg/m^2 s). This relationship was independent of pressure.

The integration to arrive at eqs. (2.44) and (2.54) assumed a linear change in x with the channel length L ($dx/dz = $ constant). However, for boiling and condensation, the local two-phase heat transfer coefficient can be a strong, non-linear function of vapour quality. This situation can be mollified by dividing the channel into short sections. In addition, in the region away from that influenced by lack of thermal equilibrium but below the dryout transition the value of the local friction multiplier and the local void fraction are little affected by the presence of an imposed heat flux.

References and further reading

Allen, W. F. (1951). 'Flow of a flashing mixture of water and steam through pipes and valves'. Trans. ASME, 73, 257–265.

Baroczy, C. J. (1965). 'A systematic correlation for two-phase pressure drop'. AIChE reprint 37 presented at 8th National Heat Transfer Conference, Los Angeles, August.

Beggs, H. D. and Brill, J. P. (1973). 'A study of two-phase flow in inclined pipes'. J. Petroleum Technology, 25, 607–617.

Benjamin, M. W. and Miller, J. G. (1942). 'The flow of a flashing mixture of water and steam through pipes'. Trans. ASME, 64, 657–664.

Bottomley, W. T. (1936–37). 'Flow of boiling water through orifices and pipes'. Trans. North East Coast Inst. of Engrs. and Ship Builders, 53, 65–100.

Bouré, J. A. (1978). 'Constitutive equations for two-phase flows'. In *Two-phase flows and heat transfer with application to nuclear reactor design problems*, Chapter 9, von Karman Inst. Book, Hemisphere, New York.

Bouré, J. A. and Reocreux, M. (1972). 'General equations of two-phase flows—applications to critical flows and to non-steady flows'. 4th All-Union Heat Transfer Conference, Minsk.

Charvonia, D. A. (1961). 'An experimental investigation of the mean liquid film thickness and the characteristics of the interfacial surface in annular, two-phase flow'. Paper 61-WA-243 presented at ASME Winter Annual Meeting in New York, 26 November–December.

Chisholm, D. (1963). 'The pressure gradient due to friction during the flow of boiling water'. NEL Report No. 78.

Chisholm, D. (1967). 'A theoretical basis for the Lockhart–Martinelli correlation for two-phase flow'. NEL Report No. 310.

Chisholm, D. (1968). 'The influence of mass velocity on friction pressure gradients during steam–water flow'. Paper 35 presented at 1968 Thermodynamics and Fluid Mechanics Convention, Institute of Mechanical Engineers (Bristol), March.

Chisholm, D. (1973). 'Pressure gradients due to friction during the flow of evaporating two-phase mixtures in smooth tubes and channels'. *Int. J. Heat Mass Transfer*, 16, 347–348.

Chisholm, D. (1983). 'Two-phase flow in pipelines and heat exchangers'. Longmans.

Chisholm, D. and Sutherland, L. A. (1969). 'Prediction of pressure gradients in pipeline systems during two-phase flow'. Paper 4 presented at Symposium on Fluid Mechanics and Measurements in Two-phase Flow Systems, University of Leeds, September.

Cicchitti, A., Lombardi, C., Silvestri, M., Soldaini, G., and Zavattarelli, R. (1960). 'Two-phase cooling experiments—pressure drop, heat transfer and burnout measurements', *Energia Nucleare*, 7(6), 407–425.

Davidson, W. F., Hardie, P. H., and Humphreys, C. G. R. (1943). 'Studies of heat transmission through boiler tubing at pressures from 500 to 3300 pounds'. *Trans. ASME*, 65, 553–591.

Delhaye, J. M. (1990). 'Basic equations for two-phase flow modelling'. In *Two-phase flows and heat transfer in the power and process industries*, Hemisphere, New York.

Dittus, F. W. and Hildebrand, A. (1942). 'A method of determining the pressure drop for oil–vapour mixtures flowing through furnace coils'. *Trans. ASME*, 64, 185–192.

Dukler, A. E. (1949). 'An investigation of pressure drop for isothermal two-phase film flow in a vertical tube'. MS Thesis, University of Delaware.

Dukler, A. E., Wicks, M., and Cleveland, R. G. (1964). 'Pressure drop and hold-up in two-phase flow Part A—A comparison of existing correlations' and 'Part B—An approach through similarity analysis'. Paper presented at AIChE meeting held Chicago 2–6 December 1962, also *AIChE Journal*, 10(1), 38–51.

Friedel, L. (1979). 'Improved friction pressure drop correlations for horizontal and vertical two-phase pipe flow'. Presented at the European Two-phase Flow Group Meeting, Ispra, Italy, Paper E2, June.

Gazley, C. Jr and Bergelin, O. P. (1949). Discussion of Lockhart and Martinelli (1949). *Chem. Eng. Prog.*, 45.

Hughmark, G. A. and Pressburg, B. S. (1961). 'Hold-up and pressure drop with gas–liquid flow in a vertical pipe'. *AIChE J.*, 7(4), 677–682.

Isbin, H. S., Moen, R. H., and Wickey, R. C. (1959). 'Two-phase steam–water pressure drops'. Nuclear Engineering Part VI. *Chem. Eng. Symp. Series* No. 23, **55**, 75–84.

Ishii, M. (1990). 'Two fluid model for two phase flow'. In *Multiphase Science and Technology*, Vol. 5, Chapter 1, Hemisphere, New York.

Kegel, P. K. (1948). 'Two-phase flow in a vertical column'. BChE Thesis, University of Delaware.

Lau, V., Yiang, Y., and Rezkallah (1992). 'Pressure drop during upwards and downward two-phase gas–liquid flow in a vertical tube'. Paper presented in *FED*, **144**, *Multiphase Flows in Wells & Pipelines*, pp. 81–7, ASME Winter Annual Meeting, Anaheim, CA.

Lewis, W. Y. and Robertson, S. A. (1940). 'The circulation of water and steam in water-tube boilers and the rational simplification of boiler design'. *Proc. Inst. Mech. Engrs. (London)*, **143**, 147–181.

Lockhart, R. W. and Martinelli, R. C. (1949). 'Proposed correlation of data for isothermal two-phase two-component flow in pipes'. *Chem. Eng. Prog.*, **45**, 39.

McAdams, W. H., Woods, W. K., and Bryan, R. L. (1942). 'Vaporization inside horizontal tubes—II—Benzene–oil mixtures'. *Trans. ASME*, **64**, 193.

Markson, A. A., Raverse, T., and Humphreys, C. G. R. (1942). 'A method of estimating the circulation in steam boiler furnace circuits'. *Trans. ASME*, **64**, 275–286.

Martinelli, R. C. and Nelson, D. B. (1948). 'Prediction of pressure drop during forced circulation boiling of water'. *Trans. ASME*, **70**, 695.

Miropolskii, E. L., Shitsman, M. E., and Shneenova, R. I. (1965). 'Influence of heat flux and velocity on hydraulic resistance with steam–water mixture flowing in tubes'. *Thermal Engng.*, **12**(5), 80.

Muscettola, M. (1963). 'Two-phase pressure drop—comparison of the "momentum exchange model" and Martinelli–Nelson's correlation with experimental measurements'. AEEW-R.284.

Sher, N. E. and Green, S. J. (1959). 'Boiling pressure drop in thin rectangular channels'. Nuclear Engineering Part VI, *Chem. Eng. Symp* No. 23, **55**, 61–73.

Sze-Foo Chien and Ibele, W. (1962). 'Pressure drop and liquid film thickness of two-phase annular and annular mist flows'. Paper presented at ASME Winter Annual Meeting, New York, 25–30 November, Paper 62-WA-170.

Tarasova, N. V. and Leont'ev, A. I. (1965). 'Hydraulic resistance with a steam–water mixture flowing in a vertical heated tube'. *Teplo. Vyso Temp.*, **3**(1), 115–123.

Tarasova, N. V., Leont'ev, A. I., Hlopuskin, V. L., and Orlov, V. M. (1966). 'Pressure drop of boiling subcooled water and steam–water mixture flowing in heated channels'. Paper 133 presented at 3rd International Heat Transfer Conference. Chicago, **4**, 178–183.

Thom, J. R. S. (1964). 'Prediction of pressure drop during forced circulation boiling of water'. *Int. J. Heat Mass Transfer*, **7**, 709–724.

Whalley, P. B. (1980). See Hewitt, G. F. (1983). 'Multiphase flow and pressure drop'. *Heat Exchanger Design Handbook*, Hemisphere, Washington, DC, Vol. 2, 2.3.2–11.

Zuber, N., Staub, F. W., Bijwaard, G., and Kroeger, P. G. (1967). 'Steady state and transient void fraction in two-phase flow systems'. GEAP 5417.

Example

Example 1

A vertical tubular test section is to be installed in an experimental high pressure water loop. The tube is 10.16 mm i.d. and 3.66 m long heated uniformly over its

EXAMPLE 73

length. An estimate of the pressure drop across the test section is required as a function of the flow-rate of water entering the test section at 204°C and 68.9 bar.

(1) Calculate the pressure drop over the test section for a water flow of 0.108 kg/s with a power of 100 kW applied to the tube using

(i) the homogeneous model
(ii) the Martinelli–Nelson model
(iii) The Thom correlation
(iv) the Baroczy correlation

(2) Estimate the pressure drop versus flow-rate relationship over the range 0.108 to 0.811 kg/s (2–15 USGPM) for a power of 100 kW and 200 kW applied to the tube using

(i) the Martinelli–Nelson model
(ii) the Baroczy correlation

Solution: Part 1

(a) The enthalpy rise across the tube $\Delta i = \dfrac{100 \times 10^3}{0.108} = 0.925$ MJ/kg

A length of tube (z_{sc}) is required to preheat the water to the saturation temperature given by

$$\frac{z_{sc}}{L} = \frac{i_f - i_{fi}}{\Delta i} = \frac{1.26 - 0.872}{0.925}, \qquad z_{sc} = \underline{1.54 \text{ m}}$$

The outlet mass quality x_o is given by

$$x_o = \frac{\Delta i + i_{fi} - i_f}{i_{fg}} = \frac{0.925 + 0.872 - 1.26}{1.51} = \underline{0.356}$$

i_{fi}—the enthalpy at the inlet temperature $= 0.872$ MJ/kg
i_f—the enthalpy of saturated water at 68.9 bar $= 1.26$ MJ/kg
i_{fg}—the latent heat of vaporization at 68.9 bar $= 1.51$ MJ/kg.

(b) *Pressure drop in preheating section*

Mass velocity $(G) = \dfrac{4W}{\pi D^2} = \dfrac{4 \times 0.108}{3.142 \times 10.16^2 \times 10^{-6}} = \underline{1335 \text{ kg/m}^2 \text{ s}}$

The Reynolds number $(Re_f) = \left[\dfrac{GD}{\mu_f}\right] = \dfrac{1335 \times 10.16 \times 10^{-3}}{\mu_f}$

at 204°C (T_{fi}) $\mu_{fi} = 1.35 \times 10^{-4}$ Ns/m^2, $v_{fi} = 1.165 \times 10^{-3}$ m^3/kg,

$$Re_{fi} = 1.00 \times 10^5, \quad f_{fo} = 0.0046$$

at 285°C (T_{SAT}) $\mu_f = 0.972 \times 10^{-4}$ Ns/m^2, $v_f = 1.35 \times 10^{-3}$ m^3/kg,

$$Re_f = 1.40 \times 10^5, \quad f_{fo} = 0.0044$$

The frictional pressure drop in the preheating section, taking average property values, is given as:

$$\Delta p_F = \frac{2 f_{fo} G^2 v_f z_{sc}}{D} = \frac{2 \times 0.0045 \times (1335)^2 \times 1.255 \times 10^{-3} \times 1.54}{10.16 \times 10^{-3}}$$

$$= 3.06 \text{ kN/m}^2$$

The acceleration of the water due to a change in specific volume produces a small pressure drop:

$$\Delta p_s = [v_f - v_{fi}] G^2 = [1.35 - 1.165] \times 10^{-3} \times (1335)^2 = 0.330 \text{ kN/m}^2$$

The gravitational head pressure drop:

$$\Delta p_z = \frac{g z_{sc}}{v_f} = \frac{9.807 \times 1.54}{1.255 \times 10^{-3}} = 12.05 \text{ kN/m}^2$$

(c) *Pressure drop in the two-phase region*
(i) *The homogeneous model*
Equation (2.44) may be used if f_{TP} is constant over the channel length and if $G^2 x (dv_g/dp)$ is very much less than unity.

$$\left(\frac{dv_g}{dp}\right) \text{ at 68.9 bar} = -4.45 \times 10^{-9} \left[\frac{m^3/kg}{N/m^2}\right]$$

$$G^2 x \left(\frac{dv_g}{dp}\right) = (1335)^2 \times 0.353 \times (-4.45 \times 10^{-9}) = -0.0028$$

(which is much less than -1).

Evaluation of f_{TP}
Various assumptions may be made about the two-phase viscosity thus:

$$\bar{\mu} = \mu_f = 0.972 \times 10^{-4} \text{ Ns/m}^2$$

Equation (2.38) $\dfrac{1}{\bar{\mu}} = \dfrac{x}{\mu_g} + \dfrac{(1-x)}{\mu_f}$ $\mu_g = 1.89 \times 10^{-5} \text{ Ns/m}^2,$

$$\bar{\mu} = 3.93 \times 10^{-5} \text{ Ns/m}^2$$

Equation (2.39) $\bar{\mu} = x\mu_g + (1-x)\mu_f$ $\bar{\mu} = 6.93 \times 10^{-5} \text{ Ns/m}^2$

Equation (2.40) $\bar{\mu} = \dfrac{1}{\bar{v}}[(1-x)v_f\mu_f + xv_g\mu_g]$

$$\bar{v} = 1.07 \times 10^{-2} \text{ m}^3/\text{kg},$$

$$\bar{\mu} = 2.52 \times 10^{-5} \text{ Ns/m}^2$$

EXAMPLE 75

Equation (2.45) $\bar{\mu} = \mu_f\left[1 + \dfrac{v_{fg}}{v_f}x\right]$ $v_{fg} = 2.65 \times 10^{-2}\ \text{m}^3/\text{kg},$

$$\bar{\mu} = 7.73 \times 10^{-4}\ \text{Ns/m}^2$$

The frictional pressure drop

$$\Delta p_f = \frac{2f_{TP}G^2 v_f(L - z_{sc})}{D}\left[1 + \frac{v_{fg}}{v_f}\left(\frac{x_o}{2}\right)\right]$$

$$= \frac{2f_{TP} \times (1335)^2 \times 1.35 \times 10^{-3} \times 2.12}{10.16 \times 10^{-3}}\left[1 + \left(\frac{2.65 \times 10^{-2} \times 0.356}{1.35 \times 10^{-3} \times 2}\right)\right]$$

$$= 4.5 \times 10^3 f_{TP}\ \text{kN/m}^2$$

Table 2.5

Equation for $\bar{\mu}$	$\bar{\mu}$ (Ns/m²)	$\dfrac{GD}{\bar{\mu}}$	f_{TP} at $x = x_o$	Average value of f_{TP}	Δp_f kN/m²
$\bar{\mu} = \mu_f$	0.972×10^{-4}	1.4×10^5	0.0044	0.0044	19.7
(2.38)	0.393×10^{-4}	3.45×10^5	0.0038	0.0040	18.0
(2.39)	0.693×10^{-4}	1.96×10^5	0.0042	0.0043	19.3
(2.40)	0.252×10^{-4}	5.38×10^5	0.00355	0.00398	17.91
(2.45)	7.73×10^{-4}	0.176×10^5	0.0062	0.0053	23.8

The acceleration pressure drop

$$\Delta p_a = G^2 v_{fg} x_o = (1335)^2 \times 2.65 \times 10^{-2} \times 0.356 = 16.7\ \text{kN/m}^2$$

The static head pressure gradient

$$\Delta p_z = \frac{g \sin \theta(L - z_{sc})}{v_{fg}x_o}\ln\left[1 + \left(\frac{v_{fg}}{v_f}\right)x\right] = \frac{9.807 \times 1 \times 2.12}{2.65 \times 10^{-2} \times 0.352}\ln(7.95)$$

$$= 4.6\ \text{kN/m}^2$$

(ii) *The Martinelli–Nelson correlation*
Equation (2.54) may be used since

$$\left[G^2\left\{\frac{x^2}{\alpha}\left(\frac{dv_g}{dp}\right) + \left(\frac{\partial \alpha}{\partial p}\right)_x\left(\frac{(1-x)^2 v_f}{(1-\alpha)^2} - \frac{x^2 v_g}{\alpha^2}\right)\right\}\right]$$

is much less than unity.
The frictional pressure drop:

$$\Delta p_f = \frac{2f_{fo}G^2 v_f(L - z_{sc})}{D}\left[\frac{1}{x_o}\int_0^{x_o}\phi_{fo}^2\,dx\right]$$

The value of

$$\left[\frac{1}{x_o} \int_0^{x_o} \phi_{fo}^2 \, dx \right]$$

for $p = 68.9$ bar, $x_o = 0.356$, from Fig. 2.5

$$= 7.05$$

$$\Delta p_f = \frac{2 \times 0.0044 \times (1335)^2 \times 1.35 \times 10^{-3} \times 2.12}{10.16 \times 10^{-3}} \times 7.05 = \underline{30.9 \text{ kN/m}^2}$$

The acceleration pressure drop:

$$\Delta p_a = G^2 v_f r_2 = (1335)^2 \times 1.35 \times 10^{-3} \times 4.55 = \underline{10.95 \text{ kN/m}^2}$$

The value of r_2 at 68.9 bar, $x_o = 0.356$ from Fig. 2.7 $= \underline{4.55}$.
   The gravitational pressure drop:

$$\Delta p_z = \frac{g \sin \theta (L - z_{sc})}{v_f x_o} \int_0^{x_o} \left[(1 - \alpha) + \left(\frac{v_f}{v_g} \right) \alpha \right] dx$$

The integral is evaluated graphically using the known values of α as a function of x evaluated from Fig. 2.6 for the specified pressure.

$$\frac{1}{x_o} \int_0^{x_o} \left[(1 - \alpha) + \left(\frac{v_f}{v_g} \right) \alpha \right] dx = 0.38 \qquad \theta = 90° \qquad \sin \theta = 1$$

$$\Delta p_z = \frac{9.807 \times 1 \times 2.12 \times 0.38}{1.35 \times 10^{-3}} = \underline{5.85 \text{ kN/m}^2}$$

(iii) *The Thom correlation*

The frictional pressure drop: the value of $\left\{ \dfrac{1}{x_o} \displaystyle\int_0^{x_o} \phi_{fo}^2 \, dx \right\}$ is interpolated from

Table 2.3 $= 4.5$ $\Delta p_f = \underline{19.85 \text{ kN/m}^2}$

The accelerational pressure drop: the value of r_2 is interpolated from

Table 2.3 $= 5.5$ $\Delta p_a = \underline{13.25 \text{ kN/m}^2}$

The gravitational pressure drop: the term

$$\frac{1}{x_o} \int_0^{x_o} \left[(1 - \alpha) + \left(\frac{v_f}{v_g} \right) \alpha \right] dx$$

can be evaluated using the following relationship for α as a function of x

$$\alpha = \frac{Mx}{1 + x(M - 1)} \quad \text{where for } p = 68.9 \text{ bar } M = 12.2$$

alternatively values of the term $\dfrac{1}{x_o} \displaystyle\int_0^{x_o} \left[(1 - \alpha) + \left(\frac{v_f}{v_g} \right) \alpha \right] dx$ are given in Thom's original paper $= 0.37$

$$\Delta p_z = \frac{9.807 \times 1 \times 2.12 \times 0.37}{1.35 \times 10^{-3}} = 5.69 \text{ kN/m}^2$$

(iv) *The Baroczy correlation*

The frictional pressure drop: the value of $\left\{ \dfrac{1}{x_o} \displaystyle\int_0^{x_o} \Omega \phi_{fo\,G=1356}^2 \, dx \right\}$ is evaluated using

Figs. 2.10 and 2.11. $\Omega = 1$ for all values of x and $\left\{ \dfrac{1}{x_o} \displaystyle\int_0^{x_o} \phi_{fo}^2 \, dx \right\}$ obtained by integration from Fig. 2.10 $= 5.75$

$$\Delta p_f = 25.4 \text{ kN/m}^2$$

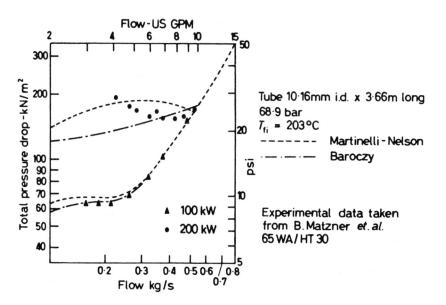

Fig. 2.13.

Table 2.6

Pressure drop components	Homogeneous model					Martinelli–Nelson model	Thom model	Baroczy model
	$\bar{\mu} = \mu_f$	$\bar{\mu}$ Eq. (2.38)	$\bar{\mu}$ Eq. (2.39)	$\bar{\mu}$ Eq. (2.40)	$\bar{\mu}$ Eq. (2.45)			
Single-phase kN/m²								
Δp_f	3.06	3.06	3.06	3.06	3.06	3.06	3.06	3.06
Δp_a	0.330	0.330	0.330	0.330	0.330	0.330	0.330	0.330
Δp_z	12.05	12.05	12.05	12.05	12.05	12.05	12.05	12.05
Two-phase kN/m²								
Δp_f	19.7	18.0	19.3	17.91	23.8	30.9	19.85	25.4
Δp_a	16.7	16.7	16.7	16.7	16.7	10.95	13.25	10.95
Δp_z	4.6	4.6	4.6	4.6	4.6	5.85	5.69	5.85
Total pressure drop kN/m²	56.4	54.7	56.0	54.7	60.5	63.1	54.2	57.64

Average of all models = 57.2 kN/m²
Highest value = 63.1 (+10.4 per cent)
Lowest value = 54.2 (−5.2 per cent)

Table 2.7 Pressure drop/flow characteristics at 100 kW

Flow rate (USGPM)	2	3	4	6	8	10	15
Flow rate (kg/s)	0.108	0.162	0.216	0.324	0.432	0.540	0.8111
Total heat added to fluid (MJ/kg)	0.925	0.615	0.461	0.307	0.230	0.185	0.123
Preheating length (z_{sc}), m	1.54	2.32	3.10	3.66	3.66	3.66	3.66
Exit quality (x_o)	0.353	0.149	0.047	0	0	0	0
Mass velocity (kg/m² s)	1335	2000	2670	4000	5340	6670	10,000
Reynolds number at							
204°C ($\times 10^{-5}$)	1.000	1.505	2.007	3.011	4.015	5.018	7.527
285°C ($\times 10^{-5}$)	1.397	2.094	2.793	4.190	5.587	6.983	10.474
$G^2 \times 10^{-6}$	1.782	4	7.13	16	28.5	44.5	100
Friction factor at 204°C	0.0046	0.0043	0.0041	0.0039	0.0037	0.0036	0.0035
285°C	0.0044	0.0041	0.0039	0.0037	0.0036	0.0035	0.0033
Single phase							
Frictional Δp_f (kN/m²)	3.06	9.65	21.8	55.1	94.3	143	308
Accelerational Δp_a (kN/m²)	0.330	0.737	1.32	2.98	5.28	8.24	18.5
Gravitational Δp_z (kN/m²)	12.05	18.1	24.1	28.6	28.6	28.6	28.6
Two phase							
Martinelli–Nelson							
($L - z_{sc}$), m	2.12	1.34	0.57	0	0	0	0
Frictional Δp_f (kN/m²)	30.9	23.5	9.45	0	0	0	0
r_2	4.55	1.95	0.93	0	0	0	0
Accelerational Δp_a (kN/m²)	10.95	10.5	8.91	0	0	0	0
Gravitational Δp_z (kN/m²)	5.85	5.65	3.21	0	0	0	0
Martinelli–Nelson total pressure drop (kN/m²)	63.1	68.2	68.9	86.7	128	180	356
Baroczy frictional Δp_f (kN/m²)	25.4	17.65	6.55	0	0	0	0
Baroczy total pressure drop (kN/m²)	57.6	62.3	65.9	86.7	128	180	356

Table 2.8 Pressure drop/flow characteristics at 200 kW

Flow rate (USGPM)	2	3	4	6	8	10	15
Flow rate (kg/s)	0.108	0.162	0.216	0.324	0.432	0.540	0.811
Total heat added to fluid (MJ/kg)	1.843	1.23	0.925	0.615	0.461	0.369	0.246
Preheating length (z_{sc}), m	0.772	1.19	1.54	2.33	3.06	3.66	3.66
Exit quality (x_o)	0.964	0.556	0.353	0.149	0.047	0	0
Mass velocity (kg/m² s)	1335	2000	2670	4000	5340	6670	10,000
Reynolds number at							
$\quad$204°C ($\times 10^{-5}$)	1.000	1.505	2.007	2.011	4.015	5.018	7.527
$\quad$285°C ($\times 10^{-5}$)	1.397	2.094	2.793	4.190	5.587	6.983	10.474
$\quad G^2 \times 10^{-6}$	1.782	4	7.13	16	28.5	44.5	100
Friction factor at 204°C	0.0046	0.0043	0.0041	0.0039	0.0037	0.0036	0.0035
$\quad$285°C	0.0044	0.0041	0.0039	0.0037	0.0036	0.0035	0.0033
Single phase							
Frictional Δp_f (kN/m²)	1.61	4.82	10.92	35.0	79.5	143.5	308.5
Accelerational Δp_a (kN/m²)	0.330	0.737	1.32	2.97	5.28	8.24	18.5
Gravitational Δp_z (kN/m²)	6.03	9.06	12.1	18.1	24.1	28.6	28.6
Two phase							
Martinelli–Nelson							
$\quad (L - z_{sc})$, m	2.89	2.50	2.11	1.34	0.57	0	0
$\quad$Frictional Δp_f (kN/m²)	87.4	105	110	84.5	35.6	0	0
$\quad r_2$	18.1	8.04	4.55	1.95	0.93	0	0
$\quad$Accelerational Δp_a (kN/m²)	43.5	43.3	43.6	42.1	35.6	0	0
$\quad$Gravitational Δp_z (kN/m²)	4.19	5.09	5.84	5.65	3.21	0	0
Martinelli–Nelson total pressure drop (kN/m²)	143	168.5	184	188	183.5	181	353
Baroczy frictional Δp_f (kN/m²)	66.9	67.6	67.5	52.8	22.25	0	0
Baroczy total pressure drop (kN/m²)	122.5	130.5	141.3	156.7	170	181	353

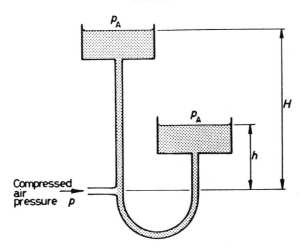

Fig. 2.14. Air lift pump for Problem 1.

The other components, accelerational and gravitational, are taken to be the same as for the Martinelli–Nelson model.

(d) *The overall pressure drop*
The total pressure drops across the entire heated length can now be evaluated. (See Table 2.6.)

Problems

1. Derive the energy equations for each phase separately for the case of an air lift pump (Fig. 2.14). Assume that the expansion of the gas is isothermal and that there is no heat or mass transfer between the phases. Assume the losses in the foot piece are negligible and that the liquid is incompressible. Derive an expression for the mechanical efficiency of the pump.

2. A novel design of desalination plant makes use of the flashing process to lift the fluid to the next stage. Assuming homogeneous flow write down the energy equation for the combined phases. Discuss the shape of the duct if the internal energy of the superheated fluid is to be converted into potential rather than kinetic energy.

3. Derive an expression for the acceleration pressure gradient dp_a/dz for the case where a mass fraction e of the liquid stream in a separated flow is entrained in the gas stream and flows at the gas velocity u_g. The remaining liquid ($W_f(1 - e)$) flows at a velocity u_f. Assume that α is the fractional cross-sectional area occupied by gas, γ is the fractional area occupied by separated liquid and $(1 - \alpha - \gamma)$ is the fractional area occupied by entrained liquid.

$$\text{Answer:} \quad -\left(\frac{dp_a}{dz}\right) = G^2 \frac{d}{dz}\left[\frac{x^2 v_g}{\alpha} + \frac{(1-e)^2(1-x)^2 v_f}{\gamma} + \frac{e^2(1-x)^2 v_f}{(1-\alpha-\gamma)}\right]$$

where

$$\gamma = 1 - \alpha - \left[\alpha e\left(\frac{1-x}{x}\right)\left(\frac{v_f}{v_g}\right)\right]$$

4. Liquid metal M.H.D. devices often make use of a jet condenser where internal energy in a vapour state is converted to kinetic energy in the liquid state. Write down the equation for the combined phases describing the energy transfer. Assume that pure saturated vapour enters the horizontal constant area channel and that this vapour is totally condensed on to a liquid stream at the exit. Assume the condensation process occurs at constant pressure and that no heat is lost from the channel walls. Discuss the values of the initial velocities of the separate phases necessary to give the maximum increase in liquid phase kinetic energy across the device.

5. Estimate the friction pressure gradient in a 10.15 cm bore unheated horizontal pipe for the following conditions:

Fluid—propylene
Pressure—8.175 bar
Temperature—7°C
Mass flow of liquid—2.42 kg/s. Density of liquid—530 kg/m³
Mass flow of vapour—0.605 kg/s. Density of vapour—1.48 kg/m³

Answer: 210 (N/m²)/m

6. Estimate the friction pressure gradient in a 13.3 cm bore return tube for the following conditions:

Fluid—steam–water
Pressure—180 bar
Mass flow of steam—12.23 kg/s
Mass flow of water—2.78 kg/s

Answer: 275 (N/m²)/m

7. Estimate the friction pressure gradient in a 5.08 cm bore evaporating tube for the following conditons:

Fluid—steam–water
Pressure—180 bar
Inlet mass flow of saturated water—2.14 kg/s
Outlet mass quality x_o—0.1825

Answer: 319 (N/m²)/m

8. Estimate the gravitational pressure gradient assuming homogeneous flow for an adiabatic flow of refrigerant R-123 in a vertical 10 mm bore tube for the following conditions:

Pressure	0.38 bar
Temperature	3°C
Flow rate	0.05 kg/s
Vapour quality	0.20

(Properties: vapour density $= 2.60$ kg/m^3, liquid density $= 1518$ kg/m^3).

Answer: 127 (N/m^2)/m

9. Determine the frictional pressure drop for R-123 for the conditions described in Problem 8 using the homogeneous flow model. (Additional properties: vapour viscosity $= 0.0126$ kN s/m^2, liquid viscosity $= 0.5856$ kN s/m^2).

Answer: 27269 (N/m^2)/m

10. Determine the frictional pressure drop for R-123 for the conditions described in Problem 8 using the Friedel separated flow model. (Additional properties: vapour viscosity $= 0.0126$ kN s/m^2, liquid viscosity $= 0.5856$ kN s/m^2, surface tension $= 0.018$ N/m).

Answer: 35,389 (N/m^2)/m

11. Prove eq. (2.44).

3

EMPIRICAL TREATMENTS OF
TWO-PHASE FLOW

3.1 Introduction

The 'homogeneous' model and the 'separated flow' model are the two most widely used and tested treatments of two-phase flow at present available. However, a large number of other empirical and semi-empirical methods have been suggested from time to time and a few of these approaches are outlined in the first half of this chapter. The most relevant and useful of these methods are those which attempt to model the particular hydrodynamic features of the flow and therefore the treatments are loosely grouped under the various flow patterns discussed in Chapter 1. Needless to say, these methods have to be used in conjunction with some method of defining the extent of the flow regime such as a flow pattern map.

The second half of the chapter is taken up with a discussion of two-phase flow through various features such as bends, enlargements, contractions, orifices, and valves.

3.2 The drift flux model

This model has been developed principally by Zuber and Findlay (1965), Wallis (1969) and Ishii (1975, 1977) together with their co-workers. The model is fully developed in Wallis (1969) and only the essential relationships will be given here. Using eq. (1.10) the relative velocity between the phases, u_{gf}, can be expressed as

$$u_{gf} = (u_g - u_f) = \frac{j_g}{\alpha} - \frac{j_f}{(1 - \alpha)} \tag{3.1}$$

Clear fractions and define a *drift flux*, j_{gf}, such that

$$j_{gf} = u_{gf}\alpha(1 - \alpha) = j_g(1 - \alpha) - j_f\alpha = j_g - \alpha j \tag{3.2}$$

The drift flux, j_{gf}, physically represents the volumetric rate at which vapour is passing forwards (in up-flow) or backwards (in down-flow) through unit area of a plane normal to the channel axis already travelling with the flow at a velocity j. To preserve continuity an equal and opposite drift flux of liquid (j_{fg}) must also pass across this same plane. Rearranging eq. (3.2),

$$j_g = \alpha j + j_{gf} \qquad (3.3a)$$

The above relationship is true for one-dimensional flow or at any local point in the flow. It is often desirable to relax the restriction of one-dimensional flow. Denoting the average properties of the flow by a bar, thus, ($\bar{u}$), then,

$$(\bar{j}_g) = (\overline{\alpha j}) + (\bar{j}_{gf}) \qquad (3.3b)$$

Dividing by ($\bar{\alpha}$)

$$\frac{(\bar{j}_g)}{(\bar{\alpha})} = (\bar{u}_g) = \frac{(\overline{\alpha j})}{(\bar{\alpha})} + \frac{(\bar{j}_{gf})}{(\bar{\alpha})} \qquad (3.4)$$

Define a parameter, C_0, such that

$$C_0 = \frac{(\overline{\alpha j})}{(\bar{\alpha})(\bar{j})} \qquad (3.5)$$

It is also found convenient to define *a weighted mean drift velocity*, $\bar{u}_{gj}$, such that,

$$\bar{u}_{gj} = \frac{(\bar{j}_{gf})}{(\bar{\alpha})} \qquad (3.6)$$

Then

$$(\bar{u}_g) = \frac{(\bar{j}_g)}{(\bar{\alpha})} = C_0(\bar{j}) + \bar{u}_{gj} \qquad (3.7)$$

Dividing through by ($\bar{j}$)

$$\frac{(\bar{u}_g)}{(\bar{j})} = \frac{(\beta)}{(\bar{\alpha})} = C_0 + \frac{(\bar{u}_{gj})}{(\bar{j})} \qquad (3.8)$$

or

$$(\alpha) = \frac{(\beta)}{C_0 + \bar{u}_{gj}/(\bar{j})} \qquad (3.9)$$

Now return to the notation defined in Section 1.4 (i.e. ($\bar{\alpha}$) = α).

It will be seen that if there is no local relative motion between the phases ($\bar{u}_{gj} = 0$) then $\alpha = \beta/C_0$; for one-dimensional homogeneous flow, $\alpha = \beta$. In other words the parameter C_0 represents an empirical factor correcting the one-dimensional homogeneous theory to account for the fact that the concentration and velocity profiles across the channel can vary independently of one another.

The drift flux model can be used with or without reference to any particular flow regime (Ishii 1977). Experimental data plotted as j_g/α versus j are used to yield expressions for C_0 and j_{gf} (or alternatively $\bar{u}_{gj}$). The particular values these parameters take up will vary depending upon whether or not the chosen data are restricted to a particular flow pattern. For example the data (Zuber et al. 1967) shown in Fig. 2.8 and similar measurements for steam–water flow at elevated pressures can be well correlated by the use of the following values in eq. (3.9) without reference to flow pattern,

$$C_0 = 1.13, \ \bar{u}_{gj} = 1.41 \left[\frac{\sigma g(\rho_f - \rho_g)}{\rho_f^2} \right]^{1/4} \tag{3.10}$$

The drift flux approach, therefore, satisfactorily accounts for the influence of mass velocity on the void fraction as seen in the separated flow model and eqs. (3.9) and (3.10) may be used to provide the required relationship between void fraction and the independent flow parameters.

The drift flux model is valuable only when the drift velocity is significant compared with the total volumetric flux (say $\bar{u}_{gj} > 0.05j$). This limits its usefulness to the bubbly, slug, and churn flow patterns.

3.3   Bubbly flow

3.3.1   *The Bankoff (1960) variable density model*

An alternative name for this approach would be the 'homogeneous model with correction for two-dimensional effects'. The model treats the case where radial gradients exist in the concentration of vapour bubbles across the channel. The concentration is assumed to be a maximum at the centre of the channel and zero at the walls. No relative motion is assumed between the vapour bubbles and the liquid at any radial position. A power law distribution is assumed for both the velocity and the void fraction

$$u/u_m = (y/r)^{1/m}$$

$$\alpha/\alpha_m = (y/r)^{1/n}$$

Where u and α are the local velocity and void fraction respectively, u_m and α_m are the maximum (axial) values, y is the distance from the tube wall and r is the tube radius; m and n are arbitrary exponents. Integration and manipulation of these equations leads directly to the result that the mean void fraction α is related to the mean volumetric quality β by the relationship

$$\alpha = C_A \bar{\beta} \tag{3.11}$$

where C_A is a function of m and n as follows:

$$C_A = \frac{2(m + n + mn)(m + n + 2mn)}{(n + 1)(2n + 1)(2m + 1)(m + 1)}$$

For n from 0.5 to 1.0 and m from 2 to 7, C_A had an effective range of 0.5–1. Bankoff found good agreement with the Martinelli–Nelson values of α ($\alpha < 0.85$) for steam–water flow with $C_A = 0.89$.

Comparison of eq. (3.11) with eq. (3.9) shows that, for the case of no local relative motion between the phases, $C_A = 1/C_0$. Expressions of the form of eq. (3.11) have been reported widely in the Russian literature (Armand 1946; Armand and Treschev 1947; Armand 1954). Armand (1946) suggested a value of 0.833 for C_A when $\beta < 0.9$. Both Armand and Treschev (1947) and Bankoff (1960) suggest that C_A increases with pressure; C_A may also be a weak function of mass velocity.

3.3.2 Slip ratio correlations

Although not strictly applicable solely to bubbly flow, both Smith (1969) and Chisholm (1983) have derived simple and useful expressions for slip ratio and it is convenient to describe them in this section.

Smith (1969) assumed a separated flow consisting of a liquid phase and a gaseous phase with a fraction e of the liquid entrained as droplets. He assumed that the momentum fluxes of the two separated phases were equal. This leads to the following expression for the slip ratio:

$$\frac{u_g}{u_f} = e + (1 - e)\left[\frac{\rho_f/\rho_g + e(1/x - 1)}{1 + e(1/x - 1)}\right]^{1/2} \tag{3.12}$$

A value of $e = 0.4$ gave the best fit to the data.

Chisholm (1983) derived the following expression:

$$\frac{u_g}{u_f} = \left(\frac{\rho_f}{\bar{\rho}}\right)^{1/2} = \left(\frac{\bar{v}}{v_f}\right)^{1/2} = \left[1 + x\left(\frac{\rho_f}{\rho_g} - 1\right)\right]^{1/2} \tag{3.13a}$$

on the basis that this slip ratio results in simple annular flow theory (see Section 3.5.1) and homogeneous theory, producing approximately equal friction pressure gradients. Equations (3.12) and (3.13a) give similar values for the predicted slip ratios, except at high qualities, when eq. (3.13a) gives the higher value. Chisholm (1983) also found eq. (3.13a) to be applicable for flows at changes of section (see Section 3.8.4) provided the Martinelli parameter X was greater than unity (see eq. (3.82)). For $X < 1$,

$$\frac{u_g}{u_f} = \left(\frac{v_g}{v_f}\right)^{1/4} = \left(\frac{\rho_f}{\rho_g}\right)^{1/4} \tag{3.13b}$$

For $X \geqslant 1$ using various identities derived in Section 1.4, the slip ratio given by eq. (3.13a) can also be expressed as

$$\frac{u_g}{u_f} = \left[1 - \beta\left(1 - \frac{v_f}{v_g}\right)\right]^{-1/2} \tag{3.13c}$$

or alternatively

$$\frac{1}{C_A} = \beta + \frac{1 - \beta}{\left[1 - \beta\left(1 - \frac{v_f}{v_g}\right)\right]^{1/2}} \tag{3.13d}$$

If $v_f/v_g \gg 1$,

$$C_A = \frac{1}{\beta + (1 - \beta)^{1/2}}$$

or

$$\alpha = \frac{\beta}{\beta + (1 - \beta)^{1/2}} \tag{3.14}$$

3.3.3   *The Hughmark (1962) correlation*

Following a suggestion by Bankoff, Hughmark developed a void fraction correlation in which C_A was related to a parameter Z which was defined as follows:

$$Z = \left[\frac{DG}{(1 - \alpha)\mu_f + \alpha\mu_g}\right]^{1/6}\left[\frac{j^2}{gD}\right]^{1/8}[1 - \beta]^{-1/4} \tag{3.15}$$

C_A was represented as a non-linear function of Z and the co-ordinates of this function are given in Table 3.1.

Table 3.1   Values of C_A a function of Z (Hughmark 1962)

Z	1.3	1.5	2.0	3.0	4.0	5.0	6.0	8.0	10.0	15.0	20.0	40.0	70.0	130.0
C_A	0.185	0.225	0.325	0.49	0.605	0.675	0.72	0.767	0.78	0.808	0.83	0.88	0.93	0.98

The inter-relationships between the various correlations may be seen more easily (NEL 1969) on a plot of (β/α) verus $(\bar{u}_{gj}/j)$ (Fig. 3.1). For those models where relative motion is ignored horizontal lines are drawn. It is clear that the dependence of C_A upon Z in the Hughmark correlation reflects the variation of (β/α) with $(\bar{u}_{gj}/j)$ and points to the fact that the assumption of no relative motion between the phases is too great a simplification.

3.3.4   *The drift flux model*

For the case of one-dimensional vertical up-flow of isolated small vapour bubbles with little coalescence ($\alpha < 0.2$) Wallis (1969) has suggested that the

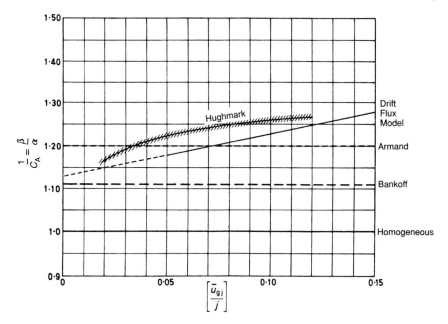

Fig. 3.1. Comparison of various void fraction models.

following values should be used in eq. (3.9):

$$C_0 = 1.0, \quad \bar{u}_{gj} = 1.53(1 - \alpha)^2 \left[\frac{\sigma g(\rho_f - \rho_g)}{\rho_f^2} \right]^{1/4} \tag{3.16}$$

Alternative values of the constant 1.53 are 1.18 (Wallis 1969) and 1.41 (Zuber et al. 1967).

For the case of vertical up-flow in the bubbly-slug flow region (referred to by Zuber et al. 1967, as the churn-turbulent region) the recommended value of $\bar{u}_{gj}$ is given by

$$\bar{u}_{gj} = 1.41 \left[\frac{\sigma g(\rho_f - \rho_g)}{\rho_f^2} \right]^{1/4} \tag{3.17}$$

A plot of $\bar{u}_{gj}$ (m/s) against pressure for the steam–water system is shown in Fig. 3.2 for convenience. The value of C_0 is dependent upon channel diameter and reduced pressure (P_r).

$$\begin{aligned}
\textit{Tubes} \quad &\text{(a)} \ D > 5 \text{ cm}, \ C_0 = 1.5 - 0.5P_r \\
&\text{(b)} \ D < 5 \text{ cm}, \ P_r < 0.5, \ C_0 = 1.2 \\
&\text{(c)} \ D < 5 \text{ cm}, \ P_r > 0.5, \ C_0 = 1.2 - 0.4(P_r - 0.5)
\end{aligned} \tag{3.18}$$

$$\textit{Rectangular ducts} \quad C_0 = 1.4 - 0.4P_r$$

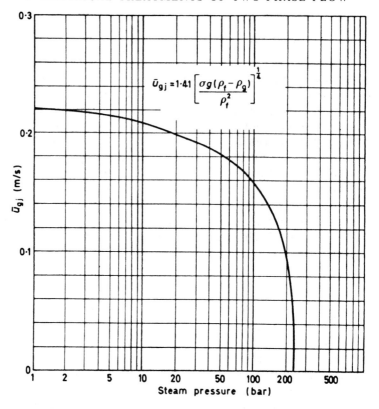

Fig. 3.2. Variation of $\bar{u}_{gj}$ for bubbly-slug flow of steam–water mixtures.

Alternative expressions have been given by Ishii (1977). For vertical down-flow the sign of $\bar{u}_{gj}$ in eq. (3.9) is changed. For horizontal bubbly flow $\bar{u}_{gj}$ is set to zero in eq. (3.9) and C_0 takes on the value given in eq. (3.18).

It is recommended that the pressure drop in bubbly flow be calculated using the separated flow model with the values of void fraction calculated from the above relationships. The wall shear force in bubbly flow is usually small and the frictional pressure gradient can be evaluated using eq. (2.35) with little error.

3.4 Slug flow

3.4.1 *The drift flux model*

For the case of vertical fully developed turbulent slug flow ($[jD\rho_f/\mu_f] > 8,000$), the following values should be used in eq. (3.9)

$$C_0 = 1.2, \qquad \bar{u}_{gj} = 0.35 \left[\frac{g(\rho_f - \rho_g)D}{\rho_f} \right]^{1/2} \qquad (3.19)$$

This expression for $\bar{u}_{gj}$ represents the bubble rise velocity in a tube of stagnant liquid for the case where inertia forces dominate. Alternative expressions are given by Wallis (1969) for $\bar{u}_{gj}$ when viscous and surface tension forces are important. Expressions for C_0 and $\bar{u}_{gj}$ for annular, rectangular, and rod cluster channels have been developed by Griffith (1963). For the case where slug flow is developing the rise velocity is somewhat increased (Moissis and Griffith 1962) above that given by eq. (3.19) and Griffith recommends that in a channel where boiling is taking place the value of $\bar{u}_{gj}$ calculated from eq. (3.19) should be increased by a factor of 1.6.

For horizontal turbulent slug flow ($[jD\rho_f/\mu_f] > 3,000$), $\bar{u}_{gj}$ can be set to zero and C_0 equals 1.2. This leads to the simple relationship first obtained by Armand (1946)

$$\alpha = 0.833\beta \qquad (3.20)$$

Incidentally, this relationship infers that the influence of mass velocity on the void fraction in horizontal flow will be absent.

Stable downwards slug flow (Martin 1976) only occurs when $[\rho_f g D^2/\sigma] < 100$, i.e., small tube diameters. For larger tube diameters the bubbles are unstable and eccentrically located off the tube axis in regions of lower liquid velocities. Consequently the value of C_0 is less than unity ($\simeq 0.9$). The value of the coefficient in the expression for $\bar{u}_{gj}$ in eq. (3.19) is higher than 0.35 ($\simeq 0.6$) because the unstable bubbles ride the wall.

The pressure drop in slug flow can be calculated using the separated flow model. The static head and acceleration terms may be estimated directly from the void fraction calculated from the above relationships. The wall shear stress is more difficult to estimate since in vertical up-flow the liquid velocity at the wall will be downwards during the passage of the vapour bubble and upwards during the passage of the subsequent liquid plug. The time average wall shear stress may therefore be either positive or negative. Negative wall shear stress results in 'negative' frictional pressure gradients in the separated flow momentum balance. Wallis (1969) has proposed that the wall shear stresses around the vapour bubble should be ignored. This is strictly true only for horizontal slug flow. This assumption results in the following expression for the frictional pressure gradient in slug flow:

$$-\left(\frac{dp}{dz} F\right) = \frac{2f_{fo} G^2 v_f}{D} (1 - \alpha) \qquad (3.21)$$

For long vapour bubbles ($L_b/D > 15$) in vertical slug flow a correction to the pressure drop must be made for the amount of liquid held up in the film around the bubble.

3.5   Annular flow

Annular flow is a particularly important flow pattern since for a wide range of pressure and flow conditions it occurs over the major part of the mass quality range (x from 0.1 and below, up to unity). In other words, in a vertical tube evaporator as much as 90 per cent of the tube length may be in annular flow.

The two-dimensional analysis of the annular flow pattern by applying the normal methods of single-phase flow would, at first sight, appear relatively straightforward. In practice, the range of flow conditions, over which the entire liquid mass flow-rate is contained as a symmetrical film on the channel wall with a smooth interface between the film and the vapour core, is very small indeed. In the general case, an unknown fraction (e) of the liquid flow will be entrained in the vapour core and the interface between the film and the vapour core will be highly disturbed. A further fundamental feature of annular flow is that the method by which the annular flow is formed greatly influences the subsequent behaviour of the system even at considerable distances downstream. This is due to the fact that different types of phase mixing device produce different amounts of initial entrainment and the mass exchange processes between the vapour core and the liquid film (those of entrainment and deposition of droplets) are relatively slow. As a result, an 'equilibrium' situation where there is no net transfer of liquid to, or from, the film is very seldom reached in isothermal systems and never in a system where evaporation is occurring.

An understanding of the physical processes of entrainment and deposition of liquid occurring in a 'non-equilibrium' (i.e., not 'fully-developed') annular flow has been developed primarily as a result of the efforts of Hewitt and Hall-Taylor (1970), Hewitt (1978), and Govan et al. (1988). This under-standing has led to a quantitative method for the calculation of 'non-equilibrium' annular flows and this will be discussed below. The physical processes of entrainment and deposition govern the occurrence of the important dryout transition in heated systems (Govan et al. 1988) and are discussed further in Chapter 9.

Only the essential relationships allowing the computation of pressure drop, film thickness, film flow-rate, and the rates of entrainment and deposition of liquid will be reviewed in this section and the reader is referred to the comprehensive treatise on annular flow by Hewitt and Hall-Taylor (1970) for fuller details of this particular flow pattern.

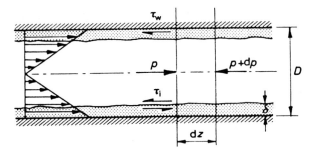

Fig. 3.3. Shear stress distribution in horizontal annular flow.

3.5.1 *The liquid film*

Despite the limited range of application, it is convenient to start by considering the entire liquid flow to be in the film and the interface to be smooth. In addition it is assumed that gravitational and accelerational forces are absent, i.e., a steady isothermal axisymmetric horizontal flow (Fig. 3.3). From a force balance on the combined phases and then on the vapour phase alone

$$-\left(\frac{dp}{dz}\,F\right) = \frac{4\tau_w}{D} \tag{3.22}$$

$$-\left(\frac{dp}{dz}\,F\right) = \frac{4\tau_i}{(D-2\delta)} \tag{3.23}$$

It is convenient to develop the present analysis in terms of already familiar parameters such as ϕ_f^2 and α. Dividing eq. (3.22) by $(dp/dz\,F)_f$, the pressure gradient for the liquid phase considered to flow alone in the channel,

$$\phi_f^2 = \frac{\tau_w}{\tau_f} \tag{3.24}$$

where τ_f is the wall shear stress when the liquid phase flows alone in the channel,

$$\tau_w = f_{TP}\left(\frac{\rho_f u_f^2}{2}\right); \qquad \tau_f = f_f\left(\frac{\rho_f j_f^2}{2}\right) \tag{3.25}$$

Substituting eq. (3.25) into eq. (3.24) and remembering that $u_f = j_f/(1-\alpha)$ (eq. (1.10)), then,

$$\phi_f^2 = \frac{1}{(1-\alpha)^2}\left(\frac{f_{TP}}{f_f}\right) \tag{3.26}$$

It is an interesting fact that the Reynolds number of the liquid film on the channel wall ($4 \delta u_f \rho_f / \mu_f$) is identical with the Reynolds number when the liquid is assumed to flow alone in the channel ($D j_f \rho_f / \mu_f$). Since the friction factor f_{TP} can be expected to be related to the liquid film Reynolds number in the same manner as the friction factor f_f is related to the liquid-alone Reynolds number then (f_{TP}/f_f) can be taken as unity. Therefore,

$$\phi_f^2 = \frac{1}{(1 - \alpha)^2} \tag{3.27}$$

This simple result is important since, knowing the liquid film flow-rate, it is possible to obtain the liquid film thickness from the pressure gradient or vice versa. A number of empirical correlations including that of Lockhart and Martinelli (Section 2.4.4.1) correspond closely with this relationship.

More refined analyses which relate the liquid film flow-rate with the frictional pressure gradient and the film thickness have been published by Calvert and Williams (1958), Mantzouranis (1959), Dukler (1960) (down-flow), Hewitt (1961) (up-flow), and Kunz and Yerazunis (1967). The general procedure in these treatments has been to write down the distribution of shear stress in the liquid film, relate this to the velocity gradient using one of the empirical relationships for eddy viscosity, integrate to get the velocity profile and integrate once again to obtain the film flow-rate. The earlier treatments imply either a constant shear stress across the film (Mantzouranis 1959) or incorrectly allowed for the effect of gravitational forces (Calvert and Williams 1958). A series of improvements have been systematically incorporated in subsequent theories including the use of the correct shear stress distribution (Hewitt 1961), the effect of curvature (Dukler 1960; Hewitt 1961) the influence of shear stress distribution on the eddy viscosity relation (Kunz and Yerazunis 1967) and the inclusion of molecular viscosity in the turbulent region (Kunz and Yerazunis 1967).

For the case where the influence of gravitational forces can be ignored these improved theories all give results very close to eq. (3.27). A plot of the liquid film friction factor f_{TP} versus the liquid film Reynolds number confirms that at low (Re < 200) and high (Re > 8000) Reynolds numbers the conventional single-phase laminar and turbulent (Blasius) friction factor relationships are also valid for film flow. At a Reynolds number of 1000 the value of f_{TP} lies about a factor 1.5 above the conventional laminar relationship on a smooth curve having the laminar and turbulent friction factor relationships as asymptotes.

In the general case a mass fraction, e, of the total liquid flow may be entrained in the vapour stream and the liquid film flow-rate will then be $W_f(1 - e)$. In many cases the entrained droplets can be assumed to travel at velocities close to the vapour velocity and for large values of the ratio (v_g/v_f) their contribution to the liquid volume fraction ($1 - \alpha$) is small. In this case

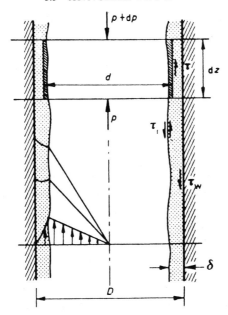

Fig. 3.4. Shear stress distribution in vertical annular flow.

eq. (3.27) can be modified so that in the evaluation of ϕ_f^2 the pressure drop for the total liquid flow (W_f) flowing alone in the channel is replaced by the pressure drop for the liquid film flow-rate ($W_f(1 - e)$) flowing alone in the channel.

In vertical annular flow it is necessary to take into account the influence of gravitational forces on the shear stress distribution in the liquid film (Fig. 3.4). Carrying out a force balance on the shaded element of liquid film it can be shown that the shear stress at any diameter d within the liquid film is given by

$$\tau = \tau_i \left[\frac{(D - 2\delta)}{d} \right] + \frac{1}{4} \left(\frac{dp}{dz} + \rho_f g \right) \left[\frac{(D - 2\delta)^2 - d^2}{d} \right] \qquad (3.28)$$

where

$$\tau_i = \frac{1}{4} \left(\frac{dp}{dz} + \rho_g g \right) [D - 2\delta] \qquad (3.29)$$

Analytical expressions for the velocity distribution in the film and the film flow-rate may be derived (Hewitt and Hall-Taylor 1970) for the case of laminar flow ($\tau = \mu_f (du_f/dy)$; $u_f = 0$ at $d = D$) by integration of eq. (3.28).

In vertical annular flow as the vapour velocity is continuously decreased for a given liquid film flow-rate a minimum is observed in the pressure gradient. This minimum corresponds closely with the interesting special condition of zero wall shear stress. The shear stress distribution in this case is illustrated in Fig. 3.4. Ignoring any changes in momentum of the streams the total pressure gradient $(\mathrm{d}p/\mathrm{d}z)$ consists only of the static head term (eq. (2.11)) since the frictional pressure gradient (based on the wall shear force) is zero. This condition highlights the fundamental differences between the momentum and energy balance approaches to two-phase pressure drop. In this special case, although no frictional losses take place at the channel wall, irreversible energy losses still occur at the interface. This condition corresponds, at low liquid flow-rates, to a value of $j_g^* \approx 1.1$. If the vapour velocity is reduced below this value the pressure gradient increases sharply, the wall shear stress becomes negative and the liquid close to the wall will flow downwards whilst that near the interface will travel upwards. Finally, the interfacial shear stress can no longer ensure the transport of the liquid flow entering the tube, 'flow reversal' takes place and the lower limit of co-current upwards annular flow is reached.

The shear stress distribution given by eq. (3.28) (or simplifications of it) forms the starting point of the analytical treatments of vertical annular flow (Hewitt 1961; Kunz and Yerazunis 1967). A considerable amount of tabulated information in dimensionless form for velocity profiles and film flow-rates as a function of wall shear stress, film thickness, and the balance between interfacial and gravitational forces has been presented by Hewitt (1961). Wallis (1969) has plotted some of these results in terms of the liquid film friction factor f_{TP} for vertical flow against film Reynolds number. The friction factor (f_{TP}) for vertical flow, at constant pressure drop, follows that for horizontal flow (i.e., the usual single-phase flow relationships) up to some particular value of film Reynolds number after which point, as might be expected, it falls dramatically. Wallis has given empirical equations relating this critical value of Reynolds number with the pressure drop.

A simple but accurate correction to the constant shear stress (horizontal flow) approach to allow for the influence of gravitational forces in vertical flow has been suggested by Collier and Hewitt (1961). Thus

$$\frac{W_{\mathrm{fv}}}{W_{\mathrm{fc}}} = \frac{(3\tau_i - 3\rho_f\,\delta\mathrm{g})}{(3\tau_i - 2\rho_f\,\delta\mathrm{g})} \tag{3.30}$$

where W_{fv} is the film flow-rate allowing for gravitational forces and W_{fc} is the film flow-rate for constant shear stress. Thus, for a given pressure drop and film thickness, the film flow-rate decreases below that implicit in eq. (3.27) as $(\rho_f\,\delta\mathrm{g})$ becomes a significant fraction of τ_i.

3.5.2 *The vapour core*

Now consider the vapour core. Returning to eq. (3.23) and dividing by $(dp/dz\ F)_g$ the pressure gradient for the vapour phase flowing alone in the channel,

$$\phi_g^2 = \frac{\tau_i}{\tau_g} \frac{D}{(D - 2\delta)} = \frac{\tau_i}{\tau_g} \frac{1}{\alpha^{1/2}} \tag{3.31}$$

where τ_g is the wall shear stress when the gas phase flows alone in the channel

$$\tau_i = f_i \left[\frac{\rho_g (u_g - u_i)^2}{2} \right]; \qquad \tau_g = f_g \left[\frac{\rho_g j_g^2}{2} \right] \tag{3.32}$$

where f_i is the interfacial friction factor, u_g is the mean vapour core velocity, and u_i is the interfacial liquid velocity. Usually the interface velocity is very small compared with the vapour core velocity and can be neglected. Substituting (3.32) in (3.31) and remembering that $u_g = j_g/\alpha$ (eq. 1.10) then

$$\phi_g^2 = \frac{1}{\alpha^{5/2}} \frac{f_i}{f_g} \tag{3.33}$$

If the interface is smooth then (f_i/f_g) will be close to unity, and therefore

$$\phi_g^2 \approx \frac{1}{\alpha^{5/2}} \tag{3.34}$$

This equation is a poor representation of the experimental variation of ϕ_g^2 with α except at very low liquid rates ($Re_f < 100$). At higher liquid rates the interface becomes disturbed by small ripples and the interfacial friction factor (f_i) increases progressively above the smooth tube value (f_g). Shearer and Nedderman (1965) have given recommendations for calculating the interfacial shear stress in this 'ripple' region. At higher liquid rates ($Re_f > 400$) much larger disturbance waves occur at the interface with the onset of turbulence in the film. A number of attempts have been made to correlate the variation of the interfacial friction factor or effective interface roughness in this 'disturbance wave' region with the other flow parameters. These studies include those of Wallis (1969), Calvert and Williams (1958), Mantzouranis (1959), Hewitt and Hall-Taylor (1970), Asali *et al.* (1985), and Lopes and Dukler (1986). The main finding from these studies has been that the interfacial friction factor or roughness is a direct function of the liquid film thickness. The simplest of the relationships proposed is that of Wallis (1969) which states that for the case of no entrainment and ignoring the interface velocity u_i,

$$\frac{f_i}{f_g} = 1 + 300\ \delta/D = 1 + 75(1 - \alpha) \tag{3.35}$$

Substitution of eq. (3.35) into (3.33) gives

$$\phi_{\rm g}^2 = \left[\frac{1 + 75(1 - \alpha)}{\alpha^{5/2}} \right] \tag{3.36}$$

This relationship, the smooth interface relationship (eq. (3.34)) and the empirical curve of Lockhart–Martinelli are shown in Fig. 3.5 together with a typical variation of $\phi_{\rm g}^2$ with liquid rate at constant vapour velocity. Wallis has considered the effect of interface velocity and entrainment upon the interfacial shear stress and has suggested empirical corrections for these effects. Assuming that the interface velocity $u_{\rm i}$ is twice the mean film velocity $u_{\rm f}$, then,

$$\tau_{\rm i} = f_{\rm i} \left[\frac{\rho_{\rm c}(u_{\rm g} - 2u_{\rm f})^2}{2} \right] \tag{3.37}$$

where $\rho_{\rm c}$ is the average density of the vapour core including entrainment, given by

$$\rho_{\rm c} = \left[\frac{W_{\rm g} + eW_{\rm f}}{W_{\rm g}} \right] \rho_{\rm g} \tag{3.38}$$

Making use of eq. (1.12) for the ratio $(u_{\rm g}/u_{\rm f})$ then

$$\left[\frac{u_{\rm g} - 2u_{\rm f}}{u_{\rm g}} \right]^2 = \left[1 - 2\left(\frac{u_{\rm f}}{u_{\rm g}} \right) \right]^2 = \left[1 - 2\left(\frac{\alpha}{1 - \alpha} \right)\left(\frac{\rho_{\rm g}}{\rho_{\rm f}} \right)\left(\frac{W_{\rm f}(1 - e)}{W_{\rm g}} \right) \right]^2 \tag{3.39}$$

Applying these corrections to eq. (3.36)

$$\phi_{\rm g}^2 = \left[\frac{1 + 75(1 - \alpha)}{\alpha^{5/2}} \right]\left[\frac{W_{\rm g} + eW_{\rm f}}{W_{\rm g}} \right]\left[1 - 2\left(\frac{\alpha}{1 - \alpha} \right)\left(\frac{\rho_{\rm g}}{\rho_{\rm f}} \right)\left(\frac{W_{\rm f}(1 - e)}{W_{\rm g}} \right) \right]^2 \tag{3.40}$$

Lopes and Dukler (1986) have shown that the momentum exchange due to the entrainment and deposition of droplets does provide a significant proportion of the pressure gradient. These measurements have enabled this term to be separately evaluated from that due to interfacial roughness.

3.5.3 Deposition and entrainment

Use can be made of the foregoing equations given in Sections (3.5.1) and (3.5.2) only if the fraction of liquid entrained in the vapour core (e) can be specified. However, in any general case it may be expected that entrainment of liquid from the film may be attempting to increase e or alternatively, deposition of liquid droplets on to the film may be causing a reduction in

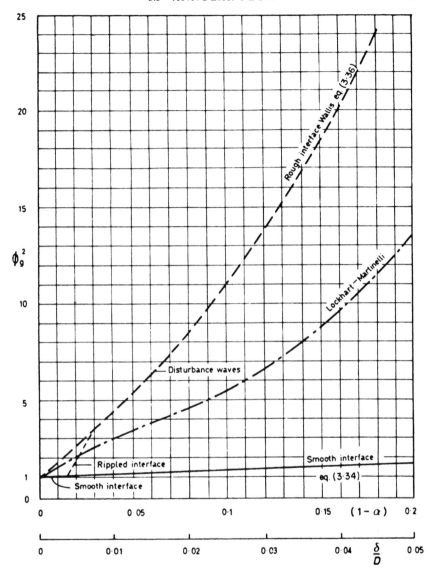

Fig. 3.5. The variation of ϕ_g^2 with film thickness.

e. In special circumstances the rate of entrainment (E) may be exactly equal to the rate of deposition (D_d) and, in this case, the fraction of liquid entrainment (e) does not change with length and the flow may be said to be in *hydrodynamic equilibrium* (i.e., 'fully developed').

The rate of deposition (D_d) of droplets from the vapour core on to the liquid film may be given by

$$D_d = kC \tag{3.41}$$

where k is the deposition or mass transfer coefficient (m/s) and C is the homogeneous mean droplet concentration in the vapour core (kg/m^3) which can be calculated provided that e is known. Measurements of k have been made and it is clear that droplet concentration has an important effect on k, as do surface tension, gas density and tube diameter. Govan et al. (1988) have correlated the available data as a relationship between $k \, (\rho_g D/\sigma)^{1/2}$ and C/ρ_g

$$\left.\begin{array}{ll} k\left(\dfrac{\rho_g D}{\sigma}\right)^{1/2} = 0.18 & \text{if } C/\rho_g < 0.3 \\[4mm] k\left(\dfrac{\rho_g D}{\sigma}\right)^{1/2} = 0.083\,(C/\rho_g)^{-0.65} & \text{if } C/\rho_g > 0.3 \end{array}\right\} \tag{3.42}$$

This is plotted in Fig. 3.6; there is considerable scatter but approximately two-thirds of the points lie within a deviation of ± 30 per cent.

Entrainment from a liquid film is associated with the onset of disturbance waves at the interface and, in general, depends on both the vapour and liquid flow-rates. At low liquid flow-rates corresponding to the 'smooth' and 'rippled' film regions (liquid film Reynolds numbers below 200) little or no entrainment will take place even at very high vapour velocities. Govan et al. (1988) provide an expression for the critical liquid film Reynolds number for the onset of entrainment

$$\text{Re}_{f_{CRIT}} = \exp\left[5.8504 + 0.4249\left(\frac{\mu_g}{\mu_f}\right)\left(\frac{\rho_f}{\rho_g}\right)^{1/2}\right] \tag{3.43}$$

With the onset of turbulence in the film, for Reynolds numbers between 200 and 3000, the amount of entrainment is a function of both vapour and liquid flow-rates. In fully turbulent flow, above a film Reynolds number of 3000 the conditions for the onset of entrainment depend mainly upon the vapour velocity. For the case of a fully turbulent film ($\text{Re}_f > 3000$) the critical vapour velocity for the start of entrainment from the film may be given by

$$j_g = 1.5 \times 10^{-4}\left(\frac{\rho_f}{\rho_g}\right)^{1/2}\frac{\sigma}{\mu_g}$$

This relationship with a different constant (2.42×10^{-4}) was suggested by Steen and recommended by Wallis (1969). In practice there is some difficulty in defining the precise vapour velocity at which entrainment starts.

At high liquid rates, the wispy-annular region is entered, the concept of a critical vapour velocity for the onset of entrainment breaks down, and

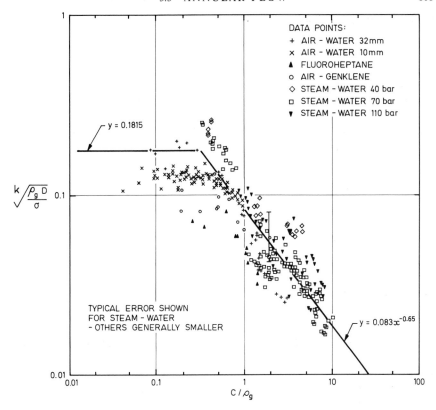

Fig. 3.6. Deposition correlation of Govan *et al.* (1988).

considerable entrainment occurs at all values of vapour velocity. The form of the criterion governing this change has yet to be ascertained (see Section 1.6.4.1) but gross entrainment at all vapour velocities is likely at values of Re_f greater than 20,000 or alternatively j_f^* greater than 1.5.

There is a lack of satisfactory experimental techniques for measuring *the rate of entrainment* (E). One approach is to deduce E from experimental measurements of the entrained liquid flow-rate under conditions of hydrodynamic equilibrium. Under these circumstances

$$E = D_d = kC_E$$

where C_E is the equilibrium homogeneous concentration of droplets in the gas core. Govan *et al.* (1988) have given the following correlation to predict the equilibrium rate of entrainment for $G_{f_F} > G_{f_{CRIT}}$

$$\frac{E}{G_g} = 5.75 \times 10^{-5} \left[(G_{f_F} - G_{f_{CRIT}})^2 \left(\frac{D\rho_f}{\sigma \rho_g^2} \right) \right]^{0.316} \tag{3.44}$$

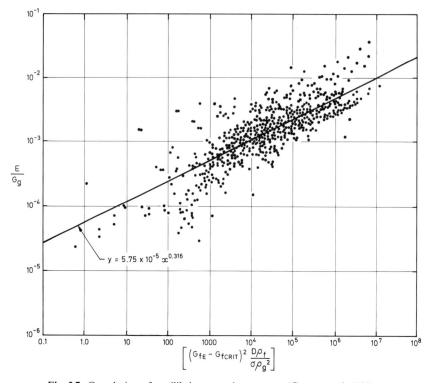

Fig. 3.7. Correlation of equilibrium entrainment rate (Govan *et al.* 1988).

where G_g is the gas mass velocity, G_{f_F} is the liquid film mass velocity ($= j_{f_F}\rho_f$), and $G_{f_{CRIT}}$ is the critical film flow-rate for the onset of entrainment (from eq. (3.43)). This correlation is illustrated in Fig. 3.7.

3.5.4   *Inter-relationships in annular flow*

The individual relationships discussed in Sections 3.5.1, 3.5.2, and 3.5.3 can be combined in various ways to provide a complete method of analysis for axisymmetric annular flow. The particular method chosen for this combination of the individual relationships depends on whether a prediction is to be made from known independent variables or, alternatively, whether one particular relationship is to be tested using actual experimental data.

In the case where a prediction is to be made, the known independent variables are usually the individual phase flow-rates, the physical properties of the phases and the channel geometry. In addition some idea of the initial amount of entrainment ($e = 0$, $e = 1$, or 'equilibrium') may be possible from a knowledge of the design of the phase mixing device and of its location

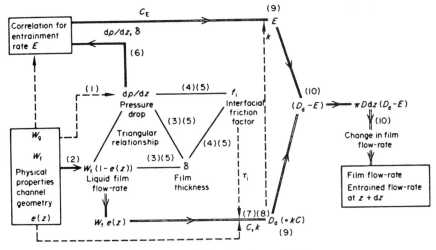

Fig. 3.8. Inter-relationships in annular flow.

relative to the point of interest (i.e., close to or well upstream). The procedure for the calculation of the film flow-rate at $(z + dz)$ given the conditions at z is shown in diagrammatic form in Fig. 3.8 and is as follows:

(1) Use known values of the phase flow-rates, physical properties and channel geometry to obtain an initial estimate of the combined frictional and static head pressure gradient from the homogeneous or separated flow models using the methods given in Chapter 2.

(2) Use the information on the amount of liquid entrainment at z to obtain the liquid film flow-rate $W_f(1 - e(z))$.

(3) Use the pressure gradient and the liquid film flow-rate to obtain the liquid film thickness.

(4) Use the vapour phase flow-rate, physical properties and channel geometry together with the calculated film thickness and entrained liquid fraction $e(z)$ to obtain the value of ϕ_g^2 (eq. (3.40)) and a new estimate of the pressure gradient.

(5) Return to step (3) and iterate between steps (3) and (5) to obtain consistent values for the film thickness and pressure gradient.

(6) Use these values of film thickness and pressure gradient to calculate C_E from Fig. 3.7.

(7) Use the calculated value of interfacial shear stress τ_i (eq. (3.37)) and the physical properties to calculate k from eq. (3.42).

(8) Use the entrained liquid flow-rate ($W_f e(z)$) and the gas flow-rate (W_g) to calculate the homogeneous mean droplet concentration in the vapour core, ($C = W_f e(z) \rho_g / W_g$).

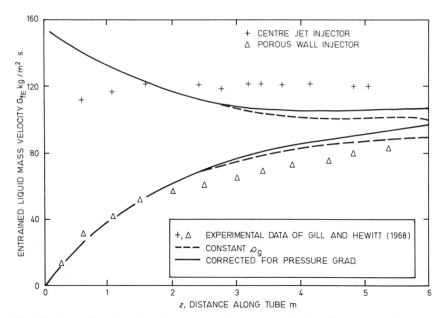

Fig. 3.9. Comparison of predicted entrainment liquid mass velocity with experimental values observed in an air–water flow with two different injectors (Govan *et al.* 1988).

(9) Use k and C to calculate D_d; use k and C_E to calculate E.

(10) Calculate $(D_d - E)$; multiply by the tube surface area $(\pi D\, dz)$ to get the change in film flow-rate (positive or negative) over the distance dz. Calculate value of $e(z + dz)$ and repeat (1)–(10) for the next increment along the channel.

The pressure gradient referred to in the previous paragraphs comprises the combined frictional and static head terms only. The use of any directly measured pressure gradient will require that this be corrected for any accelerational effects. The acceleration pressure gradient is a function of the level of entrainment (e) (see Problem 3, Chapter 2) and thus an iterative solution is often necessary.

One application of this analysis of annular flow is the prediction of the dryout condition, and some examples are given in Chapter 9. Another example given by Govan *et al.* (1988) is shown in Fig. 3.9. The variation of entrained liquid mass velocity (for an air–water flow) is shown as a function of tube length for a 31.75 mm (1.25 in) bore vertical tube. Two different phase mixing devices were used; in one instance the liquid was injected through a jet placed at the tube axis and, in the second instance, the liquid was introduced at the tube wall via a porous sinter. The phase flow-rates were identical in the two cases. the closed-form analysis above was used by Hewitt and co-workers to predict their experimental results. Good agreement is seen, especially when allowance is made for the varying gas density along the tube.

3.6 Wispy-annular flow

No attempts have so far been made to develop specific methods of analysis for this flow pattern. In general, the major fraction of the liquid is entrained in agglomerates in the vapour core and the homogeneous model has been found reasonably satisfactory for a first approximation.

In this region under conditions of hydrodynamic equilibrium (i.e., far downstream of any phase-mixing device) it is observed (Collier and Hewitt 1961) that at a given vapour flow-rate the liquid film flow-rate remains sensibly constant and that any additional increase in liquid flow-rate serves only to increase the entrained liquid flow. This observation may be used to provide a tentative basis for an analysis method in the wispy-annular flow pattern. It is recommended that the methods found satisfactory for the annular flow region be used to evaluate the liquid film flow-rate for conditions corresponding to the actual vapour flow-rate and a liquid phase flow-rate corresponding to a value of $j_f^* = 1.0$. The difference between the calculated film flow-rate and the actual liquid flow-rate may then be assumed to be entrained in the vapour core.

3.7 Chenoweth and Martin (1955) correlation for large-bore pipes

Pressure drop data for pipe diameters above 5 cm bore are very few and the data obtained in New Zealand by Lester *et al.* (1958) on 10.3 cm and 15.4 cm pipes with steam–water mixtures at pressures of 2–7 bar are valuable in this respect. When compared with the Lockhart–Martinelli correlation these data fall below the correlating line by 25–50 per cent. A similar discrepancy was noted by Baker (1954), who presented results for gas–oil flow in 10 to 25 cm bore pipes at pressures around 70 bar.

The reason for this discrepancy can be found in the fact that the Lockhart–Martinelli correlation is only valid for conditions close to atmospheric pressure ($v_g/v_f \approx 800$, C_{tt} in eq. (2.68) = 20). At higher pressures where the value of the specific volume ratio drops rapidly the value of C from eq. (2.72) also decreases. A value of $C = 10$ is given by this equation for the conditions of Chenoweth and Martin's tests with air–water flow in a 7.5 cm bore pipe at a pressure of 7 bar. This value of C substituted into eq. (2.68) very satisfactorily correlates the experimental results.

Chenoweth and Martin (1955) also developed a correlation which has been very successful in correlating data from these larger pipe sizes (both single and two-component systems). This is shown in Fig. 3.10 where the abscissa denotes the superficial liquid volumetric fraction $(1 - \beta)$ and the ordinate is the two-phase multiplier ϕ_{fo}^2. The family of lines on the plot refers to different values of the ratio $[(dp/dz\ F)_{go}/(dp/dz\ F)_{fo}]$ which in fact can be shown to be identical to the inverse of the property index $[(v_f/v_g)(\mu_f/\mu_g)^{0.2}]$ used by Baroczy. The term $(dp/dz\ F)_{go}$ is defined in the same manner as $(dp/dz\ F)_{fo}$ (cf. eq. (2.36)) and is the frictional pressure gradient that would exist if the total flow (vapour and liquid) was assumed to be vapour and was flowing in the channel. The data of Lester, when compared with the Chenoweth and Martin correlation, were represented within ± 35 per cent; considerably better than the original Lockhart–Martinelli correlation. This correlation is therefore recommended for pipe bore sizes greater than 5 cm.

3.8   Pressure losses through enlargements, contractions, orifices, bends, and valves

One of the least-studied aspects of two-phase vapour–liquid flow is that of the change in static pressure and the energy loss associated with such features as enlargements, contractions, orifices, bends, and valves. These features occur in practical pipe runs and in the vaporizing sections of boilers and evaporators. Their influence can be particularly marked on the characteristics of natural circulation units.

One common way of calculating frictional losses at local features in single-phase flow is to express the resistance of the feature in terms of the equivalent length of straight pipe. This same procedure can be applied to two-phase flows, although in general the equivalent length of pipe tends to be somewhat longer than for single-phase flows. In single-phase flow the velocity profile becomes fully developed a maximum of 10–12 pipe diameters downstream of the feature. In two-phase flow the distance downstream over which the influence of the feature is felt is greatly increased (by up to ten times).   If the change in channel cross-section is gradual and no flow separation occurs (nozzle or diffuser angles 5°–7°) a reduction of pressure energy may be recovered as a comparable increase in kinetic energy or vice versa. If wall shear forces and gravitational effects are ignored the changes in static pressures may be obtained by modifying eq. (2.10):

$$-\left(\frac{dp}{dz}\,a\right) = \frac{d}{dz}\left(\frac{W_g u_g + W_f u_f}{A}\right) = \frac{1}{A}\frac{d}{dz}(W_g u_g + W_f u_f) - \left(\frac{W_g u_g + W_f u_f}{A^2}\right)\frac{dA}{dz}$$

$$(3.45)$$

The inclusion of the term in (dA/dz) results in an additional term in the

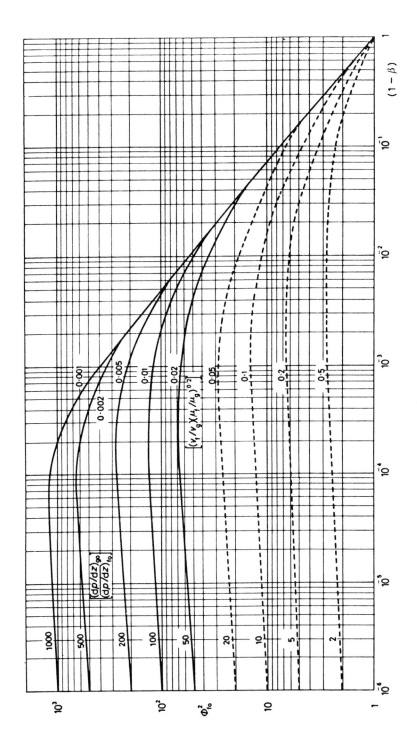

Fig. 3.10. Correlation of turbulent two-phase friction pressure drop in large horizontal pipes (Chenoweth and Martin 1955).

numerator of eqs. (2.34) and (2.53) respectively. For the homogeneous model eq. (2.31) becomes

$$-\left(\frac{dp}{dz} a\right) + G^2 \frac{d(\bar{v})}{dz} - \frac{G^2 \bar{v}}{A} \frac{dA}{dz}$$

(3.46)

and the additional term in the numerator of eq. (2.34) is

$$-\frac{G^2 v_f}{A}\left[1 + x\left(\frac{v_{fg}}{v_f}\right)\right]\frac{dA}{dz}$$

For the separated flow model eq. (2.51) becomes

$$-\left(\frac{dp}{dz} a\right) = G^2 \frac{d}{dz}\left[\frac{x^2 v_g}{\alpha} + \frac{(1-x)^2 v_f}{(1-\alpha)}\right] - \frac{G^2}{A}\left[\frac{x^2 v_g}{\alpha} + \frac{(1-x)^2 v_f}{(1-\alpha)}\right]\frac{dA}{dz}$$

(3.47)

and the additional term in the numerator of eq. (2.53) is

$$-\frac{G^2}{A}\left[\frac{x^2 v_g}{\alpha} + \frac{(1-x)^2 v_f}{(1-\alpha)}\right]\frac{dA}{dz}$$

3.8.1 *Sudden enlargement*

In the case of a sudden enlargement in the cross-sectional area of the pipe flow, separation will occur and the above equations are no longer valid. The general method used in single-phase flow is, however, still applicable to a one-dimensional separated two-phase flow. Consider a two-phase flow passing through a sudden enlargement (Fig. 3.11). Wall shear and gravitational forces are ignored and subscripts 1 and 2 will be used to denote the

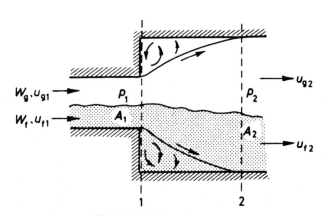

Fig. 3.11. Sudden enlargement.

conditions at planes 1 and 2. The pressure just downstream of the enlargement (plane 1) is found to be equal to that at the end of the smaller section of pipe and a simplified momentum balance for the combined flow yields

$$p_1 A_2 - p_2 A_2 = W_g(u_{g2} - u_{g1}) + W_f(u_{f2} - u_{f1}) \tag{3.48}$$

Substituting for W_g, W_f, u_g, and u_f in terms of G, x, α, v_f, and v_g using the continuity relationships eqs. (1.4) and (1.7) and defining σ as the ratio (A_1/A_2), eq. (3.48) becomes

$$p_2 - p_1 = G_1^2 \sigma v_f \left[\left\{ \frac{(1-x)^2}{(1-\alpha_1)} + \left(\frac{v_g}{v_f}\right) \frac{x^2}{\alpha_1} \right\} - \sigma \left\{ \frac{(1-x)^2}{(1-\alpha_2)} + \left(\frac{v_g}{v_f}\right) \frac{x^2}{\alpha_2} \right\} \right] \tag{3.49}$$

This equation was first derived by Romie (1958). If the void fraction remains unchanged across the feature ($\alpha_1 = \alpha_2 = \alpha$), eq. (3.49) becomes

$$p_2 - p_1 = G_1^2 \sigma (1 - \sigma) v_f \left[\frac{(1-x)^2}{(1-\alpha)} + \left(\frac{v_g}{v_f}\right) \frac{x^2}{\alpha} \right] \tag{3.50}$$

Comparing this result with that for the total flow assumed to be single-phase liquid reveals that the term in the square brackets acts as a two-phase multiplier. For homogeneous flow eq. (3.50) reduces to

$$p_2 - p_1 = G_1^2 \sigma (1 - \sigma) v_f \left[1 + \left(\frac{v_{fg}}{v_f}\right) x \right] \tag{3.51}$$

Equations (3.49), (3.50), and (3.51) relate the change in static pressure across the feature to the flow parameters. In the case of a sudden enlargement only a fraction of the decrease in kinetic energy can be recovered as an increase in pressure energy; the remainder is dissipated as heat. This frictional dissipation dE can be evaluated from a simplified energy balance on the combined phases between planes 1 and 2.

$$-(p_2 - p_1)[W_g v_g + W_f v_f] = W \, dE + \frac{W_g}{2}(u_{g2}^2 - u_{g1}^2) + \frac{W_f}{2}(u_{f2}^2 - u_{f1}^2) \tag{3.52}$$

Using the continuity equations (1.4) and (1.7), assuming that $\alpha_1 = \alpha_2 = \alpha$, and remembering that $\sigma = A_1/A_2$, eq. (3.52) can be rearranged as

$$p_2 - p_1 = -\frac{dE}{[xv_g + (1-x)v_f]} + \frac{G_1^2 (1 - \sigma^2)\left[\dfrac{x^3 v_g^2}{\alpha^2} + \dfrac{(1-x)^3 v_f^2}{(1-\alpha)^2}\right]}{2[xv_g + (1-x)v_f]} \tag{3.53}$$

The first term on the right-hand side of eq. (3.53) represents the frictional pressure change across the enlargement. This can be evaluated by substituting eq. (3.50) for $(p_2 - p_1)$ into eq. (3.53). For homogeneous flow this frictional pressure loss becomes

$$\Delta p_f = \frac{G_1^2}{2}(1 - \sigma)^2 v_f \left[1 + \left(\frac{v_{fg}}{v_f} \right) x \right] \qquad (3.54)$$

Measurements of void fraction made in the vicinity of sudden enlargements by Richardson (1958) (horizontal flow) and by Petrick and Swanson (1959) (vertical flow) indicate that the assumption $\alpha_1 = \alpha_2 = \alpha$ is only approximately true. Large variations in void fraction take place across the enlargement and the initial value is only recovered some considerable distance downstream. Ferrel and McGee (1966) found that for steam–water flow at a pressure of 8.2 bar the homogeneous model (eq. (3.51)) over-predicted the rise in static pressure across an enlargement by a factor 1.5–2. The use of measured void fractions in eq. (3.50), however, gave good agreeement. Fitzsimmons (1964) on the other hand found that for steam–water flow through expansions at a pressure of 82.6 bar and high mass velocities ($G > 2700 \text{ kg/m}^2 \text{ s}$) the homogeneous model gave satisfactory predictions. Chisholm and Sutherland (1969) have applied the method discussed in Section 2.5.2 to evaluate the static pressure change at an enlargement,

$$p_2 - p_1 = G_1^2 \sigma (1 - \sigma) v_f (1 - x)^2 \left[1 + \frac{C}{X} + \frac{1}{X^2} \right] \qquad (3.55)$$

The value of C is given by eq. (2.72) with $\lambda = 1$ and $C_2 = 0.5$. In fact the value of C_2 is found to vary slightly with pressure and a better value for C_2 is $(v_f/v_g)^{1/6}$. Equation (3.55) or alternatively eq. (3.50) (with the value of α for pipe flow) is recommended for the calculation of the static pressure change across an enlargement.

3.8.2 Sudden contraction

The general approach used for single-phase flow can again be applied to a two-phase separated flow passing through a sudden contraction (Fig. 3.12). As the fluid passes from plane 1 to plane c at the vena contracta it is accelerated and pressure energy is converted to kinetic energy with little or no frictional dissipation. Beyond plane c, however, the conditions are similar to those for a sudden enlargement and frictional dissipation occurs. Equations (3.50) and (3.53) derived in Section 3.8.1 can therefore be applied to this 'enlargement' from plane c to plane 2 to obtain an expression for the frictional pressure loss. The coefficient of contraction C_c is used to denote the ratio A_c/A_2 and replaces σ; likewise the mass velocity G_c ($= G_2/C_c$) replaces G_1 in eqs. (3.50) and (3.53). Rearranging and simplifying the equations yields the following expression for the frictional pressure loss

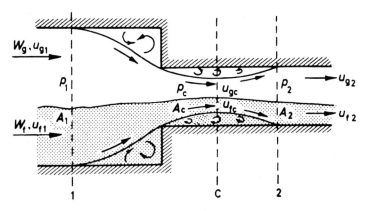

Fig. 3.12. Sudden contraction.

across a sudden contraction:

$$\Delta p_f = \frac{dE}{[xv_g + (1-x)v_f]} = \left(\frac{G_2}{C_c}\right)^2 (1 - C_c)$$

$$\times \left[\frac{(1+C_c)\left\{\dfrac{x^3 v_g^2}{\alpha^2} + \dfrac{(1-x)^3 v_f^2}{(1-\alpha)^2}\right\}}{2[xv_g + (1-x)v_f]} - C_c\left\{\dfrac{x^2 v_g}{\alpha} + \dfrac{(1-x)^2 v_f}{(1-\alpha)}\right\} \right] \quad (3.56)$$

For homogeneous flow this frictional pressure loss becomes

$$\Delta p_f = \frac{G_2^2 v_f}{2}\left[\frac{1}{C_c} - 1\right]^2\left[1 + \left(\frac{v_{fg}}{v_f}\right)x\right] \quad (3.57)$$

The change in static pressure at a sudden contraction is given by the sum of the frictional dissipation and the theoretical kinetic energy change. For the homogeneous model,

$$p_1 - p_2 = \frac{G_2^2 v_f}{2}\left[\left(\frac{1}{C_c} - 1\right)^2 + \left(1 - \frac{1}{\sigma^2}\right)\right]\left[1 + \left(\frac{v_{fg}}{v_f}\right)x\right] \quad (3.58)$$

In the above equation $\sigma = A_1/A_2$ and the coefficient of contraction C_c is a function of σ. For turbulent single-phase flow Perry (1963) gives

$1/\sigma$	0	0.2	0.4	0.6	0.8	1.0
C_c		0.586	0.598	0.625	0.686	0.790	1.0
$\left(\dfrac{1}{C_c} - 1\right)^2$		0.5	0.45	0.36	0.21	0.07	0

Both Ferrel and McGee (1966) and Geiger and Rohrer (1966) found that the homogeneous model (eqs. (3.57) and (3.58)) satisfactorily represented their experimental data for steam–water flow through sudden contractions. Fitzsimmons (1964) measured the pressure change across three contraction geometries (5.08/3.81 cm, 5.08/2.54 cm, 7.62/5.08 cm) for steam–water flow at high mass velocities and a pressure of 82.6 bar. The measured pressure drops were found to lie within 20 per cent of the homogeneous model predictions in the first two cases but to be up to a factor of two greater in the third case. It seems likely that the placing of the downstream pressure tapping in this case did not allow the full pressure recovery, bearing in mind that a considerably greater length of pipe is required in two-phase flow. Again measurements of void fraction made by Richardson (1958) (horizontal flow) and by Petrick and Swanson (1959) (vertical flow) in the vicinity of sudden contractions indicate that the assumption implicit in the above equations, than $\alpha_1 = \alpha_2 = \alpha$, is only approximately true. The measured void fractions are considerably different from those predicted from the homogeneous model and the fact that this model gives the best predictions for the change in static pressure must be regarded as coincidence (Chisholm 1969).

3.8.3   Orifices

The pressure drop caused by the passage of a vapour–liquid flow through a sharp-edged orifice is of considerable interest in the field of flow metering. It is also important in the design of steam and refrigeration plant and in assessing the consequences of the failure of high-pressure circuits.

Again it is possible to derive an expression for the static pressure change across an orifice by extending the conventional single phase incompressible flow treatment to the case of a separated two-phase flow (Fig. 3.13) and writing the energy equation for the combined phases between plane 1 and the vena contracta plane 2. The expressions obtained can be progressively simplified by assuming the void fraction is unchanged on passing through the orifice and, subsequently, if desired, by assuming homogeneous flow. In this latter case, the change in static pressure between planes 1 and 2 will be

$$p_1 - p_2 = \frac{G_1^2 v_{\mathrm{f}}}{2C_{\mathrm{D}}^2}\left[\left(\frac{A_1}{A_{\mathrm{o}}}\right)^2 - 1\right]\left[1 + \left(\frac{v_{\mathrm{fg}}}{v_{\mathrm{f}}}\right)x\right] \qquad (3.59)$$

In the above equation, the subscript o refers to the plane of the orifice plate and C_{D} is the coefficient of discharge which equals the contraction coefficient C_{c} modified to allow for the small amount of frictional dissipation and velocity profile effects. In single-phase flow, C_{D} is a complex function of Reynolds number and (A_{o}/A_1).

In practice, it is found that the use of eq. (3.59) is very inaccurate and

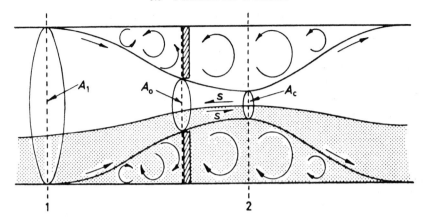

Fig. 3.13. Flow through an orifice.

considerably overestimates the pressure drop (James 1965–66; Hoopes (1957), particularly at low qualities. James (1965–66) has proposed that the two-phase multiplier should be evaluated using a modified mass quality x_m where x_m is related empirically to the true mass quality, x, by the expression

$$x_m = x^{1.5} \tag{3.60}$$

This simple modification should, however, be used to provide rough estimates only. Hoopes (1957) and Thom (1966) found that an alternative form of the two-phase multiplier was an improvement over the unmodified homogeneous model,

$$p_1 - p_2 = \frac{G_1^2 v_f}{2 C_D^2} \left[\left(\frac{A_1}{A_o} \right)^2 - 1 \right] \left[\frac{(1-x)^2}{(1-\alpha)} + \left(\frac{v_g}{v_f} \right) \frac{x^2}{\alpha} \right] \tag{3.61}$$

where α the void fraction is evaluated from a straight pipe flow correlation for conditions at the vena contracta. Although this represents an improvement over the unmodified homogeneous model, the accuracy of prediction (± 50 per cent) is still relatively poor.

A completely different and superior approach is that in which the flow processes in the orifice itself are considered. Such a method has been proposed by Chisholm (Chisholm 1958, 1966, 1967a, c; Chisholm and Watson 1966; Watson *et al.* 1967). The following assumptions are made:

(a) the two-phase flow is incompressible
(b) the upstream momentum is negligible
(c) there is no phase change in the orifice

(d) wall shear forces are neglected in relation to interfacial forces between phases
(e) the proportion of the flow cross-section occupied by each phase remains constant through the orifice.

Consider the orifice shown in Fig. 3.13. In single-phase flow it is found that the contraction to the vena contracta occurs with little or no frictional loss. Therefore, following assumption (b),

$$p_1 A_1 - p_2 A_c = W u_c \tag{3.62}$$

A more convenient form of the momentum balance across the orifice in single-phase flow is

$$(p_1 - p_2) A_o + F = W u_c \tag{3.63}$$

where u_c is the single-phase velocity at the vena contracta and F is the force due to the variation of pressure over the upstream face of the orifice plate. Jobson (1965–66) has shown that

$$F = f \frac{W^2 v}{A_o} = \left[\frac{1}{C_D} - \frac{1}{2C_D^2} \right] \frac{W^2 v}{A_o} \tag{3.64}$$

where C_D is the discharge coefficient ($\approx A_c/A_o$).

Chisholm (1967a) has applied eq. (3.63) to each phase assuming a shear force S exists between the phases. Thus for the liquid phase

$$(p_1 - p_2) A_{fo} + f \frac{W_f^2 v_f}{A_{fo}} + S = W_f u_{fc} \tag{3.65}$$

and for the vapour phase

$$(p_1 - p_2) A_{go} + f \frac{W_g^2 v_g}{A_{go}} - S = W_g u_{gc} \tag{3.66}$$

Dividing eq. (3.65) by (A_{fo}/v_f) and substituting the relationship $W_f = A_{fc} u_{fc}/v_f$ and eq. (3.64)

$$(p_1 - p_2) v_f \left[1 + \frac{S C_D}{A_{fc}(p_1 - p_2)} \right] = \frac{u_{fc}^2}{2} \tag{3.67}$$

Similarly for the vapour phase

$$(p_1 - p_2) v_g \left[1 - \frac{S C_D}{A_{gc}(p_1 - p_2)} \right] = \frac{u_{gc}^2}{2} \tag{3.68}$$

For convenience a 'shear force ratio' S_R is defined as

$$S_R = \frac{S C_D}{A_{gc}(p_1 - p_2)} \qquad (3.69)$$

and the velocity ratio (u_{gc}/u_{fc}) is denoted as K.

Substituting eq. (3.69) and the definition of K into (3.67) and (3.68) and combining gives

$$K = \frac{1}{Z}\left(\frac{v_g}{v_f}\right)^{0.5} \qquad (3.70)$$

where

$$Z = \left\{\frac{1 + S_R(A_{gc}/A_{fc})}{1 - S_R}\right\}^{0.5} \qquad (3.71)$$

Also

$$\frac{A_{gc}}{A_{fc}} = Z\left(\frac{x}{1-x}\right)\left(\frac{v_g}{v_f}\right)^{0.5} \qquad (3.72)$$

Let

$$\Delta p_f = \frac{W_f^2 v_f}{2A_c^2} = \frac{W^2(1-x)^2 v_f}{2C_D^2 A_o^2} \qquad (3.73)$$

and

$$\Delta p_g = \frac{W_g^2 v_g}{2A_c^2} = \frac{W^2 x^2 v_g}{2C_D^2 A_o^2} \qquad (3.74)$$

where Δp_f and Δp_g are the pressure drops for the liquid and vapour phases flowing alone through the orifice.

From eqs. (3.73) and (3.74)

$$\left(\frac{x}{1-x}\right)\left(\frac{v_g}{v_f}\right)^{0.5} = \left(\frac{\Delta p_g}{\Delta p_f}\right)^{0.5} = \frac{1}{X} \qquad (3.75)$$

Dividing eq. (3.67) by (3.73) and remembering that $(p_1 - p_2) = \Delta p_{TP}$ and $A_c = A_{fc} + A_{gc}$

$$\frac{\Delta p_{TP}}{\Delta p_f} = \frac{[1 + (A_{gc}/A_{fc})]^2}{[1 + S_R(A_{gc}/A_{fc})]} \qquad (3.76)$$

Substituting eqs. (3.71), (3.72), and (3.75) into (3.76)

$$\frac{\Delta p_{TP}}{\Delta p_f} = 1 + C\left(\frac{\Delta p_g}{\Delta p_f}\right)^{0.5} + \frac{\Delta p_g}{\Delta p_f} = 1 + \frac{C}{X} + \frac{1}{X^2} \qquad (3.77)$$

where

$$C = Z + \frac{1}{Z}$$

$$= K\left(\frac{v_f}{v_g}\right)^{0.5} + \frac{1}{K}\left(\frac{v_g}{v_f}\right)^{0.5} \tag{3.78}$$

A number of alternative derivations of eq. (3.77) are possible and this equation can be used quite generally for the correlation of two-phase pressure drop data (Chisholm and Sutherland 1969) (see Section 2.5.2). The alternative form in terms of the pressure drop with the vapour flowing alone is

$$\frac{\Delta p_{TP}}{\Delta p_g} = 1 + CX + X^2 \tag{3.79}$$

If there is no shear force between the phases, $S = 0$, $S_R = 0$ and from eq. (3.71) $Z = 1$ and $C = 2$. Thus, from eq. (3.79)

$$\left(\frac{\Delta p_{TP}}{\Delta p_g}\right)^{0.5} = 1 + X \tag{3.80}$$

This equation was originally derived by Murdock (1962) and was used by him as a basis for analysis of data for steam–water and gas–liquid flow through orifices. The data were actually fitted better by the slightly modified form:

$$\left(\frac{\Delta p_{TP}}{\Delta p_g}\right)^{0.5} = 1 + 1.26X \tag{3.81}$$

Chisholm (1967a, 1977) has taken all available data for flow through orifices (excluding those parts which do not comply with the assumptions stated earlier) and has correlated them in terms of eqs. (3.77) and (3.78). Table 3.2 summarizes his findings.

Clearly C is considerably different from 2 and phase interaction must be considered. A satisfactory procedure (Chisholm 1977) for predicting C for steam–water flow through orifice plates is to use eq. (3.78) with

$$X \geqslant 1, \qquad K = \left(\frac{\bar{v}}{v_f}\right)^{1/2} \tag{3.82a}$$

where $\bar{v}$ is the homogeneous mean specific volume of the steam–water mixture given by eq. (2.23)

$$X < 1, \qquad K = \left(\frac{v_g}{v_f}\right)^{1/4} \tag{3.82b}$$

Values of C calculated from this recommendation are also given in Table 3.2 for comparison.

Table 3.2 Chisholm's (1967a, 1977) values of C for various studies of two-phase flow through orifices

Data	Bizon (1965)		Chisholm and Watson (1966)	Collins and Gacesa (1971)	James (1965–66)	Murdock (1962)	Silberman (1960)	Thom (1966)
Orifice diameter (d) cm (in.)	1.14 (0.45)	1.78 (0.7)	0.95 (0.375)– 2.54 (1)	4.1 (1.61)– 5.4 (2.12)	14.2 (5.591)– 16.8 (6.615)	2.54 (1)– 3.18 (1.25)	1.27 (0.5)– 1.9 (0.75)	2.59 (1.02)
Pipe diameter (D) cm (in.)	2.54 (1)		5.08 (2)	5.9 (2.32)– 7.4 (2.91)	20.05 (7.9)	5.84 (2.5)– 10.15 (4)	2.54 (1)	3.86 (1.52)
d/D	0.45	0.7	0.187–0.5	—	0.707, 0.837	0.26–0.5	0.5–0.75	0.67
Range of X	0.6–2.7	0.25–2.4	0.5–5.0	1–16	0.033–3.3	0.041–0.25	0.125–∞	0.33–8.25
Range of $\left(\dfrac{v_g}{v_f}\right)^{0.5}$	4		11.2–20.0	4.6	9.1–22	3.87–34.7	~29	8.9–14.5
$C(X<1)$ experiment	2.48	2.34	5.3	2.92	2.41	2.66	5	3.79
C, eq. (3.82)	2.5	2.5	5.57	2.61	3.35–5.9	2.48–6.06	5.57	3.79

Chisholm (1967d) has also suggested a modified form of eq. (3.77) thus

$$\frac{\Delta p_{TP}}{\Delta p_f} = 1 + 5.3y + y^2 \tag{3.83}$$

where

$$y = \frac{1}{X}\frac{C_f}{C_g} \tag{3.84}$$

and C_f and C_g are the contraction coefficients for the liquid and the vapour phase corresponding to the two-phase pressure drop. This single equation satisfactorily correlates the data for the three orifices tested by Chisholm and Watson (1966).

Two-phase pressure drop data across thin plate ($t < 10$ mm) restrictions with rectangular slots have been reported by Janssen and Kervinen (1966). Chisholm and Sutherland (1969) recommend that the value of C for use in eq. (3.77) be evaluated from eq. (2.71) with $\lambda = 1$ and $C_2 = 0.5$. Pressure drop data across thick plates ($t > 10$ mm) have been reported by Cermak et al. (1964) and by Janssen and Kervinen (1966). Chisholm and Sutherland (1969) recommend that the value of C for use in equation (3.77) for this feature be evaluated from eq. (2.71) with $\lambda = 1$ and $C_2 = 1.5$.

3.8.4   Nozzles and venturis

Interest in two-phase flow through nozzles and venturis (Moissis and Radovcich (1963)) is increasing as a result of the possible application to jet pumps, M.H.D. devices, and flow metering. Equations (3.77) and (3.79) can also be derived for nozzles and venturis. Graham (1967) has given the results of measurements for the flow of air–water mixtures through nozzles 1.59, 2.54, and 3.50 cm diameter mounted in a 5.08 cm bore pipe. A single value of $C = 14$ was found to correlate all the data taken. Data taken by Bizon (1965) for a venturi with a throat to upstream pipe diameter ratio of 0.58 with steam–water flow were correlated with $C = 2.6$. It is recommended that for nozzles, C be evaluated using eq. (2.71) with $\lambda = 1$ and $C_2 = 0.5$.

To allow for the expansion of the vapour phase (Chisholm 1967d), eqs. (3.70), (3.72), and (3.75) require modification. If the expansion satisifes the equation $pv^n = $ constant then

$$K = \frac{1}{Z}\left(\frac{v_{gl}}{v_f}\right)^{0.5}\frac{1}{Br^{1/n}} \tag{3.85}$$

where v_{gl} is the vapour specific volume upstream of the feature, r is the ratio of the downstream-to-upstream static pressure (p_2/p_1) and B is given by

$$B = \left[\left(\frac{n-1}{n}\right)\left(\frac{1-r}{1-r^{[(n-1)/n]}}\right)\left(\frac{1}{r^{2/n}}\right)\right]^{0.5} \tag{3.86}$$

Table 3.3 Values of B

		r	
n	1.0	0.75	0.55
1.0	1.0	1.245	1.575
1.1	1.0	1.212	1.514
1.2	1.0	1.199	1.496
1.3	1.0	1.182	1.422
1.4	1.0	1.168	1.387

also

$$\frac{A_{gc}}{A_{fc}} = Z\left(\frac{x}{1-x}\right)\left(\frac{v_{gl}}{v_f}\right)^{0.5} B \qquad (3.87)$$

and

$$\left(\frac{\Delta p_g}{\Delta p_f}\right)^{0.5} = \frac{1}{X} = \left(\frac{x}{1-x}\right)\left(\frac{v_{gl}}{v_f}\right)^{0.5} B \qquad (3.88)$$

Table 3.3 gives values of B for various values of r and n. The validity of this method was demonstrated by Chisholm (1967d) using the data of Vogrin (1963) for the flow of air–water mixtures through a venturi having an inlet diameter of 2.54 cm and throat diameter of 0.514 cm. A value of $C = 5.3$ correlates the data using eqs. (3.77) and (3.88) and this value is, in turn, consistent with the measurements reported by Vogrin of the void fraction at the throat and the local slip ratio. Watson et al. (1967) found that for air–water flow through various sharp edged orifices with $X > 0.1$ the term B was not required in eq. (3.88).

3.8.5 Bends, tees, coils, and valves

Experimental data on flow around bends and other fittings are given by Chenoweth and Martin (1955) and Fitzsimmons (1964). In the latter work it was reported that the pressure drop around bends was up to 2.5 times higher than that predicted using single-phase pressure loss measurements and multiplying by an experimentally evaluated straight pipe multiplier (this latter term was found to agree reasonably well with the homogeneous model at the high mass velocities used).

Chisholm (1967b) has examined the data of Fitsimmons (1964) for 90° bends using eq. (3.77). He found that the coefficient C_2 in eq. (2.71) for C was a function of relative radius R/D where R is the radius of the bend and D is the pipe diameter. Table 3.4 gives values of C_2 for 90° bends as a function

Table 3.4   Values of C_2

	R/D				
	0	1	3	5	7
Normal bend	2.0	4.35	3.4	2.2	1.0
Bend with upstream disturbance within 50 D	1.7	3.1	2.5	1.75	1.0

of R/D. The high value of C_2 at an R/D of unity presumably results from the fact that sharp bends act as efficient phase separators. The case with R/D equal to zero corresponds to a tee which appears to act as an efficient phase mixing device. Chisholm and Sutherland (1969) recommend the following values of C_2 and λ for use with eqs. (2.71) and (3.77):

90° bends $\qquad\qquad\qquad\qquad$ $C_2 = 1 + 35\,D/L \qquad \lambda = 1$ (3.89)

180° bends and 90° bends with
upstream disturbance $\qquad\qquad$ $C_2 = 1 + 20\,D/L, \qquad \lambda = 1$ (3.90)

Tees $\qquad\qquad\qquad\qquad\qquad$ $C_2 = 1.75, \qquad\qquad \lambda = 1$ (3.91)

The ratio L/D is the equivalent length to diameter ratio of the bend.
A general expression given by Chisholm (1980) is

$$C_2 = 1 + \frac{2.2}{k_{fo}(2 + R/D)} \tag{3.92}$$

where k_{fo} is the loss coefficient for all the flow assumed to be single-phase liquid.

Pressure drop and void fraction measurements for flow in helical coils have been reported by Rippel *et al.* (1966), by Banerjee *et al.* (1968), and by Boyce *et al.* (1969). In general, it has been found that for estimating the frictional component of the pressure drop the two-phase multiplier to be used should be that for straight pipe flow. The influence of coiling on two-phase frictional pressure drops is therefore correctly allowed for by evaluating the single-phase pressure gradients taking account of the increased friction factors resulting from coiling (Srinivasan *et al.*, 1968). For laminar flow the ratio of the coil to straight pipe friction factor is given by

$$f/f_s = \left\{ 1 - \left[1 - \left(\frac{11.6}{K} \right)^{0.45} \right]^{2.22} \right\}^{-1} \tag{3.93}$$

where $K = [\mathrm{Re}(D/d)^{0.5}]$, d is the coil diameter, D the tube internal diameter, and f_s the straight pipe friction factor. The transition Reynolds number is increased in a coil

$$(Re)_{trans} = 2300[1 + 8.6(D/d)^{0.45}] \tag{3.94}$$

For turbulent flow the increase in friction factor is given by

$$f/f_s = [Re(D/d)^2]^{0.05} \quad \text{for} \quad [Re(D/d)^2] > 6 \tag{3.95}$$

Despite the presence of strong secondary flows in a coil the value of void fraction measured by Boyce *et al.* (1969) appeared to remain unchanged from that in a straight pipe. It is recommended that the frictional and accelerational components of the pressure drop be evaluated using straight pipe values for void fraction. Equation (3.27) remains valid for annular flow in helical coils.

For valves, Chisholm and Sutherland (1969) recommend the following values of C_2 (based on the measurements of Fitzsimmons 1964) for the evaluation of C in eq. (2.71) and its subsequent use in eq. (3.77):

$$\text{gate valve} \quad C_2 = 1.5$$
$$\text{globe valve} \quad C_2 = 2.3$$

3.9 Conclusions

None of the empirical methods for pressure drop prediction is suitable for every case. There is a continuing need to assess the available correlations to establish which is best over certain ranges of conditions and what the expected accuracy may be. Large data banks containing upward of 25,000 individual measurements of pressure drop have been assembled. Caution, however, must be used even for large data bases; easily measured data points tend to dominate at the expense of those for difficult test conditions.

Over the years a number of 'best buy' comparisons have been made (ESDU 1977; Idsinga 1975; Friedel 1976; Bryce 1977). The detailed results are interesting, but even more significant are the large standard deviations (typically 25–50 per cent) found with the methods examined. Based on detailed comparisons Whalley (1980) has made some tentative recommendations with respect to the published correlations for friction multiplier ϕ_{fo}^2 (see Section 2.5.3). As regards void fraction, the Smith and Chisholm correlations (see Section 3.3.2) are both simple and accurate.

A note of caution is however required; the reader should treat with reservations any claims of accuracy made for particular correlations. Whenever possible, efforts should be made to ascertain the flow pattern in order to examine the more specific correlations covering just one region.

References and further reading

Armand, A. A. (1946). 'The resistance during the movement of a two-phase system in horizontal pipes'. *Izvestia Vses. Teplo. Inst.* (1), 16–23. AERE-Lib/Trans 828.

Armand, A. A. (1954). 'Investigation of the mechanism of two-phase flow in vertical pipes'. In Styrikovich, M. A. (ed.), *Hydrodynamics and heat transfer with boiling*, Akad. Nauk SSSR, Moscow.

Armand, A. A. and Treschev, G. G. (1947). 'Investigation of the resistance during the movement of vapour–water mixtures in a heated boiler tube at high pressure'. *Izvestia Vses. Teplo. Inst.* (4), 1–5. AERE-Lib/Trans 816.

Aslia, J. C., Hanratty, T. J. and Andreussi, P. (1985). 'Interfacial drag and film height for vertical annular flow'. *AIChE J.*, **31**(6), 895–902.

Baker, O. (1954). 'Simultaneous flow of oil and gas'. *The Oil and Gas J.*, 26 July, 185–195.

Banerjee, S., Rhodes, E., and Scott, D. S. (1968). 'Pressure drops, flow patterns and hold-up for co-current gas liquid flow in helically coiled tubes'. Paper presented at AIChE meeting, Tampa, Florida, May.

Bankoff, S. G. (1960). 'A variable density single-fluid model of two-phase flow with particular reference to steam–water flow'. *Trans. ASME J. Heat Transfer*, Series C, **82**.

Baroczy, C. J. (1965). 'A systematic correlation for two-phase pressure drop'. AIChE reprint 37 presented at 8th National Conference, Los Angeles, August.

Bizon, E. (1965). 'Two-phase flow measurement with sharp-edged orifices and venturis'. AECL-2273.

Boyce, B. E., Collier, J. G., and Levy, J. (1969). 'Hold-up and pressure drop measurements in the two-phase flow of air–water mixtures in helical coils'. In *Co-current gas liquid flow*, Plenum Press.

Bryce, W. M. (1977). 'A new flow dependent slip correlation which gives hyperbolic steam–water mixture flow equations'. AEEW-R 1099.

Calvert, S. and Williams, B. (1958). *AIChE J.*, **1**, 78.

Cermak, J. O., Jicha, J. J., and Lightner, R. G. (1964). 'Two-phase pressure drop across vertically mounted thick plate restrictions'. *Trans. ASME J. of Heat Transfer* Series C, **86**, 227–239.

Chenoweth, J. M. and Martin, M. W. (1955). 'Turbulent two-phase flow'. *Petroleum Refiner*, **34**(10), 151–155, October.

Chisholm, D. (1958). 'The flow of steam/water mixtures through sharp-edged orifices'. *Engng. and Boiler House Review*, August, 252–256.

Chisholm, D. (1966). 'A brief review of the literature of flow of two-phase mixtures through orifices, nozzles and venturimeters'; and Watson, G. G. 'The flow of steam–water mixtures through orifices'. Papers presented at meeting held at NEL 6 Jan. 1965, 'Metering of two-phase mixtures'. NEL Report No. 217, February.

Chisholm, D. (1967a). 'Flow of incompressible two-phase mixtures through sharp-edged orifices'. *J. Mech. Engng. Sci.*, **9**(1), 72–78.

Chisholm, D. (1967b). 'Pressure losses in bends and tees during steam–water flow'. NEL Report No. 318 (1967); see also *Engng. and Boiler House Review*, August.

Chisholm, D. (1967c). 'Pressure gradients during the flow of incompressible two-phase mixtures through pipes, venturis and orifice plates'. *Brit. Chem. Engng.*, **12**(9), 454–457.

Chisholm, D. (1967d). 'Flow of compressible two-phase mixtures through throttling devices'. *Chem. and Process Engng.*, December, 73–78.

Chisholm, D. (1969). 'Theoretical aspects of pressure changes at changes of section during steam–water flow'. NEL Report No. 418, June.

Chisholm, D. (1977). 'Research Note: Two-phase flow through sharp-edged orifices'. *J. Mech. Engng. Sci.*, **19**(3), 127–129.

Chisholm, D. (1980). 'Two-phase pressure drop in bends'. *Int. J. Multiphase Flow*, **6**(4), 363–367.

Chisholm, D. (1983). *Two-phase flow in pipelines and heat exchangers*, George Godwin, London and New York.

Chisholm, D. and Sutherland, L. A. (1969). 'Prediction of pressure gradients in pipeline systems during two-phase flow'. Paper 4 presented at Symposium on Fluid Mechanics and Measurements in Two-phase Flow Systems, Leeds, 24–25 September.

Chisholm, D. and Watson, G. G. (1966). 'The flow of steam/water mixtures through sharp-edged orifices'. NEL Report No. 213, January.

Collier, J. G. and Hewitt, G. F. (1961). 'Data on the vertical flow of air–water mixtures in the annular and dispersed flow regions. Part II. Film thickness and entertainment data and analysis of pressure drop measurements'. *Trans. Instn. Chem. Engrs.*, **39**(2), 127–136.

Collins, D. B. and Gacesa, M. (1971). 'Measurement of steam quality in two-phase up flow with venturis and orifice plates'. *J. Basic Engng.*, **93**(1), 11–21.

Cousins, L. B. and Hewitt, G. F. (1968). 'Liquid phase mass transfer in annular flow droplet deposition and liquid entrainment'. AERE-R 5657.

Dukler, A. E. (1960). 'Fluid mechanics and heat transfer in vertical falling film systems'. *Chem. Engng. Prog. Symp. Series* No. 30, **56**, 1–5.

Dukler, A. E., Wicks, M., and Cleveland, R. G. (1962). 'Pressure drop and hold up in two phase flow—Part A. A comparison of existing correlations' and 'Part B. An approach through similarity analysis'. Paper presented at AIChE meeting, Chicago, 2–6th December.

ESDU (1976). 'The frictional component of pressure gradient for two-phase gas or vapour/liquid flow through straight pipes'. Engineering Sciences Data Unit, London, September.

ESDU (1977). 'The gravitational component of pressure gradient for two-phase gas or vapour/liquid flow through straight pipes'. Engineering Sciences Data Unit, London.

Ferrel, J. K. and McGee, J. W. (1966). 'Two-phase flow through abrupt expansions and contractions'. TID-23394. Vol. 3. Rayleigh N.C. Department of Chemical Engineering. North Carolina State University.

Fitzsimmons, D. E. (1964). 'Two-phase pressure drop in piping components'. HW 80970 Rev. 1.

Friedel, L. (1976). 'Momentum exchange and pressure drop in two-phase flow'. *Proceedings NATO Advanced Study Institute*, Istanbul, Turkey, 16–27 August, I, 239–312. Hemisphere Pub. Corp.

Geiger, G. E. and Rohrer, W. M. (1966). 'Sudden contraction losses in two phase flow'. *J. of Heat Transfer*, February, 1–9.

Gill, L. E. and Hewitt, G. F. (1968). 'Sampling probe studies of the gas core in annular two-phase flow III. Distribution of velocity and droplet flowrate after injection through axial jet. *Chem. Eng. Sci.*, **23**, 677–686.

Govan, A. H., Hewitt, G. F., Owen, D. G., and Bott, T. R. (1988). 'An improved CHF modelling code', *2nd UK National Conference on Heat Transfer*, **1**, 33–48 Inst. of Mechanical Engineers, London, September.

Graham, E. J. (1967). 'The flow of air–water mixtures through nozzles'. NEL Report No. 308.

Griffith, P. (1963). A.S.M.E. paper 63-HT-20, presented at National Heat Transfer Conference, Boston, Mass.

Hewitt, G. F. (1961). 'Analysis of annular two-phase flow: Application of the Dukler analysis to vertical upward flow in a tube'. AERE-R 3680.

Hewitt, G. F. (1978). 'Liquid mass transport in annular two-phase flow'. Paper given in 1978 Seminar of the International Centre for Heat and Mass Transfer, Dubrovnik, Yugoslavia, September.

Hewitt, G. F. and Hall-Taylor, N. (1970). *Annular two-phase flow*, Pergamon Press.

Hoopes, J. W. (1957). 'Flow of steam–water mixtures in a heated annulus and through orifices'. *AIChE Journal*, **3**(2), 268–275.

Hughmark, G. A. (1962). 'Hold-up in gas liquid flow'. *Chemical Engineering Progress*, **58**(4), 62–65.

Idsinga, W. (1975). 'An assessment of two-phase pressure drop correlations for steam–water systems'. MSc Thesis, MIT, May.

Ishii, M. (1975). *Thermo-fluid dynamic theory of two-phase flow*. Chapters IX and X, Eyrolles, Paris, Scientific and Medical Publication of France, NY.

Ishii, M. (1977). 'One dimensional drift-flux model and constitutive equations for relative motion between phases in various two-phase flow regimes'. Argonne Nat. Lab. ANL-77-47, October.

James, R. (1965–66). 'Metering of steam–water two-phase flow by sharp-edged orifices'. *Proc. Inst. Mech. Engrs.*, **180**, Part I, 23, 549–572.

Janssen, E. and Kervinen, J. A. (1966). 'Two-phase pressure drop across contractions and expansions: water and steam mixtures at 600 to 1400 p.s.i.a.' GEAP 4622 (1964); see also *Proc. 3rd Int. Heat Transfer Conf. 1966*, **5**, 13–25.

Jobson, D. A. (1965–66). 'On the flow of a compressible fluid through orifices'. *Proc. Inst. Mech. Engrs.* **180**, Part I, 23, 549–572.

Kunz, H. R. and Yerazunis, S. (1967). 'An analysis of film condensation, film evaporation and single phase heat transfer'. Paper 67-HT-1 presented at U.S. National Heat Transfer Conference, Seattle, August.

Lester, G. W. (1958). 'Correlation of two-phase pressure drop measurement for steam/water mixtures in a 4.06-in diameter and a 6.06-in diameter horizontal pipeline'. AERE CE/M 217.

Lopes, J. C. B. and Dukler, A. E. (1986). 'Droplet Entrainment in Vertical Annular Flow and its Contribution to Momentum Transfer'. *AIChE J.*, **32**(9), 1500–1515.

Lottes, P. A. (1961). 'Expansion losses in two-phase flow'. *Nuc. Science & Engrg.*, **9**, 26–31.

Mantzouranis, B. G. (1959). PhD Thesis 1959, University of London, see also Anderson, G. H. and Mantzouranis, B. G. (1960). *Chemical Engineering Science*, **12**, 109.

Martin, C. S. (1976). 'Vertically downward two-phase slug flow'. *J. of Fluids Engng.*, **98**, Series 1, 4, 715–722, December.

Mendler, O. J. (1961). 'Sudden expansion losses in single and two-phase flow'. PhD Thesis, University of Pittsburgh (1963); see also *Trans. ASME J. of Heat Transfer*, **83c**(3), 261.

Moissis, R. and Griffith, P. (1962). 'Entrance effects in two phase slug flow'. *Trans ASME J. of Heat Transfer*, **84**, Series C, 29.

Moissis, R. and Radovcich, N. A. (1963). 'Two-phase flow through a vertical venturi'. Paper 63-HT-42, presented at Heat Transfer Conference AIChE–ASME, Boston, August.

Murdock, J. W. (1962). 'Two-phase flow measurement with orifices'. *J. Basic Engineering*, **84**(4), 419–433.

NEL (1969). 'Designing for Two-Phase Flow—Report of a Meeting at NEL 17 Jan. 1968, Part II: Void fraction prediction under saturated conditions'. NEL Report No. 386.

Perry's chemical engineers handbook (4th edn) (1963). McGraw-Hill, pp. 5–30.

Petrick, M. and Swanson, B. S. (1959). 'Expansion and contraction of an air–water mixture in vertical flow'. *AIChE Journal*, **5**(4), 440–445.

Richardson, B. E. (1958). 'Some problems in horizontal two-phase two-component flow'. ANL-5949.

Rippel, G. R. *et al.* (1966). 'Two-phase flow in a coiled tube'. *I. and E. C. Process Design and Develop.*, **5**(1), 32–38.

Romie, F. (1958). Private communication to P. Lottes (see Lottes, 1961), American Standard Co.

Shearer, C. J. and Nedderman, R. M. (1965). 'Pressure gradient and liquid film thickness in co-current upwards flow of gas liquid mixtures; application to film cooler design'. *Chem. Engng. Sci.*, **20**, 671–682.

Silberman, E. (1960). 'Air/water flow through orifices, bends and other fittings in a horizontal pipe'. AD-253–494, Project No. 63, St. Anthony Falls Hydraulic Laboratory,' University of Minnesota.

Smith, S. L. (1969). 'Void fractions in two-phase flow. A correlation based on an equal velocity head model'. *Proc. Inst. Mech. Engng.*, **184**(36), 647–664.

Srinevasan, P. S., Nandapurkar, S. S., and Holland, F. A. (1968). 'Pressure drop and heat transfer in coils', *The Chemical Engineer*, CE 113–119, May (1968).

Thom, J. R. S. (1966). 'Some experiences on the two-phase flow of water and steam through a sharp-edged orifice'. Paper presented at meeting held at NEL 6 Jan. 1965, 'Metering of two-phase mixtures'. NEL Report No. 217, February.

Vogrin, J. A. (1963). 'An experimental investigation of two-phase two component flow in a horizontal converging–diverging nozzle'. ANL-6754.

Wallis, G. B. (1969). *One dimensional two-phase flow*, McGraw-Hill.

Watson, G. G., Vaughan, V. E., and McFarlane, M. W. (1967). 'Two-phase pressure drop with a sharp-edged orifice'. NEL Report No. 290, May.

Whalley, P. B. (1980). Private Communication to G. F. Hewitt.

Zuber, N. and Findlay, J. (1965). 'Average volumetric concentration in two-phase flow systems'. *Trans. ASME J. Heat Transfer*, **87**, 453.

Zuber, N., Staub, F. W., Bijwaard, G., and Kroeger, P. G. (1967). 'Steady state and transient void fraction in two-phase flow systems'. *GEAP* 5417.

Examples

Example 1
Calculate the film thickness, film flow-rate, and pressure drop for vertical up-flow of an air–water mixture in a 30 mm bore tube. The air flow is 0.1 ks/s and the water flow 0.2 kg/s. The estimated entrained liquid fracton (e) is 0.615. The relevant physical properties are:

$$\rho_f = 1000 \text{ kg/m}^3 \qquad \rho_g = 1.64 \text{ kg/m}^3$$

$$\mu_f = 10^{-3} \text{ N s/m}^2 \qquad \mu_g = 1.8 \times 10^{-5} \text{ N s/m}^2$$

$$\sigma = 0.072 \text{ N/m}$$

Solution
The solution of this problem follows that recommended in Section 3.5.4 and illustrated in Fig. 3.8.

(a) *Initial estimate of pressure gradient*

$$\text{Reynolds number of liquid phase} = \frac{W_f D}{A \mu_f} = \frac{4 \times 0.2}{\pi \times 0.03 \times 10^{-3}} = \underline{8500}$$

$G_f = 283.5 \text{ kg/m}^2 \text{ s}$
$f_f = 0.0082$

$$-\left(\frac{dp}{dz} F\right)_f = \frac{2 f_f G_f^2}{D \rho_f} = \frac{2 \times 0.0082 \times 283.5^2}{0.03 \times 10^3} = \underline{43.9 \text{ (N/m}^2)/ \text{ m}}$$

$$\text{Reynolds number of gas phase} = \frac{W_g D}{A \mu_g} = \frac{4 \times 0.1}{\pi \times 0.03 \times 1.8 \times 10^{-5}} = \underline{2.36 \times 10^5}$$

$G_g = 141.7 \text{ kg/m}^2 \text{ s}$
$f_g = 0.0036$

$$-\left(\frac{dp}{dz} F\right)_g = \frac{2 f_g G_g^2}{D \rho_g} = \frac{2 \times 0.0036 \times 141.7^2}{0.03 \times 1.64} = \underline{2.93 \times 10^3 \text{ (N/m}^2)/\text{m}}$$

$$X_{tt} = \left[\frac{(dp/dz \, F)_f}{(dp/dz \, F)_g}\right]^{1/2} = \left[\frac{43.9}{2.93 \times 10^3}\right]^{1/2} = \underline{0.1225}$$

From eq. (2.69)
$\phi_g^2 = 1 + CX + X^2$ where $C = 20$

$$= 1 + 20 \times 0.1225 + 0.015 = \underline{3.465}$$

$$-\left(\frac{dp}{dz} F\right) = -\left(\frac{dp}{dz} F\right)_g \times \phi_g^2 = 2.93 \times 10^3 \times 3.465 = \underline{10.15 \text{ (kN/m}^2)/\text{m}}$$

(b) *Entrainment*
The entrained liquid fracton $(e) = 0.615$
Therefore the liquid film flow-rate $W_{fF} = \underline{0.077 \text{ kg/s}}$.

(c) *Estimate of film thickness*
As discussed in Section (3.5.1) eq. (3.27) is modified when entrainment occurs
such that

$$\phi_{fF}^2 = \frac{1}{(1-\alpha)^2}$$

where ϕ_{fF}^2 is evaluated using the pressure drop for the liquid film flow-rate
(W_{fF}) flowing alone in the tube.
Reynolds number $Re_{fF} = Re_f(1-e) = 8500 \times 0.385 = \underline{3270}$

$$f_{fF} = 0.01045$$

$$-\left(\frac{dp}{dz}F\right)_{fF} = \frac{2f_{fF}G_{fF}^2}{D\rho_f} = \frac{2 \times 0.01045 \times 109^2}{0.03 \times 10^3} = \underline{8.3 \text{ (N/m}^2\text{)}/\text{ m}}$$

$$\phi_{fF}^2 = \left[\frac{(dp/dz\ F)}{(dp/dz\ F)_{fF}}\right] = \frac{10.15 \times 10^3}{8.3} = \underline{1.225 \times 10^3}$$

$$(1-\alpha)^2 = 0.815 \times 10^{-3} \qquad (1-\alpha) = \frac{4\delta}{D} = \underline{2.855 \times 10^{-2}}$$

$$\delta, \text{ the liquid film thickness} = \underline{0.214 \text{ mm}}$$

(d) *Revised estimate of pressure gradient*
Use can now be made of eq. (3.40) to obtain a new estimate of the value of
ϕ_g^2 and hence the pressure gradient

$$\phi_g^2 = \left[\frac{1 + 75(1-\alpha)}{\alpha^{5/2}}\right]\left[\frac{W_g + eW_f}{W_g}\right]\left[1 - 2\left(\frac{\alpha}{1-\alpha}\right)\left(\frac{\rho_g}{\rho_f}\right)\left(\frac{W_f(1-e)}{W_g}\right)\right]^2$$

$$= \left[\frac{1 + 75 \times 2.855 \times 10^{-2}}{0.97145^{5/2}}\right]\left[\frac{0.1 + 0.615 \times 0.2}{0.1}\right]$$

$$\times \left[1 - 2\left(\frac{0.97145}{2.855 \times 10^{-2}}\right)\left(\frac{1.64}{10^3}\right)\left(\frac{0.077}{0.1}\right)\right]^2$$

$$= \left[\frac{3.14}{0.929}\right][2.23][1 - 0.086]^2 = \underline{6.58}$$

$$-\left(\frac{dp}{dz}F\right) = -\left(\frac{dp}{dz}F\right)_g \times \phi_g^2 = 2.93 \times 10^3 \times 6.58 = \underline{19.3 \text{ (kN/m}^2\text{)}/\text{m}}.$$

(e) *Iteration to obtain accurate values of film thickness and pressure drop*
It is now possible to return to step (c) and obtain new estimates of δ, the film

thickness. Iteration finally obtains consistent values of the film thickness and two-phase pressure drop

$$\delta = 0.175 \text{ mm}, \qquad \phi_g^2 = 5.19, \qquad -\left(\frac{dp}{dz} F\right) = \underline{15.2 \text{ (kN/m}^2)/\text{m}}$$

The film thickness therefore is 0.175 mm
The film flow-rate   is 0.077 kg/s
The pressure drop   is 15.2 kN/m² m

Example 2
Calculate the static pressure change and the irreversible pressure loss at a sudden enlargement in a horizontal pipe from 25 mm to 50 mm diameter for the flow of a steam–water mixture. The system pressure is 10 bar, the mass quality is 5 per cent and the mass flow-rate is 0.7 kg/s.

Solution
From the steam table $v_f = 1.127 \times 10^{-3} \text{ m}^3/\text{kg}$ $\qquad v_g = 0.156 \text{ m}^3/\text{kg}$

The upstream cross-sectional area $= \dfrac{\pi}{4} \times 25^2 \times 10^{-6} = 4.91 \times 10^{-4} \text{ m}^2$

The upstream mass velocity $G_1 = \dfrac{0.7}{4.91 \times 10^{-4}} = \underline{1425 \text{ kg/m}^2 \text{ s}}$

The upstream to downstream area $\sigma = \frac{1}{4}$

From eq. (3.55)

$$p_2 - p_1 = G_1^2 \sigma (1 - \sigma) v_f (1 - x)^2 \left[1 + \frac{C}{X} + \frac{1}{X^2} \right]$$

where

$$C = [\lambda + (C_2 - \lambda)(v_{fg}/v_g)^{0.5}][(v_g/v_f)^{0.5} + (v_f/v_g)^{0.5}] \qquad (2.71)$$

and

$$\lambda = 1 \qquad C_2 = 0.5$$

Therefore

$$C = 0.5[(138.5)^{0.5} + (1/138.5)^{0.5}] = 5.935$$

$$X^2 = \left[\frac{1-x}{x} \right]^2 \frac{v_f}{v_g} = \left[\frac{0.95}{0.05} \right]^2 \frac{1}{138.5} = 2.61$$

$$p_2 - p_1 = 1425^2 \times \tfrac{1}{4} \times (1 - \tfrac{1}{4}) \times 1.127 \times 10^{-3}(0.95)^2 \left[1 + \frac{5.935}{\sqrt{2.61}} + \frac{1}{2.61} \right]$$

$$= \underline{1.95 \text{ kN/m}^2} \text{ (pressure rise)}$$

The irreversible pressure loss from eq. (3.64)

$$\Delta p_F = \frac{G_1^2}{2}(1 - \sigma)^2 v_f \left[1 + \left(\frac{v_{fg}}{v_f}\right)x\right]$$

$$= \frac{1.425^2 \times 10^6}{2}(0.75)^2 \times 1.127 \times 10^{-3}[1 + 137.5 \times 0.05]$$

$$= 5.06 \text{ kN/m}^2$$

Problems

1. Starting with eq. (3.9) derive an expression for the void fraction in a heated channel in which evaporation is occurring in terms of the mass quality x, the inlet velocity u_{fo} of liquid to be evaporated, the drift velocity $\bar{u}_{gj}$, and the phase densities.

Answer:
$$\bar{\alpha} = \frac{x}{C_0 x[(\rho_f - \rho_g)/\rho_f] + [C_0 + u_{gj}/u_{fo}]\rho_g/\rho_f}$$

2. Using the equation derived from Problem 1 and values of C_0 and $\bar{u}_{gj}$ given by eq. (3.10) verify that the drift flux model does satisfactorily account for the effect of mass velocity seen in Fig. 2.8.

3. Calculate the presssure drop around 20 cm radius 90° bend in a horizontal 50 mm bore pipe for the flow of a steam–water mixture. The system pressure is 10 bar, the mass quality is 5 per cent, and the mass flow-rate 0.7 kg/s.

4. Starting with the shear stress distribution given by eq. (3.28) for a liquid film in vertical annular flow obtain the expressions for the velocity profile $(u_f(d))$ and the film flow-rate (W_{fF}) for the case of laminar flow.

Answer:
$$u_f(d) = \frac{1}{\mu_f}\left[Y\left(\frac{d^2 - D^2}{16}\right) - X\ln\left(\frac{d}{D}\right)\right]$$

where $Y = \left(\frac{dp}{dz} + \rho_f g\right)$ and $X = \frac{\tau_i}{2}(D - 2\delta) + \frac{Y}{8}(D - 2\delta)^2$

$$W_{fF} = \frac{2\pi\rho_f}{\mu_f}\left[X\left\{\frac{\delta(D - \delta)}{4} - \frac{(D - 2\delta)^2}{8}\ln\left(\frac{D}{D - 2\delta}\right)\right\} - \frac{Y}{16}(\delta(D - \delta))^2\right]$$

5. Repeat the calculation discussed in Example 1 for the case where the liquid flow-rate is increased to 1.0 kg/s.
   (*Note:* in this case the flow pattern is 'wispy annular', the initial pressure drop calculation should be made using the homogeneous model, and the recommendations listed in Section 3.6 should be used to compute the entrained liquid flow-rate.)

6. Determine the slip ratio using Smith's method as a function of local vapour quality for water boiling at 10.03 bar inside a tube. The flow is all liquid at the inlet

and all vapour at the exit. Also comment on this method as the vapour quality approaches 0.0 and 1.0.

7. Repeat Problem 6 using the Chisholm method.

8. For water evaporating inside a boiler tube, determine the Reynolds number for the onset of entrainment of liquid droplets in the vapour flow for temperatures ranging from 50 to 350°C in Table A.3 in the Appendix. Can you give a physical explanation for the trend you find?

4

INTRODUCTION TO POOL AND CONVECTIVE BOILING

4.1 Introduction

The object of this chapter is to introduce the subject of boiling, particularly for the case where the fluid is evaporated in a stagnant pool or a heated channel or duct. The elementary thermodynamic aspects of the liquid–vapour phase change are briefly reviewed to provide a background to a discussion of the basic aspects of bubble nucleation and growth. This, in turn, leads to a qualitative description of the heat transfer regimes present when a stagnant liquid is boiled in a container such as an electric kettle. Finally, the various modes of heat transfer encountered when a liquid is circulated through a heated channel are described and classified by means of a boiling map which enables the detailed features of convective boiling to be dealt with in sequence in the succeeding chapters.

4.2 Elementary thermodynamics of vapour/liquid systems

Figure 4.1 shows diagrammatically the pressure–volume–temperature (PVT) surface for unit mass of a pure substance. The various stable equilibrium phase states can be clearly seen. The stable regions of present interest are those of liquid alone, liquid and vapour, and vapour alone. This limits the temperature range of interest to T_1 (the triple point) at the lower end and T_c (the critical point) at the upper end. For a constant temperature T the pressure and volume vary along a line such as ABCD. Liquid only exists along the line AB and vapour only exists along the line CD. Liquid and vapour co-exist along the line BC. The saturation curve is the locus of points such as B and C. Along the line BC the pressure remains constant and the volume of the unit mass (specific volume, v) is determined by its heat content (enthalpy, i). The projection of the liquid + vapour surface in the pressure–temperature plane (viewed in the direction of the arrow) results in a single curve—the vapour pressure curve—relating pressure and temperature between T_1 and T_c. Corresponding values of the pressure and temperature taken from this curve are known as the saturation pressure (p_{SAT}) and saturation temperature (T_{SAT}) respectively.

Experimentally, the vapour pressure curve may be determined by partly

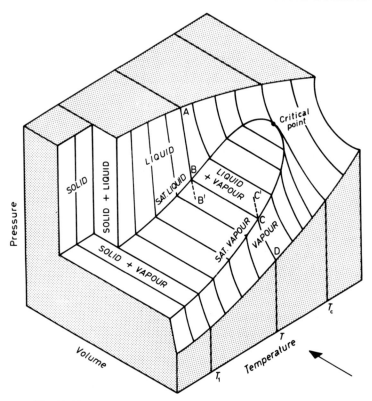

Fig. 4.1. Pressure–volume–temperature surface for a pure substance.

filling a large diameter container with the chosen substance under vacuum, mounting the container in a controlled uniform temperature field and recording the pressure of the vapour above the liquid for each desired temperature level. At equilibrium, the number of molecules striking and being absorbed by the interface from the vapour phase is exactly equal to the number of molecules being emitted through the interface from the liquid phase. For a planar interface, the pressures in the liquid and vapour immediately adjacent to the interface are equal at equilibrium.

So far only stable equilibrium phase states have been considered. Other metastable or unstable states can occur where the co-ordinates of pressure, volume, and temperature (PVT) do not lie in any of the surfaces shown in Figs. 4.1–4.4. For example, it is possible with care to reduce the pressure imposed on a liquid at constant temperature along a line AB without the formation of vapour at point B. Likewise it is possible to increase the pressure imposed on a vapour along a line DC without the formation of liquid at C. The PVT co-ordinates of these metastable states lie along an extrapolation

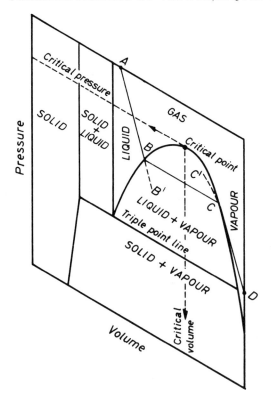

Fig. 4.2. Pressure–volume surface for a pure substance.

of AB to B′ or DC to C′. Point B′ may also be reached by carefully increasing the liquid temperature above the saturation temperature corresponding to the imposed static pressure; this process is referred to as superheating and the metastable liquid state is referred to as superheated liquid. Similar statements can be made about the vapour state at C′ where the equivalent cooling process is termed supersaturation and the metastable vapour state is referred to as a supersaturated vapour.

Vapour and liquid phases can co-exist in unstable equilibrium states along lines such as BB′ or CC′. In this instance the pressures in the liquid and vapour in the vicinity of the interface are no longer equal at equilibrium. If the interface is concave with the centre of curvature in the vapour phase then the vapour pressure (p_g) will be greater than the liquid pressure (p_f) by an amount given by the following relationship:

$$p_g - p_f = \sigma\left(\frac{1}{r_1} + \frac{1}{r_2}\right) \tag{4.1}$$

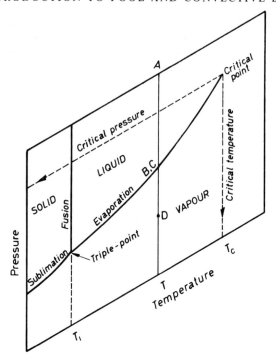

Fig. 4.3. Pressure–temperature surface for a pure substance.

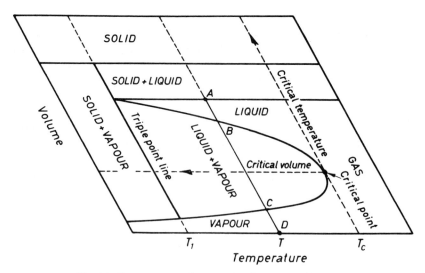

Fig. 4.4. Temperature–volume surface for a pure substance.

where σ is the surface tension and r_1 and r_2 are the orthogonal radii of curvature of the interface. Because of the increased vapour pressure the number of molecules striking and being absorbed by the interface from the vapour phase is greater than when the interface was planar. To maintain equilibrium the number of molecules emitted through the interface from the liquid phase must increase correspondingly. This increase can only be accomplished by increasing the temperature of the system (liquid *and* vapour) above that necessary for equilibrium with a planar interface, i.e., the saturation temperature (T_{SAT}) corresponding to the liquid pressure (p_f). Therefore the liquid adjacent to the curved interface is superheated with respect to the imposed liquid pressure.

A further feature of vapour–liquid equilibrium at curved interfaces is that the relationship between liquid temperature and vapour pressure (the vapour pressure curve) is itself somewhat changed as a consequence of the curvature. For a concave interface the vapour pressure for a given liquid temperature is slightly reduced below that for a planar interface; for a convex interface the reverse is the case.

This latter equilibrium state is unstable since any change in the curvature of the interface in either sense will result in a divergent departure from the equilibrium state.

4.3 The basic processes of boiling

4.3.1 *Vapour formation*

Vapour may form in one of three ways corresponding to the departure from a stable, metastable, or unstable equilibrium state. The formation of vapour at a planar interface occurs when the liquid temperature is increased fractionally above the corresponding saturation temperature. The term 'evaporation' will be reserved for this process and will be used to denote vapour formation at a continuous liquid surface such as the interface between the liquid film and vapour core in annular flow. Evaporation and condensation at a planar interface can be described in terms of the imbalance of molecular fluxes passing through the interface from the vapour and liquid phases respectively. The application of kinetic theory to the evaluation of the interfacial heat transfer resistance is fully discussed in Section 10.3.

4.3.1.1 *Superheat requirements for vapour nucleation*   Initial considerations relating to the formation of vapour from a metastable liquid or an unstable equilibrium state invariably start from the equation defining the mechanical equilibrium of a spherical vapour nucleus (radius r^*) in a liquid at constant temperature (T_g) and pressure (p_f). Thus

$$p_g - p_f = \frac{2\sigma}{r^*} \tag{4.2}$$

where p_g is the vapour pressure inside the nucleus and p_f is the imposed liquid pressure corresponding to a saturation temperature T_{SAT}. Curvature of the interface fractionally lowers the vapour pressure (p_g) inside the nucleus compared with that above a planar interface (p_∞) for the same liquid temperature (Thomson 1871). It may be shown that

$$p_g = p_\infty \exp(-2\sigma v_f M / r^* RT) \approx p_\infty \left(1 - \frac{2\sigma v_f}{p_\infty r^* v_g}\right) \tag{4.3}$$

Alternatively, from (4.2) and (4.3),

$$p_\infty - p_f = \frac{2\sigma}{r^*}\left(1 + \frac{v_f}{v_g}\right) \tag{4.4}$$

To calculate the liquid superheat ($T_g - T_{SAT}$) corresponding to the pressure difference ($p_\infty - p_f$), use can be made of the Clausius–Clapeyron equation and the perfect gas law

$$\frac{dp}{dT} = \frac{Ji_{fg}}{T(v_g - v_f)} \tag{4.5}$$

If the specific volume of the liquid is small ($v_g \gg v_f$) and $Mpv_g = RT$, then

$$\frac{1}{p}\,dp = \frac{Ji_{fg}M}{RT^2}\,dT \tag{4.6}$$

Equation (4.6) can be integrated from p_f to p_∞ and from T_{SAT} to T_g:

$$\ln\left(\frac{p_\infty}{p_f}\right) = -\frac{Ji_{fg}M}{R}\left(\frac{1}{T_g} - \frac{1}{T_{SAT}}\right) = \frac{Ji_{fg}M}{RT_g T_{SAT}}(T_g - T_{SAT}) \tag{4.7}$$

Substituting eq. (4.4) and rearranging,

$$(T_g - T_{SAT}) = \frac{RT_{SAT} T_g}{Ji_{fg}M}\ln\left[1 + \frac{2\sigma}{p_f r^*}\left(1 + \frac{v_f}{v_g}\right)\right] \tag{4.8}$$

If $v_g > v_f$ and $(2\sigma/p_f r^*) \ll 1$ then eq. (4.8) simplifies to

$$(T_g - T_{SAT}) = \Delta T_{SAT} = \frac{RT_{SAT}^2}{Ji_{fg}M}\frac{2\sigma}{p_f r^*} \tag{4.9}$$

Equation (4.9) can be used without significant error for reduced pressures in the range $0.01 < P_r < 1$.

4.3.1.2 *Homogeneous nucleation*   It will be observed from eq. (4.9) that the size of the equilibrium vapour nucleus ($r*$) becomes smaller as the superheat (ΔT_{SAT}) is increased, i.e., as a result of a decrease in system pressure (p_f) along a line BB' (Fig. 4.1). Close to B' the size of the equilibrium nucleus approaches molecular dimensions. Thermal fluctuations occur in the metastable liquid and there is a small but finite probability of a cluster of molecules with vapour-like energies coming together to form a vapour embryo of the size of the equilibrium nucleus. This process of vapour formation in a metastable liquid is referred to as 'homogeneous nucleation'. The probability of the formaton of vapour nuclei of the necessary size can be estimated from the Boltzmann equation for the distribution of molecular clusters of size r. The number of nuclei of radius r, $N(r)$ per unit volume, is given by

$$N(r) = N \, e^{-\Delta G(r)/kT_g}$$

where N is a constant approximately equal to the number of molecules per unit volume provided $N(r)$ is small. $\Delta G(r)$ is the free energy of formation of a nucleus of radius r and k is the Boltzmann constant. The free energy of formation of a nucleus of radius r, $\Delta G(r)$ is given by

$$\Delta G(r) = 4\pi r^2 \sigma - \tfrac{4}{3}\pi r^3 (p_g - p_f) \tag{4.10}$$

Substituting eq. (4.2) into (4.10),

$$\Delta G(r) = 4\pi r^2 \sigma \left(1 - \frac{2r}{3r*} \right) \tag{4.11}$$

The free energy $\Delta G(r)$ is a maximum when $r = r*$

$$\Delta G(r*) = \tfrac{4}{3}\pi r*^2 \sigma = \frac{16}{3} \frac{\pi \sigma^3}{(p_g - p_f)^2} \tag{4.12}$$

Since the free energy of formation is less, bubbles smaller than $r*$ will collapse and bubbles larger than $r*$ will grow spontaneously. Homogeneous vapour nucleation will therefore occur if one further molecule collides with an equilibrium embryo. The rate of nucleation dn/dt in a metastable liquid at temperature T_g is given by the product of the number of equilibrium nuclei per unit volume and a collision frequency λ,

$$\frac{dn}{dt} = \lambda N(r*) = \lambda N \, e^{-\Delta G(r*)/kT_g} \tag{4.13}$$

Expressions for the collision frequency λ have been given by Westwater (1958) $[\lambda = kT_g/h]$ and Bernath (1952) $[\lambda = (2\sigma/\pi m)^{1/2}]$ where m is the mass of one molecule and h is Planck's constant. For water at 100°C, λ is approximately 10^{12}–10^{13} s^{-1}.

The rate of homogeneous nucleation dn/dt is an extremely sensitive function of the superheated liquid temperature T_g. At low superheats the rate is insignificantly small but it increases very rapidly as the superheat is increased. Simpson and Walls (1965) have suggested that significant nucleation occurs for values of dn/dt between 10^9 and 10^{13} m^{-3} s^{-1}. For benzene these values correspond to a very narrow range of temperature (T_g) from 224°C to 225.2°C respectively.

Equation (4.13) is in a rather awkward form to evaluate T_g and a rather simple expression sufficiently accurate for most purposes has been produced by Lienhard (1976) on the basis of the measurements of Skripov and Pavlov (1970):

$$(T_{rg} - T_{rSAT}) = 0.905 - T_{rSAT} + 0.095 T_{rSAT}^8 \qquad (4.14)$$

where the subscript 'r' refers to a reduced quantity, i.e., an absolute temperature divided by the critical temperature. For water at atmospheric pressure the liquid temperature corresponding to a nucleation rate of 10^{13} m^{-3} s^{-1} is 320.7°C, i.e., a superheat (ΔT_{SAT}) of 220.7°C. This value of superheat is much greater than any experimental value which has been measured for water (Knapp 1958) even under very carefully controlled conditions. It is possible to state that, for water at least, homogeneous nucleation from a metastable liquid state can be discounted as a mechanism for vapour formation. However, homogeneous nucleation can and does occur in organic liquids. The reader is referred to the review by Blander and Katz (1975) for further details.

4.3.1.3 *Heterogeneous nucleation*   Foreign bodies and container surfaces normally provide ample nuclei to act as centres of vapour formation. This third method of vapour generation from pre-existing nuclei is termed 'heterogeneous nucleation'. Examples of pre-existing nuclei are non-condensible gas bubbles held in suspension in the liquid and gas- or vapour-filled cracks or cavities in container surfaces (known as nucleation 'sites') or in suspended foreign bodies in the bulk liquid. The presence of a dissolved gas, say air, in the liquid necessitates the gas partial pressure being taken into account when considering the mechanical equilibrium of the vapour nucleus. Thus eq. (4.2) is modified to

$$p_g + p_a - p_f = \frac{2\sigma}{r*} \qquad (4.15)$$

and eq. (4.8) to

$$T_g - T_{SAT} = \frac{R T_{SAT} T_g}{J i_{fg} M} \ln\left[1 + \left(\frac{2\sigma}{p_f r*}\right)\left(1 + \frac{v_f}{v_g}\right) - \frac{p_a}{p_f}\right] \qquad (4.16)$$

The presence of dissolved gas reduces the superheat required to maintain a bubble of radius r^* in unstable equilibrium. In practice the evaluation of the gas partial pressure may be complicated by the presence of temperature gradients and by finite rates of diffusion of the gas through the liquid.

4.3.1.4 *Nucleation at solid surfaces*   The free energy of formation $\Delta G(r^*)$ of an equilibrium vapour embryo is reduced by a factor ϕ in the presence of a flat surface. Kast (1964) and Bankoff (1957) have shown that ϕ is a function of the contact angle θ between the surface and the liquid (measured through the liquid, Fig. 4.5(a)). The factor ϕ is given by

$$\phi = \frac{(2 + 2\cos\theta + \cos\theta\sin^2\theta)}{4} \tag{4.17}$$

If the liquid completely wets the surface $\theta = 0°$, $\phi = 1$ and there is no reduction of the free energy of formation of the embryo. If the surface is completely non-wetting ($\theta = 180°$) then $\phi = 0$ and no superheat is required for nucleation at the surface. In practice most solid–liquid systems lie in the range between $\theta = 0°$ and $90°$ and ϕ lies in the range 1 to 0.5. Consequently this reduction of $\Delta G(r^*)$ is still insufficient to account for the very much lower superheats found in practical situations, particularly with water, compared with those required for homogeneous nucleation.

The solution to the paradox lies in the presence of pits or cavities in the surface (shown schematically in Fig. 4.5(b), (c), and (d)). As the included angle of the cavity (β) decreases the 'effective' contact angle (θ') increases. For the particular case shown in Fig. 4.5(c) the interface in the cavity is flat, the 'effective' contact angle, θ', is 180° so that ϕ and $\Delta G(r^*)$ are zero. In the case of Fig. 4.5(d), $\Delta G(r^*)$ is negative; that is to say that the vapour pressure in the cavity is lower than the static liquid pressure. Thus it is possible for

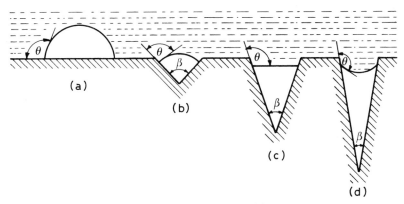

Fig. 4.5. Interfaces in cavities.

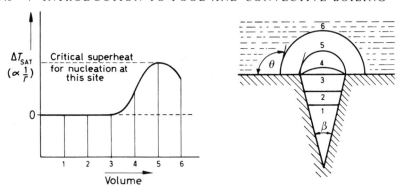

Fig. 4.6. Nucleation from cavities.

vapour to continue to exist in cavities in contact with subcooled liquids provided the contact angle (θ) is greater than 90°. The effective contact angle, θ', is related to the flat plate contact angle, θ, and the included angle of the cavity by

$$\theta' = \theta + \frac{(180 - \beta)}{2} \tag{4.18}$$

Since on any surface there are pits or cavities with a wide range of included angles, there will always be situations like those illustrated in Fig. 4.5(c) and 4.5(d), with any liquid–surface combination, having a contact angle θ greater than 90°. At first sight, it would appear that nucleation in these situations should occur at (in the case of Fig. 4.5(c)) or slightly below (in the case of Fig. 4.5(d)) the corresponding saturation temperature, i.e., no superheat. In actual fact when considering the motion of the interface in the cavity as the liquid temperature is slowly raised above the saturation temperature, it will be seen (Fig. 4.6) that the interface rapidly travels up to the lip of the cavity and then, as the vapour pressure in the cavity increases, the interface becomes concave. For the nucleus to continue to grow, the liquid tempera-ture must now be progressively increased above the saturation temperature to exceed the equilibrium superheat corresponding to the radius of curvature of the interface. This radius of curvature decreases as the nucleus grows until the contact angle θ with the flat surface is established. Further growth then tends to increase the radius of curvature and the bubble grows spontaneously in the superheated liquid. Thus it can be stated that the size of the mouth of the cavity or pit determines the superheat at which a vapour bubble will be nucleated at that site. This statement has been confirmed experimentally (Griffith and Wallis 1960).

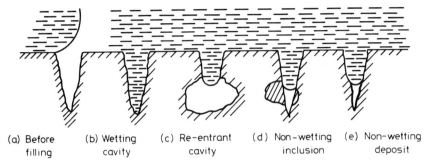

(a) Before filling   (b) Wetting cavity   (c) Re-entrant cavity   (d) Non-wetting inclusion   (e) Non-wetting deposit

Fig. 4.7. Formation of an active site.

4.3.1.5 *Sizing of active nucleation sites*   Only a very small fraction of the crevices and cavities in a surface are able to act as effective nucleation sites. To explain this it is necessary to consider what happens when an initially dry surface is flooded with liquid (Bankoff 1959) (Fig. 4.7). The advancing liquid will trap a mixture of air and vapour in the crevice (Fig. 4.7(a)). The air will rapidly dissolve and, if the liquid wets the cavity walls ($\theta < 90°$) then the remaining vapour pressure will be insufficient to balance the surface tension forces leading to complete penetration of the liquid to the base of the crevice (Fig. 4.7(b)). A cavity that is completely filled with liquid cannot act as a nucleation site. If, however, the walls of the cavity are poorly wetted or irregular in shape, the curvature of the interface may reverse so that surface tension forces are now resisting further penetration even when the vapour pressure in the cavity is negligible. During subsequent heating the vapour pressure rises sharply, drawing the interface back up to the mouth of the cavity. Stabilization of the interface within a crevice may occur as a result of an initial enlargement (Fig. 4.7(c)), as a result of a non-wetting inclusion, in, say, a metal surface (Fig. 4.7(d)), or as a result of a non-wetting surface film or deposit (Fig. 4.7(e)). In this latter case the entire surface may be covered by such a film or deposit. When the surface is flooded liquid is forced into the cavity by the imposed liquid pressure. If heat is applied to the surface subsequently, 'wetting' may occur as a result of the dissolution of grease films in a solvent liquid or in the case of liquid metals, a chemical reaction between a non-wetting surface oxide and the liquid metal. Nucleation will then be initiated from the size of vapour embryo given by the minimum radius to which the liquid has penetrated into the cavity. Thus,

$$r = \frac{2\sigma}{p_f - (p_g + p_a)} \tag{4.19}$$

This relationship also gives the minimum neck radius of an active re-entrant cavity for the case where $\theta \leqslant 90°$. A number of experimental confirmations

Table 4.1   Contact angles for various
fluids (Shakir and Thome 1986)

Liquid	Surface	Contact angle (degrees)
Water	Copper	86
Water	Brass	84
Benzene	Copper	25
Benzene	Brass	23
Ethanol	Copper	14–19
Ethanol	Brass	14–18
Methanol	Copper	25
Methanol	Brass	22
n-Propanol	Copper	13
n-Proponol	Brass	8

of this method (Holtz 1966; Fabic 1964; Griffith and Snyder 1964; Dwyer 1969) of characterizing the size of active sites have been attempted with varying success.

Lorentz et al. (1974) have developed a model which takes into account both the wettability of the surface, in terms of the contact angle θ, and the geometrical shape of the cavity, in terms of the included angle β. The model assumes a conical cavity being inundated by an advancing liquid front (Fig. 4.8(a)). Once the vapour is trapped by the liquid, the interface readjusts to form a vapour embryo and radius r (Fig. 4.8(b)). Conservation of vapour volume requires that r is a function of θ and β (Fig. 4.8(c)). This model is useful in that if the size of active cavities on a surface is known for one liquid, then the equivalent value of r for other liquids with different contact angles can be derived.

The contact angle θ is not generally known for most fluids and surface combinations. Table 4.1 shows contact angles measured by Shakir and Thome (1986) at 20°C for some common fluids on copper and brass discs polished with emery paper. Water has a large contact angle while organic fluids have smaller ones in the range from 8 to 25°.

4.3.2   Simple bubble dynamics

4.3.2.1   *Bubble growth*   Once the vapour nucleus has attained a size greater than that for unstable equilibrium it will grow spontaneously. The growth may be limited initially by the inertia of the surrounding liquid or latterly by the rate at which the latent heat of vaporization can be conducted to the vapour–liquid interface.

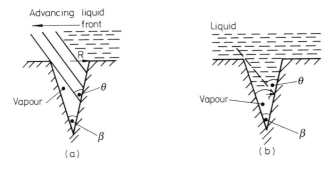

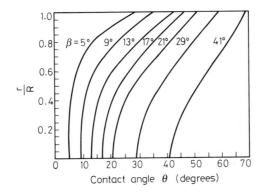

Fig. 4.8. Vapour trapping model of Lorentz *et al.* (1974), for sizing active cavities.

Consider a spherical element of a non-viscous incompressible liquid of radius r and thickness δr surrounding a spherical cavity of radius R held at a constant excess pressure Δp. The mechanical energy equation for this situation can be expressed as

$$\tfrac{1}{2}\rho_f \int_R^\infty 4\pi r^2 \dot{r}^2 \, \mathrm{d}r = \frac{4\pi}{3}(R^3 - R_0^3)\,\Delta p \tag{4.20}$$

where $\dot{r}$ is the velocity of the spherical liquid element and R_0 is the initial radius of the cavity. From continuity, $\dot{r}$ is related to the velocity of the bubble wall, $\dot{R}$,

$$\dot{r} = \left(\frac{R^2}{r^2}\right)\dot{R} \tag{4.21}$$

Substitution of eq. (4.21) into (4.20) yields

$$2\pi\rho_f R^3 \dot{R}^2 = \frac{4\pi}{3}(R^3 - R_0^3)\,\Delta p \tag{4.22}$$

Differentiating eq. (4.22) with respect to time and dividing through by $R^2 \dot{R}$ yields

$$RR\ddot{} + \tfrac{3}{2}\dot{R}^2 = \frac{\Delta p}{\rho_f} \tag{4.23}$$

This equation was first derived by Rayleigh (1917). The pressure in the liquid adjacent to the bubble wall is $[p_g(R, t) - 2\sigma/R]$ where $p_g(R, t)$ is the vapour pressure inside the bubble. This pressure is a function of both radius and time. The excess pressure Δp is therefore given by

$$\Delta p = \left[p_g(R, t) - \frac{2\sigma}{R} \right] - p_f \tag{4.24}$$

where p_f is the imposed static pressure well away from the bubble. Equation (4.23) then becomes

$$R\ddot{R} + \tfrac{3}{2}\dot{R}^2 = -v_f \left[p_f - p_g(R, t) + \frac{2\sigma}{R} \right] \tag{4.25}$$

It is usual to assume complete thermodynamic equilibrium at the interface and therefore $p_g(R, t)$ is taken as the vapour pressure corresponding to the instantaneous liquid temperature at the interface $T_g(R, t)$. It is known that this assumption is not strictly correct and Bornhorst and Hatsopoulos (1967) have considered the importance of non-equilibrium effects on the bubble growth problem.

·For water, the bubble growth process very rapidly becomes governed by conduction of heat to the interface and Plesset and Zwick (1954) have obtained a solution assuming this limiting condition, viz.,

$$\dot{R} = [T_g - T_g(R, t)] \left(\frac{k_f}{i_{fg}\rho_g} \right) \left(\frac{\pi}{3} \alpha_f t \right)^{-1/2} \tag{4.26}$$

where α_f is the thermal diffusivity of the liquid and k_f the thermal conductivity.

The asymptotic solution, valid at sufficiently large values of bubble radius, is obtained by assuming that the bubble wall temperature $T_g(R, t)$ has rapidly fallen to the saturation temperature; T_g in eq. (4.26) is the uniform temperature (superheated) of the liquid far from the bubble.

Thus

$$R = \frac{2\Delta T_{SAT} k_f}{i_{fg}\rho_g} \left(\frac{3t}{\pi\alpha_f} \right)^{1/2} \tag{4.27}$$

Mikic et al. (1970) have examined vapour bubble growth within a liquid close to a surface analytically. For a bubble growing in an initially uniformly superheated liquid from an initial radius greater than the critical

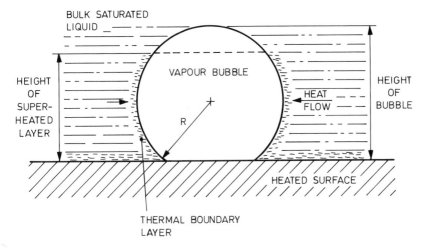

Fig. 4.9. Relaxation microlayer model of Van Stralen (1970).

radius (r^*) given by eq. (4.8), they gave the following expression:

$$R^+ = \tfrac{2}{3}[(t^+ + 1)^{3/2} - (t^+)^{3/2} - 1] \qquad (4.28)$$

where $R^+ = RA/B^2$; $t^+ = tA^2/B^2$

$$A = \left[b\,\frac{\Delta T_{SAT} i_{fg} \rho_g}{T_{SAT} \rho_f} \right]^{1/2} ; \qquad B = \left[\frac{12}{\pi} Ja^2 \alpha_f \right]^{1/2}$$

For bubble growth in an infinite medium $b = \tfrac{2}{3}$ and $\Delta T_{SAT} = (T_g - T_{SAT})$. In the inertia-controlled region $t^+ \ll 1$ eq. (4.28) becomes

$$R^+ = t^+ \qquad (4.29)$$

which is the Rayleigh solution. For the diffusion-controlled region $t^+ \gg 1$ eq. (4.28) becomes

$$R^+ = (t^+)^{1/2} \qquad (4.30)$$

which is the asymptotic solution eq. (4.27). The validity of eq. (4.28) was confirmed experimentally by Lien and Griffith (1969) who measured the growth of vapour bubbles in superheated water over the pressure range 0.012–0.38 bar, superheats in the range 8.3–15.5°C and $58 < Ja < 2690$.

Van Stralen (1970) has extended bubble growth theory to a bubble growing at a heated wall with an exponential radial temperature profile in the liquid. He also assumed that only part of the bubble was in the superheated thermal boundary layer as shown in Fig. 4.9, which he calls the *relaxation* microlayer. The ratio of the height of the superheated layer to the height of the bubble is the new parameter b. This bubble growth model leads to the following expression:

$$R = \left(\frac{12\alpha_f}{\pi}\right)^{1/2} b\left(\frac{\rho_f c_{pf}}{\rho_g i_{fg}}\right)[\Delta T_{SAT} \exp(-t/t_g)^{1/2}]t^{1/2} \tag{4.31}$$

where t_g is the bubble growth time.

For bubble growth at a heated wall, Moore and Mesler (1961) postulated that a thin liquid layer (approximatey 1 μm thick) is formed underneath a rapidly growing vapour bubble, based on microthermocouple temperature measurements. This *evaporation* microlayer is distinct from the relaxation microlayer. Subsequently, Van Stralen *et al.* (1975) developed a new comprehensive bubble growth model for a bubble growing at a heated wall with an evaporation microlayer and a relaxation microlayer. Their model describes the entire growth stage of hydrodynamic and asymptotic growth in the following bubble growth equation

$$R(t) = \frac{R_1(t)R_2(t)}{R_1(t) + R_2(t)} \tag{4.32}$$

where the initial hydrodynamic controlled growth stage, including the exponential radial temperature profile, is given by the modified Rayleigh solution:

$$R_1(t) = \left[\frac{2\rho_g i_{fg}\,\Delta T_{SAT}\,\exp(-t/t_g)^{1/2}}{3\rho_f T_{SAT}}\right]^{1/2} t \tag{4.33}$$

and the asymptotic diffusion controlled growth stage for the combined relaxation and evaporation microlayers is determined as a sum of their contributions as follows:

$$R_2(t) = R_{rm}(t) + R_{em}(t) \tag{4.34}$$

The relaxation microlayer bubble growth radius $R_{rm}(t)$ is given by eq. (4.31) and the evaporation microlayer contribution $R_{em}(t)$ was derived to be

$$R_{em}(t) = 0.373 Pr_f^{-1/6}\left(\frac{\alpha_f^{1/2}\rho_f c_{pf}}{\rho_g i_{fg}}\right)[\Delta T_{SAT}\exp(-t/t_g)^{1/2}]t^{1/2} \tag{4.35}$$

Thome and Davey (1981) and Preston *et al.* (1969) have experimentally investigated bubble growth rates for liquid nitrogen and argon using high-speed cine films and a computerized image analysis system. Contrary to previous studies that measured only one or two bubble growth sequences that supported one of the above-mentioned bubble growth theories, measurement of as many as 30 consecutive bubbles showed that there is a wide variation in growth rates from bubbles at the same boiling site and also in comparison to other remote boiling sites on the same surface. For instance, it was observed that the wake of a rapidly growing large bubble could prematurely 'pull' the next bubble off the surface while it was just beginning to grow. Thus, experimental verification of bubble growth and departure theories must utilize data for statistically mean bubble growth rates and departure diameters.

4.3.2.2 *Bubble detachment and frequency* The size and shape of vapour bubbles departing from a heated surface are a strong function of the way in which they are formed. The prime forces acting on a vapour bubble during the later phases of its growth are buoyancy and hydrodynamic drag forces attempting to detach it from the surface and surface tension and liquid inertia forces acting to prevent detachment. The liquid inertia force is a dynamic force resulting from the displacement of liquid during bubble growth. The growth velocity of a bubble and hence the inertial force is a strong function of the liquid superheat which, in turn, is inversely proportional to the size of the active cavity. A small cavity thus forms a bubble with a faster growth rate than from a large cavity. Hatton and Hall (1966) have considered all the forces acting on a growing bubble and have concluded that for small cavity sizes ($r_c < 10 \, \mu m$ for water at atmospheric pressure) the bubble size at departure is dictated mainly by a balance between buoyancy and liquid inertia forces. For larger cavity sizes, the growth rate decreases, the dynamic forces become small, and the bubble size at departure is set by a balance between buoyancy and surface tension forces (Howell and Siegel 1966). Fritz (1935) and Ende first considered this latter case and proposed the equation

$$D_d = 0.0208\theta \left[\frac{\sigma}{g(\rho_f - \rho_g)} \right]^{1/2} \tag{4.36}$$

Other more complete equations including inertia, buoyancy and surface tension forces and the 'pinching' effect of the vapour neck joining the bubble to the surface have been developed (Hatton and Hall 1966; Han and Griffith 1965; Keshock and Siegel 1964). When surface tension forces are dominant, the departing bubbles tend to be spherical. With inertia forces dominant the bubble tends to be hemispherical and when both forces are significant the bubble has an elongated oblate shape (Johnson et al. 1966).

Cooper et al. (1978) more recently completed an experimental study on bubble growth and departure in a stagnant uniformly superheated liquid under both normal gravity and reduced gravity conditions. Contrary to the above results and theories, they concluded that the surface tension force can either assist or resist bubble departure depending on the circumstances. They offered some new predictive equations for bubble departure diameters.

Individual nucleation sites emit bubbles with a certain mean frequency, which varies from site to site. The bubble departure frequency is equal to the reciprocal of the sum of the bubble growth and waiting periods. The bubble growth time can be obtained by solving a bubble growth equation, such as eq. (4.27), with a bubble departure equation, such as eq. (4.31). The waiting period between the departure of a bubble and the initiation of growth of the next bubble is difficult to predict, depending upon the formation of a new boundary layer and the nucleation characteristics of the boiling site.

Jakob (1958) observed that the product of bubble departure frequency and diameter was a constant which Zuber (1963) has evaluated in the following form:

$$f D_\mathrm{d} = 0.59 \left[\frac{\sigma g (\rho_\mathrm{f} - \rho_\mathrm{g})}{\rho_\mathrm{f}^2} \right]^{1/4} \tag{4.37}$$

Under conditons where the bubble departure process is governed by dynamic forces the relationship becomes $f^2 D_\mathrm{d} = $ constant. For water at atmospheric pressure bubble departure diameters are in the range 1–2.5 mm and bubble frequencies in the range 20–40 s^{-1}.

4.4   Pool boiling

4.4.1   *Pool boiling curve*

Pool boiling is defined as boiling from a heated surface submerged in a large volume of stagnant liquid. This liquid may be at its boiling point, in which case the term *saturated pool boiling* is employed or below its boiling point, in which case the term *subcooled pool boiling* is used. The results of investigations into heat transfer rates in pool boiling are usually plotted on a graph of surface heat flux (ϕ) against heater wall surface temperature (T_w)—the 'boiling curve'. Such a curve for the boiling of water at atmospheric pressure is shown diagrammatically in Fig. 4.10. An alternative presentation is to use the wall superheat ($T_\mathrm{w} - T_\mathrm{SAT}$) rather than the wall temperature itself.

The component parts of the boiling curve are well known—Fig. 4.11 shows a representation of each region.

(a) The natural convection region AB where temperature gradients are set up in the pool and heat is removed by natural convection to the free surface and thence by evaporation to the vapour space.

(b) The onset of nucleate boiling (ONB) where the wall superheat becomes sufficient to cause vapour nucleation at the heating surface. This may occur close to the point where the curves AB and B′C meet as is usually the case for water at atmospheric pressure and above. Alternatively it may occur at much larger superheats than those required to support fully developed nucleate boiling, resulting in a sharp drop in surface temperature from B to B′ for the case of a constant surface heat flux. This latter behaviour is associated with fluids at very low reduced pressures, e.g. water at below atmospheric pressure and liquid metals in particular.

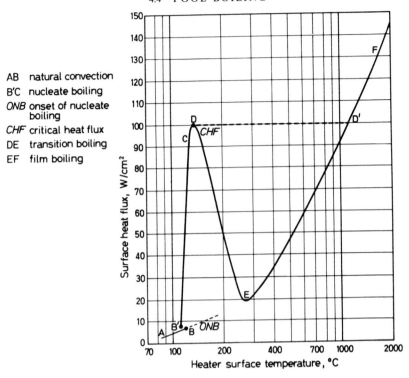

AB   natural convection
B'C   nucleate boiling
ONB onset of nucleate boiling
CHF critical heat flux
DE   transition boiling
EF   film boiling

Fig. 4.10. Pool boiling curve for water at atmospheric pressure.

(c) The nucleate boiling region (B'C) where vapour nucleation occurs at the heating surface. Starting with a few individual sites at low heat fluxes the vapour structure changes, as the heat flux is increased, as a result of bubble coalescence and finally, at high heat fluxes, vapour patches and columns are formed close to the surface.

(d) The critical heat flux (CHF or point D) marks the upper limit of nucleate boiling where the interaction of the liquid and vapour streams causes a restriction of the liquid supply to the heating surface.

(e) The transition boiling region (DE) is characterized by the existence of an unstable vapour blanket over the heating surface that releases large patches of vapour at more or less regular intervals. Intermittent wetting of the surface is believed to occur. This region can only be studied under conditions approximating to constant surface temperature.

(f) The film boiling region (EF) where a stable vapour film covers the entire heating surface and vapour is released from the film periodically in the form of regularly spaced bubbles. Heat transfer is accomplished

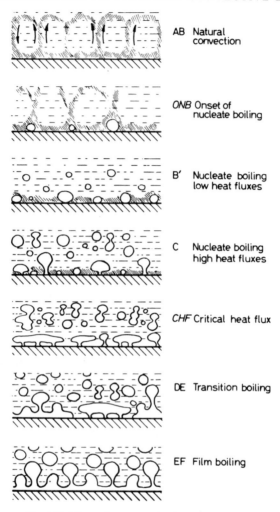

AB Natural convection

ONB Onset of nucleate boiling

B′ Nucleate boiling low heat fluxes

C   Nucleate boiling high heat fluxes

CHF Critical heat flux

DE Transition boiling

EF Film boiling

Fig. 4.11. The various stages in the pool boiling curve.

principally by conduction and convection through the vapour film with radiation becoming significant as the surface temperature is increased.

Each region will now be considered in more detail.

4.4.1.1 *The natural convection region*   In the natural convection region the liquid may be at or below the saturation temperature. The temperature gradient away from the surface may be established from work on single-phase natural convection. For example, for turbulent natural convection from a horizontal flat surface, one well-known correlation (Fishenden and Saunders

1950) is

$$\left[\frac{hD}{k_f}\right] = 0.14\left[\left(\frac{\beta g\,\Delta TD^3\rho_f^2}{\mu_f^2}\right)\left(\frac{c_p\mu}{k}\right)_f\right]^{1/3} \tag{4.38}$$

where D is the diameter of the surface and β is the coefficient of expansion.

4.4.2 Nucleation in a temperature gradient

As the surface heat flux is increased the surface temperature exceeds the saturation temperature. Earlier discussion of vapour nucleation has considerered only uniform temperature fields and it is now necessary to discuss the influence on nucleation of a temperature gradient away from a surface. Consider Fig. 4.12 which shows a conical active nucleation site with a hemispherical vapour nucleus, radius r_c (corresponding to a contact angle, θ, of 90°), sitting at the mouth of the cavity. For other contact angles a simple geometrical relationship exists between the radius of the vapour nucleus and the radius of the cavity mouth. Also shown is the liquid temperature profile away from the surface. The liquid temperature well away from the surface is $T_{f\infty}$ and the wall temperature is T_W. The temperature gradient is assumed to be essentially linear through the thermal boundary layer, thickness δ. If the heat transfer coefficient in single-phase natural convection is h, then δ is given approximately by the expression

$$\delta = \frac{k_f}{h} \tag{4.39}$$

Typically δ may be of the order of 0.1 mm for water.

Hsu (1962) postulated that the criterion for nucleation from this site is that the temperature of the liquid surrounding the top of the bubble should exceed that necessary for the nucleus to remain in equilibrium (eq. (4.8)). A convenient way of representing this criterion diagrammatically is shown in Fig. 4.12, taking into account distortion of the temperature isotherm due to the presence of the bubble. If the line representing the liquid temperature profile intersects the equilibrium bubble curve then nucleation occurs. The first nucleation site to be activated corresponds to the point of tangency between the equilibrium bubble curve and the liquid temperature profile. The wall temperature corresponding to this condition will be $(T_W)_{ONB}$. The remaining variable is the relation of the location of the liquid temperature isotherm for T_g to the critical bubble radius. Hsu (1962) assumed this distance was $2r_c$ but Han and Griffith (1965) suggested $1.5r_c$ from potential flow theory. It is often convenient to assume that no distortion of the isotherm occurs and the distance is then r_c. With this latter assumption, if the liquid pool is at the saturation temperature, i.e., $T_{f\infty} = T_{SAT}$, then the size of the

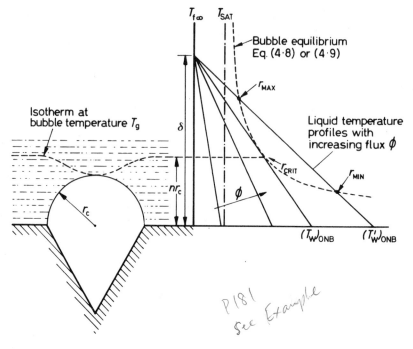

Fig. 4.12. Onset of nucleation in pool boiling.

cavity satisfying the condition of tangency will be $r_c = \delta/2$, i.e., approximately 50 μm for water at atmospheric pressure. The superheat corresponding to this size of cavity is about 1°C for water. In practice some 10–15°C of superheat is normally required to initiate boiling of water at atmospheric pressure from a flat metallic surface.

The discrepancy lies in the fact that cavities of radius 50 μm are not normally active sites (that is they do not contain a vapour embryo). Returning to Fig. 4.12, if an active site of size r_c does not exist on the heating surface then the wall temperature must be increased until the liquid temperature profile intersects the equilibrium bubble curve at a point where active cavities do exist, e.g., to $(T'_W)_{ONB}$. Brown (1967) has measured the active site distribution on various surfaces and reasonable populations of active sites (greater than 1 site/cm^2) occur only for cavity radii below 10 μm.

The size of active nucleation sites may be estimated from the methods given in Section 4.3.1 particularly eq. (4.19). Alternatively, as a rough guide the following figures may be taken for the maximum size of active nucleation sites on smooth metallic surfaces; for water ~ 5 μm, for organics and refrigerants ~ 0.5 μm, for cryogenic fluids on aluminium or copper

~ 0.1–0.3 µm. These figures may be considerably increased on rough surfaces, specially prepared porous surfaces or porous deposits caused by fouling (see Chapter 11).

4.4.3 Nucleate boiling correlations

In the nucleate boiling region the surface temperature increases very slowly for a relatively large change in surface heat flux (Fig. 4.10). The surface temperature is often expressed in the form of an empirical power law relationship,

$$T_W - T_{SAT} = \psi \phi^n \tag{4.40}$$

where ψ and n are constants depending upon the physical properties of the liquid and vapour and upon the nucleation properties of the surface. n will normally be in the range 0.25–0.5. Very many dimensionless relationships have been published over the years which attempt to seek a relationship of the arbitrary form

$$\mathrm{Nu} = C_1 \, \mathrm{Re}^x \, \mathrm{Pr}^y \tag{4.41}$$

by analogy with the relationships found satisfactory for forced convection of single phase fluids. Various characteristic parameters have been inserted into these dimensionless parameters (Zuber and Fried 1962). For example, as one of many alternatives, the bubble diameter at departure from the surface given by eq. (4.36) can be used as the characteristic length in forming the Nusselt and Reynolds numbers

$$\mathrm{Nu} = \frac{h}{k_f} \left[\frac{\sigma}{g(\rho_f - \rho_g)} \right]^{1/2} \tag{4.42}$$

Rohsenow (1952) chose the superficial liquid velocity towards the surface as the velocity in the Reynolds number

$$\mathrm{Re} = \frac{\phi}{i_{fg} \rho_f} \left[\frac{\sigma}{g(\rho_f - \rho_g)} \right]^{1/2} \frac{\rho_f}{\mu_f} \tag{4.43}$$

The complete equation proposed by Rohsenow (1952) contains an arbitrary constant C_{sf} included to account for the differing nucleation properties of any particular liquid–surface combination thus

$$\mathrm{Nu} = \frac{1}{C_{sf}} \, \mathrm{Re}^{(1-n)} \, \mathrm{Pr}_f^{-m} \tag{4.44}$$

This can be arranged to give

$$\left[\frac{c_{pf} \Delta T_{SAT}}{i_{fg}} \right] = C_{sf} \left[\frac{\phi}{\mu_f i_{fg}} \left(\frac{\sigma}{g(\rho_f - \rho_g)} \right)^{1/2} \right]^n \left[\frac{c_p \mu}{k} \right]_f^{m+1} \tag{4.45}$$

Table 4.2   Values of C_{sf} for the Rohsenow equation for various liquid–surface combinations

Liquid–surface combination	C_{sf}
n-Pentane on polished copper	0.0154
n-Pentane on polished nickel	0.0127
Water on polished copper	0.0128
Carbon tetrachloride on polished copper	0.0070
Water on lapped copper	0.0147
n-Pentane on lapped copper	0.0049
n-Pentane on emery rubbed copper	0.0074
Water on scored copper	0.0068
Water on ground and polished stainless steel	0.0080
Water on Teflon pitted stainless steel	0.0058
Water on chemically etched stainless steel	0.0133
Water on mechanically polished stainless steel	0.0132

Whenever possible it is recommended that a pool boiling experiment be carried out to determine the value of C_{sf} applicable to the particular conditions of interest. In the absence of such information a value of C_{sf} of 0.013 may be used as a first approximation.

The original equation had values of $n = 0.33$ and $m = 0.7$. Rohsenow has, however, recommended that *for water only* the value of m be changed to zero. Values of C_{sf} for various liquid–surface combinations are given in Table 4.2. A detailed study of the values of C_{sf} and n for various liquid–surface combinations and various surface preparation techniques has been reported by Vachon *et al.* (1967a, b).

Forster and Zuber (1955) used the bubble radius (eq. (4.27)) and the bubble growth velocity (eq. (4.26)) and obtained the following correlation

$$\frac{\phi}{\rho_g i_{fg}}\left(\frac{\pi}{\alpha_f}\right)^{1/2}\left[\frac{\rho_f r^{*3}}{2\sigma}\right]^{1/4} = 0.0015\left[\frac{\rho_f}{\mu_f}\left(\frac{\Delta T_{SAT}k_f}{\rho_g i_{fg}}\right)^2\frac{\pi}{\alpha_f}\right]^{5/8}\left[\frac{c_p\mu}{k}\right]_f^{1/3} \quad (4.46)$$

where r^* is the equilibrium bubble radius given by eq. (4.9). The coefficient (0.0015) was evaluated from data for water at 1 and 50 bar pressure, n-butyl alcohol at 3.4 bar, aniline at 2.4 bar and mercury at 1 and 3 bar pressure. It is presumably influenced by the surface condition in the same way as C_{sf} but there has been no systematic examination published so far.

The above correlations have *not* been effective for predicting independently obtained data. Stephan and Abdelsalam (1980) have developed accurate nucleate pool boiling correlations for several classes of fluids (water, organics, refrigerants and cryogens) using statistical regression techniques. Their organic fluid correlation is given as

$$\frac{hD_d}{k_f} = 0.0546 \left[\left(\frac{\rho_g}{\rho_f}\right)^{1/2} \left(\frac{\phi D_d}{k_f T_{SAT}}\right) \right]^{0.67} \left(\frac{i_{fg} D_d^2}{\alpha_f^2}\right)^{0.248} \left(\frac{\rho_f - \rho_g}{\rho_f}\right)^{-4.33} \quad (4.47)$$

where the bubble departure diameter D_d is obtained from equation (4.36).

The contact angle θ is a fixed value of $35°$, irrespective of the fluid. This correlation has proved to be very reliable, accurately predicting independent data for numerous light to heavy hydrocarbons (Sardesai *et al.* 1986) and a wide range of alcohols and solvents (Thome and Shakir 1987; Bajorek *et al.* 1990). For organic fluids, this is the recommended method.

The dimensionless correlations discussed so far have a number of disadvantages; they require accurate physical properties, they are complicated to evaluate and the inherent uncertainty induced by the condition of the surface is considerable.

As an alternative, simple dimensional equations can be prescribed for individual fluids using data from experimental studies. One basis for such equations is the work of Borishanski (1969) which makes use of the law of corresponding states. The heat transfer coefficient is evaluated from

$$h = A^* \phi^{0.7} F(p) \quad (4.48)$$

where $F(p)$ is a function of reduced pressure P_r and A^* is a constant evaluated at the reference reduced pressure of $P_r^* = 0.0294$. Mostinski (1963) proposed the following expressions for A^* and $F(p)$

$$\left.\begin{array}{l} A^* = 0.1011 p_{cr}^{0.69} \\ F(p) = 1.8 p_r^{3.17} + 4 P_r^{1.2} + 10 P_r^{10} \end{array}\right\} \quad (4.49)$$

The Mostinski method is at least as accurate as any of the physical property based correlations and is a good deal simpler to use.

Cooper (1984) has developed a more accurate reduced pressure correlation based on an extensive study. His correlation is given as

$$h = 55 P_r^{0.12 - 0.4343 \ln R_P} (-0.4343 \ln P_r)^{-0.55} M^{-0.5} \phi^{0.67} \quad (4.50)$$

where h is in $W/m^2 °C$ and the heat flux ϕ is in W/m^2. P_r is the reduced pressure and M is the molecular weight of the liquid. R_p is the surface roughness parameter in μm. For an unspecified surface, R_p is set equal to 1.0 μm. Cooper also recommended multiplying h by a factor of 1.7 for boiling on horizontal copper cylinders. The correlation covers reduced pressures from 0.001 to 0.9 and molecular weights from 2 to 200. This is the recommended method for water and refrigerants and for organic fluids with poorly defined physical properties.

The Gorenflo (1990, 1993) method is an alternative approach for predicting nucleate pool boiling coefficients. It is based on a reference heat transfer coefficient h_0 at the following standard conditions: reduced pressure $P_{ro} = 0.1$, surface roughness $R_{po} = 0.4\,\mu m$ and heat flux $\phi_0 = 20,000\ W/m^2$. The reference values of h_0 are given in Table 4.3 for a selection of fluids. To obtain coefficients at other conditions, the following expression is used

$$h = h_0 F_{PF} [\phi/\phi_0]^{nf} [R_p/R_{po}]^{0.133} \tag{4.51}$$

where F_{PF} is the pressure correction factor correlated as

$$F_{PF} = 1.2P_r^{0.27} + 2.5P_r + \frac{P_r}{1 - P_r} \tag{4.52}$$

which approaches a value of 1.0 at $P_r = 0.1$. The heat flux correction term has an exponent nf given as

$$nf = 0.9 - 0.3P_r^{0.3} \tag{4.53}$$

which equals 0.75 at $P_r = 0.1$. The surface roughness of the actual surface R_p is in micrometres and can be set to 0.4 for an unknown surface. The above expression is for all fluids except water and helium. For water the corresponding expressions are

$$F_{PF} = 1.73P_r^{0.27} + \left(6.1 + \frac{0.68}{1 - P_r}\right)P_r^2 \tag{4.54}$$

and

$$nf = 0.9 - 0.3P_r^{0.15} \tag{4.55}$$

4.4.4   Heat transfer mechanisms

The mechanism of nucleate pool boiling has been the subject of considerable debate during the past twenty or so years. Detailed observations and plausible theoretical models abound (Han and Griffith 1965). None, however, have become generally accepted as the correct explanation for the great improvement in heat transfer in this region.

In the case of boiling of a liquid considerably below its saturation temperature (Fig. 4.13), no bubbles are visible at the point of bubble

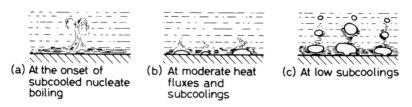

(a) At the onset of subcooled nucleate boiling

(b) At moderate heat fluxes and subcoolings

(c) At low subcoolings

Fig. 4.13. Representation of subcooled nucleate boiling.

Table 4.3 Reference values of h_0 (Gorenflo 1993)

Fluid		P_{CRIT} (bar)	h_0 (W/m^2 °C)
Methane		46.0	7000
Ethane		48.8	4500
Propane		42.4	4000
Butane		38.0	3600
n-Pentane		33.7	3400
i-Pentane		33.3	2500
Hexane		29.7	3300
Heptane		27.3	3200
Benzene		48.9	2750[a]
Toluene		41.1	2650[a]
Diphenyl		38.5	2100
Ethanol		63.8	4400
n-Propanol		51.7	3800
i-Propanol		47.6	3000
n-Butanol		49.6	2600
i-Butanol		43.0	4500
Acetone		47.0	3950[a]
R11		44.0	2800
R12		41.6	4000
R13		38.6	3900
R13B1		39.8	3500
R22		49.9	3900
R23		48.7	4400
R113		34.1	2650
R114		32.6	3800
R115		31.3	4200
R123		36.7	2600[b]
R134a		40.6	4500
R152a		45.2	4000[b]
R226		30.6	3700
R227		29.3	3800
RC318		28.0	4200
R502		40.8	3300
Chloromethane		66.8	4400
Tetrafluoromethane		37.4	4750
Water		220.64	5600
Ammonia		113.0	7000
Carbon dioxide		73.8	5100[c]
Sulfur hexafluoride		37.6	3700
Oxygen	(on Cu)	50.5	9500
	(on Pt)	50.5	7200
			(*Continued*)

Table 4.3   (*Continued*)

Fluid		P_{CRIT} (bar)	h_0 (W/m^2 °C)
Nitrogen	(on Cu)	34.0	10000
	(on Pt)	34.0	7000
Argon	(on Cu)	49.0	8200
	(on Pt)	49.0	6700
Neon	(on Cu)	26.5	20000
Hydrogen	(on Cu)	13.0	24000
Helium		2.28	2000[d]

[a] Mean value from scattered data; [b] Preliminary estimated value; [c] At triple point pressure; [d] At standard heat flux of 1000 W/m^2 at $P_r = 0.3$.

nucleation (ONB). Hot jets of liquid are projected from the superheated boundary layer into the cold liquid. As the heat flux or the pool temperature is increased so vapour bubbles appear at the heating surface but condense before they are released. This process induces microconvection in the thermal boundary layer as illustrated by the Schlieren photographs of Bähr (1965) which are reproduced in Fig. 4.14. Finally, at temperatures close to the saturation temperature, bubbles at the heating surface grow and detach only to be condensed in the cold liquid as they rise to the free surface. When the liquid is at the saturation temperature, bubbles no longer condense in the liquid pool but rise to the free surface.

Hui and Thome (1985) have also shown that the boiling nucleation process and hence active boiling site density on a heated surface are adversely affected by subcooling of the liquid pool. This in turn decreases the boiling heat transfer coefficient with increasing subcooling as shown by their data for benzene in Fig. 4.15.   *vertical food dish*

Detailed visual observations of the bubble growth process synchronized with studies of the temperature variation of the heating surface reveal an interesting picture. Figure 4.16 illustrates, in an idealized form, the behaviour during nucleation at a single site on the surface. After detachment of the previous bubble there is a period of time (the waiting period) while the bubble nucleus attains a critical condition. As soon as the critical condition is reached, the bubble grows very rapidly pushing superheated liquid away from the surface and leaving a thin layer of liquid only a few micrometres thick (the 'microlayer') adhering to the surface below the bubble. This liquid is rapidly vaporized with the consequent rapid removal of heat from the surface leading to a sharp reduction in surface temperature. While the bubble is growing rapidly inertial forces hold it to the surface. When the rapid

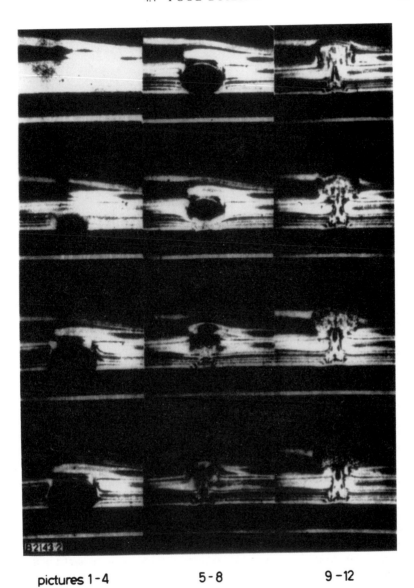

pictures 1-4   5-8   9-12

Fig. 4.14. Sequence of pictures demonstrating the bubble generation, the bubble condensation and the induced micro-convection. Time interval between successive photographs 0.7×10^{-3} s (Bähr 1965).

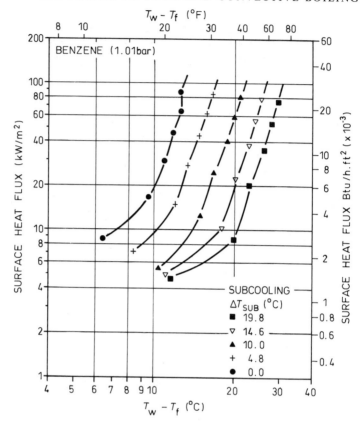

Fig. 4.15. Subcooled boiling of benzene on a vertical disk (25.4 mm in diameter) at 1.01 bar.

growth rate is concluded by the evaporation of the film, buoyancy forces cause the bubble to detach from the surface. The disturbance zone around the bubble is about 1.5 times the bubble departure diameter.

An important feature of the bubble cycle is the waiting period. In the nucleation model of Hsu (1962) this period is occupied by reforming the superheated boundary layer by transient conduction as illustrated in Fig. 4.17. It is assumed that the detachment of the bubble completely removes the thermal boundary layer and that at the start of the waiting period ($t = 0$) the bulk liquid $T_{f\infty}$ extends up to the wall. The thermal boundary layer is slowly built up again by transient conduction until at a time $t = t_1$ the temperature at a distance nr_c from the surface becomes equal to that required for equilibrium of a bubble of radius r_c. If there is an active cavity of r_c then bubble growth will occur after this time t_1, the end of the waiting period. The onset of nucleation (ONB) corresponds to an infinite waiting period. The

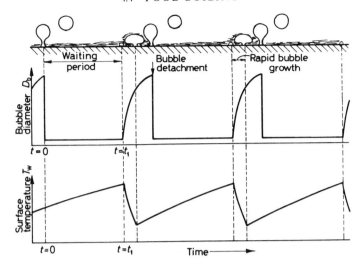

Fig. 4.16. Behaviour at a single nucleation site.

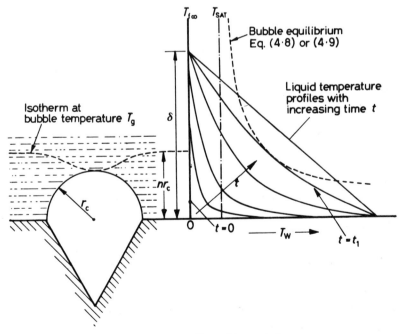

Fig. 4.17. The waiting period.

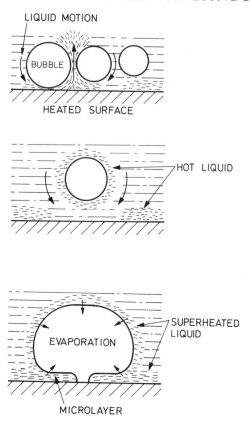

Fig. 4.18. Plain surface nucleate pool boiling mechanisms: (a) bubble agitation; (b) thermal boundary layer stripping; (c) evaporation.

theory of Hsu (1962) assumes transient conduction to the liquid from a surface held at constant temperature. In actual fact, a considerable amount of heat is removed from the heating surface in the microlayer vaporization process and the temperature of the heating surface itself must be recovered by transient conduction of heat from material surrounding the site.

Hsu and Graham (1976) surveyed the heat transfer mechanisms active in the nucleate pool boiling regime. As illustrated in Fig. 4.18, they concluded that the principal heat transfer mechanisms responsible for the increase in heat transfer relative to the single-phase natural convection regime are as follows:

Bubble agitation. The motion of the liquid induced by the growing and departing bubbles modifies the liquid-phase *natural* convection process, creating a *forced* convection heat transfer process. Heat is transported in the form of sensible heat in the superheated liquid.

Thermal boundary layer stripping. The thermal boundary layer formed by transient heat conduction into the liquid in the vicinity of the heated wall is periodically removed by the hydrodynamic drag of departing vapour bubbles. Heat is transported as sensible heat in the superheated liquid.

Evaporation. Vapour bubbles growing on the heated wall at active nucleation sites are formed by vaporization of superheated liquid surrounding the bubbles and by thin-film evaporation in the microlayer trapped beneath the bubbles. Heat is transported in the form of latent heat.

The actual boiling process is a combination of these mechanisms, which are not mutually exclusive of one another, i.e., heat passes through the liquid thermal boundary layer in all of them.

4.4.5 *Critical heat flux (CHF)*

The curve, shown in Fig. 4.10 for the pool boiling of water at atmospheric pressure from a flat surface, can only be obtained in its entirety under circumstances where the temperature of the heating surface is controlled to a specified value. In many practical cases, however, the surface heat flux is the independently controlled variable. In this case, the boiling curve in the natural convection (AB) and nucleate boiling (B′C) regions remains basically unaltered. However, if an attempt is made to increase the value of the surface heat flux above D the surface temperature will jump from that corresponding to D ($\sim 135°C$) to that corresponding to point D′ ($\sim 1150°C$) the next stable operating point in the film boiling region. In many practical cases this large temperature jump is sufficient to cause failure of the heating surface. Hence the colloquial name 'burnout' which is often used to refer to this phenomenon.

It is now generally accepted that the critical heat flux in pool boiling occurs as a result of a hydrodynamic flow pattern transition close to the heating surface. The mechanism is one in which insufficient liquid is able to reach the heating surface due to the rate at which vapour is leaving the surface. Zuber (1958) idealized the flow pattern as a square array of vapour jets, leaving the heated surface with a spacing corresponding to the most rapidly growing Taylor instability wavelength λ_d. Interaction of the adjacent vapour jets due to Helmholtz instability limits the vapour volumetric flux from the surface to

$$(j_g)_{\text{MAX}} = \frac{\pi}{24}\left[\frac{\sigma g(\rho_f - \rho_g)}{\rho_g^2}\right]^{1/4} \tag{4.56}$$

For the case where the liquid is at the saturation temperature and all the vapour is produced at the heating surface (for liquid metals and also for certain two-component systems this latter assumption may not be valid) then

$$j_g = \frac{\phi}{\rho_g i_{fg}} \tag{4.57}$$

and thus

$$\phi_{CRIT} = K i_{fg} \rho_g^{1/2} [\sigma g(\rho_f - \rho_g)]^{1/4} \tag{4.58}$$

where, according to Zuber, $K = \pi/24$. This theory thus confirmed the earlier correlation proposed by Kutateladze (1951) in which K was 0.16 ± 0.03. More recently, Lienhard and Dhir (1973a) have re-examined the Zuber theory in the light of up-to-date experimental evidence and have concluded that the constant K given by Zuber should be increased by a factor 1.14, i.e., K should be 0.149. The Zuber analysis is strictly valid only for boiling of pure fluids on large thick well-wetted horizontal surfaces facing upwards.

Lienhard and Dhir (1973a, b), Lienhard et al. (1973), Sun and Lienhard (1970), and Ded and Lienhard (1972) have extended this theory to finite surfaces, such as small plates, cylinders and spheres:

$$\frac{\phi_{CRIT}}{(\phi_{CRIT})_{Zuber}} = fn(L') \tag{4.59}$$

The ratio of $\phi_{CRIT}/(\phi_{CRIT})_{Zuber}$ in this equation is given in terms of a constant or as a function of a dimensionless characteristic dimension. Thus,

$$L' = 2\pi\sqrt{3}\,\frac{L}{\lambda_d} \tag{4.60}$$

where the Taylor instability wavelength is

$$\lambda_d = 2\pi\sqrt{3}\left[\frac{\sigma}{g(\rho_f - \rho_g)}\right]^{1/2} \tag{4.61}$$

The characteristic dimension in eq. (4.60) might be a radius R, a length L or a perimeter P. Table 4.4 lists the various relationships. Figure 4.19 compares the predictions with experimental data. The relationships are valid within about 20 per cent. They are not valid if the heater dimension becomes too small. When $L' < 0.38$ mm (0.15 in), the instability process is no longer the Taylor–Helmholtz instability and surface tension forces dominate. As L' decreases further, the critical and minimum heat fluxes move together and the boiling curve becomes monotonic. For very thin wires, the heater is immediately enveloped in vapour, such that nucleate and transition boiling are excluded, and the heat transfer process passes directly from natural convection to film boiling.

Liaw and Dhir (1986) have investigated CHF on partially and well-wetted surfaces for water. By oxidizing the test surface they were able to reduce the

Table 4.4 Critical heat flux for finite bodies (adapted from Lienhard 1981)

Geometry	$\dfrac{\phi_{\text{CRIT}}}{(\phi_{\text{CRIT}})_{\text{Zuber}}}$	Dimension	Range
1. Infinite flat plate	1.14	Width or diameter	$L' \geqslant 2.7$
2. Small flat heater	$1.14 A_{\text{heater}}/\lambda_{\text{d}}^2$	Width or diameter	$0.07 \leqslant L' \leqslant 0.2$
3. Horizontal cylinder	$0.89 + 2.27\,e^{-3.44\sqrt{R'}}$	Radius R	$R' \geqslant 0.15$
4. Large horizontal cylinder	0.90	Radius R	$R' \geqslant 1.2$
5. Small horizontal cylinder	$0.94/(R')^{1/4}$	Radius R	$0.15 \leqslant R' < 1.2$
6. Large sphere	0.84	Radius R	$4.26 \leqslant R'$
7. Small sphere	$1.734/(R')^{1/2}$	Radius R	$R' \leqslant 4.26$
8. Small horizontal ribbon oriented vertically*	$1.18/(H')^{1/4}$	Height of side H	$0.15 \leqslant H' \leqslant 2.96$
9. Small horizontal ribbon oriented vertically†	$1.4/(H')^{1/4}$	Height of side H	$0.15 \leqslant H' \leqslant 5.86$
10. Any large finite body	about 0.9	Length L	about $L' \geqslant 4$
11. Small slender cylinder of any cross section	$1.4/(P')^{1/4}$	Transverse perimeter P	$0.15 \leqslant P' \leqslant 5.86$
12. Small bluff body	$\text{Constant}/(L')^{1/2}$	Length L	about $L' \geqslant 4$

* Heated on both sides
† One side insulated.

contact angle from 90° to 18°. The CHF increased with decreasing contact angle as shown in Fig. 4.20. The Zuber and Lienhard–Dhir theories predict the low contact angle data well. Using local void fraction measurements, Dhir (1992) explained the effect of contact angle as follows: (a) for well-wetted surfaces it is the vapour removal rate that controls the CHF phenomenon consistent with hydrodynamic theory and (b) on partially wetted surfaces it is the rate of vapour generation that sets the upper limit on CHF.

Small amounts of impurity in the liquid can also produce large changes in the critical heat flux. In the case of a liquid subcooled below its saturation temperature the critical heat flux is increased

$$\phi_{\text{CRIT}} = K i_{\text{fg}} \rho_{\text{g}}^{1/2} [\sigma g (\rho_{\text{f}} - \rho_{\text{g}})]^{1/4} [1 + B\,\Delta T_{\text{SUB}}] \qquad (4.62)$$

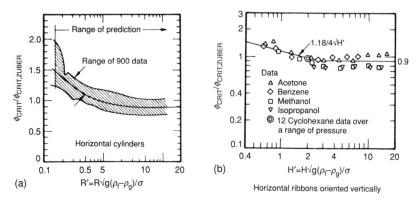

(a)

(b)

Horizontal ribbons oriented vertically

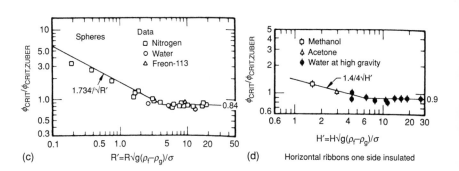

(c)

(d)

Horizontal ribbons one side insulated

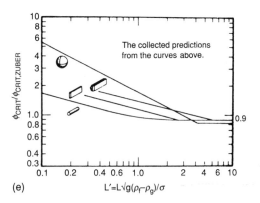

(e)

Fig. 4.19. The critical pool boiling heat flux on several heaters (Lienhard *et al.*).

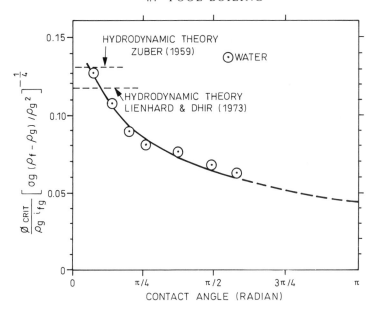

Fig. 4.20. Dependence of CHF on contact angle (Liaw and Dhir 1986).

Ivey and Morris (1962) recommended the value of B be given by

$$B = 0.1\left(\frac{\rho_f}{\rho_g}\right)^{0.75}\left(\frac{c_{pf}}{i_{fg}}\right) \qquad (4.63)$$

4.4.6 Transition boiling

In this little-studied region of the boiling curve, liquid periodically contacts the heating surface with the result that the formation of large amounts of vapour forces the liquid away from the surface and a vapour film or blanket is formed. This, in turn, collapses, allowing liquid to contact the surface once more. The region is normally only obtainable by controlling the surface temperature to a predetermined value. From Fig. 4.10 it will be seen that for water at atmospheric pressure this temperature range is about 140–250°C. Because of the periodic nature of the process the surface heat flux and temperature undergo large variations with time. No adequate theory or model of this region exists at this time.

Liaw and Dhir (1986) have measured transition boiling curves for water and R113 for surfaces of different wettabilities. The contact angles of the fluids were varied by oxidizing the surface or by applying a fluoro-silicone sealant to it. They observed that the critical heat flux decreases with increasing contact angle; this thereby shifts the position of the transition boiling curve, which depends on the critical heat flux as one of its end points. They presented a new method for predicting the transition boiling curve based on their work. In another study, Dhuga and Winterton (1986) carried out transient pool boiling curve tests on an aluminium block with an anodized heat transfer surface. By measuring the electrical current and thus obtaining the impedance between the bulk liquid and the surface, the proportion of the surface wetted by the liquid was ascertained. Interestingly, they observed substantial liquid contact with the heated surface at the critical heat flux and negligible or no contact at the minimum film boiling heat flux (point E in Fig. 4.10).

4.4.7   Film boiling

At high temperature differences a continuous vapour film blankets the heater surface. The major resistance to heat transfer is confined to this vapour film and, because of the lack of liquid–solid contact, this region is the most tractable to analyse. Vapour is removed from the film in the form of bubbles released regularly in both time and space (Fig. 4.11). The wavelength is that for Taylor instability given by

$$\lambda_c = 2\pi \left[\frac{\sigma}{g(\rho_f - \rho_g)} \right]^{1/2} \tag{4.64}$$

The relationships for the heat transfer coefficient in laminar or turbulent film boiling in various geometrical situations can be established by direct analogy with the identical relationships derived for film-wise condensation (see Chapter 10). For example, for laminar film boiling from a horizontal tube, Bromley (1950) gives, by analogy with eq. (10.71),

$$h = 0.62 \left[\frac{g(\rho_f - \rho_g)\rho_g k_g^3 i'_{fg}}{D\mu_g \Delta T} \right]^{1/4} \tag{4.65}$$

where i'_{fg} is an effective latent heat of vaporization allowing for the effect of superheat

$$i'_{fg} = i_{fg} \left[1 + 0.68 \left(\frac{c_{pg}\Delta T}{i_{fg}} \right) \right] \tag{4.66}$$

For very large tubes, horizontal flat surfaces, and very thin wires, eq. (4.65) is inaccurate.

Berenson (1961) developed a new model by incorporating the hydro-dynamic wave instability theory of Taylor into the Bromley model, obtaining the following expression:

$$h = 0.425 \left[\frac{g(\rho_f - \rho_g)\rho_g k_g^3 i'_{fg}}{\mu_g \, \Delta T_{SAT} [\sigma/g(\rho_f - \rho_g)]^{1/2}} \right]^{1/4} \tag{4.67}$$

where the physical properties of the vapour are evaluated at the mean film temperature, the liquid properties at the saturation temperature and a value of 0.5 is used in eq. (4.66) rather than 0.68. This equation correctly predicts a wide variety of independently obtained data. Breen and Westwater (1962) have also modified eq. (4.65) empirically to include the wavelength of instability.

Surface temperatures are usually high in film boiling and heat may be transferred by radiation. Bromley (1950) proposes the following approximation for combining the effects of convection and radiation:

$$h = h_c + 0.75h_r \tag{4.68}$$

where h_c is the convective coefficient (for example from eq. (4.65)) and h_r is the radiation coefficient. The value of h_r is calculated by assuming radiation between infinite parallel planar surfaces with the liquid acting as a perfect black body:

$$h_r = \sigma \varepsilon \left[\frac{T_W^4 - T_f^4}{(T_W - T_f)} \right] \tag{4.69}$$

where σ is the Stefan–Boltzman constant and ε is the emissivity of the heating surface.

4.5   Convective boiling

In Section 1.6.1 the various hydrodynamic conditions encountered when a liquid is boiled in a confined channel were described in general qualitative terms. These earlier comments will provide a starting point for the more detailed discussion of the various heat transfer topics to be dealt with in this section and the following chapters.

4.5.1   *The regimes of heat transfer*

Consider again a vertical tube heated uniformly over its length with a low heat flux and fed with subcooled liquid at its base at such a rate that the liquid is totally evaporated over the length of the tube. Figure 4.21 shows, in diagrammatic form, the various flow patterns encountered over the length of the tube, together with the corresponding heat transfer regions.

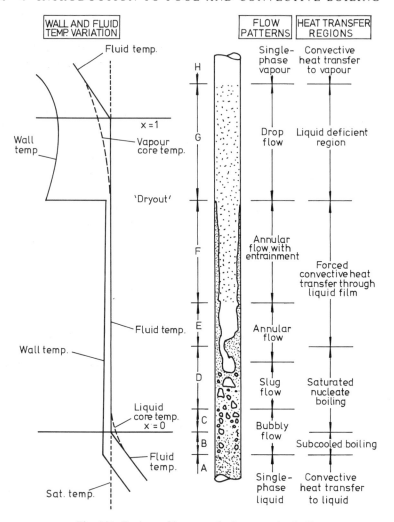

Fig. 4.21. Regions of heat transfer in convective boiling.

Whilst the liquid is being heated up to the saturation temperature and the wall temperature remains below that necessary for nucleation, the process of heat transfer is *single phase convective heat transfer to the liquid phase* (region A). At some point along the tube, the conditions adjacent to the wall are such that the formation of vapour from nucleation sites can occur. Initially vapour formation takes place in the presence of subcooled liquid (region B) and this heat transfer mechanism is known as *subcooled nucleate boiling*. In the subcooled boiling region, B, the wall temperature remains essentially constant a few degrees above the saturation temperature, whilst the mean bulk fluid temperature is increasing to the saturation temperature. The amount by which the wall temperature exceeds the saturation tempera- ture is known as the 'degree of superheat', ΔT_{SAT}, and the difference between the saturation and local bulk fluid temperature is known as the 'degree of subcooling', ΔT_{SUB}.

The transition between regions B and C, the *subcooled nucleate boiling region* and the *saturated nucleate boiling region* is clearly defined from a thermodynamic viewpoint. It is the point at which the liquid reaches the saturation temperature ($x = 0$) found on the basis of simple heat balance calculations. However, as explained in Section 1.6.2, subcooled liquid can persist in the liquid core even in the region defined as *saturated nucleate boiling*. Vapour generated in the subcooled region is present at the transition between regions B and C ($x = 0$); thus some of the liquid must be subcooled to ensure that the liquid mixed mean (mixing cup) enthalpy equals that of saturated liquid (i_f). This effect occurs as a result of the radial temperature profile in the liquid and the subcooled liquid flowing in the centre of the channel will only reach the saturation temperature at some distance down- stream of the point $x = 0$. The non-equilibrium processes which occur around this transition will be discussed in detail in Chapter 6.

In the regions C to G, the variable characterizing the heat transfer mechanism is the thermodynamic mass 'quality' (x) of the fluid. The 'quality' of the vapour–liquid mixture at a distance, z is given on a thermodynamic basis as

$$x(z) = \frac{i(z) - i_f}{i_{fg}} \tag{4.70}$$

or in terms of heat flux and length

$$x(z) = \frac{4\phi}{DGi_{fg}} (z - z_{SC}) \tag{4.71}$$

where z_{SC} is the length of tube required to bring the enthalpy of the liquid up to the saturated liquid enthalpy, i_f. In the region $0 < x < 1$ and for complete thermodynamic equilibrium, x represents the ratio of the vapour mass flow-rate to the total mass flow-rate. From the thermodynamic definition, eq. (4.70), x may have both negative values and values greater than unity. Although these are sometimes used for convenience, particularly in Russian publications, negative values or values greater than unity have no practical significance other than to signify that in the former case the condition is that of a subcooled liquid and, in the latter case, that of a superheated vapour. The variable x is also often referred to as the 'vapour weight fraction'.

As the quality increases through *the saturated nucleate boiling region* a point may be reached where a fundamental transition in the mechanism of heat transfer takes place. The process of 'boiling' is replaced by the process of 'evaporation'. This transition is preceded by a change in the flow pattern from bubbly or slug flow to annular flow (regions E and F). In the latter regions the thickness of the thin liquid film on the heating surface is often such that the effective thermal conductivity is sufficient to prevent the liquid in contact with the wall being superheated to a temperature which would allow bubble nucleation. Heat is carried away from the wall by forced convection in the film to the liquid–vapour core interface, where evaporation occurs. Since nucleation is completely suppressed, the heat transfer process can no longer be called 'boiling'. The region beyond the transition has been referred to as *the two-phase forced convective region* of heat transfer (regions E and F).

At some critical value of the quality the complete evaporation of the liquid film occurs. This transition is known as 'dryout' and is accompanied by a rise in the wall temperature for channels operating with a controlled surface heat flux. The area between the dryout point and the transition to *dry saturated vapour* (region H) has been termed the *liquid deficient region* (corresponding to the drop flow pattern) (region G). This condition of 'dryout' often puts an effective limit on the amount of evaporation that can be allowed to take place in a tube at a particular value of heat flux. It is extremely important in the design of evaporators, pipe stills, steam boilers, nuclear reactors, and other units cooled by forced convection boiling.

4.5.2   *The boiling map*

It is useful at this stage to describe, at least qualitatively, the progressive variation of the local heat transfer coefficient along the length of the tube as evaporation proceeds. The local heat transfer coefficient can be established by dividing the surface heat flux (constant over the tube length) by the difference between the wall temperature and the bulk-fluid temperature. Typical variations of these two temperatures with length along the tube are shown in Fig. 4.21. The variation of heat transfer coefficient with length along the tube for the conditions represented in Fig. 4.21 is given in Fig. 4.22 (curve (i), solid line). In the *single phase convective heat transfer region*, the heat transfer coefficient is relatively constant, changing only slightly due to the influence of temperature on the liquid physical properties. In the *subcooled nucleate boiling region* the temperature difference between the wall and the bulk fluid decreases linearly with length up to the point where $x = 0$. The heat transfer coefficient, therefore, increases linearly with length in this region. In the *saturated nucleate boiling region* the temperature difference and therefore the heat transfer coefficient remain constant. Because of the

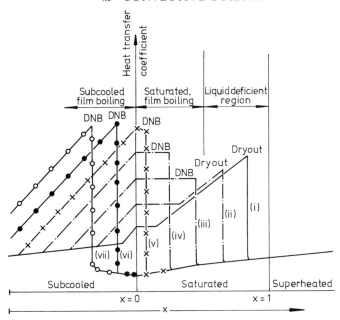

Fig. 4.22. Variation of heat transfer coefficient with quality with increasing heat flux as parameter.

reducing thickness of the liquid film in the *two-phase forced convective* region heat transfer in this region is characterized by an increasing coefficient with increasing length or mass quality. At the dryout point the heat transfer coefficient is suddenly reduced from a very high value in the forced convective region to a value near to that expected for heat transfer by forced convection to dry saturated vapour. As the quality increases through the *liquid deficient region* so the vapour velocity increases and the heat transfer coefficient rises correspondingly. Finally, in the *single-phase vapour region* ($x > 1$) the heat transfer coefficient levels out to that corresponding to convective heat transfer to a single-phase vapour flow.

The above comments have been restricted to the case where a relatively low heat flux is supplied to the walls of the tube. The effect of progressively increasing the surface heat flux whilst keeping the inlet flow-rate constant will now be considered with reference to Figs. 4.22–24. Figure 4.22 shows the heat transfer coefficient plotted against mass quality with increasing heat flux as parameter (curves (i)–(vii)). Figure 4.23 shows the various regions of two-phase heat transfer in forced convection boiling on a three-dimensional diagram with heat flux, mass quality and temperature as co-ordinates—'the boiling surface'.

Figure 4.24 shows the regions of two-phase forced convective heat transfer

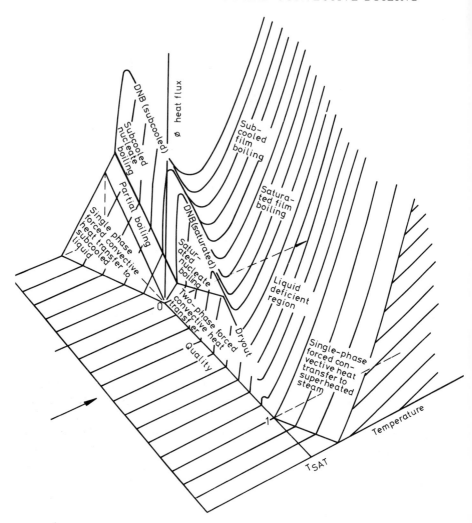

Fig. 4.23. Forced convection boiling surface.

as a function of quality with increasing heat flux as ordinate (an elevation view of Fig. 4.23 taken in the direction of the arrow).

Curve (i) of Fig. 4.22 relates to the conditions shown in Fig. 4.21 for a low heat flux being supplied to the walls of the tube. The temperature pattern shown in Fig. 4.21 will be recognized as the projection in plan view (temperature–quality co-ordinates) of Fig. 4.23.

Curve (ii) shows the influence of increasing the heat flux. Subcooled boiling is initiated sooner, the heat transfer coefficient in the nucleate boiling region

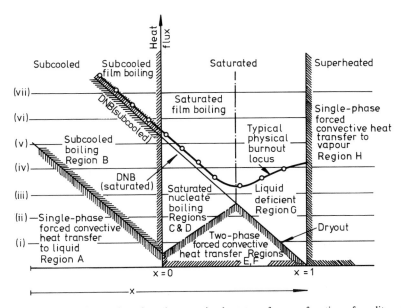

Fig. 4.24. Regions of two-phase forced convective heat transfer as a function of quality with increasing heat flux as ordinate.

is higher but is unaffected in the two-phase forced convective region. Dryout occurs at a lower mass quality. Curve (iii) shows the influence of a further increase in heat flux. Again, subcooled boiling is initiated earlier and the heat transfer is again higher in the nucleate boiling region. As the mass quality increases, before the two-phase forced convective region is initiated and while bubble nucleation is still occurring, an abrupt deterioration in the cooling process takes place. This transition is essentially similar to the critical heat flux phenomenon in saturated pool boiling and will be termed 'departure from nucleate boiling' (DNB).

The mechanism of heat transfer under conditions where the critical heat flux (DNB or dryout) has been exceeded is dependent on whether the initial condition was the process of 'boiling' (i.e., bubble nucleation in the subcooled or low mass quality regions) or the process of 'evaporation' (i.e., evaporation at the liquid film–vapour core interface in the higher mass quality areas). In the latter case, the 'liquid deficient region' is initiated; in the former case the resulting mechanism is one of 'film boiling' (Fig. 4.24).

Returning to Figs. 4.22 and 4.24 it can be seen that further increases in heat flux (curves (vi) and (vii)) cause the condition of 'departure from nucleate boiling' (DNB) to occur in the subcooled region with the whole of the saturated or 'quality' region being occupied by firstly 'film boiling' and, in the latter stages, by the 'liquid deficient region'—both relatively inefficient modes of heat transfer.

In Figs. 4.23 and 4.24 the film boiling region has been arbitrarily divided into two regions: 'subcooled film boiling' and 'saturated film boiling'. 'Film boiling' in forced convective flow is essentially similar to that observed in pool boiling. An insulating vapour film covers the heating surface through which the heat must pass. The heat transfer coefficient is orders of magnitude lower than in the corresponding region before the critical heat flux was exceeded, due mainly to the lower thermal conductivity of the vapour.

4.5.3   *The critical heat flux condition in forced convection boiling*

Two separate and distinct processes—'departure from nucleate boiling' (DNB) and 'dryout'—have been discussed which can cause an abrupt temperature rise in a heat flux-controlled situation. In the present text, a distinction will be made between the precise point at which a more or less rapid surface temperature rise from a temperature close to the saturation value is initiated and the point at which failure of the heating surface due to rupture or melting occurs. The term 'critical heat flux' will be used to cover the former case and will serve as a generic term covering the three processes; DNB (subcooled), DNB (saturated), and dryout. The failure of the heating surface will be termed 'physical burnout' and the heat flux which causes this condition, 'the physical burnout heat flux'. The term 'burnout heat flux', which also suggests this latter condition, has been abused by many who use it to denote the heat flux at which the rapid deterioration of the cooling process occurs. This ambiguity can be illustrated by reference to Fig. 4.24. Low heat transfer coefficients coupled with the relatively high heat flux values required to initiate film boiling in the subcooled or low quality regions results in extremely high temperature differences at the critical heat flux condition. Failure of the heating surface usually occurs and thus the heat flux to initiate DNB is often identical with that to cause 'physical burnout'— the physical burnout heat flux. However, this is not the case in the liquid deficient region where higher heat transfer coefficients and lower critical heat fluxes cause only modest temperature excursions at 'dryout'. In this region the physical burnout locus denotes a particular isotherm representing the failure criterion for the chosen heating surface. The physical burnout heat flux may be many times the critical flux in this region.

In the following chapters criteria will be established whereby the boundaries delineated in Figs. 4.23 and 4.24—'the boiling surface'—can be fixed. In addition methods for calculating heat transfer rates and other relevant information in each heat transfer region will be discussed in detail.

References and further reading

Bähr, A. (1965). 'The microconvection in boiling water induced by single bubbles'. Paper presented at Symposium on Two-phase Flow, Exeter, 21–23 June.

Bajorek, S. M., Lloyd, J. R., and Thome, J. R. (1990). 'Evaluation of multicomponent pool boiling heat transfer coefficients'. Presented at 9th Int. Heat Transfer Conf., Jerusalem, Paper 1-BO-07.

Bankoff, S. G. (1957). 'Ebullition from solid surfaces in the absence of a pre-existing gaseous phase'. *Trans. ASME*, **79**, 735.

Bankoff, S.G. (1959). 'The prediction of surface temperature at incipient boiling'. *Chem. Engng. Prog. Symp. Series* No. 29, **55**, 87.

Berenson, P. J. (1961). 'Film-boiling heat transfer from a horizontal surface'. *J. Heat Transfer*, **83**, 351–358.

Bernath, L. (1952). *Ind. Engng. Chem.*, **44**, 1310.

Blander, M. and Katz, J. L. (1975). 'Bubble Nucleation in Liquids'. *AIChE J.*, **21**(5), 833–848.

Borishanski, V. M. (1969). 'Correlation of the effect of pressure on the critical heat flux and heat transfer rates using the theory of thermodynamic similarity'. In *Problems of Heat Transfer and Hydraulics of Two-phase Media*, pp. 16–37, Pergamon Press.

Bornhorst, W. J. and Hatsopoulos, G. N. (1967). *Trans. ASME J. Appl. Mechanics*, December, 847.

Breen, B. P. and Westwater, J. W. (1962). 'Effect of diameter of horizontal tubes on film boiling heat transfer'. AIChE preprint 19 presented at 5th National U.S. Conference, Houston, Texas, August.

Bromley, L. A. (1950). 'Heat transfer in stable film boiling'. *Chem. Engng. Prog.*, **46**, 221–227.

Brown, W. T. (1967). 'A study of flow surface boiling'. PhD Thesis, MIT Mech. Eng. Dept.

Cooper, M. G. (1984). 'Heat flow rates in saturated nucleate pool boiling—a wide-ranging examination using reduced properties.' *Advances in Heat Transfer*, Academic Press, Orlando, **16**, 157–239.

Cooper, M. G., Judd, A. M., and Pike, R. A. (1978). 'Shape and departure of single bubbles growing at a wall.' Presented at 6th Int. Heat Transfer Conf., Toronto, **1**, 115–120.

Ded, J. S. and Lienhard, J. H. (1972). 'The peak pool boiling heat flux from a sphere.' *AIChE J.*, **18**(2), 337–342.

Dhir, V. K. (1992). 'Some observations from maximum heat flux data obtained on surfaces having different degrees of wettability.' *Pool and External Flow Boiling*, Proceedings of Engineering Foundation Conference, Santa Barbara, California, March 22–27.

Dhuga, D. S. and Winterton, R. H. S. (1986). 'The pool boiling curve and liquid–solid contact.' Presented at 8th Int. Heat Transfer Conf., San Francisco, **4**, 2055–2059.

Dwyer, O. E. (1969). 'On incipient boiling wall superheats in liquid metals'. *Int. J. Heat Mass Transfer*, **12**, 1403–1419.

Fabic, S. (1964). 'Vapour nucleation on surfaces subject to transient heating'. Report NE-64-1, University of California (Berkeley).

Fishenden, M. and Saunders, O. (1950). *An Introduction to Heat Transfer*, Oxford University Press.

Forster, H. K. and Zuber, N. (1955). 'Bubble dynamics and boiling heat transfer'. *AIChE J.*, **1**, 532.

Fritz, W. (1935). 'Berechnung des Maximal Volume von Dampfblasen'. *Phys. Z.*, **36**, 379.

Gorenflo, D. (1993). 'Pool boiling'. *VDI-Heat Atlas*, VDI-Verlag, Düsseldorf.

Gorenflo, D., Sokol, P. and Caplanis, S. (1990) 'Pool boiling heat transfer from single plain tubes to various hydrocarbons'. *Int. J. Refrig.*, **13**, 286–292.

Griffith, P. and Snyder, G. (1964). 'The mechanism of void formation in initially subcooled systems'. Report 9041-26, Mech. Engng. Dept., MIT.

Griffith, P. and Wallis, J. D. (1960). 'The role of surface conditions in nucleate boiling'. *Chem. Engng. Prog. Symp. Series* No. 30, 49.

Han, C. Y. and Griffith, P. (1965). 'The mechanism of heat transfer in nucleate pool boiling. Pt. I: Bubble initiation, growth and departure'. *Int. J. Heat Mass Transfer*, **8**, 887.

Hatton, A. P. and Hall, I. S. (1966). 'Photographic study of boiling on prepared surfaces'. Paper presented at 3rd International Heat Transfer Conference, Chicago, August.

Holtz, R. E. (1966). 'The effect of the pressure-temperature history upon incipient boiling superheats in liquid metals'. ANL-7184.

Howell, J. R. and Siegel, R. (1966). 'Incipience, growth and detachment of boiling bubbles in saturated water from artificial nucleation sites of known geometry and size'. Paper presented at 3rd International Heat Transfer Conference, Chicago, August.

Hsu, Y. Y. (1962). 'On the size of range of active nucleation cavities on a heating surface'. *Trans. ASME J. of Heat Transfer*, **84**, 207.

Hsu, Y. Y. and Graham, R. W. (1976). 'Transport processes in boiling and two-phase systems', Hemisphere, Washington, DC.

Hui, T. O. and Thome, J. R. (1985). 'A study of binary mixture boiling: boiling site density and subcooled heat transfer'. *Int. J. Heat Mass Transfer*, **28**, 919–928.

Ivey, H. J. and Morris, D. J. (1962). 'On the relevance of the vapour–liquid exchange mechanism for sub-cooled boiling heat transfer at high pressure'. AEEW-R 137.

Jakob, M. (1958). *Heat transfer*, Vol. 1, Wiley, New York.

Johnson, M. A., de la Pena, J., and Mesler, R. B. (1966). 'Bubble shapes in nucleate boiling'. *Chem. Engng. Prog. Symp. Series*, **62**, 1.

Kast, W. (1964). 'Bedeutung der Keimbildung und der instationanen Wärmeübertragung für den Wärmeübergang bei Blasenverdampfung und Tropfenkondensation'. *Chemie Ing. Teckn.*, 36 Jahrg. 1964/N9, 933–940.

Keshock, E. G. and Siegel, R. (1964). 'Forces acting on bubbles in nucleate boiling under normal and reduced gravity conditions'. NASA TN-D-2299.

Knapp, R. T. (1958). 'Cavitation and nuclei'. *Trans. ASME*, **80**, 1321.

Kutateladze, S. S. (1951). 'A hydrodynamic theory of changes in the boiling process under free convection'. *Izv. Akad. Nauk SSSR Otd. Tekh. Nauk.*, **4**, 529.

Liaw, S. P. and Dhir, V. K. (1986). 'Effect of surface wettability on transition boiling heat transfer from a vertical surface'. Presented at 8th Int. Heat Transfer Conf., San Francisco, **4**, 2031–2036.

Lien, Y. and Griffith, P. (1969). 'Bubble growth rates at reduced pressure'. ScD Thesis, Mech. Engng. Dept., MIT.

Lienhard, J. H. (1976). 'Correlation of the limiting liquid superheat'. *Chem. Engng. Sci.*, **31**, 847–849.

Lienhard, J. H. (1981). 'A heat transfer textbook'. Prentice-Hall, Englewood Cliffs, 409.

Lienhard, J. H. and Dhir, V. K. (1973a). 'Extended hydrodynamic theory of the peak and minimum pool boiling heat fluxes'. NASA-CR-2270, 194 pages, July.

Lienhard, J. H. and Dhir, V. K. (1973b). 'Hydrodynamic prediction of peak pool boiling heat fluxes from finite bodies'. *J. Heat Transfer*, **95**, 152–158.

Lienhard, J. H., Dhir, V. K., and Riherd, D. M. (1973). 'Peak pool boiling heat-flux measurements on finite horizontal flat plates'. *J. Heat Transfer*, **95**, 477–482.

Lorentz, J. J., Mikic, B. B., and Rohsenow, W. M. (1974). 'The effect of surface conditions on boiling characteristics'. *Heat Transfer*, Vol. 4, *Proceedings of 5th International Heat Transfer Conference*, Tokyo, Japan.

Mikic, B. B., Rohsenow, W. M., and Griffith, P. (1970). 'On bubble growth rates'. *Int. J. Heat Mass Transfer*, **13**, 657–665.

Moore, F. D. and Mesler, R. B. (1961). *AIChE J.*, **7**, 620.

Mostinski, I. L. (1963). *Teploenergetika*, **4**, 66. English Abstract, *Brit. Chem. Engng.*, **8**(8), 580.

Plesset, M. and Zwick, S. A. (1954). *J. Appl. Phys.*, **25**, 493–500.

Preston, G., Thome, J. R., Bald, W. B., and Davey, G. (1979). 'The measurement of growing bubbles on a heated surface using a computerized image analysis system'. *Int. J. Heat Mass Transfer*, **22**, 1457–1459.

Rayleigh, Lord (1917). *Phil. Mag.*, **34**(94) (1917); *Sci. Papers* (1920), Cambridge University Press, **6**, 504.

Rohsenow, W. M. (1952). 'A method of correlating heat transfer data for surface boiling of liquids'. *Trans. ASME*, **74**, 969–975.

Sardesai, R. G., Palen, J. W., and Thome, J. R. (1986). 'Nucleate pool boiling of hydrocarbon mixtures'. Presented at AIChE National Conf., Miami Beach, November 2–7, Paper 127a.

Shakir, S. and Thome, J. R. (1986). 'Boiling Nucleation of Mixtures on Smooth and Enhanced Surfaces'. Presented at 8th Int. Heat Transfer Conf., San Francisco, **4**, 2081–2086.

Simpson, H. C. and Walls, A. S. (1965). 'A study of nucleation phenomena in transient pool boiling', Paper 18 presented at Symposium on Boiling Heat Transfer in Steam Generating Units and Heat Exchangers, Manchester, 15–16 September.

Skripov, V. P. and Pavlov, P. A. (1970). *Teplo. Vyso. Temp.*, **8**(4), 833.

Stephen, K. and Abdelsalam, M. (1980). 'Heat transfer correlations for natural convection boiling'. *Int. J. Heat Mass Transfer*, **23**, 73–87.

Sun, K. H. and Lienhard, J. H. (1970). 'The peak pool boiling heat flux on horizontal cylinders'. *Int. J. Heat Mass Transfer*, **13**, 1425–1439.

Thome, J. R. and Davey, G. (1981). 'Bubble growth rates in liquid nitrogen, argon and their mixtures'. *Int. J. Heat Mass Transfer*, **24**, 89–97.

Thome, J. R. and Shakir, S. (1987). 'A new correlation for nucleate pool boiling of aqueous mixtures'. *AIChE Symp. Ser.*, **83**(257), 46–51.

Thomson, W. (1871). *Phil. Mag.*, **42**(4), 448.

Vachon, R. I., Nix, G. H., and Tangor, G. E. (1967). 'Evaluation of constants for the Rohsenow pool boiling correlation'. Paper 67-HT-33 presented at US National Heat Transfer Conference, Seattle, August.

Vachon, R. I., Tanger, G. E., Davis, D. L., and Nix, G. H. (1967). 'Pool boiling on polished and chemically etched stainless steel surfaces'. Paper 67-HT-34 presented at US National Heat Transfer Conference, Seattle, August.

Van Stralen, S. J. D. (1970). Presented at 4th Int. Heat Transfer Conf., Paris, Paper B7.6.

Van Stralen, S. J. D., Sohal, M. S., Cole, R., and Sluyter, W. M. (1975). 'Bubble growth rates in pure and binary systems: combined effect of relaxation and evaporation microlayers. *Int. J. Heat Mass Transfer*, **18**, 453.

Westwater, J. W. (1958). 'Boiling of liquids'. In Drew, T. B. and Hooper, J. P. (eds.), *Advances in Chemical Engineering*, Vol. 2, Academic Press, New York.

Zuber, N. (1958). 'On the stability of boiling heat transfer'. *Trans. ASME*, **80**, 711.

Zuber, N. (1963). 'Nucleate boiling—the region of isolated bubbles—similarity with natural convection'. *Int. J. Heat Mass Transfer*, **6**, 53.

Zuber, N. and Fried, E. (1962). 'Two phase flow and boiling heat transfer to cryogenic liquids'. *ARS Journal*, 1332–1341, September.

Example

Example 1
Determine the wall superheat required to cause nucleation from a horizontal flat plate immersed in water at atmospheric pressure,

(a) when active cavities of all sizes are present,
(b) when active cavities of sizes below 5 μm only are present.

Solution
The relevant physical properties for water are:

$$T_{SAT} = 100°C = 373 \text{ K} \qquad M = 18 \text{ kg/kmol}$$

$$\sigma = 0.059 \text{ N/m} \qquad i_{fg} = 2.255 \text{ MJ/kg}$$

$$\beta \text{ (coefficient of expansion)} = 7.2 \times 10^{-4} \text{ °C}^{-1}$$

$$v_f = \left(\frac{\mu_f}{\rho_f}\right) = 3.1 \times 10^{-7} \text{ m}^2/\text{s} \quad Pr_f = 1.9$$

(a) *The thermal boundary layer thickness*
From eq. (4.38)

$$\frac{hD}{k_f} = 0.14\left[\left(\frac{\beta g\, \Delta T_w D^3 \rho_f^2}{\mu_f^2}\right)\left(\frac{c_p \mu}{k}\right)_f\right]^{1/3}$$

$$\frac{h}{k_f} = \frac{1}{\delta} = 0.14\left[\left(\frac{7.2 \times 10^{-4} \times 9.806 \times \Delta T_w}{(3.1 \times 10^{-7})^2}\right) \times 1.9\right]^{1/3}$$

$$= 0.14 \times 5.192 \times 10^3\, \Delta T_w^{1/3}$$

$$\delta = 1.38 \times 10^{-3}/\Delta T_w^{1/3} \text{ m} \qquad\qquad (1)$$

(b) *Local superheat relationship*
Use can be made of eq. (4.9) since $v_g > v_f$ and

$$\left(\frac{2\sigma}{p_f r^*}\right) = \frac{2 \times 0.059}{10^5 \times 5 \times 10^{-6}} = 0.24(<1)$$

Equation (4.9) gives

$$\Delta T_{SAT} = \frac{RT_{SAT}^2}{Ji_{fg}M} \frac{2\sigma}{p_f r^*} \tag{4.9}$$

The method used is that by Hsu for the determination of the nucleation condition within a temperature gradient. No distortion of the isotherms by the bubble nucleus is assumed.
   Thus, $r_c = r^*$

$$\Delta T_{SAT} = \frac{8314.3 \times 373^2}{1 \times 2.255 \times 10^6 \times 18} \left(\frac{2 \times 0.059}{10^5 \times r_c}\right) = \left(\frac{2.96 \times 10^{-5}}{r_c}\right)°C \tag{2}$$

(c) *Wall superheat requirement*
Now with all cavities present

$$r_c = \delta/2 \qquad (\Delta T_W)_{ONB} = 2\Delta T_{SAT} \qquad \text{(Fig. 4.12)}$$

For condition (a) combining eqs. (1) and (2)

$$(\Delta T_W)_{ONB} = \frac{4.92 \times 10^{-5}}{r_c} = \frac{9.84 \times 10^{-5}}{\delta} = \left(\frac{9.84 \times 10^{-5}}{1.38 \times 10^{-3}}\right) \Delta T_W^{1/3}$$

$$= 7.12 \times 10^{-2} \, \Delta T_W^{1/3}$$

Therefore $(\Delta T_W)_{ONB} = \underline{0.019°C}$

$$\delta = \underline{5.18 \text{ mm}}$$

For condition (b) because the intersection of the liquid temperature profile and the equilibrium bubble curve takes place close to the horizontal axis in Fig. 4.12,

$$(\Delta T_W)_{ONB} \approx \Delta T_{SAT}$$

$$\approx \frac{2.96 \times 10^{-5}}{r_c} = \frac{2.96 \times 10^{-5}}{5 \times 10^{-6}} = \underline{5.9°C}$$

Problems

1. Compute the relationship between superheat and equilibrium bubble radius for liquid sodium at one atmosphere pressure ($T_{SAT} = 881°C$) using eq. (4.8). Compare this with that obtained from eq. (4.9) and also the value for water at one atmosphere pressure. Does sodium require a higher or lower superheat to cause

nucleation than water?

$$M = 23 \text{ kg/kmol} \qquad R = 8314.3 \text{ N m/kmol K} \qquad i_{fg} = 3.78 \text{ MJ/kg}$$

$$\sigma = 0.113 \text{ N/m} \qquad \rho_f = 742 \text{ kg/m}^3 \qquad \rho_g = 0.274 \text{ kg/m}^3$$

2. Compute the most likely size of active cavity on a stainless steel surface immersed in liquid sodium at one atmosphere. Assume that the vapour pressure of sodium within the cavity when 'wetting' takes place is negligibly small.

3. Modify eq. (4.23) to account for the influence of viscous forces within the liquid on the growth rate of a spherical bubble in a uniformly superheated fluid.

Answer:
$$\left[R\ddot{R} + \tfrac{3}{2}\dot{R}^2 = \frac{1}{\rho_f}\left(\Delta p - \frac{4\mu\dot{R}}{R} \right) \right]$$

4. For nucleate pool boiling, determine the boiling heat transfer coefficient for a fluid at saturation at the following conditions:

Pressure	7.5 bar
Critical pressure	50 bar
Molecular weight	100
Heat flux	10 kW/m^2
Surface roughness	2 μm

5. Repeat Problem 4 using the Gorenflo method if the standard heat transfer coefficient is 2850 W/m^2 °C.

6. Determine the wall temperature difference for *p*-xylene boiling at the following condition on a horizontal steel tube using the Stephan–Abdelsalam correlation:

Pressure	1.01 bar
Temperature	138.4°C
Heat flux	20,000 W/m^2

(Properties: liquid density = 762 kg/m^3, vapour density = 3.3 kg/m^3, surface tension = 0.016 N/m, liquid thermal conductivity = 0.123 W/m °C, latent heat = 340.1 kJ/kg, liquid specific heat = 2.093 kJ/kg °C, critical pressure = 34.47 bar).

7. Calculate the critical heat flux for the situation in Problem 6 for a 25.4 mm diameter horizontal tube.

5

SUBCOOLED BOILING HEAT TRANSFER

5.1 Introduction

After a brief comment concerning heat transfer to a single-phase liquid flowing in a channel, the remainder of this chapter is taken up with the topic of heat transfer to a flowing subcooled liquid under conditions where nucleate boiling takes place at the channel wall. The criteria necessary for the onset of nucleation with forced convection are considered in detail. This is followed by a description of alternative methods for calculating the change of heat transfer rate and surface temperature with the independent variables such as velocity, heat flux, and liquid temperature. Finally, the various possible mechanisms of subcooled boiling are considered.

Although a very large proportion of the studies on forced convection boiling have been carried out using the water–steam system, whenever justified the relationships derived in this and the following chapters have been kept as general as possible so as to apply to all single-component systems.

5.2 Single-phase liquid heat transfer

Returning to Fig. 4.21, the first heat transfer regions encountered when a liquid is fed to a vertical evaporator tube are those of 'single-phase heat transfer to the liquid phase' (region A) and 'subcooled boiling' (region B). Figure 5.1 shows in an idealized form the flow pattern and the variation of surface and liquid temperatures in the regions designated by A, B, and C. In the following discussion it is assumed that the pressure drop along the channel is small compared with the applied static pressure and that the saturation temperature is constant along the channel. In any practical analysis, however, the variation of saturation pressure and temperature with length can be accounted for and the contribution of 'flashing' to the vapour flow-rate taken into consideration when calculating the mass quality (x).

Consider initially the region of the tube in which the bulk liquid temperature is rising to the saturation temperature (regions A and B, $z = 0$ to $z = z_{sc}$). A simple heat balance up to distance z from the tube inlet

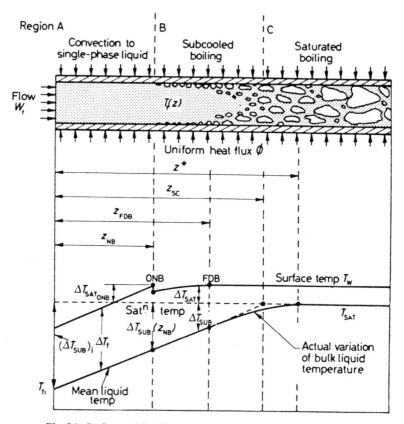

Fig. 5.1. Surface and liquid temperature distribution in subcooled boiling.

assuming a uniform heat flux along the tube gives

$$\phi \pi D z = W_f c_{pf}(T_f(z) - T_{fi}) \tag{5.1}$$

where c_{pf} is the liquid specific heat and $T_f(z)$ is the mean liquid temperature at a length z from the tube inlet. Rearranging and substituting $G = 4W_f/\pi D^2$ then $T_f(z)$ is given by

$$T_f(z) = T_{fi} + \frac{4\phi z}{G c_{pf} D} \tag{5.2}$$

Region C, saturated nucleate boiling, starts when $T_f(z)$ equals T_{SAT}, the saturation temperature. The combined length of regions A and B, the subcooled regions, z_{SC}, is given by

$$z_{SC} = \frac{G c_{pf} D}{4\phi}(T_{SAT} - T_{fi}) \tag{5.3}$$

The tube surface temperature in region A, convective heat transfer to single-phase liquid, is given by

$$T_W = T_f(z) + \Delta T_f \tag{5.4}$$

and

$$\Delta T_f = \phi/h_{fo} \tag{5.5}$$

where ΔT_f is the temperature difference between the tube inside surface and the mean bulk liquid temperature at a length z from the tube inlet, h_{fo} is the heat transfer coefficient to single phase liquid under forced convection. The liquid in the channel may be in laminar or turbulent flow. The laws governing laminar or turbulent convective heat transfer are well established. For laminar flow a variety of theoretical relationships are available depending on the boundary conditions, i.e., constant surface heat flux or surface temperature, developing velocity profile or fully developed flow. The following empirical equation based on experimental data takes into account the effect of varying physical properties across the flow stream and the influence of free convection.

$$\left[\frac{h_{fo}D}{k_f}\right] = 0.17 \left[\frac{GD}{\mu_f}\right]^{0.33} \left[\frac{c_p\mu}{k}\right]_f^{0.43} \left[\frac{Pr_f}{Pr_W}\right]^{0.25} \left[\frac{D^3\rho_f^2 g\beta\,\Delta T}{\mu_f^2}\right]^{0.1} \tag{5.6}$$

This relationship is valid for heating in vertical upflow or cooling in vertical downflow for $z/D > 50$ and $[GD/\mu_f] < 2000$.

For turbulent flow the well-known Dittus–Boelter equation has been found satisfactory.

$$\left[\frac{h_{fo}D}{k_f}\right] = 0.023 \left[\frac{GD}{\mu_f}\right]^{0.8} \left[\frac{c_p\mu}{k}\right]_f^{1/3} \tag{5.7}$$

This relationship is valid for heating in vertical upflow for $z/D > 50$ and $[GD/\mu_f] > 10,000$. The heating surface temperature, T_W, is thus given by

$$T_W = T_{fi} + \phi \left[\frac{4z}{Gc_{pf}D} + \frac{1}{h_{fo}}\right] \tag{5.8}$$

Now, region B, the subcooled boiling region, is initiated when T_W equals $(T_{SAT} + (\Delta T_{SAT})_{ONB})$. Thus the length of tube under non-boiling conditions, z_{NB}, is given by

$$z_{NB} = \frac{Gc_{pf}D}{4}\left[\frac{(\Delta T_{SUB})_i + (\Delta T_{SAT})_{ONB}}{\phi} - \frac{1}{h_{fo}}\right] \tag{5.9}$$

or

$$z_{SC} - z_{NB} = \frac{Gc_{pf}D}{4}\left[\frac{1}{h_{fo}} - \frac{(\Delta T_{SAT})_{ONB}}{\phi}\right] \tag{5.10}$$

$(\Delta T_{SUB})_i$ is the inlet subcooling $(T_{SAT} - T_{fi})$; $(\Delta T_{SAT})_{ONB}$, the amount of superheat required to initiate nucleation, is a function of the heat flux ϕ. This relationship between $(\Delta T_{SAT})_{ONB}$ and ϕ is discussed in Section 5.3. Region B will disappear if $(z_{SC} - z_{NB})$ becomes zero. Examination of eq. (5.10) shows this will occur if $\Delta T_f = (\Delta T_{SAT})_{ONB}$. From eq. (5.2).

$$T_f(z_{NB}) = T_{fi} + \frac{4\phi z_{NB}}{Gc_{pf}D} \tag{5.11}$$

and $\Delta T_{SUB}(z_{NB})$, the subcooling at the onset of subcooled boiling is given by

$$\Delta T_{SUB}(z_{NB}) = \frac{4\phi}{Gc_{pf}D}(z_{SC} - z_{NB}) \tag{5.12}$$

5.3   The onset of subcooled nucleate boiling

Consider now the conditions under which boiling will be initiated in the vertical heated tube. No boiling can occur whilst the temperature of the heating surface remains below the saturation temperature of the fluid at that particular location, that is $T_{SAT} > T_W$ and from eqs. (5.5) and (5.8) this minimum limiting condition for nucleation is given by

$$T_{SAT} \leqslant T_W, \qquad T_{SAT} \leqslant \left[T_f(z) + \frac{\phi}{h_{fo}} \right] \tag{5.13}$$

and

$$T_{SAT} \leqslant T_W, \qquad T_{SAT} \leqslant \left\{ \phi \left[\frac{4z}{Gc_{pf}D} + \frac{1}{h_{fo}} \right] + T_{fi} \right\} \tag{5.14}$$

or since $(T_{SAT} - T_{fi}) = (\Delta T_{SUB})_i$, and $(T_{SAT} - T_f(z)) = \Delta T_{SUB}(z)$,

$$\Delta T_{SUB}(z) \leqslant \phi/h_{fo} \tag{5.15}$$

$$(\Delta T_{SUB})_i \leqslant \phi \left[\frac{4z}{Gc_{pf}D} + \frac{1}{h_{fo}} \right] \tag{5.16}$$

In a plot of ϕ, the surface heat flux, against $(\Delta T_{SUB})_i$, the inlet subcooling (Fig. 5.2), the inequality (5.16) is, in fact, obeyed in an area ('subsaturation zone') bounded by the abscissa and a line of slope

$$\left[\frac{4z}{Gc_{pf}D} + \frac{1}{h_{fo}} \right]^{-1}$$

Within this region there is no possibility whatsoever of nucleation. As the distance, z, along the channel increases, so the slope decreases, while increases

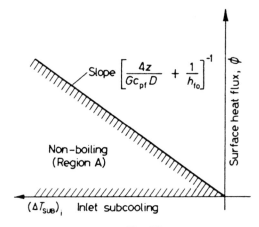

Slope $\left[\dfrac{4z}{Gc_{pf}D} + \dfrac{1}{h_{fo}}\right]^{-1}$

Surface heat flux, ϕ

Non-boiling
(Region A)

$(\Delta T_{SUB})_i$ Inlet subcooling

Fig. 5.2.

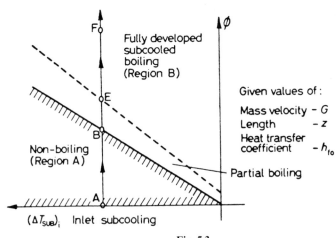

F

Fully developed
subcooled
boiling
(Region B)

ϕ

E

B

Given values of :

Mass velocity – G
Length   – z
Heat transfer
coefficient   – h_{fo}

Partial boiling

Non-boiling
(Region A)

A

$(\Delta T_{sub})_i$ Inlet subcooling

Fig. 5.3.

in the mass velocity, G, the diameter, D, or the liquid phase heat transfer coefficient, h_{fo}, will yield an increased slope. Usually in any problem the surface heat flux and inlet subcooling are given together with the desired flow, and it is simple to calculate at what position along the tube the surface temperature first exceeds the saturation value.

It should be realized that fully developed subcooled boiling does not exist immediately where the surface exceeds the saturation temperature. A region of 'partial boiling' exists between the subsaturation zone and fully developed subcooled boiling (Fig. 5.3). This 'partial boiling' zone consists of comparatively few nucleation sites and, in this zone, a proportion of the

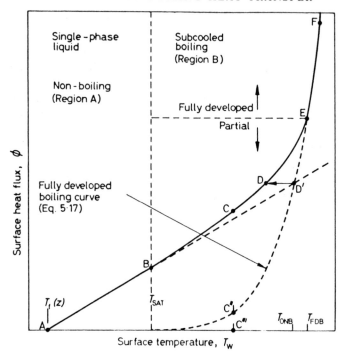

Fig. 5.4. Regions of heat transfer in convective boiling.

heat would be transferred by normal single-phase processes between patches
of bubbles.

Consider the variation of the temperature of the tube inner surface at a
point, z, from the inlet as the heat flux is steadily increased at a given inlet
subcooling and mass velocity. Figure 5.4 shows the relationship in a
qualitative form. The three regions delineated in Fig. 5.3 are shown as the
single-phase (subsaturation) region AB, the 'partial boiling' region BCDE
and the fully developed subcooled boiling region EF. It will be shown later
that the heat flux/surface temperature relationship in the fully developed
region is, to a first approximation, independent of velocity and subcooling
and may be correlated by the equations of the type obtained from pool
boiling experiments. Many such correlations exist but in order to retain
simplicity for the present the following equation is written in the form of a
power law relationship,

$$(T_W)_{SCB} = T_{SAT} + \psi \phi^n \tag{5.17}$$

ψ is a parameter usually containing physical property values and, in many
cases, also containing a further function characterizing the particular heating

surface/fluid combination, n is an exponent having a value in different correlations ranging from 0.25 to 0.5.

As the heat flux is increased at constant subcooling the relationship between the surface temperature, T_W, and heat flux will follow the line ABD′ until the first bubbles nucleate. This line is, of course, given by eq. (5.4). A higher degree of superheat is necessary to initiate the first bubble nucleation sites at a given heat flux than indicated by the curve ABCDE. When nucleation first occurs the surface temperature drops (Bertoletti et al. 1964) from D′ to D and for further increases in heat flux, follows the line DEF.

The position of D′ could be either before or after the intersection of ABD′ and C″EF; Bowring (1962), however, suggested that the condition for nucleation and thus the onset of subcooled boiling is

$$(T_W)_{SPL} = (T_W)_{SCB}$$

or

$$T_f(z) + \left[\frac{\phi}{h_{fo}} \right] = T_{SAT} + \psi \phi^n \tag{5.18}$$

and thus Bowring's expression for the local subcooling for onset of nucleate boiling is

$$\Delta T_{SUB}(z_{NB}) = \left[\frac{\phi}{h_{fo}} - \psi \phi^n \right] \tag{5.19}$$

Equation (5.19) is based on an arbitrary assumption and, as has been pointed out earlier, nucleation is a statistical process, depending upon the surface characteristics, so the relationship is only approximate.

A more refined treatment of the onset of nucleation can be derived by considering the temperature profile in the region adjacent to the heated wall. This treatment was originally proposed by Hsu for pool boiling (see Chapter 4) and has been used to predict both the onset of subcooled boiling and the suppression of saturated nucleate boiling. If the liquid is at a uniform temperature T_g, then bubble nuclei of radius r_c will grow if this temperature exceeds that given by

$$T_g = \frac{R T_{SAT} T_g}{J i_{fg} M} \ln(1 + \zeta) + T_{SAT} \tag{5.20}$$

where

$$\zeta = \left(\frac{2\sigma}{p_f r_c} \right)$$

In a heated system, there is a temperature gradient away from the wall which is approximated in the linear form:

$$T_f(y) = T_W - \left(\frac{\phi y}{k_f}\right) \tag{5.21}$$

The postulate of Hsu is that the bubble nuclei on cavities in the heated wall will only grow if the lowest temperature on the bubble surface (i.e., that furthest away from the wall) is greater than T_g. Allowing for the distortion of the liquid temperature profile by the bubble by plotting T_g against nr_c and T_f against y on the same ordinates (see Fig. 4.12) it will be seen that only that range of bubbles for which $T_f > T_g$ (i.e., radii r_{MIN} to r_{MAX}) can grow. If, over the whole field $T_f < T_g$, then no nuclei will grow. When eq. (5.21) is just tangent to eq. (5.20) then nuclei of a critical radius, r_{CRIT} will grow. Thus, for nucleation of the critical nucleus ($n = 1$)

$$T_f(y) = T_g \quad \text{and} \quad \frac{dT_f(y)}{dy} = \frac{dT_g}{dr} \tag{5.22}$$

Bergles and Rohsenow (1963) obtained a graphical solution of eqs. (5.20) and (5.21) for the conditions set by eq. (5.22).

Their equation, *which is valid for water only* over the pressure range 15–2000 psia (1–138 bar) is

$$\phi_{ONB} = 15.60 p^{1.156} (\Delta T_{SAT})_{ONB}^{(2.30/p^{0.0234})} \tag{5.23a}$$

where ϕ_{ONB} (Btu/h ft^2) is the heat flux to cause nucleation at a wall superheat $(\Delta T_{SAT})_{ONB}$ ($^\circ$F) and at a system pressure p (psia). Alternatively, in SI units

$$(\Delta T_{SAT})_{ONB} = 0.556 \left[\frac{\phi_{ONB}}{1082 p^{1.156}}\right]^{0.463 p^{0.0234}} \tag{5.23b}$$

where $(\Delta T_{SAT})_{ONB}$ is in $^\circ$C, ϕ_{ONB} is in W/m^2 and p in bar. Analytical solution of the equations was carried out by Davis and Anderson (1966), who proceeded as follows:

$$\frac{dT_f(y)}{dy} = -\frac{\phi}{k_f} \tag{5.24}$$

$$\frac{dT_g}{dr} = -\frac{2RT_{SAT}^2 \sigma}{JMi_{fg} p_f r^2 (1 + \zeta)} \left[1 - \frac{RT_{SAT}}{JMi_{fg}} \ln(1 + \zeta)\right]^{-2} \tag{5.25}$$

Equation (5.25) can be simplified provided the wall superheat is not too high if it is assumed that

$$1 \gg \left[\frac{RT_{SAT}}{JMi_{fg}} \ln(1 + \zeta)\right]$$

and hence

$$\frac{dT_g}{dr} = -\frac{B}{r^2}\frac{1}{(1+\zeta)} \tag{5.26}$$

where

$$B = \left[\frac{2\sigma T_{SAT} v_{fg}}{Ji_{fg}}\right] \tag{5.27}$$

Equating eqs. (5.24) and (5.26) and remembering that, at the point of tangency $(y = r)$, since n is taken as unity,

$$r_{CRIT} = -\frac{\sigma}{p_f} + \left[\left(\frac{\sigma}{p_f}\right)^2 + \left(\frac{Bk_f}{\phi}\right)\right]^{1/2} \tag{5.28}$$

Substitution of r_{CRIT} for r and eq. (5.21) for $T_f(y)$ $(=T_g)$ in eq. (5.20) yields the wall superheat required to initiate nucleate boiling

$$(T_W - T_{SAT})_{ONB} = \frac{[(RT_{SAT}^2)/JMi_{fg}]\ln(1+\zeta_{CRIT})}{[1-(RT_{SAT})\ln(1+\zeta_{CRIT})/JMi_{fg}]} + \frac{\phi_{ONB}r_{CRIT}}{k_f} \tag{5.29}$$

where

$$\zeta_{CRIT} = \left(\frac{2\sigma}{p_f r_{CRIT}}\right)$$

If $\zeta \ll 1$ then eq. (5.26) approximates to

$$\frac{dT_g}{dr} = -\frac{B}{r^2} \tag{5.30}$$

Equation (5.28) then becomes

$$r_{CRIT} = \sqrt{\frac{Bk_f}{\phi}} \tag{5.31}$$

and eq. (5.29) becomes

$$(T_W - T_{SAT})_{ONB} = \frac{B}{r_{CRIT}} + \frac{\phi_{ONB}r_{CRIT}}{k_f} \tag{5.32}$$

Substituting eq. (5.31) into eq. (5.32)

$$\phi_{ONB} = \frac{k_f}{4B}(T_W - T_{SAT})_{ONB}^2 = \frac{k_f}{4B}(\Delta T_{SAT})_{ONB}^2 \tag{5.33a}$$

Alternatively eq. (5.33a) may be expressed as

$$(T_W - T_{SAT})_{ONB} = (\Delta T_{SAT})_{ONB} = \left[\frac{8\sigma\phi_{ONB}T_{SAT}}{Ji_{fg}k_f\rho_g}\right]^{0.5} \tag{5.33b}$$

Equations (5.23) and (5.33) are in good agreement with each other and adequately predict the onset of nucleation in the experiments carried out by Bergles and Rohsenow (1963) with water flowing at velocities up to 17.5 m/s at low pressures and temperatures and also in the experiments of Rohsenow and Clark (1951) at high pressure.

Frost and Dzakowic (1967) have extended this treatment to cover other liquids. In their study nucleation was assumed to occur when the liquid temperature $T_f(y)$ was matched to the temperature for bubble equilibrium T_g at a distance nr_c where $n = [c_p \mu / k]_f^2$ rather than r_c as assumed in earlier works. No justification for this assumption was presented. Thus eqs. (5.33a) and (5.33b) become respectively,

$$\phi_{ONB} = \frac{k_f}{4B} \left(\frac{(\Delta T_{SAT})_{ONB}}{Pr_f} \right)^2 \qquad (5.34a)$$

$$(T_W - T_{SAT})_{ONB} = (\Delta T_{SAT})_{ONB} = \left[\frac{8\sigma \phi_{ONB} T_{SAT}}{J i_{fg} k_f \rho_g} \right]^{0.5} Pr_f \qquad (5.34b)$$

This equation was compared with experimental data for the onset of boiling for a variety of different fluids. Rearranging (5.34) a parameter X was defined thus:

$$X = \left[\frac{(\Delta T_{SAT})_{ONB}}{\phi_{ONB}^{0.5} Pr_f} \right] = \left[\frac{4B}{k_f} \right]^{0.5} = \left[\frac{8\sigma T_{SAT}}{J i_{fg} k_f \rho_g} \right]^{0.5} \qquad (5.35)$$

Figure 5.5 shows values of (X_{EXP}/X_{REF}) plotted against reduced pressure. X_{EXP} was calculated from reported values of $(\Delta T_{SAT})_{ONB}$; X_{REF} was evaluated from fluid physical properties at a constant reduced pressure of 0.05. Values of X_{REF} for the various fluids are given in Table 5.1.

Equation (5.35) represents the lower bound of the experimental results shown in Fig. 5.5; i.e., boiling in most instances was not initiated until higher values of surface temperature than estimated from eq. (5.35). This is due mainly to the fact that this treatment assumes a complete range of active cavity sizes which most surfaces do not have.

Davis and Anderson (1966) point out that the above treatments can only predict the onset of nucleate boiling accurately if there is a sufficiently wide range of 'active' cavity sizes available. At low fluxes and low pressures the values of r_{CRIT} predicted from eqs. (5.28) and (5.31) may be so large that no 'active' sites of this size are present on the heating surface. In this case an estimate of the largest 'active' cavity size available on the heating surface must be made (see Chapter 4) and substituted into eq. (5.29) or eq. (5.32). Davis and Anderson (1966) found reasonable agreement with experimental data for water and benzene when a maximum 'active' cavity size of 1 μm radius was used.

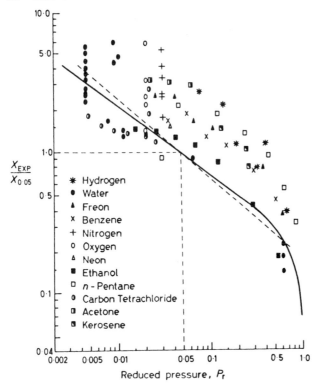

Fig. 5.5. Experimental data for the onset of boiling compared with eq. (5.35) (Frost and Dzakowic 1967).

For the subcooled boiling region eq. (5.33) must be solved simultaneously with the heat transfer equation:

$$\phi = h_{fo}(\Delta T_{SAT} + \Delta T_{SUB}(z)) \;=\; h_{fo}\left(T_w - T_f\right) \tag{5.36}$$

to give the heat flux ϕ_{ONB} and $(\Delta T_{SAT})_{ONB}$ required for the onset of boiling. The boundary between region A and region B shown on Fig. 4.23 can be derived from such a simultaneous solution and it is noted that at the condition $\Delta T_{SUB}(z) = 0$

$$(\phi_{ONB})_{\Delta T(z)=0} = \frac{4h_{fo}^2 B}{k_f} \tag{5.37}$$

For heat fluxes below that given by eq. (5.37) the bulk liquid must be superheated before vapour generation can take place at the surface, i.e., there is no subcooled boiling region. The conditions present in the tube under these circumstances are shown diagrammatically in Fig. 5.6. Once a vapour

Table 5.1 Reference values of $X[=(4B/k_f)^{0.5}]$ evaluated at a reduced pressure of 0.05

Fluid	X_{REF} $K/(W/m^2)^{0.5}$	X_{REF} $°R/(Btu/h\ ft^2)^{0.5}$
Ammonia	0.00432	0.0138
Carbon dioxide	0.01186	0.0379
Carbon tetrachloride	0.01364	0.0436
Mercury	0.00397	0.0127
Neon	0.00241	0.0077
n-Pentane	0.01214	0.0388
Para-hydrogen	0.00338	0.0108
Nitrogen	0.00519	0.0166
Propane	0.00960	0.0307
Oxygen	0.00507	0.0162
Benzene	0.01339	0.0428
Water	0.00441	0.0141
Freon 12	0.01417	0.0453
Ethanol	0.00891	0.0285
Acetone	0.01120	0.0358
Kerosene (JP-4)	0.01267	0.0405
Helium-4	0.00231	0.0074
Argon	0.00601	0.0192

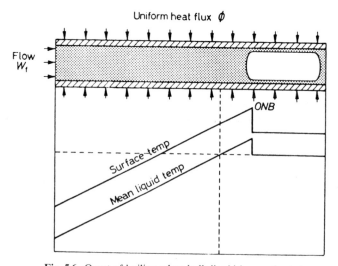

Fig. 5.6. Onset of boiling when bulk liquid is superheated.

bubble forms in the superheated liquid its growth rate is very rapid and it quickly fills the tube cross-section. This explosive formation of vapour is often a source of instability since it is accompanied by a sharp local increase in static pressure which may reduce, stop, or even reverse the flow in the upstream section of the tube. A liquid film initially remains between the vapour bubble and the heated wall. If the conditions are severe this film may be completely evaporated and a form of 'dryout' with consequent overheating of the tube wall will occur (Collier 1968). Superheating is particularly common with the liquid alkali metals such as sodium and potassium and also occurs in glass apparatus (Hewitt et al. 1965) where the range of active cavity sizes is severely restricted. For well wetting fluids (those with small contact angles), Robertson and Clarke (1981) have observed superheating for liquid nitrogen and Celata et al. (1992) obtained similar effects with refrigerants R-12 and R-114 at low pressures.

The nucleation model discussed in the preceding paragraphs has been applied by Fiore and Ricque (1964) to the prediction of the onset of nucleation under forced convection conditions of high boiling point multi-component organic liquids. The analysis was extended to cover the condition of degassing where the liquid contains appreciable amounts of dissolved gases. In this case the relationship (5.20) is modified to allow for the partial pressure of the gas in the bubble nucleus. The term ζ is then given by

$$\zeta = \left(\frac{2\sigma}{p_f r} - \frac{p_a}{p_f} \right) \tag{5.38}$$

where p_a is the partial pressure of gas in the bubble nucleus in equilibrium with the liquid at a temperature T_g.

The nucleation models discussed so far ignore the convective effect of the local flow pattern around the nucleus on the bubble surface temperature. Kenning and Cooper (1965) have reported an investigation in which the flow patterns near vapour bubble nuclei in forced convection boiling heat transfer were modelled on a large scale by the flow past an air bubble. The experimental observations were used to deduce the surface temperature of the bubble nuclei when the local heat transfer was predominantly by convection. The bubble surface temperature was taken to correspond to that of the dividing streamline. The position of the dividing streamline was determined experimentally and expressed as a function of the bubble Reynolds number (Fig. 5.7). The resulting criterion for the inception of boiling was given by the equation

$$(\Delta T_{SAT})_{ONB} \left[\frac{k_f u^* \rho_f}{\phi \mu_f} \right] = 1.33 \left[\frac{B k_f u^{*2} \rho_f^2}{\phi \mu_f^2} \right]^{0.6} \tag{5.39}$$

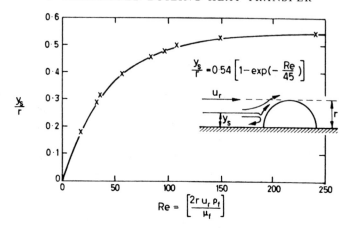

Fig. 5.7. Position of dividing streamline around a bubble nucleus (Kenning and Cooper 1965).

over the range

$$0.6 < \left[\frac{B k_f u^{*2} \rho_f^2}{\phi \mu_f^2} \right] < 20$$

where u^* is the 'friction' velocity ($=\sqrt{(\tau_w/\rho_f)}$) and B is given by eq. (5.27). This expression is more complex than eq. (5.33) and on the present evidence there is no reason to recommend the use of eq. (5.39) in preference to eq. (5.33):

A photographic study of incipient subcooled boiling for forced convection flow of pressurized water past a heated plate has been presented by Jiji and Clark (1962). This work suggests that a considerably higher surface heat flux (two and one-half times) is required to permit visual detection of bubbles than to cause an improvement in heat transfer coefficient at the same point.

The subcooled boiling process for the refrigerants R-123 and R-11 has been observed by Kattan et al. (1992) using video tapes and still photographs for flow in a horizontal annulus with an outer glass tube and a heated inner copper tube. Figure 1.9(a) shows the active boiling sites are situated nearly all on the top half of the tube at a low flow rate (liquid Reynolds number about 3000) because of thermal stratification in the *mixed* convection flow regime, i.e. combined effects of forced and natural convection. Further along the tube, active boiling sites occurred all around the tube as shown in Fig. 1.9(b).

5.4   Partial subcooled boiling

In the 'partial boiling' region, once boiling has been initiated, only comparatively few nucleation sites are operating so that a proportion of the heat will
be transferred by normal single-phase processes between patches of bubbles.
As the surface temperature increases, so the number of bubble sites also
increases and thus the area for single-phase heat transfer decreases.

As the surface temperature is increased further, the whole surface is covered
by bubble sites, boiling is 'fully developed' and the single-phase component
reduces to zero. In the 'fully developed' boiling region, velocity and
subcooling, both of which have a strong influence on single-phase heat
transfer, have little or no effect on the surface temperature as observed
experimentally. Throughout the partial boiling region Bowring (1962)
suggests that

$$\phi = \phi_{SPL} + \phi_{SCB} \tag{5.40}$$

where ϕ is the total average surface heat flux, ϕ_{SPL} is the average surface
heat flux transferred by single-phase convection, and ϕ_{SCB} is the average
surface heat flux transferred by bubble nucleation.

Consider first the case of constant inlet subcooling and velocity, but
decreasing heat flux, so that the line FEDCB in Fig. 5.4 is followed. In
saturated pool boiling, with zero subcooling and velocity, i.e., a very small
single-phase heat transfer component ($\phi_{SPL} \rightarrow 0$), the line FED'C'' would be
followed. The difference between the two curves therefore represents the
single-phase component. For example, at a surface temperature corresponding to C'''

$$\phi_{SCB} = C'''C''$$

$$\phi_{SPL} = C''C$$

If the two heat flux components, ϕ_{SCB} and ϕ_{SPL} from Fig. 5.4 are plotted
separately the curves shown in Fig. 5.8 are obtained. The shape of the ϕ_{SPL}
curve is not known quantitatively, hence, it is necessary to approximate its
form.

Three different empirical solutions will now be discussed. In the first
method, that elaborated by Bowring (1962), the single-phase component of
the total heat flux, ϕ_{SPL} is given by curve (i) of Fig. 5.8, thus for

$$\left. \begin{array}{c} T_{SAT} \leqslant T_W \leqslant T_{FDB} \\ \phi_{SPL} = h_{fo}(T_{SAT} - T_f(z)) \end{array} \right\} \tag{5.41}$$

and for

$$T_w > T_{FDB}, \phi_{SPL} = 0 \tag{5.42}$$

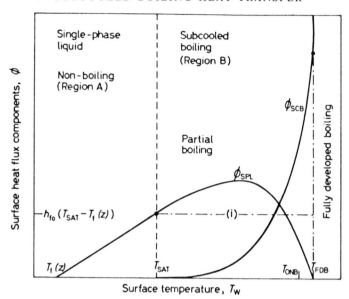

Fig. 5.8. Bowring model.

To summarize this method then, for the various regions shown in Fig. 5.1:

(a) Where $\Delta T_{\text{SUB}}(z)$ is greater than the value for the onset of subcooled boiling (given by eq. (5.19)). In this region heat transfer is entirely by single-phase convection even though T_{W} may be greater than T_{SAT}.

(b) Where $\Delta T_{\text{SUB}}(z)$ is less than the value for the onset of subcooled boiling but greater than the value for fully developed subcooled boiling, $\Delta T_{\text{SUB}}(z)_{\text{FDB}}$. In this region the single-phase component of the heat flux is given by eq. (5.41).

(c) Where $\Delta T_{\text{SUB}}(z)$ is less than the value for fully developed subcooled boiling, $\Delta T_{\text{SUB}}(z)_{\text{FDB}}$. In this case the single-phase component of the heat flux is zero.

It is necessary to define $\Delta T_{\text{SUB}}(z)_{\text{FDB}}$, the value of the local subcooling at which fully developed subcooled boiling can be said to start.

Bowring (1962) notes that Engelberg-Forster and Grief (1959), after examining experimental data covering a wide range of conditions and fluids, correlated the heat flux at which boiling becomes fully developed (at point E, Fig. 5.4), by the relation

$$\phi_{\text{E}} = 1.4\phi_{\text{D}} \tag{5.43}$$

where ϕ_{D} corresponds to the point of intersection of the line ABD′ single-phase line with the curve FEC″. Hence, from Fig. 5.4 and remembering

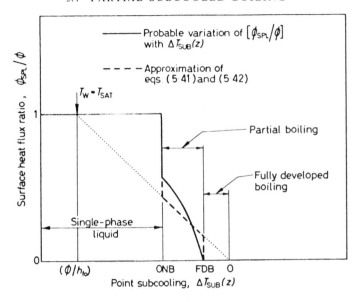

Fig. 5.9. Bowring model.

that Bowring (1962) assumed that this intersection corresponded to the onset of boiling $(T_W)_{ONB}$,

$$(T_W)_{ONB} - T_f(z) = \frac{\phi_E}{1.4 h_{fo}} \tag{5.44}$$

and also from eq. (5.17)

$$(T_W)_{ONB} - T_{SAT} = \psi \left[\frac{\phi_E}{1.4} \right]^n \tag{5.45}$$

Now at the point of onset of fully developed boiling $\phi = \phi_E$ and $T_{SAT} - T_f(z) = \Delta T_{SUB}(z)_{FDB}$, thus, on subtracting eq. (5.45) from eq. (5.44)

$$\Delta T_{SUB}(z)_{FDB} = \left[\frac{\phi}{1.4 h_{fo}} \right] - \psi \left[\frac{\phi}{1.4} \right]^n \tag{5.46}$$

Figure 5.9 shows the variation of the single-phase heat flux component ϕ_{SPL} as the local subcooling is reduced. The Bowring procedure is illustrated by an example at the end of the chapter.

The second empirical solution is that presented by Rohsenow (1953). This method again involves the superposition of a single-phase forced convection component and a subcooled boiling component (eq. (5.40)). However, in this method the single-phase component ϕ_{SPL} of the total heat flux is given by

$$\phi_{SPL} = h_{fo}(T_W - T_f(z)) \tag{5.47}$$

This differs considerably from that suggested by Bowring (eq. (5.41)). Experimental data examined by Rohsenow were reduced by subtracting the component ϕ_{SPL} from the total heat flux ϕ imposed during the experiment. The residual term $(\phi - \phi_{SPL})$ was assumed to be the subcooled boiling component, ϕ_{SCB} and was successfully correlated using the equation suggested earlier by Rohsenow (1952) for the saturated nucleate pool boiling of various liquids (see Section 4.3)

$$\left(\frac{c_{pf}\,\Delta T_{SAT}}{i_{fg}}\right) = C_{sf}\left[\frac{\phi_{SCB}}{\mu_f i_{fg}}\sqrt{\frac{\sigma}{g(\rho_f - \rho_g)}}\right]^{0.33}\left(\frac{c_{pf}\mu_f}{k_f}\right)^{1.7} \tag{4.45}$$

where C_{sf} is a constant described by the liquid–surface combination. The data of Rohsenow and Clark (1951) for subcooled boiling of water at 138 bar (2000 psia) in a 4.56 mm i.d. tube were correlated with a value of $C_{sf} = 0.006$. The value of h_{fo} used in eq. (5.47) for the reduction of these data is given by the usual Dittus–Boelter equation (eq. (5.7)) where the coefficient is ordinarily suggested to be 0.023. However, for non-boiling heat transfer data at 138 bar (2000 psia) Rohsenow and Clark (1951) found a better correlation with a coefficient of 0.019. Rohsenow similarly correlated the data of Krieth and Summerfield (1949) and Piret and Isbin (1953). Table 5.2 gives a summary of the values of C_{sf} obtained using this superposition method of correlation.

Table 5.2   Values of C_{sf} in eq. (4.45) obtained in the reduction of the forced (and natural) convection subcooled boiling data of various investigators

Reference	Heating surface	C_{sf}	Fluid–surface
Rohsenow & Clark (1951)	Vertical tube (4.56 mm i.d.)	0.006	Water–nickel
Krieth & Summerfield (1949)	Horizontal tube (14.9 mm i.d.)	0.015	Water–stainless steel
Piret & Isbin (1953)	Vertical tube (27.1 mm i.d.)	0.013	Water–copper
		0.013	Carbon tetrachloride–copper
		0.0022	Isopropyl alcohol–copper
		0.0030	n-Butyl alcohol–copper
		0.00275	50% K_2CO_3–copper
		0.0054	35% K_2CO_3–copper
Bergles & Rohsenow (1963)	Horizontal tube (2.39 mm i.d.)	0.020	Water–stainless steel

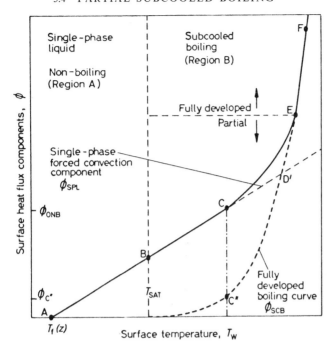

Fig. 5.10. Method of Bergles and Rohsenow (1963).

The third empirical solution is that presented by Bergles and Rohsenow (1963). The suggested procedure is summarized by reference to Fig. 5.10. The fully developed subcooled boiling curve, ϕ_{SCB}, FED′C″ is established from experimental data at high values of ΔT_{SAT} or constructed using an empirical equation of the form of eq. (5.17). The single phase forced convection curve ABCD′ is constructed from eq. (5.47).

The flux ϕ_{ONB} for the onset of nucleation (point C on Fig. 5.10) can be calculated using the Bergles and Rohsenow method (empirically expressed by eq. (5.23) for water or eq. (5.33) for other liquids).

The fixing of the point C enables a point C″ to be found on the extrapolated fully developed subcooled boiling curve at the same surface temperature as the point C (Fig. 5.10). A simple interpolation formula was given (Bergles and Rohsenow 1963) which satisfies the characteristics of the boiling curve between C and E—thus,

$$\phi = \phi_{SPL}\left[1 + \left\{\frac{\phi_{SCB}}{\phi_{SPL}}\left(1 - \frac{\phi_{c''}}{\phi_{SCB}}\right)\right\}^2\right]^{1/2} \tag{5.48}$$

Equation (5.48) is then applied by picking off values of ϕ_{SCB} and ϕ_{SPL} at various values of T_W to fill in the section CE.

This latter method is recommended for rapid calculations of the complete forced convective subcooled boiling curve when a limited number of data are available. The method of Bowring is, however, recommended for detailed calculations, particularly if other properties of the subcooled boiling system are required, such as void fraction and pressure drop.

A further method based on that suggested by Chen for saturated forced convection boiling heat transfer will be discussed in Chapter 7.

5.5   Fully developed subcooled boiling

Subcooled boiling is characterized by the formation of vapour at the heating surface in the form of vapour bubbles at preferred nucleation points. These bubbles condense in the cold liquid away from the heating surface with the result that no net vapour is produced. At high heat fluxes the onset of subcooled boiling is encountered at high degrees of subcooling, and the vapour bubbles may grow and collapse whilst still attached to, or even sliding along, the heating surface. These processes are sometimes accompanied by noise and vibration of the heating surface.

Probably the first investigators of subcooled boiling were Mosciki and Broder (1926) who studied the heat transfer from an electrically heated vertical platinum wire, submerged in water at atmospheric pressure. They found that the wire temperature at the highest obtainable heat flux was essentially independent of the water temperature. Local heat transfer coefficients considerably greater than the values predicted for non-boiling conditions were found by both McAdams et al. (1941) (preheating sections of a horizontal tube evaporator), and Davidson et al. (1943) (preheating parts of his experimental boiler tubes). Both sets of workers suggested the occurrence of subcooled boiling as the cause of the discrepancy.

Knowles (1948) studied subcooled boiling of river water on an electrically heated stainless steel tube centrally located in an outer glass tube. His experiments, although plagued by deposition of scale on the heating element, gave the first quantitative data on the subcooled boiling curve and showed that high heat fluxes (up to 7.25 MW/m^2 (2.3×10^6 Btu/h ft^2)) could be transferred by the subcooled boiling mechanism under conditions of forced convection. Experiments with similar geometry (but purer water) are reported by McAdams et al. (1946).

Typical of the results obtained by McAdams et al. (1946) is Fig. 5.11. The various aspects of the forced convective subcooled boiling curve discussed earlier are clearly visible. An alternative method of plotting the same data is to use logarithmic co-ordinates as illustrated in Figs. 5.12 and 5.13 where the combined effects of inlet subcooling and velocity are shown. The data obtained by McAdams for fully developed subcooled boiling of *water* were

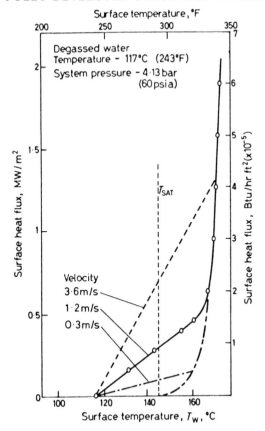

Fig. 5.11. Typical curve for subcooled boiling (McAdams 1946).

represented by eq. (5.17) with $n = 0.259$. With the heat flux ϕ expressed in MW/m^2, ψ varied from 22.62 to 28.92 as the concentration of dissolved gas decreased from 300 to 60 cm^3 of air at standard conditions per m^3 of water. Equation (5.17) represented the data satisfactorily over the following range of variables—velocity 0.3 to 11 m/s, subcooling 11° to 83°C, pressure 2 to 6 bar, and equivalent diameter 4.3 mm to 12.2 mm. The value of ψ quoted for the data of McAdams *et al.* (1946) is only valid for water and over the range quoted above. As mentioned earlier, ψ is a function of the physical properties of the fluid and of the liquid–surface combination.

Krieth and Summerfield (1949) studied subcooled boiling of water in both horizontal and vertical electrically heated tubes and also found that the surface temperature in subcooled boiling was constant and independent of the subcooling. Their experiments covered the following range of conditions—

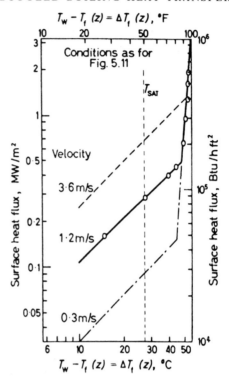

Fig. 5.12. Typical curve for subcooled boiling (McAdams 1946).

velocity 1.8 to 4 m/s, pressure 1.4 to 14 bar, tube diameters 13.6 and 14.9 mm, and heat flux from 1.14 to 4.575 MW/m². Over the pressure range studied it was observed that ΔT_{SAT} decreased with pressure at a given heat flux following approximately the relationship

$$\Delta T_{SAT} \propto p^{-0.75} \tag{5.49}$$

The data obtained from this study were later satisfactorily correlated using the superposition method of Rohsenow (1952). Krieth and Summerfield (1950) also reported a study of subcooled boiling for aniline and n-butyl alcohol in the same experiment. The data in the fully developed subcooled boiling region are similar in form to the results for water. Figure 5.14 illustrates data obtained by Rohsenow and Clark (1951) for water flowing in an electrically heated nickel tube 4.57 mm i.d. and 23.8 cm long and demonstrates very clearly that the governing temperature difference in fully developed subcooled boiling is ΔT_{SAT}, the wall 'superheat' rather than the overall temperature difference ($T_w - T_f(z)$). These data were also correlated using the superposition model of Rohsenow (1952, 1953).

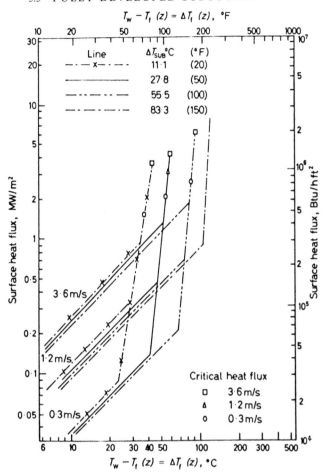

Fig. 5.13. Effect of velocity and subcooling on degassed distilled water, 4.13 bar (60 psia), $T_{SAT} = 145°C$ (293°F) (McAdams 1946).

Jens and Lottes (1951) summarized experiments on subcooled boiling of water flowing upwards in vertical electrically heated stainless steel or nickel tubes, having inside diameters ranging from 3.63 to 5.74 mm. System pressures ranged from 7 bar to 172 bar, water temperatures from 115° to 340°C, mass velocities from 11 to 1.05×10^4 kg/m² s, and heat fluxes up to 12.5 MW/m². These data were correlated by a dimensional equation *valid for water only*

$$\Delta T_{SAT} = 25\phi^{0.25} e^{-p/62} \tag{5.50}$$

where p is the absolute pressure in bar, ΔT_{SAT} is in °C, and ϕ is in MW/m².

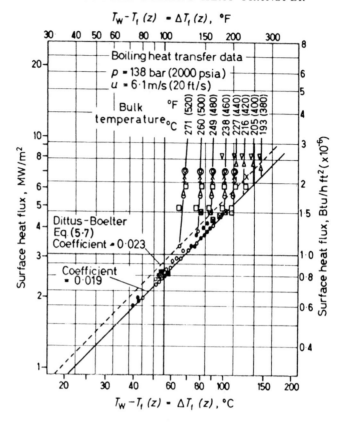

Fig. 5.14. Effect of subcooling on forced convection boiling data (Rohsenow and Clark 1951).

In British Engineering Units the equation becomes

$$\Delta T_{SAT} = 1.9\phi^{0.25} e^{-p/900} \tag{5.50a}$$

where p is the absolute pressure in psia, ΔT_{SAT} is in °F, and ϕ is in Btu/h ft². This equation holds for degassed water only and also correlates the data of McAdams *et al.* (1949).

Thom *et al.* (1965) reported that the values of ΔT_{SAT} estimated from eqs. (5.50) or (5.50a) were consistently low over the range of their experiments. A modified equation of the same form and valid only for water was suggested:

$$\Delta T_{SAT} = 22.65\phi^{0.5} e^{-p/87} \tag{5.51}$$

in SI Units. In British Engineering Units the equation becomes

$$\Delta T_{SAT} = 0.072\phi^{0.5} e^{-p/1260} \tag{5.51a}$$

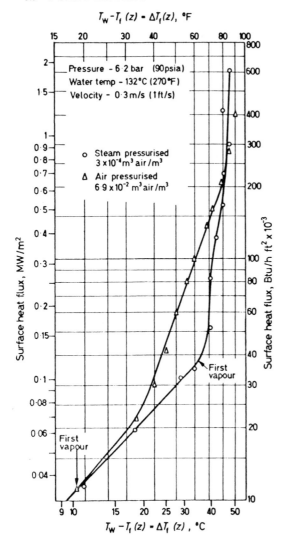

Fig. 5.15. Effect of dissolved air (McAdams 1949).

The presence of gas dissolved in the water increases the heat transfer in subcooled boiling. Gas bubbles can form at the surface and agitate the liquid in the same way that vapour bubbles do. McAdams *et al.* (1949) demonstrated the effect of dissolved gases in forced convective subcooled boiling (Fig. 5.15). It can be seen that the presence of gas has a considerable effect on the position where the first evolution of bubbles is seen and also on the curve for the boiling process at the lower heat fluxes. However, the effect

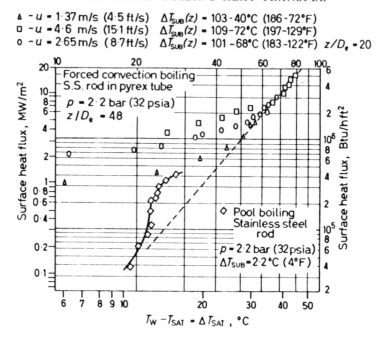

Fig. 5.16. Forced convection subcooled boiling data and pool boiling data for a stainless steel tube (Bergles and Rohsenow 1963).

dies out at the higher heat fluxes near the critical heat flux point. An exactly similar observation was made by Pike *et al.* (1955) for subcooled pool boiling on nickel wire in degassed and tap water.

It has generally been assumed that the fully developed region of the boiling curve in forced convection coincides with the extrapolation of the pool boiling curve. This suggestion would appear reasonable, at least, for low velocities. Experiments carried out by Bergles and Rohsenow (1963) were designed with the object of directly comparing data for pool boiling and forced convection boiling. Data for both boiling processes were taken for the same surface—this was accomplished by using an annular geometry for the forced convection boiling tests. Stainless steel tubing, 1.64 mm o.d. was used as the heater in the tests. The data obtained are shown in Fig. 5.16. The forced convection boiling data appear to merge into an asymptote which can be represented by a dotted line. It is apparent that forced convection boiling cannot be considered an extrapolation of pool boiling for these data, at least. However, it would appear that the *form* of the equations found suitable for the correlation of pool boiling data is also suitable for the representation of forced convective, fully developed, subcooled boiling data.

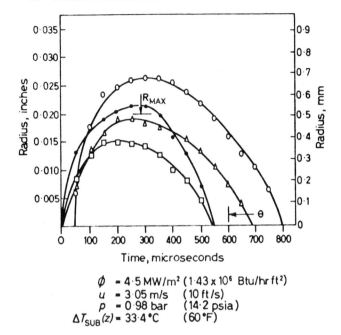

ϕ = 4·5 MW/m² (1·43 × 10⁶ Btu/hr ft²)
u = 3·05 m/s   (10 ft/s)
p = 0·98 bar   (14·2 psia)
$\Delta T_{SUB}(z)$ = 33·4 °C   (60 °F)

Fig. 5.17. Bubble radius vs time—data of Gunther (1951).

The values of the slope and intercept (n and ψ of eq. (5.17)) will, of course, need to be altered from the values obtained for pool boiling.

An early attempt to establish the mechanisms of heat transfer in subcooled forced convection boiling was the excellent photographic study carried out by Gunther (1951). The test section was a transparent channel of rectangular cross-section, with an electrically heated ribbon 3.17 mm wide suspended lengthwise to divide the channel into two flow passages. The fluid used was water at atmospheric pressure with a linear velocity between 1.5 and 12 m/s. The study showed that at higher subcoolings ($\sim 50°$C) the vapour bubbles grow and collapse whilst still attached to the heating surface. The attached bubbles, whilst growing and collapsing, simultaneously slide downstream on the heater surface with a velocity approximately 80 per cent of that of the free stream, increasing somewhat for larger bubbles.

The bubble growth and collapse process was photographed at a rate of 20,000 frames per second, using a Kerr cell shutter. Radius versus time curves were determined for selected bubbles. R_{MAX}, the radius at full growth of an 'average' bubble, was defined and evaluated as a function of the conditions. A lifetime, θ, of the 'average' bubble was also defined and evaluated. A typical set of curves for bubble radius versus time is shown in Fig. 5.17. The particular value of R_{MAX} and θ, designated for this set of conditions, is also

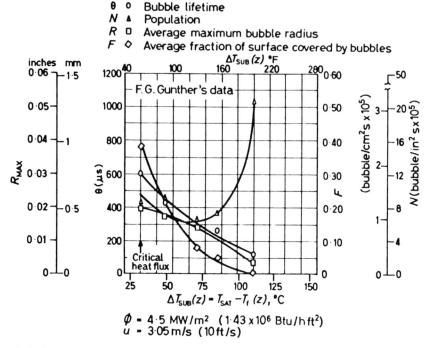

Fig. 5.18. Effect of subcooling on bubble population, lifetime, radius, and average fraction of surface covered by bubbles (constant heat flux).

shown. The number of new bubbles of 'average' size appearing per unit area per unit time (N) was also evaluated and finally the time average percentage (F) of the heating surface covered by bubbles calculated. The variation of the four bubble parameters R_{MAX}, θ, N and F with system subcooling is shown in Fig. 5.18 for a constant heat flux and velocity. Bubble size, lifetime, population, and coverage decrease as the subcooling is increased. However, at high subcoolings very large populations of minute bubbles were observed. All four parameters also decrease as the velocity is increased at constant heat flux and subcooling. Figure 5.19 shows the influence of heat flux at constant velocity and subcooling. Bubble size and lifetime decrease moderately while bubble population and coverage increase sharply as the heat flux is increased at constant velocity and subcooling.

The mechanisms of subcooled nucleate boiling have been the subject of much debate (see also Section 4.4.4). Gunther and Krieth (1950) using measurements from their photographic study (Gunther 1951) and also Clark and Rohsenow (1951), found that the rate of visible vapour evolution in subcooled forced convective boiling could only account for a very small fraction (1–2 per cent) of the total heat flow. Latent heat transport within the

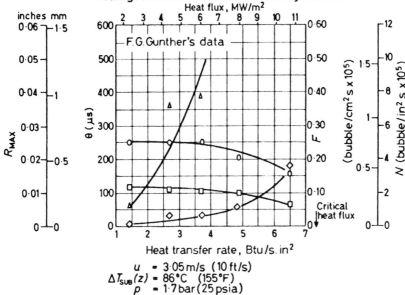

θ o = Bubble lifetime
N ▲ = Population
R □ = Average maximum bubble radius
F ◇ = Average fraction of surface covered by bubbles

Fig. 5.19. Effect of heat transfer rate of bubble population, lifetime, radius, and average fraction of surface covered by bubbles (constant subcooling).

bubble by simultaneous evaporation at the base and condensation at the roof of the bubble was also shown (Gunther and Krieth 1950) to be very small if it is assumed that the bubble is surrounded by a stagnant liquid layer. However, Bankoff and Mikesell (1959) point out that if turbulent and convective heat transport dominates in the flow of heat from the condensing bubble surfaces, the picture is considerably changed and latent heat transport might, indeed, be significant.

Ellion (1955) also reported a study of subcooled boiling using high-speed photographic methods. The influences of velocity, pressure, and temperature were studied and measurements were made of bubble radii and radial velocities in subcooled water and carbon tetrachloride.

In a study of the mechanism of subcooled nucleate boiling Bankoff (1960) proposed a so-called 'sequential rate process model'. The model postulates three steps for the transfer of heat. First, heat flows from the surface to the adjacent two-phase boundary layer. Secondly, the heat flows through the two-phase layer and, finally, passes into the main subcooled single-phase core. Simple expressions for the first and last stages are in reasonable

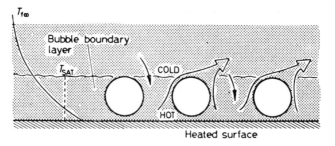

Fig. 5.20. Thermocapillarity mechanism of subcooled boiling (Brown 1967).

agreement with Gunther's data (1951), Bankoff (1960) was able to derive an approximate relation giving the temperature at the edge of the two-phase region which was applied to Gunther's data. The relationship predicted a steep rise in the temperature at the edge of the single-phase core as the heat flux increased, approaching the saturation temperature at the critical heat flux.

Measurements of the bubble boundary layer thickness and the temperature profiles associated with forced convective subcooled boiling of water in a rectangular channel, at pressures between 14 and 70 bar have been reported by Jiji and Clark (1964). The bubble boundary layer increased with distance from the start of subcooled boiling due to increases in bubble size. A correlation of boundary layer thickness in terms of the independent variables was given, both in dimensional and dimensionless forms. Temperature measurements showed that contrary to Bankoff's (1960) prediction the mean temperature at the interface between the single-phase core and the bubble boundary layer approached the inlet temperature as the heat flux is increased and is always well below the saturation temperature.

It has been noted by Brown (1967) that a vapour bubble in a temperature gradient is subjected to a variation of surface tension which tends to move the interfacial liquid film. This, in turn, drags with it adjacent warm liquid so as to produce a net flow around the bubble from the hot to the cold region which is released as a jet in the wake of the bubble (see Fig. 5.20). Brown suggests that this mechanism, known as thermo-capillarity, can transfer a considerable fraction of the heat flux. Moreover, it would appear to explain a number of observations about the bubble boundary layer made by various workers. Semeria (1964) has reported seeing jets of the type described. Other observations which can be explained include the fact that bubbles are attracted to heated walls, despite inertial effects appearing small, the temperature fluctuations in the bubble boundary layer (Jiji and Clark 1964) and the fact that the mean temperature in the boundary layer is lower than saturation (Jiji and Clark 1964). The observation that the temperature at the interface between the single-phase core and the bubble boundary layer *falls* with increased heat flux is attributed to the increased circulation caused by thermocapillarity.

Other experimental studies of subcooled forced convective boiling have been published by Hsu and Ing (1962), by Jicha and Frank (1962), and by Bernath and Begell (1959).

In recent studies Vandervort *et al.* (1992) have identified all possible mechanisms for subcooled boiling.

(1) Single phase liquid turbulent forced convection.
(2) Direct wall to vapour transport at the point of bubble attachment.
(3) Microlayer vaporization of liquid to vapour at the base of the bubble (Fig. 4.15c) followed by (4).
(4) Internal microconvection of vapour across the bubble followed by (5).
(5) Vapour condensation at the bubble tip and convection into the subcooled liquid core.
(6) Bubble-induced turbulent mixing in the liquid phase (Fig. 4.15a).
(7) Vapour–liquid exchange.
(8) Thermocapillarity or Marangoni-induced microconvection (Fig. 5.20).

The most dominant heat transfer processes in highly subcooled boiling are believed to be mechanisms (3) and (8). The mechanism by which the critical heat flux is reached in subcooled boiling is discussed in Chapter 9.

References and further reading

Bankoff, S. G. (1960). 'On the mechanism of subcooled nucleate boiling'. Parts I and II. *Heat Transfer, Buffalo, Chem. Engng. Prog. Symp. Series*, No. 32, **57**, 156–163, 164, 172.

Bankoff, S. G. and Mikesell, R. D. (1959). 'Bubble growth rates in highly subcooled nucleate boiling'. Preprint 5 presented at AIChE–ASME Heat Transfer Conference, 1958, *Chem. Engng. Prog. Symp. Series* No. 29, 95–109.

Bergles, A. E. and Rohsenow, W. M. (1963). 'The determination of forced convection surface boiling heat transfer'. Paper 63-HT-22 presented at 6th National Heat Transfer Conference of the ASME–AIChE, Boston, 11–14 August.

Bernath, L. and Begell, W. (1959). 'Forced convection local boiling heat transfer in narrow annuli'. *Heat Transfer—Chicago, Chem. Engng. Prog. Symp. Series* No. 29, **55**, 59–65.

Bertoletti, S., Lombardi, C., and Silvestri, M. (1964). 'Heat transfer to steam–water mixtures'. CISE—Report R 78, January.

Bowring, R. W. (1962). 'Physical model based on bubble detachment and calculation of steam voidage in the subcooled region of a heated channel'. OECD Halden Reactor Project Report HPR-10.

Brown, W. T. (1967). 'A study of flow surface boiling'. PhD Thesis, Mech. Engng. Dept., MIT, January.

Celata, G. P., Cumo, M., and Setaro, T. (1992). 'Hysteresis effect in the boiling incipience of well wetting fluids'. Paper presented at 10th National Heat Transfer Congress held in Genoa, Italy, UIT, 461–470, 25–27 June.

Clark, J. H. and Rohsenow, W. M. (1951). 'A study of the mechanism of boiling heat transfer'. *Trans ASME*, **73**, 609.

Collier, J. G. (1968). 'Boiling of liquid alkali metals'. *Chemical and Process Engng*, August, 167–173.

Davidson, W. F., Hardie, P. H., Humphreys, C. G. R., Markson, A. A., Mumford, A. R., and Raverse, T. (1943). 'Studies of heat transmission through boiler tubing at pressures from 500 to 3000 pounds'. *Trans. ASME*, **65**(6), 553–591.

Davis, E. J. and Anderson, G. H. (1966). 'The incipience of nucleate boiling in forced convection flow'. *AIChE Journal*, **12**(4), 774–780.

Ellion, M. E. (1955). 'Study of mechanism of boiling heat transfer'. Jet. Prop. Lab. Memo. 20–88, March.

Engelberg-Forster, K. and Grief, R. (1959). 'Heat transfer to a boiling liquid— mechanism and correlations'. *Trans. ASME J. of Heat Transfer*, Series C, **81**, 43–53.

Fiore, A. and Ricque, R. (1964). 'Propriétés thermiques des Polyphényles—détermination du début de l'ébullition nuclée du terphényl OMZ en convection forces'. EUR 2442 f.

Frost, W. and Dzakowic, G. S. (1967). 'An extension of the method of predicting incipient boiling on commercially finished surfaces'. Paper 67-HT-61 presented at ASME-AIChE Heat Transfer Conference, Seattle, August.

Gunther, F. C. (1951). 'Photographic study of surface-boiling heat transfer to water with forced convection'. *Trans. ASME*, 115–123, February.

Gunther, F. C. and Krieth, F. (1950). 'Photographic study of bubble formation in heat transfer to subcooled water'. Progress Report No. 4–120. J.P.L. California Institute of Technology, Pasadena, California.

Hewitt, G. F. *et al.* (1965). 'Burnout and film flow in evaporation of water in tubes'. AERE-R 4864.

Hsu, S. T. and Ing, P. W. (1962). 'Experiments in forced convection subcooled nucleate boiling heat transfer'. Paper 62-HT-38 presented at 5th National Heat Transfer Conference, Houston, 5–8 August.

Jens, W. H. and Lottes, P. A. (1951). 'Analysis of heat transfer burnout, pressure drop and density data for high pressure water'. ANL-4627, May.

Jicha, J. J. and Frank, S. (1962). 'An experimental local boiling heat transfer and pressure drop study of a round tube'. Paper 62-HT-48, presented at 5th National Heat Transfer Conference, Houston, 5–8 August.

Jiji, L. M. and Clark, J. A. (1962). 'Incipient boiling in forced convection channel flow'. Paper 62-WA-202 presented at the ASME Winter Annual Meeting, New York, 25–30 November.

Jiji, L. M. and Clark, J. A. (1964). 'Bubble boundary layer and temperature profiles for forced convection boiling in channel flow'. *J. Heat Transfer*, 50–58, February.

Kattan, N., Thome, J. R., and Favrat, D. (1992). 'Convective boiling and two-phase flow patterns in an annulus'. Paper presented at 10th National Heat Transfer Congress held in Genoa, Italy, UIT, 309–320, 25–27 June.

Kenning, D. B. R. and Cooper, M. G. (1965). 'Flow patterns near nuclei and the initiation of boiling during forced convection heat transfer'. Paper 11 presented at the Symposium on Boiling Heat Transfer in Steam Generating Units and Heat Exchangers held in Manchester, IMechE (London) 15–16 September.

Knowles, J. W. (1948). 'Heat transfer with surface boiling'. *Can. J. Research*, **26**, 268–278.

Krieth, F. and Summerfield, M. (1949). 'Heat transfer to water at high flux densities with and without surface boiling'. *Trans. ASME*, **71**(7), 805–815.

Krieth, F. and Summerfield, M. (1950). 'Pressure drop and convective heat transfer with surface boiling at high heat flux; data for aniline and *n*-butyl alcohol'. *Trans. ASME*, **72**, 869–879.

McAdams, W. H., Woods, W. K., and Bryan, R. L. (1941). 'Vaporization inside horizontal tubes'. *Trans. ASME*, **63**(6), 545–552.

McAdams, W. H., Kennel, W. E., Minden, C. S. L., Carl, R., Picornell, P. M., and Dew, J. E. (1949). 'Heat transfer at high rates to water with surface boiling'. *Ind. Engng. Chem.*, **41**(9), 1945–1953.

Mosciki, I. and Broder, J. (1926). 'Discussion of heat transfer from a platinum wire submerged in water'. *Recznicki Chemje*, **6**, 329–354.

Pike, F. P., Miller, P. D., and Beatty, K. O., Jr. (1955). 'The effect of gas evolution on surface boiling at wire coils'. *Heat Transfer—St. Louis, Chem. Engng. Prog. Symp. Series* No. 17, **51**, 13–19.

Piret, E. L. and Isbin, H. S. (1953). 'Two-phase heat transfer in natural circulation evaporators'. AIChE Heat Transfer Symposium, St. Louis, *Chem. Engng. Prog. Symp. Series*, No. 6, **50**, 305.

Robertson, J. M. and Clarke, R. H. (1981). 'The onset of boiling of liquid nitrogen in plate-fin heat exchangers'. *AIChE Symp. Ser.*, **77**(208), 86–95.

Rohsenow, W. M. (1952). 'A method of correlating heat transfer data for surface boiling of liquids'. *Trans. ASME*, **74**, 969.

Rohsenow, W. M. (1953). 'Heat transfer with evaporation'. *Heat Transfer—A Symposium held at the University of Michigan During the Summer of 1952.* University of Michigan Press, 101–150.

Rohsenow, W. M. and Clark, J. A. (1951). 'Heat transfer and pressure drop data for high heat flux densities to water at high subcritical pressure'. *1951 Heat Transfer and Fluid Mechanics Institute.* Stanford University Press, Stanford, California.

Semeria, R. (1964). 'Analyse fine de l'ébullition à l'échelle locale'. *C.R. Acad. Sc. Paris*, 471–476.

Thom, J. R. S., Walker, W. M., Fallon, T. A., and Reising, G. F. S. (1965). 'Boiling in subcooled water during flow up heated tubes or annuli'. Paper 6 presented at the Symposium on Boiling Heat Transfer in Steam Generation Units and Heat Exchangers held in Manchester, IMechE (London) 15–16 September.

Vandervort, C. L., Bergles, A. E., and Jensen, M. K. (1992). 'Heat transfer mechanisms in very high heat flux subcooled boiling'. Presented in *Fundamentals of Subcooled Flow Boiling*, HTD-Vol. 217, pp. 1–9. Winter Ann. Mtg ASME, Anaheim, CA.

Examples

Example 1

A vertical tubular test section is to be installed in an experimental high-pressure water loop. The tube is 10.16 mm i.d. and 3.66 m long, heated uniformly over its length. An estimate of the inside surface temperature distribution along the tube is required for a flow of water of 8 USGPM. The temperature of the water entering the test section is 203°C and the test section exit pressure is 68.9 bar. The total power applied to the tube is 200 kW.

Solution
The given conditions are:

Tube bore (D)	= 10.16 mm i.d.	Tube length (L) = 3.66 m	
Power applied	= 200 kW	Waterflow	= 8 USGPM
			= 0.432 kg/s
Pressure at exit = 68.9 bar		Inlet temp.	= 203°C

$$\text{Enthalpy rise} = \frac{200}{0.432} = 0.4625 \text{ MJ/kg}$$

(a) *Length of tube required to preheat water to the saturation temperature*

$$\text{Mass velocity } G = \frac{4W}{\pi D^2} = \frac{4 \times 0.432}{\pi \times 1.016^2 \times 10^{-4}} = 5340 \text{ kg/m}^2 \text{ s}$$

$$\text{Heat flux } \phi = \frac{200}{\pi \times 1.016 \times 10^{-2} \times 3.66} = 1720 \text{ kW/m}^2$$

Inlet enthalpy = 0.872 MJ/kg. Enthalpy of saturated liquid = 1.261 MJ/kg

$$z_{SC} = \frac{GD}{4\phi} ((i_f)_{SAT} - i_{fi})$$

$$= \frac{5340 \times 10.16 \times 10^{-3}}{4 \times 1720 \times 10^3} (1.261 - 0.872) \times 10^6 = \underline{3.06 \text{ m}}$$

(b) *Length of tube at which $T_W = T_{SAT}$*

$$z_{NB} = \frac{Gc_{pf}D}{4} \left[\frac{(\Delta T_{SUB})_i}{\phi} - \frac{1}{h_{fo}} \right]$$

$T_f(z)$	h_{fo}	ΔT_f	T_W
°C	kW/m² °C	°C	°C
203	44.0	39.1	242.1
260	47.8	36.0	296.0
271	48.4	35.5	306.5

$$(\Delta T_{SUB})_i = 80.5°C$$

$$z_{NB} = \frac{5340 \times 4.94 \times 10^3 \times 10.16 \times 10^{-3}}{4} \left[\frac{80.5}{1720 \times 10^3} - \frac{1}{47.8 \times 10^3} \right]$$

$$= 1.73 \text{ m}$$

(c) *The relationship between ΔT_{SAT} and ϕ*
Use the Jens and Lottes relationship (eq. (5.50)) (alternatively the Thom relationship (eq. (5.51)) could have been used).

$$\Delta T_{SAT} = 25\phi^{0.25} e^{-p/62}$$

$$= 25 \times \phi^{0.25} e^{-(68.9/62)} = 8.2 \phi^{0.25}$$

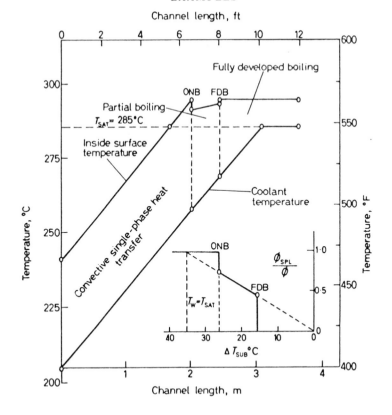

Fig. 5.21.

(d) *The value of* $\Delta T_{SUB}(z)_{ONB}$

$$\Delta T_{SUB}(z)_{ONB} = (\phi/h_{fo}) - \psi\phi^n \qquad (5.19)$$

$$= \frac{1720}{47.8} - 8.2(1.72)^{0.25} = 36.0 - 9.4 = \underline{26.6^\circ C}$$

(e) *Construction of diagram of* ΔT_{SUB} *versus* $[\phi_{SPL}/\phi]$ Fig. 5.21
From the diagram ϕ_{SPL} (at $\Delta T_{SUB}(z) = 26.6^\circ C$) $= 0.74\phi$

$$\phi_{SCB} = 0.26\phi = 0.26 \times 1720 = \underline{0.446 \text{ MW/m}^2}$$

From the Jens and Lottes relationship

$$\Delta T_{SAT} = 8.2(0.446)^{0.25} = 6.7^\circ C$$

(f) *Evaluation of* $\Delta T_{SUB}(z)_{FDB}$

$$\Delta T_{SUB}(z)_{FDB} = \left[\frac{\phi}{1.4h_{fo}}\right] - \psi\left[\frac{\phi}{1.4}\right]^n \qquad (5.46)$$

$$= 25.7 - 8.65 = \underline{17.05^\circ C}$$

From the diagram ϕ_{SPL} (at $\Delta T_{SUB}(z) = 17.05°C) = 0.46\phi$

$$\phi_{SCB} = 0.54\phi = 0.54 \times 1720 = 0.93 \text{ MW/m}^2$$

From the Jens and Lottes relationship

$$\Delta T_{SAT} = 8.2(0.93)^{1/4} = \underline{8.05°C}$$

The complete temperature distribution is shown in Fig. 5.21.

Example 2
Liquid nitrogen flows through a 0.5 in i.d. pipe at a rate of 1 lb/s. The pressure at the exit of the heated pipe is 60 psia and the fluid temperature is 140°R. Calculate the tube wall temperature and the heat flux required to initiate boiling.

Solution
This example uses British Engineering Units. The following equations must be solved simultaneously to obtain both the wall temperature and the heat flux to initiate boiling.

$$(\Delta T_{SAT})_{ONB} = X \ \text{Pr}_f \phi_{ONB}^{1/2} \tag{5.35}$$

and

$$\phi_{ONB} = h_{fo}((\Delta T_{SAT})_{ONB} + \Delta T_{SUB}(z)) \tag{5.36}$$

$$T_{SAT} = 164°R \qquad \Delta T_{SUB}(z) = 24°F$$

The heat transfer coefficient h_{fo} is given by

$$h_{fo} = 0.023\left(\frac{k_f}{D}\right)\left(\frac{GD}{\mu_f}\right)^{0.8}\left(\frac{c_p\mu}{k}\right)_f^{0.4} \tag{5.7}$$

Values of physical properties at bulk temperature 140°F

$$k_f = 0.08 \text{ Btu/h ft °F}$$

$$\mu_f = 0.4 \text{ lb/ft h} \qquad \text{Pr}_f = 1.25 \qquad \text{Pr}_f^{0.4} = 1.093$$

$$c_{pf} = 0.25 \text{ Btu/lb °F}$$

$$D = 0.0416 \text{ ft}$$

$$G = 2.64 \times 10^6 \text{ lb/h ft}^2$$

$$\left[\frac{GD}{\mu}\right] = 2.74 \times 10^5 \qquad \left[\frac{GD}{\mu}\right]^{0.8} = 2.239 \times 10^4$$

$$h_{fo} = 1025 \text{ Btu/h ft}^2 \text{ °F}$$

$$X = \left(\frac{X}{X_{REF}}\right)X_{REF}$$

From Table 5.1, X_{REF} for a reduced pressure of 0.05

$$X_{REF} = 0.0166 \,^\circ R/[Btu/h \, ft^2]^{0.5}$$

The critical pressure of nitrogen is 493 psia.

$$P_r = (p/p_{CRIT}) = 60/493 = 0.122$$

From Fig. 5.5 X/X_{REF} for a reduced pressure of $0.122 - X/X_{REF} = 0.66$

$$X = 0.0166 \times 0.66 = 0.011 \,^\circ R/[Btu/h \, ft^2]^{0.5}$$

Eqs. (5.35) and (5.36) become

$$(\Delta T_{SAT})_{ONB} = 0.011 \times 1.25 \, \phi_{ONB}^{0.5}$$

and

$$= 0.0137 \phi_{ONB}^{0.5}$$

$$\phi_{ONB} = 1025((\Delta T_{SAT})_{ONB} + 24)$$

Simultaneous solution of these equations yields

$$(\Delta T_{SAT})_{ONB} = 2.3^\circ F \qquad (T_W)_{ONB} = 166.3^\circ R \qquad \phi_{ONB} = 27{,}000 \, Btu/h \, ft^2.$$

Problems

1. Liquid oxygen passes through a 0.5 in i.d. pipe at a rate of 1 lb/s. The pressure at the end of the heated pipe is 100 psia and boiling is initiated at a wall temperature of 210°R. Calculate the heat flux and liquid subcooling at this point.

2. Calculate the superheat required to initiate boiling in example 1 using the method of Davis and Anderson (eq. (5.53)) rather than that of Bowring (eq. (5.19)). No active cavities larger than 10 μm radius exist on the tube heated surface—does this influence the calculation?

3. Water is being heated up to saturation temperature and evaporated in a vertical tube. The heat flux is $6 \, kW/m^2$. The single-phase heat transfer coefficient is $40 \, kW/m^2 \,^\circ C$. Is there any danger of instability due to the water superheating before nucleation occurs?

4. Derive the expression analogous to eq. (5.2) for a tube with a uniform wall temperature.

6

VOID FRACTION AND PRESSURE DROP IN SUBCOOLED BOILING

6.1 Introduction

In describing the process of subcooled boiling in forced convective flow of liquid in a confined channel it has been noted that vapour in various amounts is generated at the heated surfaces in the channel. A knowledge of the amount of vapour present in the channel is required to enable the accelerative and static head components of the pressure gradient to be calculated. In the particular case of the design of liquid cooled nuclear reactors information on the void fraction under subcooled conditions is often required because of its influence upon the reactivity of the system. As a result of this latter requirement the majority of experimental data is with water as the liquid.

6.2 Void fraction in subcooled boiling

Consider, again, a tube to the wall of which is applied a uniform heat flux ϕ (Fig. 6.1). Subcooled liquid is introduced at a uniform rate, W_f at the inlet. As noted in the previous chapter subcooled boiling starts after a length z_{NB} and saturated nucleate boiling after a length z_{SC}.

In the representation of subcooled boiling shown in Fig. 6.1 it will be seen that, at first, for high degrees of subcooling ($\Delta T_{SUB}(z)$ just less than $\Delta T_{SUB}(z)_{ONB}$) the vapour generated remains as discrete bubbles attached to the surface whilst growing and collapsing; voidage in this region is essentially a wall effect. At somewhat lower subcoolings, bubbles detach from the surface, condensing only slowly as they move through the slightly subcooled liquid; voidage in this region is a bulk fluid effect. In this latter region the bubbles may be very large and irregular (Kirby 1963) particularly at high heat fluxes and low pressures, and an appreciable fraction (30–50 per cent) of the channel may be occupied by vapour. Thus, the 'void fraction' will vary with length in a manner similar to curve ABCDE in Fig. 6.1. The section AB represents the wall effect in subcooled boiling; the section BCD the build-up of vapour in the slightly subcooled region, and, finally, the section DE represents the void fraction in saturated nucleate boiling; a subject already treated in Chapter 2. The point C corresponds to the point at which

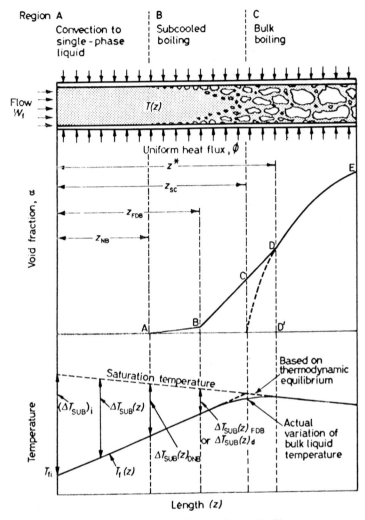

Fig. 6.1. Void fraction in subcooled boiling.

saturated nucleate boiling is calculated to begin if thermodynamic equilibrium is assumed. However, as pointed out earlier, the fact that bubbles are present in the bulk stream implies that the liquid must be below the saturation temperature, to obtain a mean enthalpy equivalent to that of saturated liquid. Not until point D is reached is the total enthalpy increase sufficient to raise all the liquid to the saturation temperature and maintain a void fraction DD'.

6.2.1   The highly subcooled region

In this region (AB) small bubbles grow and collapse whilst attached to the heated surface and do not penetrate far into the bulk subcooled flow. The conditions are those described earlier as 'partial subcooled boiling'. The fraction of the heat used in the formation of net vapour is insignificant.

A simple expression for the void fraction at point B has been suggested by Levy (1967) from a consideration of the forces exerted on a vapour bubble attached to the wall and the temperature distribution in the single-phase liquid away from the heated surface. The basis of this model is discussed in the following section. It is assumed that at point B the bubbles of radius r are spherical (i.e., on the point of detachment) and are spaced at a distance S apart. The number of bubbles around the heated perimeter P_h is then (P_h/S) and the volume of vapour in a section of channel S in length is $(P_h/S)(\frac{4}{3}\pi r^3)$. The void fraction at point B is therefore given by

$$\alpha_w = \left(\frac{P_h}{S}\right)\left(\frac{4}{3}\pi r^3\right)\left(\frac{1}{AS}\right) = \frac{16}{3}\pi \frac{r}{D_h}\left(\frac{r}{S}\right)^2 \tag{6.1}$$

where D_h is the heated equivalent diameter.

If the bubbles are assumed to be packed in a square array and to interfere with each other if $r/S \approx 0.25$, then,

$$\alpha_w \approx \left(\frac{\pi}{3}\right)\frac{r}{D_h} \approx \frac{\pi}{6}\frac{Y_B}{D_h} \tag{6.2}$$

where Y_B is the distance from the wall to the tip of the vapour bubble. It will be shown that Y_B is given by the following equation:

$$Y_B = 0.015\left[\frac{\sigma D_e}{\tau_w}\right]^{1/2} \tag{6.3}$$

where τ_w is the wall shear stress given by eq. (6.12) and D_e is the hydraulic equivalent diameter.

Typical values calculated by Levy from eqs. (6.2) and (6.3) for water flowing in a 12.7 mm diameter tube at a heat flux of 790 kW/m² (2.5 × 10⁵ Btu/h ft²) are given in Table 6.1.

6.2.2   The point of departure of vapour bubbles from the heated surface (point B)

The transition between the first and second regions, point B of Fig. 6.1, is associated with the onset of *fully developed subcooled boiling* (point E of Fig. 5.4). Bowring (1962) applied the following criterion to establish this

Table 6.1 Values of α_w for water in a 12.7 mm (0.5 in.) diameter tube

Pressure bar (psia)	Mass flow-rate kg/m² s (lb/h ft²)		
	136 (0.1 × 10⁶)	680 (0.5 × 10⁶)	1360 (1.0 × 10⁶)
4 (60)	0.055	0.013	0.007
20 (300)	0.047	0.011	0.006
40 (600)	0.040	0.010	0.005
69 (1000)	0.033	0.008	0.004
138 (2000)	0.019	0.005	0.002

point. The subcooling at point B, $\Delta T_{SUB}(z)_B$ should be the least of:

(a) $(\Delta T_{SUB})_i$ = the inlet subcooling
(b) $\Delta T_{SUB}(z)_{ONB}$ = the subcooling at the onset of subcooled boiling given previously
(c) $\Delta T_{SUB}(z)_d$ = the subcooling at which bubble detachment occurs, where

$$\Delta T_{SUB}(z)_d = \eta \frac{\phi}{Gv_f} \qquad (6.4)$$

η is an empirical factor derived from experimental data for water and found to depend only on the system pressure. The following relationship for water over the pressure range 11 to 138 bar was found,

$\eta \times 10^6 = 14.0 + 0.1p$ where p is the system pressure in bar and G, ϕ, and v_f are expressed in SI units $\qquad (6.5)$

or

$\eta = 0.067[14.0 + 0.0068p]$ where p is the system pressure in psia and G, ϕ, and v_f are expressed in British engineering units.

It has been shown (Levy 1967; Staub 1967) that the expression proposed by Bowring is in reasonable agreement with experimental data at low velocities (~ 680 kg/m² s; 0.5×10^6 lb/h ft²) only and appears to predict values of subcooling at bubble departure considerably below those observed for higher mass velocities.

A force balance can be used to establish a criterion for the point of bubble departure. Once the size of the bubble at incipient departure is known it is possible to calculate the subcooling by assuming, firstly, the single-phase temperature profile in the liquid and, secondly, that the fluid temperature at a distance corresponding to the departure bubble radius equals the

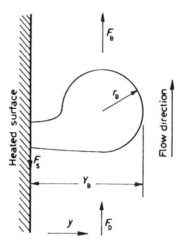

Fig. 6.2. Forces on a departing bubble (Levy 1967).

saturation temperature T_{SAT}. The treatments of subcooled void fraction by Levy (1967) and Staub (1967) both employ this line of reasoning.

Four forces are usually taken into account when considering whether a bubble will remain on or depart from the heating surface. Surface tension and inertia forces hold the bubble to the surface while buoyancy and frictional drag forces attempt to remove it. Inertia forces are usually assumed to be negligible near the point of bubble departure. Following Levy's treatment (Fig. 6.2), the force balance is

$$C_B g(\rho_f - \rho_g)r_B^3 + C_F \frac{\tau_w}{D_e} r_B^3 - C_S r_B \sigma = 0 \qquad (6.6)$$

$$\text{Buoyancy} \qquad \text{Drag} \qquad \text{Surface} \atop \text{tension}$$

Equation (6.6) can be rearranged and solved for the bubble radius r_B. If it is assumed that the distance to the tip of the bubble, Y_B is proportional to the bubble radius r_B, then

$$Y_B = C\left[\frac{\sigma D_e}{\tau_w}\right]^{1/2}\left[1 + C'\left(\frac{g(\rho_f - \rho_g)D_e}{\tau_w}\right)\right]^{-1/2} \qquad (6.7)$$

This distance can be made dimensionless in the usual way in terms of the parameter u^* $(=\sqrt{(\tau_w/\rho_f)})$. Thus

$$Y_B^+ = \left[\frac{Y_B u^* \rho_f}{\mu_f}\right] = C\frac{(\sigma D_e \rho_f)^{1/2}}{\mu_f}\left[1 + C'\left(\frac{g(\rho_f - \rho_g)D_e}{\tau_w}\right)\right]^{-1/2} \qquad (6.8)$$

In Section 5.3 it has been shown that the liquid temperature adjacent to the tip of the vapour bubble must exceed the saturation temperature by a finite amount if the bubble is either growing or in equilibrium. In order to simplify the present analysis Levy assumed that the liquid temperature at the bubble tip T_B was equal to the saturation temperature T_{SAT}.

The temperature T_B at the position Y_B^+ can also be specified in terms of the dimensionless parameter T_B^+.

$$T_B^+ = \frac{c_{pf}\rho_f u^*}{\phi}(T_W - T_B) \tag{6.9}$$

where ϕ is the wall heat flux and T_W is the wall temperature.

Using the universal temperature profile originally proposed by Martinelli (1947),

$$
\begin{aligned}
T_B^+ &= \Pr_f Y_B^+ && \text{for } (0 \leqslant Y_B^+ \leqslant 5) \\
T_B^+ &= 5\left[\Pr_f + \ln\left(1 + \Pr_f\left\{\frac{Y_B^+}{5} - 1\right\}\right)\right] && \text{for } (5 \leqslant Y_B^+ \leqslant 30) \\
T_B^+ &= 5\left[\Pr_f + \ln(1 + 5\Pr_f) + 0.5\ln\left\{\frac{Y_B^+}{30}\right\}\right] && \text{for } (Y_B^+ > 30)
\end{aligned}
\tag{6.10}
$$

Remembering that $T_B = T_{SAT}$ and that $(T_W - T_f(z)_d) = \phi/h_{fo}$, on rearranging eq. (6.9),

$$(T_{SAT} - T_f(z)_d) = \Delta T_{SUB}(z)_d = \phi\left[\frac{1}{h_{fo}} - \frac{T_B^+}{c_{pf}\rho_f u^*}\right] \tag{6.11}$$

Provided values of C and C' are known, eqs. (6.8), (6.10), and (6.11) may be solved to determine the point at which vapour bubbles first leave the heated surface. The heat transfer coefficient, h_{fo} may be calculated from the standard single-phase relationship (Chapter 5, eq. (5.7)) and the wall shear stress τ_W from the relationship

$$\tau_W = \left[\frac{f_{fo}G^2 v_f}{2}\right] \tag{6.12}$$

where f_{fo} is the single-phase Fanning friction factor corresponding to a relative roughness of $\varepsilon/D_e = 10^{-4}$ (smooth drawn tubing).

Levy evaluated the constants C and C' in eq. (6.8) from experimental data. It was found that buoyancy forces appear to play a negligible part even at low mass velocities and $C' = 0$. The best fit value of C was found to be 0.015 (see eq. (6.3)).

Figure 6.3 shows the prediction of liquid subcooling at the point of departure for subcooled boiling of water in a 12.7 mm diameter tube at a

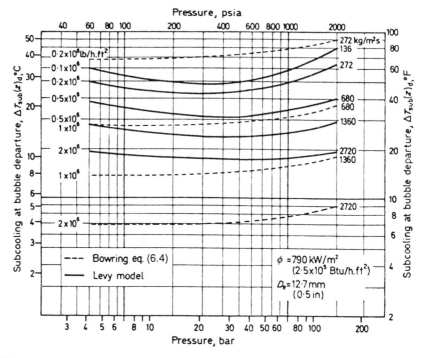

Fig. 6.3. Prediction of local subcooling at point B for steam–water mixtures (Levy 1967).

heat flux of 790 kW/m² (2.5 × 10⁵ Btu/h ft²). The values calculated from the expression proposed by Bowring (eq. (6.4)) are shown for comparison.

The predictions from the Levy model are in good agreement with the data of Egen *et al.* (1957), Maurer (1960), and Rouhani (1965). The treatment proposed by Staub (1967) is basically similar in approach, although there are a number of different assumptions. For example, in Staub's model the expression for the drag term in the force balance (eq. (6.6)) is based on the force on a bubble layer rather than an isolated bubble. The increased friction factor, f_{fo} and heat transfer coefficient h_{fo} induced by the presence of the bubble layer having an equivalent roughness (r_B/D_e) is also allowed for. In contrast to the finding of Levy, Staub concluded that at low mass velocities ($G < 270$ kg/m² s) the drag term in the force balance was negligible and the departure of the bubble was governed primarily by buoyancy forces.

Saha and Zuber (1974) have proposed a simple method to calculate the point of 'net vapour generation' which can be assumed to be coincident with point B, the point of bubble detachment, shown in Fig. 6.1. At low flow-rates the bubble detachment is assumed to be thermally controlled, occurring at a fixed value of Nusselt number $[\phi D/k_f(T_{SAT} - T_f(z)_B)]$. At high flow-rates

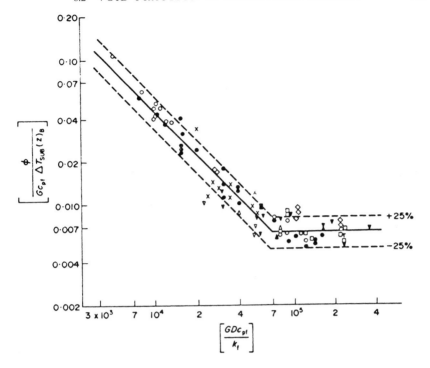

Fig. 6.4. Conditions for point of bubble detachment (point B) (Saha and Zuber 1974).

bubble departure is hydrodynamically induced and occurs at a fixed Stanton number $[\phi/Gc_{pf}(T_{SAT} - T_f(z)_B)]$. Figure 6.4 shows experimental data from three fluids plotted as Stanton number versus Peclet number $[GDc_{pf}/k_f]$. Two regions can be distinguished. Below a Pe value of 70,000 the data fall on a line of slope minus one. Above this value St is a constant.

If Pe < 70,000, the value of $\Delta T_{SUB}(z)_B$ $(= T_{SAT} - T_f(z)_B)$ at point B is given by

$$\Delta T_{SUB}(z)_B = 0.0022 \left[\frac{\phi D}{k_f}\right] \qquad (6.13)$$

for Pe > 70,000, the value of $\Delta T_{SUB}(z)_B$ at point B is given by

$$\Delta T_{SUB}(z)_B = 153.8 \left[\frac{\phi}{Gc_{pf}}\right] \qquad (6.14)$$

This correlation is simple to use and a recent critical review of models by Lee *et al.* (1992) has shown that it remains the most accurate.

6.2.3   *The low subcooling region*

Now, consider the region BCD in Fig. 6.1, the region of low subcooling. In this region bubbles grow and detach, condensing only slowly as they pass through the slightly subcooled liquid. The void fraction in this region rises sharply with length from the transition point, B.

Bowring (1962) proposed a model for the estimation of void fraction in this region. In addition, simple empirical methods of calculating the void fraction in the subcooled boiling of water have been suggested by Thom *et al.* (1965), Kroeger and Zuber (1968), and Levy (1967). The model proposed by Bowring and the empirical methods of Thom, Zuber, and Levy will be described in the following paragraphs.

Bowring's (1962) model, taken in conjunction with the detailed heat transfer relationships as set out in Chapter 5, is recommended as the best available at present, Equation (6.4) leads to a length for bubble detachment z_d of

$$z_d = \frac{Gc_{pf}D}{4}\left[\frac{(\Delta T_{SUB})_i}{\phi} - \left(\frac{\eta}{Gv_f}\right)\right] \tag{6.15}$$

Bowring shows that the fractional change in bubble volume due to recondensation averaged over the slightly subcooled region is small (<7 per cent in all the experimental cases he investigated and <2 per cent in most) and is neglected. Heat is removed from the heated surface by several simultaneous mechanisms: as latent heat, by convection caused by bubble agitation and by single-phase heat transfer between patches of bubbles, thus

$$\phi = \phi_e + \phi_a + \phi_{SPL} \tag{6.16}$$

By comparison with eq. (5.40) it is seen that

$$\phi_{SCB} = \phi_e + \phi_a \tag{6.17}$$

Writing equations for ϕ_e and ϕ_a, Bowring defines the ratio (ϕ_a/ϕ_e) as ε, thus

$$\varepsilon = \frac{\phi_a}{\phi_e} = \left(\frac{\rho_f}{\rho_g}\right)\left(\frac{c_{pf}}{i_{fg}}\right)\tau \tag{6.18}$$

where τ is an effective temperature rise of the water pulled in and pushed out by the bubble. Bowring arrived at the following empirical relationships by analysing available experimental void fraction data for water.

For the pressure range 1–9.5 bar (14.7–140 psia)

$$(1 + \varepsilon) = 1 + 3.2\left(\frac{\rho_f}{\rho_g}\right)\left(\frac{c_{pf}}{i_{fg}}\right) \tag{6.19}$$

for the pressure range 9.5–50 bar (140–735 psia)

$$(1 + \varepsilon) = 2.3 \tag{6.20}$$

and for rectangular channels 2.5 mm wide for pressures above 50 bar (735 psia)

$$(1 + \varepsilon) = 2.6 \tag{6.21}$$

Gregorio and Merlini (1968) have suggested an alternative expression for ε which overcomes the discontinuity in this value at the point of bubble detachment

$$\varepsilon = \left[1 + \frac{v_g}{v_f} \frac{c_{pf}}{i_{fg}} \frac{\Delta T_{SUB}(z)}{i_{fg}} \right] \tag{6.22}$$

The calculation of the void profile involves the addition of the wall and free stream void components

$$\alpha = \alpha_W + \alpha_B \tag{6.23}$$

The wall voidage, α_W, may be calculated from eq. (6.2)—(i.e., α_W is constant beyond B).

The free stream voidage, α_B, is given by direct analogy with the bulk boiling case (Chapter 2)

$$\frac{u_g}{u_f} = \left(\frac{x}{1 - x} \right) \left(\frac{1 - \alpha}{\alpha} \right) \left(\frac{\rho_f}{\rho_g} \right) \tag{1.12}$$

In the bulk boiling region all of the heat flux is converted into latent heat of vaporization giving the relationship between mass fraction vapour, $x(z)$, and heat flux ϕ

$$x(z) = \frac{4\phi}{DGi_{fg}} (z - z_{SC}) \tag{4.71}$$

However, in subcooled boiling, only the latent heat component of eq. (6.16) is used to produce vapour, thus

$$x'(z) = \frac{4\phi_e}{DGi_{fg}} (z - z_d) \tag{6.24}$$

Substituting eqs. (6.17) and (6.18) this equation becomes

$$x'(z) = \frac{4}{DGi_{fg}} \left[\frac{\phi - \phi_{SPL}}{(1 + \varepsilon)} \right] (z - z_d) \tag{6.25}$$

where ϕ_{SPL} is given by the equations derived previously in Chapter 5 depending on whether $\Delta T_{SUB}(z)$ is greater or less than $\Delta T_{SUB}(z)_{FDB}$. Substitution of eq. (6.25) into eq. (1.12) and rearranging, gives

$$\left(\frac{\alpha_B}{1 - \alpha_B} \right) (1 - x'(z)) = \left(\frac{\rho_f}{\rho_g} \right) \left(\frac{4}{DGi_{fg}} \right) \left(\frac{u_f}{u_g} \right) \left(\frac{\phi - \phi_{SPL}}{(1 + \varepsilon)} \right) (z - z_d) \tag{6.26}$$

$(1 - x'(z))$ may be calculated from eq. (6.25) if necessary but is usually close to unity and can be taken as such.

It is now possible to evaluate the void fraction at the position z_{SC}, i.e., point C, and at point D (Fig. 6.1). For the value of α at the bulk boiling boundary α_W is neglected and it is assumed that $(1 - x'(z))$ is equal to unity. It is also assumed that fully developed boiling occurs so that ϕ_{SPL} is zero. Therefore eqs. (6.23) and (6.26) with $z = z_{SC}$ yield

$$\left(\frac{\alpha_B}{1 - \alpha_B}\right) \approx \left(\frac{\rho_f}{\rho_g}\right)\left(\frac{4}{DGi_{fg}}\right)\left(\frac{u_f}{u_g}\right)\left(\frac{\phi}{(1 + \varepsilon)}\right)(z_{SC} - z_d) \qquad (6.27)$$

Substituting for z_{SC} and z_d respectively the following relationship is obtained:

$$\left(\frac{\alpha_B}{1 - \alpha_B}\right) \approx \left(\frac{\rho_f}{\rho_g}\right)\left[\frac{c_{pf}\eta\phi}{Gi_{fg}(u_g/u_f)(1 + \varepsilon)v_f}\right] \qquad (6.28)$$

It is also possible to evaluate the void fraction at the point D (Fig. 6.1) i.e., DD'. By solving eqs. (4.71) and (6.24) simultaneously for the intersection of the values of $x(z)$ in the subcooled and bulk boiling regions, Bowring obtained

$$x(z^*) = \frac{c_{pf}\eta\phi}{Gi_{fg}\varepsilon v_f} \qquad (6.29)$$

where $x(z^*)$ is the weight fraction of vapour at the intersection with the bulk boiling void fraction curve, (point D). Thus

$$\left(\frac{\alpha^*}{1 - \alpha^*}\right)(1 - x(z^*)) = \left(\frac{\rho_f}{\rho_g}\right)\left(\frac{u_f}{u_g}\right)\left[\frac{c_{pf}\eta\phi}{Gi_{fg}\varepsilon v_f}\right] \qquad (6.30)$$

Measurements made at the Argonne National Laboratory and reported by Marchaterre et al. (1958) were excellently correlated by the Bowring model. Void fractions were measured for subcooled boiling of water in rectangular ducts 6.35 mm × 50.8 mm and 12.7 × 50.8 mm cross-section for pressures up to 41 bar (600 psia). The model was also compared with an extensive set of data obtained by Rouhani (1965). The model seriously overestimates the void fraction at low mass velocities probably due to the fact that the assumption regarding negligible condensation of the bubbles is not valid under these conditions.

The empirical method of Thom et al. (1965) relates the subcooled void fraction, to a 'reduced' steam quality, x'. This relationship is similar to that proposed earlier by the same author (Thom 1964) to relate void fractions in saturated forced convection boiling and is of the form

$$\alpha = \frac{Mx'}{1 + x'(M - 1)} \qquad (6.31)$$

The 'reduced' steam quality, x', is defined in terms of the fluid enthalpy at the start of the second region (point B) thus

$$x' = \frac{(i(z) - i'_B)}{(i_g - i'_B)} \tag{6.32}$$

where $i(z)$ is the local fluid enthalpy, i'_B is the fluid enthalpy at the point B, and i_g is the enthalpy of the saturated vapour.

It was found that for the conditions of the experimental programme (pressurized water at 52 and 69 bar) the value of the fluid enthalpy at the point B, i'_B, could be related to the enthalpy of saturated water, i_f, by the empirical relationship

$$i'_B = i_f \left[1 - 0.0645 \frac{\phi}{G}\right] \tag{6.33}$$

The value of M in eq. (6.31) is a function of the system pressure, and has a value of 16 at 52 bar and 10 at 69 bar.

Kroeger and Zuber (1968) have proposed a more general empirical method for the prediction of the void fraction in the subcooled boiling region. The method is based on an assumed liquid temperature distribution between an arbitrarily chosen length z_0 (associated with the start of void formation) and z^* (Fig. 6.1). Two simple relationships were arbitrarily chosen to represent the temperature profile

$$T^* = \left[\frac{T_f(z) - T_f(z_0)}{T_{SAT} - T_f(z_0)}\right] = \left[\frac{\Delta T_{SUB}(z_0) - \Delta T_{SUB}(z)}{\Delta T_{SUB}(z_0)}\right] \tag{6.34}$$

where T^* can be given by

$$T^* = 1 - \exp[-Z^+] \tag{6.35}$$

or by

$$T^* = \tanh Z^+ \tag{6.36}$$

where

$$Z^+ = \left\{\frac{z - z_0}{z_{SC} - z_0}\right\} \tag{6.37}$$

The 'true' mass fraction vapour distribution is then given by

$$x'(z) = \frac{4\phi(z_{SC} - z_0)}{DGi_{fg}} \left\{\frac{Z^+ - T^*}{1 + [c_{pf} \Delta T_{SUB}(z_0)/i_{fg}](1 - T^*)}\right\} \tag{6.38}$$

$$= \frac{c_{pf} \Delta T_{SUB}(z_0)}{i_{fg} + c_{pf} \Delta T_{SUB}(z_0)(1 - T^*)} (Z^+ - T^*) \tag{6.39}$$

The void fraction $\alpha(z)$ is then given by

$$\alpha(z) = x'(z) \Big/ \left\{ \frac{C_0(\rho_f - \rho_g)}{\rho_f} x'(z) + \left[C_0 + \frac{\bar{u}_{gj}}{u_{fo}} \right] \frac{\rho_g}{\rho_f} \right\} \tag{6.40}$$

where C_0 is a distribution parameter (see Chapter 3), u_{fo} is the inlet liquid velocity and $\bar{u}_{gj}$ is the drift velocity in the churn-turbulent region of bubbly flow and given by

$$\bar{u}_{gj} = 1.41 \left\{ \frac{\sigma g(\rho_f - \rho_g)}{\rho_f^2} \right\}^{1/4} \tag{3.10}$$

It was found that the best representations of experimental void fractions were obtained with eq. (6.36) used for the variation of T^* and a value of $C_0 = 1.13$ or 1.4 used for circular or rectangular geometries respectively. The main difficulty found when applying this treatment is the estimation of z_0; it can be assumed to be equal to z_{ONB} or alternatively z_{FDB} or z_d.

The basis of the empirical method suggested by Levy (1967) is very similar to that proposed by Kroeger and Zuber. It is assumed that the 'true' mass fraction vapour $x'(z)$ is related to the 'thermodynamic' mass fraction vapour $x(z)$ by the relationship

$$x'(z) = x(z) - x(z)_d \exp\left(\frac{x(z)}{x(z)_d} - 1 \right) \tag{6.41}$$

where $x(z)_d$ is the (negative) thermodynamic quality at the point of bubble detachment given by

$$x(z)_d = -\left[\frac{c_{pf} \Delta T_{SUB}(z)_d}{i_{fg}} \right] \tag{6.42}$$

Equation (6.41) is a simple relationship which satisfies the following necessary boundary conditions,

(a) at $x(z) = x(z)_d$, $x'(z) = 0$

(b) $\dfrac{d}{dz}(x'(z))$ at point B must be zero

(c) $x'(z) \to x(z)$ for $x(z) \gg |x(z)_d|$

Substitution of the value of 'true' mass fraction vapour from eq. (6.41) into eq. (6.42) yields the void fraction versus length profile. Levy chose a value for $C_0 = 1.13$ and used the expression for the drift velocity $\bar{u}_{gj}$ given by

$$\bar{u}_{gj} = 1.18 \left[\frac{\sigma g(\rho_f - \rho_g)}{\rho_f^2} \right]^{1/4} \tag{6.43}$$

Levy also points out that the use of a value of $C_0 = 1.13$ is questionable in a developing flow where the major fraction of the vapour is close to the wall.

The treatments of both Kroeger and Zuber (1968) and Levy (1967) fit the experimental data acceptably well. Both methods can be applied to fluids other than water. Levy's method is recommended as being the simpler to use.

Further studies of subcooled void fraction have been reported by Bartolemei and Chanturiya (1967) who present measurements for water in the pressure range 14 to 41 bar in tubes 15.4 mm and 24 mm i.d. In addition to the void fraction both surface and fluid temperature distributions were recorded. Their findings agree qualitatively with the analyses of Bowring and Levy.

6.3   Pressure drop in subcooled boiling

The state of knowledge relating to pressure drop in the subcooled boiling region is very unsatisfactory indeed. It might have been expected that, using the variation of void fraction in the subcooled region as a guide (Fig. 6.1), efforts would have been made to correlate pressure drop in at least two regions, the high subcooling and low subcooling areas. However, by establishing the void fraction profile as a function of length it is possible to present a satisfactory and logical procedure at least for the low subcooling region. This method was first suggested by Sher (1957).

The total pressure gradient for a separated flow model is given by eq. (2.53). To evaluate this equation it is necessary to know the local vapour fraction, $x(z)$ (defined on a flow-rate basis), the local void fraction, α, and the local value of the two-phase friction multiplier, ϕ_{fo}^2.

6.3.1   *The slightly subcooled region*

Consider first the slightly subcooled region BCD (Figs. 6.1 and 6.5). It has been noted that there is no essential physical difference between subcooled boiling and saturated nucleate boiling and that no discontinuity exists in any characterizing parameter at the bulk boiling boundary. Therefore, it may be postulated that eq. (2.53) could be used to evaluate the local total pressure gradient. Figure 6.5 depicts the variation of the local mass fraction of vapour, $x'(z)$, and the local value of ϕ_{fo}^2. The true local vapour mass fraction is given as VWXYZ (in the region VW, the existence of a quality as defined would depend on the bubbles sliding along the wall). For the sake of simplification, with little loss of accuracy, it is assumed that the variation is VW'X'YZ which may be represented as follows:

(a) for $z < z_{FDB}$ (or z_d), $x'(z) = 0$.

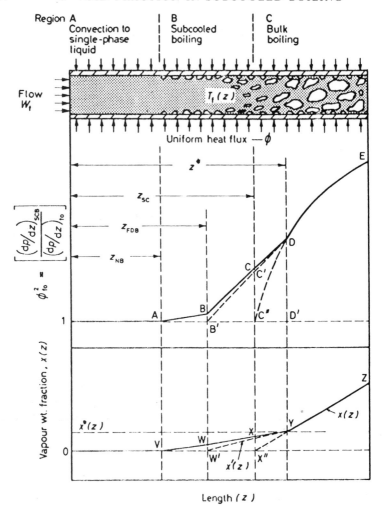

Fig. 6.5. Subcooled boiling pressure gradient.

This may also be expressed as

$$\Delta T_{\text{SUB}}(z) > \Delta T_{\text{SUB}}(z)_{\text{FDB}} \text{ (or } \Delta T_{\text{SUB}}(z)_{\text{d}}), \quad x'(z) = 0$$

(b) For z_{d} (or z_{FDB}) $< z < z^*$, (or $\Delta T_{\text{SUB}}(z)_{\text{FDB}}$ or $\Delta T_{\text{SUB}}(z)_{\text{d}}) > \Delta T_{\text{SUB}}(z) > 0$)

$$x'(z) = \frac{4}{DGi_{\text{fg}}} \left[\frac{\phi}{(1 + \varepsilon)} \right] (z - z_{\text{d}}) \tag{6.44}$$

where

$$z^* = \frac{Gc_{pf}D}{4}\left[\frac{(\Delta T_{SUB})_i}{\phi} + \frac{\eta}{G\varepsilon v_f}\right] \tag{6.45}$$

(c) For $z > z^*$, x is the normal 'thermodynamic quality' given by

$$x(z) = \frac{4\phi}{DGi_{fg}}(z - z_{SC}) \tag{4.71}$$

The value of $x(z)$ at z^* is given by eq. (6.29)

$$x(z^*) = \frac{c_{pf}\eta\phi}{Gi_{fg}\varepsilon v_f} \tag{6.29}$$

The value of ϕ_{fo}^2 may be evaluated as a function of length since the quality distribution, $x(z)$ is now known. Alternatively, the value of ϕ_{fo}^2 may be calculated directly from a knowledge of the void fraction, using the Levy equation,

$$\phi_{fo}^2 = \frac{(1 - x(z))^{1.75}}{(1 - \alpha)^2} \tag{6.46}$$

6.3.2 The highly subcooled region

The procedure outlined above is not satisfactory for the highly subcooled region (region AB) where the vapour bubbles grow and collapse at the heated surface. Hirata and Nishiwaki (1963) have attempted an analysis of the boundary layer flow over a porous plate through which gas is being bubbled. The wall shear stress for a rough plate, given by Schlichting, is

$$\tau_w = \frac{1}{2}\rho_f u_{f\infty}^2\left[2.87 + 1.58\ln\left(\frac{z}{\varepsilon}\right)\right]^{-2.5} \tag{6.47}$$

where the surface roughness ε was originally introduced by Nikuradse as the grain size of the sand glued to the plate. For simplicity it was assumed that the time-averaged diameter of the growing bubble could replace ε. It was shown that this was 0.75 times the diameter of the bubble at departure. If use is made of Levy's (1967) model this, in turn, is given approximately by eq. (6.3). Hirata and Nishiwaki (1963) show that the increase in wall shear stress is likely to be much more sensitive to the nucleation site density and the bubble departure diameter than to the heat flux.

6.3.3 Experimental studies of subcooled boiling pressure drop

Only a limited amount of experimental work has been carried out on the study of pressure gradients in the subcooled region. The higher pressure gradients associated with subcooled boiling, as compared with single-phase liquid flow, were noted by Buchberg et al. (1951), and others in the early studies of subcooled boiling. No attempts at correlation were made but observations concerning the qualitative influence of the independent variables were given. It was observed that the relative increase in pressure gradient in the subcooled boiling region over that with no boiling was largest at the lowest pressures. It was also noted that the pressure gradient increased as the subcooling was reduced and the heat flux was increased.

The first published work devoted to the measurement of local pressure gradients in the subcooled boiling region was that of Reynolds (1954). Since no measurements of the subcooled void fraction profile were obtained, correlation of the total pressure gradient only was possible. It was observed that soon after subcooled boiling was initiated there was a region where the pressure gradient was *lower* than that for single-phase liquid flow and that after this point the gradient increased rapidly.

Reynolds expressed the ratio of the total subcooled boiling pressure gradient $(dp/dz)_{SCB}$ to the single-phase isothermal pressure gradient $(dp/dz)_{fo}$ (at the same bulk temperature and mass velocity) as a function of

$$\left[1 - \frac{\Delta T_{SUB}(z)}{\Delta T_{SUB}(z_{NB})} \right]$$

thus,

$$\left[\left(\frac{dp}{dz} \right)_{SCB} \middle/ \left(\frac{dp}{dz} \right)_{fo} \right]_T = \text{fn} \left[1 - \frac{\Delta T_{SUB}(z)}{\Delta T_{SUB}(z_{NB})} \right] = \text{fn} \left(\frac{z - z_{NB}}{z_{SC} - z_{NB}} \right) \quad (6.48)$$

It was found that, by plotting the data in the form suggested by eq. (6.48), a curve was obtained which was practically independent of mass velocity and absolute pressure and a function of only the heat flux; this could be fitted with the hyperbolic cosine, as follows:

$$\left[\left(\frac{dp}{dz} \right)_{SCB} \middle/ \left(\frac{dp}{dz} \right)_{fo} \right]_T = \cosh \left[a' \left(1 - \frac{\Delta T_{SUB}(z)}{\Delta T_{SUB}(z_{NB})} \right) \right] \quad (6.49)$$

where a' is an empirical term which is a function of heat flux only, thus,

$$a' = 1.2 + 4.6 \left(\frac{\phi}{10^6} \right) \quad (6.50)$$

where ϕ is in Btu/h ft^2 or

$$a' = 1.2 + 1.46 \left(\frac{\phi}{10^3} \right) \quad (6.51)$$

where ϕ is in kW/m^2.

In order to evaluate the pressure loss from the start of the subcooled boiling region, z_{NB}, to a point, z, it is necessary to integrate eq. (6.49). Reynolds (1954) attempted to extend the range of the correlation by comparing its predictions with the data of Buchberg et al. (1951). Satisfactory agreement was achieved over a wide range of conditions provided the surface heat flux was lower than 1.58 MW/m² (5 × 10⁵ Btu/h ft²).

Lottes et al. (1962) report a correlation developed by Rohde (their reference 50) at Argonne National Laboratory. Rohde evaluated the local pressure gradients in the work by Buchberg et al. (1951). Lottes et al. point out that the method of measurement of the pressure loss was such that the reported values will include only the frictional and static head terms—complete condensation having occurred by the time the fluid arrived at the downstream pressure tap.

The Rohde correlation was expressed as

$$\left[\left(\frac{dp}{dz}\right)_{SCB} \Big/ \left(\frac{dp}{dz}\right)'_{fo}\right]_{F+S} = \exp\left[\frac{26.5}{p}\left(\frac{\phi}{\phi'_{SPL}} - 1\right)\right] \qquad (6.52)$$

where $(dp/dz)'_{fo}$ is the single-phase liquid pressure gradient* at the same heat flux, velocity, and temperature, p is the system pressure in bar and ϕ'_{SPL} is the maximum heat flux which could be sustained at that point without subcooled boiling. This flux is given by

$$\phi'_{SPL} = \frac{(\Delta T_{SUB})_i + \Delta T_{SAT}}{[(4z/Gc_{pf}D) + 1/h_{fo}]} \qquad (6.53)$$

This equation can be solved iteratively using the Jens and Lottes equation for the value of ΔT_{SAT}.

Sher (1957) reported local values of the subcooled boiling frictional pressure gradient for water flow at 138 bar (2000 psia) in rectangular channels (2.5 mm (and 1.25 mm) × 25.4 mm × 68.5 cm). Void fraction profiles were taken and the total gradients corrected for acceleration and static head components. In addition to the earlier noted effects of heat flux, pressure, and subcooling on the value of $[(dp/dz)_{SCB}/(dp/dz)_{fo}]$, Sher noted that the ratio was also increased by decreased mass velocities. Sher suggests that, when the subcooled void fraction profile is available, a method of evaluating the pressure gradient should be used which is essentially identical with that discussed in Section 6.3.1.

Tarasova et al. (1966) have derived a form of equation for subcooled boiling pressure drop similar to that used by Reynolds. The proposed

---

* Note: $\left(\dfrac{dp}{dz}\right)'_{fo}$ will be less than the isothermal gradient $\left(\dfrac{dp}{dz}\right)_{fo}$.

relationship is

$$\left\{\left[\left(\frac{dp}{dz}\right)_{\text{SCB}}\bigg/\left(\frac{dp}{dz}\right)_{\text{fo}}\right] - 1\right\}\left(\frac{\phi}{i_{\text{fg}}G}\right)^{-0.7}\left(\frac{\rho_{\text{f}}}{\rho_{\text{g}}}\right)^{-0.08} = \text{fn}(Z) \qquad (6.54)$$

where

$$Z = \left[\frac{z - z_{\text{NB}}}{z_{\text{SC}} - z_{\text{NB}}}\right] = \left[1 - \frac{\Delta T_{\text{SUB}}(z)}{\Delta T_{\text{SUB}}(z_{\text{NB}})}\right] \qquad (6.55)$$

The function $\text{fn}(Z)$ may be represented by the following equation,

$$\text{fn}(z) = 20Z/(1.315 - Z) \qquad (6.56)$$

Pressure drop data of subcooled boiling in forced convective flow of water in internally heated annuli and at pressures close to atmospheric have been reported by Jordan and Leppert (1962). In the study local friction factors were evaluated for the heated wall from overall pressure gradient measurements corrected for acceleration and static head effects. These friction factors were plotted, following Sabersky and Mulligan (1955), as a function of the Stanton number (h/Gc_{p}) in an attempt to examine the Reynolds' analogy in the subcooled boiling regime,

$$2f_{\text{SCB}} = \frac{h_{\text{SCB}}}{Gc_{\text{p}}} \qquad (6.57)$$

Jordan and Leppert showed that their own data together with those of Reynolds (1954) and Owens and Schrock (1960), were in qualitative agreement with eq. (6.57). However, 'apparent' friction factors evaluated from the *total* pressure gradients observed in these experiments increased at a rate greater than that predicted by eq. (6.57). In all probability this effect is due to the inclusion of the acceleration term in the 'apparent' friction factor. However, it is impossible to verify this because no void fraction data were taken for the experiments in question.

Experimental data on subcooled boiling pressure gradients for water at pressures higher than atmospheric have been reported by Owens and Schrock (1960) and by Jicha and Frank (1962). The experiments of Owens and Schrock were carried out in vertical stainless steel tubes having internal diameters of 3 mm and 4.6 mm and lengths of 38 cm and 40.6 cm respectively. The results appeared insensitive to pressure over the range of variables investigated in contrast to the findings of most other investigations. The total pressure gradients were correlated in terms of the ratio of the length from inception of subcooled boiling $(z - z_{\text{NB}})$ to the total subcooled boiling length $(z_{\text{SC}} - z_{\text{NB}})$ after the manner of Reynolds

$$\left[\left(\frac{dp}{dz}\right)_{\text{SCB}}\bigg/\left(\frac{dp}{dz}\right)'_{\text{fo}}\right]_{\text{T}} = 0.97 + 0.28 \exp[6.13Z] \qquad (6.58)$$

where $(dp/dz)'_{fo}$ is the single-phase liquid pressure gradient at the same heat flux, velocity, and temperature.

Jicha and Frank (1962) report data in the pressure range 55 to 103 bar (800–1500 psia) taken from a vertical stainless steel tube 10.6 mm i.d. and 76 cm in length for heat fluxes up to 4.1 MW/m^2 (1.3×10^6 Btu/h ft^2). Again no void fraction data were taken and an empirical correlation of the combined frictional and accelerational pressure gradient terms was presented. Further studies of subcooled boiling pressure drop for water have been given by Miller (1954) and by Poletavkin (1959). For liquids other than water and refrigerants there appears to be a dearth of information.

An investigation of pressure drop for subcooled boiling of water in small diameter tubes has been reported by Dormer and Bergles (1964). A satisfactory correlation of a large amount of data was obtained by plotting the results in the manner shown in Fig. 6.6. The parameters chosen were $(\Delta p_{SCB}/\Delta p_{SPL})$, the ratio of the subcooled boiling pressure drop across the test section to the pressure drop obtained in a similar unheated tube at the same velocity and inlet temperature, and (ϕ/ϕ_{SAT}), the ratio of the actual heat flux to that required to produce just saturated water at the exit of the tube. The use of these parameters eliminated all effects except geometry. The diameter is seen to have a small but measurable effect, whereas the length-to-diameter ratio is of prime importance. The correlation was found to be suspect when the velocities and heat fluxes covered by Dormer were greatly exceeded.

It will be observed from Fig. 6.6 that the rapid increase in subcooled pressure drop occurs at a particular value of ϕ/ϕ_{SAT} depending on the length-to-diameter ratio (z/D) of the test section. It might be expected that the sharp increase in pressure drop above that experienced for single-phase isothermal flow, would be caused by the transition from the highly subcooled region, where the bubbles remain attached to the heating surface, to the slightly subcooled region, where the bubbles detach from the wall (Whittle and Forgan 1967). In a uniformly heated channel this transition will occur first at the exit of the test section. For a fixed geometry and mass velocity

$$\frac{\phi}{\phi_{SAT}} = \frac{(T_f(z) - T_{fi})}{(T_{SAT} - T_{fi})} = \left[\frac{(\Delta T_{SUB})_i - \Delta T_{SUB}(z)}{(\Delta T_{SUB})_i} \right] \qquad (6.59)$$

The value of (ϕ/ϕ_{SAT}) corresponding to the point of bubble detachment occurring just at the end of the test section $(\phi/\phi_{SAT})^*$ is given by

$$\left(\frac{\phi}{\phi_{SAT}} \right)^* = \left[\frac{(\Delta T_{SUB})_i - \Delta T_{SUB}(z)_d}{(\Delta T_{SUB})_i} \right] \qquad (6.60)$$

Using the Bowring model of bubble detachment the value of $\Delta T_{SUB}(z)_d$ is given by eq. (6.4). The value of ϕ_{SAT} and $(\Delta T_{SUB})_i$ are related by eq. (5.3)

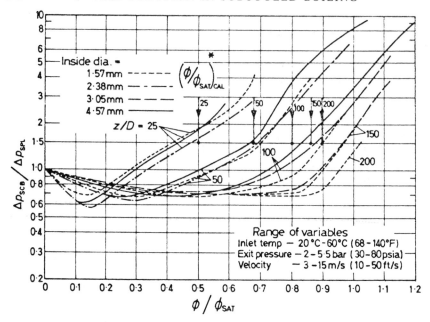

Fig. 6.6. Pressure drop correlation for subcooling boiling of low pressure water (Dormer and Bergles 1964).

which rewritten becomes

$$\phi_{\text{SAT}} = \frac{G c_{\text{pf}} D}{4z} (\Delta T_{\text{SUB}})_i \tag{6.61}$$

Substituting for $(\Delta T_{\text{SUB}})_i$ and $\Delta T_{\text{SUB}}(z)_d$ from eq. (6.61) and (6.4) into eq. (6.60) and rearranging gives

$$\left(\frac{\phi}{\phi_{\text{SAT}}}\right)^* = \frac{1}{1 + \left[\left(\dfrac{\eta c_{\text{pf}}}{4 v_f}\right) \Big/ \left(\dfrac{z}{D}\right)\right]} \tag{6.62}$$

For water at atmospheric pressure $(c_{\text{pf}}/v_f) \approx 4 \times 10^6 \text{ J/m}^3 \text{ °C}$, and the Bowring value of η for the same conditions (given by eq. (6.5) is 14.0×10^{-6} $(\text{m}^3 \text{ °C/J})$. Poor agreement is observed between the calculated and experimental values of $(\phi/\phi_{\text{SAT}})^*$ when these values are used in eq. (6.62). Better agreement with the value of (ϕ/ϕ_{SAT}) corresponding to a constant value of $(\Delta p_{\text{SCB}}/\Delta p_{\text{SPL}})$ of 1.5 (Fig. 6.6) is obtained when η is taken as 25×10^{-6} $(\text{m}^3 \text{ °C/J})$.

References and further reading

Bartolomei, C. C. and Chanturiya, V. M. (1967). 'Experimental study of true void fraction when boiling subcooled water in vertical tubes'. *Thermal Engineering*, **14**(2), 123–128.

Bowring, R. W. (1962). 'Physical model based on bubble detachment and calculation of steam voidage in the subcooled region of a heated channel'. *OECD Halden Reactor Project Report HPR-10*.

Buchberg, H., Romie, F., Lipkis, R., and Greenfield, M. (1951). 'Heat transfer, pressure drop and burnout studies with and without surface boiling for de-aerated and gassed water at elevated pressures in a forced flow system'. *1951 Heat Transfer and Fluid Mechanics Institute*, 171–191, Stanford University, 20–22 June.

Dormer, J. Jr. and Bergles, A. E. (1964). 'Pressure drop with surface boiling in small diameter tubes'. Report no. 8767–31, Dept. of Mech. Engng., MIT, 1 September.

Egen, R. A., Dingee, D. A., and Chastain, J. W. (1957). 'Vapour formation and behaviour in boiling heat transfer'. ASME Paper No. 57-A-74; also BMI-1163, February.

Gregorio, P. and Merlini, C. (1968). 'Voids, friction and vibrations associated with localized boiling'. *AERE Trans.* 1114.

Hirata, M. and Nishiwaki, N. (1963). 'Skin friction and heat transfer for liquid flow over a porous wall with gas injection'. *Int. J. Heat Mass Transfer*, **6**, 941–949.

Jicha, J. J. and Frank, S. (1962). 'An experimental local boiling heat transfer and pressure drop study of a round tube'. Paper 62-HT-48 presented at 5th National Heat Transfer Conference, Houston, August.

Jordan, D. P. and Leppert, G. (1962). 'Pressure drop and vapour volume with subcooled nucleate boiling'. *Int. J. Heat Mass Transfer*, **5**, 751–761.

Kirby, G. (1963). 'Film of subcooled boiling at atmospheric pressure at high heat fluxes'. Private communication, A. E. E. Winfrith.

Kroeger, P. G. and Zuber, N. (1968). 'An analysis of the effects of various parameters on the average void fractions in subcooled boiling'. *Int. J. Heat Mass Transfer*, **11**, 211–233.

Lee, S. C., Dorra, H., and Bankoff, S. G. (1992). 'A critical review of predictive models for the onset of significant void in forced convection subcooled boiling'. Presented in *Fundamentals of Subcooled Flow Boiling* HTD-Vol. 217, pp. 33–39. Paper presented at ASME Winter Ann. Mtg, Anaheim, CA, Nov.

Levy, S. (1967). 'Forced convection subcooled boiling prediction of vapour volumetric fraction'. *Int. J. Heat Mass Transfer*, **10**, 951–965.

Lottes, P. A. *et al.* (1962). *Boiling water reactor technology, status of the art report.* Vol. 1, *Heat transfer and hydraulics*, ANL 6561.

Marchaterre, J. F. *et al.* (1958). 'Natural and forced-convection boiling studies'. ANL 5735.

Martinelli, R. C. (1947). 'Heat transfer to molten metals'. *Trans. ASME*, **69**, 947.

Maurer, G. W. (1960). 'A method of predicting steady state boiling vapour fractions in reactor coolant channels'. *Bettis Technical Review*, WARD-BT-19.

Miller, M. S. (1954). 'Pressure in forced circulation flow of subcooled water with and without surface boiling'. AM Thesis, MIT, August.

Owens, W. L. and Schrock, V. E. (1960). 'Local pressure gradients for subcooled boiling of water in vertical tubes'. ASME Paper No. 60-WA-249.

Poletavkin, P. G. (1959). 'Hydraulic resistance with surface boiling of water'. *Teploenergetika*, **12**, 13–18, Trans. RTS-1513.

Reynolds, J. B. (1954). 'Local boiling pressure drop'. ANL 5178.

Rouhani, S. Z. (1965). 'Void measurements in the region of subcooled and low quality boiling'. Paper E5 presented at Symposium on Two-phase Flow, Exeter University, 21–23 June.

Sabersky, R. H. and Mulligan, H. E. (1955). 'On the relationship between fluid friction and heat transfer in nucleate boiling'. *Jet Propulsion*, **25**(1), 9–12.

Saha, P. and Zuber, N. (1974). 'Point of net vapour generation and vapour void fraction in subcooled boiling'. *Proc. of the 5th International Heat Transfer Conference*, Tokyo. Paper B4.7.

Sher, N. C. (1957). 'Estimation of boiling and non-boiling pressure drop in rectangular channels at 2000 psia'. WAPD-TH-300.

Staub, F. W. (1967). 'The void fraction in subcooled boiling—prediction of the initial point of net vapour generation'. ASME preprint 67-HT-36 presented at 9th National Heat Transfer Conference, Seattle, August.

Tarasova, N. V., Leontiev, A. I., Hlopushin, V. I., and Orlov, V. M. (1966). 'Pressure drop of boiling subcooled water and steam–water mixture flowing in heated channels'. *Proc. of 3rd Int. Heat Transfer Conf.*, Vol. 4, Paper 113.

Thom, J. R. S. (1964). 'Prediction of pressure drop during forced circulation boiling of water'. *Int. J. Heat Mass Transfer*, **7**, 709–724.

Thom, J. R. S., Walker, W. W., Fallon, T. A., and Reising, G. F. S. (1965). 'Boiling in subcooled water during flow up heated tubes or annuli'. Paper 6 presented at Symposium on Boiling Heat Transfer in Steam Generating Units and Heat Exchangers held in Manchester, IMechE (London), 15–16 September.

Whittle, R. H. and Forgan, R. (1967). 'A correlation for the minima in the pressure drop versus flow-rate curves for subcooled water flowing in narrow heated channels'. *Nuc. Engng. and Design*, **6**, 89–99.

Examples

Example 1
Estimate the void fraction profile along a 10.16 mm i.d. tube heated uniformly over a 3.66 m length for the conditions of Example 1 of Chapter 5 using

(a) the approximate method of Thom
(b) the method proposed by Bowring.

Solution
Only the last 1.93 m of the tube is of interest. Figure 6.7 shows the important locations and variables already determined previously.

(a) *Thom's method*
The 'reduced' steam quality x' is calculated using eq. (6.32) thus,
The enthalpy at the onset of the low subcooling region (point B) is denoted by i'_B.

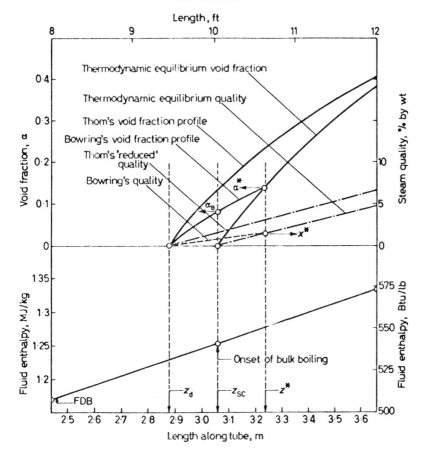

Fig. 6.7.

From eq. (6.33)

$$i'_B = i_f\left[1 - 0.0645\frac{\phi}{G}\right]$$

$$i_f = 1.261 \text{ MJ/kg} \qquad \phi = 1720 \text{ kW/m}^2 \qquad G = 5340 \text{ kg/m}^2 \text{ s}$$

$$i'_B = 1.261\left[1 - 0.0645\left(\frac{1720}{5340}\right)\right] = 1.261 \times 0.9793$$

$$= 1.234 \text{ MJ/kg}$$

The value of x' at the tube exit is then given by eq. (6.32)

$$x' = \frac{i(z) - i'_B}{i_g - i'_B} = \frac{1.334 - 1.234}{2.77 - 1.234} = \underline{0.065}$$

The void profile may then be calculated from eq. (6.31)

$$\alpha = \frac{Mx'}{1 + x'(M - 1)} \quad \text{where at 69 bar } M = 10$$

$$\alpha = \frac{10x'}{1 + 9x'}$$

Values calculated from this equation are:

x'	0.01	0.02	0.03	0.04	0.05	0.06
α	0.092	0.169	0.236	0.294	0.345	0.390

Figure 6.7 shows the estimated void profile using Thom's method.

(b) *Bowring's method*
Calculate $\Delta T_{SUB}(z)_d$ from eq. (6.4):

$$\Delta T_{SUB}(z)_d = \eta \frac{\phi}{Gv_f}$$

$\eta = [14.0 + 0.1p] \times 10^{-6} [\text{m}^3 \,^\circ\text{C/J}]$
$p = 68.9$ bar $\eta = 20.89 \times 10^{-6}$ m^3 deg/J
$v_f = 1.35 \times 10^{-3}$ m^3/kg

$$\Delta T_{SUB}(z)_d = \frac{20.89 \times 10^{-6} \times 1.720 \times 10^6}{5340 \times 1.35 \times 10^{-3}} = 5^\circ\text{C}$$

Now $\Delta T_{SUB}(z)_B$ is the least of

$$(\Delta T_{SUB})_i = 80.5^\circ\text{C} \qquad \Delta T_{SUB}(z)_{ONB} = 26.6^\circ\text{C} \qquad \Delta T_{SUB}(z)_d = 5^\circ\text{C}$$

Therefore, $\Delta T_{SUB}(z)_B = 5^\circ\text{C}$.

From eq. (6.15)

$$z_d = \frac{5340 \times 4.82 \times 10^3 \times 10.16 \times 10^{-3}}{4} \left[\frac{80.5}{1720 \times 10^3} - \frac{20.89 \times 10^{-6}}{5340 \times 1.35 \times 10^{-3}} \right]$$

$$= \underline{2.88 \text{ m}}$$

Assume $1 + \varepsilon = 2.4$. Evaluate $x(z^*)$ from eq. (6.29):

$$x(z^*) = \frac{c_{pf}\eta\phi}{Gi_{fg}\varepsilon v_f} = \frac{4.82 \times 10^3 \times 20.89 \times 10^{-6} \times 1720 \times 10^3}{5340 \times 1.51 \times 10^6 \times 1.4 \times 1.35 \times 10^{-3}}$$

$$= \underline{0.0114}$$

$$z^* = \frac{Gc_{pf}D}{4} \left[\frac{(\Delta T_{SUB})_i}{\phi} + \frac{\eta}{G\varepsilon v_f} \right]$$

$$= \frac{5340 \times 4.82 \times 10^3 \times 10.16 \times 10^{-3}}{4} \left[\frac{80.5}{1720 \times 10^3} + \frac{20.89 \times 10^{-6}}{5340 \times 1.35 \times 10^{-3} \times 1.4} \right]$$

$$= \underline{3.23 \text{ m}}$$

For lengths greater than this value the thermodynamic equilibrium quality is used.

A slip ratio (u_g/u_f) of 1.69 is assumed for 68.9 bar steam pressure. Equation (6.30) is used to calculate α^*.

$$\left(\frac{\alpha^*}{1-\alpha^*}\right)(1-x(z^*)) = \left(\frac{\rho_f}{\rho_g}\right)\left(\frac{u_f}{u_g}\right)\left(\frac{c_{pf}\eta\phi}{Gi_{fg}\varepsilon v_f}\right)$$

$$= \left(\frac{0.02795}{1.35 \times 10^{-3}}\right)\left(\frac{1}{1.69}\right) \times 1.14 \times 10^{-2}$$

$$= \underline{0.14}$$

$$\alpha^* = \underline{0.1242}$$

Equation (6.27) is used to calculate α_B at $z = z_{SC}$:

$$\left(\frac{\alpha_B}{1-\alpha_B}\right) = \left(\frac{\rho_f}{\rho_g}\right)\left(\frac{c_{pf}\eta\phi}{Gi_{fg}(u_g/u_f)(1+\varepsilon)v_f}\right)$$

$$= \left(\frac{0.02795}{1.35 \times 10^{-3}}\right)\left(\frac{4.82 \times 10^3 \times 20.89 \times 10^{-6} \times 1.720 \times 10^6}{5340 \times 1.51 \times 10^6 \times 1.69 \times 2.4 \times 1.35 \times 10^{-3}}\right)$$

$$= 20.65 \times 0.004 = \underline{0.082}$$

$$\alpha_B \text{ at } z = z_{SC} = \underline{0.0758}$$

Equation (1.12) can be used to provide values of void fraction from thermodynamic quality with $u_g/u_f = 1.69$.

$$\left(\frac{\alpha}{1-\alpha}\right) = \left(\frac{x}{1-x}\right)\left(\frac{u_f}{u_g}\right)\left(\frac{\rho_f}{\rho_g}\right)$$

x	0.005	0.01	0.02	0.03	0.04	0.047
α	0.058	0.100	0.199	0.274	0.337	0.378

The void profiles given by Bowring's method are also shown on Fig. 6.7.

Example 2
Using Dormer's correlation calculate the subcooled boiling pressure drop across a vertical heated test section 2.39 mm i.d. $L/D = 50$ for flows of up to 204 kg/h and heat fluxes of 0, 3.15, 6.30, 9.45, and 12.60 MW/m² . The water temperature is 21°C and the exit pressure from the channel is 2.06 bar.

Solution
Calculate the single-phase pressure drop for no heat applied.

$$\Delta p_{fo} = \frac{2f_{fo}L}{D}\frac{G^2}{\rho_f}$$

Flow area $= \dfrac{\pi D^2}{4} = \dfrac{\pi}{4} \times (2.39 \times 10^{-3})^2 = 4.49 \times 10^{-6} \text{ m}^2$

Flow-rate $= \dfrac{204}{3600} = 0.0566 \text{ kg/s}$

$$\text{Mass velocity} = \frac{5.66 \times 10^{-2}}{4.49 \times 10^{-6}} = 12{,}630 \text{ kg/m}^2 \text{ s}$$

Viscosity μ_f for water at $21°C = 0.98 \times 10^{-3}$ N s/m^2

$$\text{Reynolds number } \frac{GD}{\mu_f} = \frac{12{,}630 \times 2.39 \times 10^{-3}}{0.98 \times 10^{-3}} = 3.08 \times 10^4$$

Assume $\varepsilon/D = 0.0006$

$$f_{fo} = 0.006$$

$$\Delta p_{fo} = \frac{2 \times 0.006 \times 50 \times 12.63^2 \times 10^6}{997.5} = 9.63 \times 10^4 \text{ N/m}^2$$

Table 6.2 gives the pressure drop values between 0 and 204 kg/h. The pressure drop versus flow characteristic is plotted in the figure. ϕ_{SAT}, the heat flux to cause a saturated condition at the test section exit, is given by eq. (6.61).

$$\phi_{SAT} = \frac{G c_{pf} D}{4z} (\Delta T_{SUB})_i$$

$$T_{SAT} \text{ at } 2.06 \text{ bar} = 121.0°C \qquad (\Delta T_{SUB})_i = 100°C$$

Table 6.2

W Flow kg/h	G Mass velocity kg/m^2 s	Reynolds number Re($\times 10^{-4}$)	Friction factor f_{fo}	Δp_{fo} kN/m^2	ϕ_{SAT} MW/m^2	ϕ/ϕ_{SAT} $\phi=3.15$ MW/m^2	ϕ/ϕ_{SAT} $\phi=6.30$ MW/m^2	ϕ/ϕ_{SAT} $\phi=9.5$ MW/m^2
22.7	1405	0.343	0.0109	2.29	2.94	1.070	—	—
45.4	2810	0.685	0.0087	6.89	5.89	0.534	1.069	—
68	4220	1.028	0.0078	13.85	8.8	0.356	0.713	1.069
90.6	5620	1.370	0.0072	22.8	11.8	0.267	0.534	0.802
113	7020	1.713	0.0069	34.1	14.7	0.214	0.428	0.641
136	8440	2.056	0.0066	47.0	17.7	0.178	0.356	0.534
159	9850	2.398	0.0064	62.1	20.6	0.153	0.305	0.458
181.5	11300	2.741	0.0062	78.6	23.6	0.1336	0.2673	0.4009
204	12630	3.084	0.0060	96.3	26.5	0.1188	0.238	0.356

ϕ/ϕ_{SAT} $\phi=12.6$ MW/m^2	$\Delta p/\Delta p_{fo}$ $\phi=3.15$ MW/m^2	$\Delta p/\Delta p_{fo}$ $\phi=6.30$ MW/m^2	$\Delta p/\Delta p_{fo}$ $\phi=9.45$ MW/m^2	$\Delta p/\Delta p_{fo}$ $\phi=12.60$ MW/m^2	Δp kN/m^2 $\phi=3.15$ MW/m^2	Δp kN/m^2 $\phi=6.30$ MW/m^2	Δp kN/m^2 $\phi=9.45$ MW/m^2	Δp kN/m^2 $\phi=12.60$ MW/m^2
—	10	—	—	—	22.9	—	—	—
—	0.93	10	—	—	6.40	68.9	—	—
—	0.70	1.64	10	—	9.70	22.7	138.5	—
1.069	0.64	0.93	2.6	10	14.60	21.2	59.2	228
0.855	0.67	0.75	1.3	3.5	22.85	25.6	44.3	119.4
0.713	0.69	0.70	0.93	1.64	32.4	32.9	43.7	77.0
0.611	0.71	0.65	0.80	1.18	44.1	40.4	49.7	73.4
0.5346	0.74	0.64	0.71	0.93	58.1	50.3	55.8	73.1
0.475	0.78	0.65	0.70	0.84	75.0	62.5	67.3	80.8

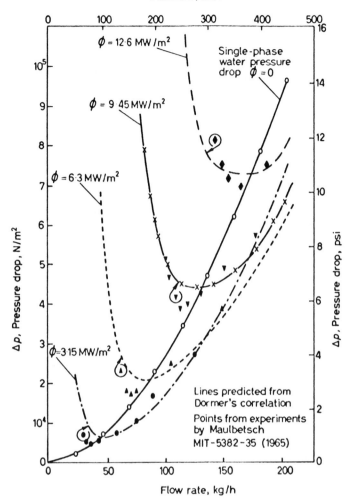

Fig. 6.8.

$$\phi_{\text{SAT}} = \frac{12,630 \times 4.187 \times 10^3 \times 100}{4 \times 50}$$

$$= \underline{26.5 \text{ MW/m}^2}$$

$$\phi = 3.15 \text{ MW/m}^2$$

$$\frac{\phi}{\phi_{\text{SAT}}} = \frac{3.15}{26.5} = 0.119$$

From Fig. 6.6 $\dfrac{\Delta p_{SCB}}{\Delta p_{SPL}} = 0.78$

Therefore $\Delta p_{SCB} = 9.63 \times 10^4 \times 0.78 = \underline{75.0 \text{ kN/m}^2}$

Table 6.2 gives the pressure drop values for the various flow-rates at heat flux values of 3.15, 6.30, 9.45, and 12.6 MW/m². Figure 6.8 shows the pressure drop/flow characteristics for each heat flux together with experimental results obtained by Maulbetsch.

Problems

1. For the conditions given in Example 1 (Chapters 5 and 6) calculate the pressure gradient profile over the region of subcooled and saturated boiling, i.e., the last 1.93 m. Calculate the overall pressure drop across the tube accounting for the increased pressure gradient in the subcooled boiling region and compare the result with that given in Example 1 of Chapter 2.
2. Tabulate the friction factor to account for subcooled boiling at the wall of a heated tube corresponding to the conditions used in Table 6.1. Assume that the bubble layer acts as a sand roughness equivalent to 0.75 times the diameter of the bubble at departure. Use the model developed by Levy to provide this departure diameter.
3. Evaluate $\Delta T_{SUB}(z)_B$ using the equations given by Saha and Zuber (eqs. (6.13) and (6.14)) for the conditions shown in Fig. 6.3 and compare with the predictions of Bowring and Levy.

7

SATURATED BOILING HEAT TRANSFER

7.1 Introduction

In this chapter the present state of knowledge regarding the heat transfer characteristics of saturated forced convective boiling and evaporation will be discussed. Returning to Fig. 4.21, it can be seen that this covers the regions C and D, where nucleate boiling is occurring at the wall and where the flow pattern would be typically bubbly, slug, or low vapour velocity annular and regions E and F where there is no nucleation at the wall; this latter region is normally associated with annular flow.

For the regions under consideration, it would appear that, for most practical purposes, the assumption of thermodynamic equilibrium between the phases of a single component fluid is acceptable. However, there are some areas, notably at small values of the reduced pressure, P_r, where significant non-equilibrium between the phases does occur. This has been observed with refrigerants (Gouse and Dickson 1966) and liquid metals (Chen 1965). The methods outlined in Chapter 2 for evaluating pressure drop and void fraction may be employed, provided the effects of possible changes in phase distribution due to the presence of a heat flux are allowed for (see Section 9.7.2).

Research work from many fields of engineering has contributed to the knowledge of forced convective boiling inside tubes and other shaped conduits. Notably, work has been published in the technical journals covering mechanical, chemical, refrigeration, petroleum, boiler plant, and nuclear engineering amongst others. Studies in some of these fields date back into the nineteenth century. However, it is only in the last twenty or so years that important contributions have been made towards understanding the processes involved in progressive vaporization along a tube. The lack of progress in the earlier work arises, in part, from the large number of experimental variables which are present but, in the main, it is the consequence of the fact that early workers in the field reported only effects on the performance of the evaporator resulting from independent variables altered by the operator. It is clear from a brief inspection of the complex flow patterns which exist when progressive evaporation takes place along a tube that it is essential to study the local or point conditions in the evaporator. No such study was published prior to 1952.

In the early experiments on evaporators condensing steam was used as the heating medium. In this case the heat flux along the length of the evaporator is a dependent variable, the magnitude of which, at any point, is set by the *local overall* heat transfer coefficient and overall temperature difference at that point. In addition, the mass quality, $x(z)$, or weight fraction vaporized at any point is a function of the total heat added to the fluid up to that point. It has been established that the heat transfer coefficient in forced convective boiling may be strongly dependent upon either the heat flux or the mass quality, $x(z)$, in certain regions. Obviously the case of a long evaporator tube heated by condensing steam is a complex one to analyse fully.

Two experimental approaches have been used which considerably simplify the situation. Firstly, the use of electrically heated tubular test sections has allowed a uniform heat flux distribution and level to be specified as an independently controlled variable resulting in a linear variation of the mass quality, $x(z)$, along the tube length. Secondly, by considering only a very short section of the evaporator, it may be assumed that conditions are approximately constant. Methods of measuring the rate of condensate production over a number of short lengths along a steam-heated evaporator have been introduced.

The following sections do not attempt to review all the available literature on this subject. The intention is rather to provide a brief account of the present state of knowledge. A review giving details of a large part of the literature published in the period 1935–57 has been prepared previously (Collier 1958). In contrast to the situation for subcooled boiling there are a considerable number of published studies for liquids other than water.

7.2 Saturated forced convective boiling in a round tube

As noted earlier, one of the prime variables in saturated forced convective boiling is $x(z)$, the 'mass quality' or the 'weight fraction vaporized'. This variable replaces the subcooling, $\Delta T_{SUB}(z)$, used in the subcooled region. The mass quality at any point for a tube with a uniform heat flux is given by

$$x(z) = \frac{4\phi}{DGi_{fg}} (z - z_{sc}) \tag{7.1}$$

Should a region of subcooled boiling precede the saturated forced convective boiling region then for values of $z < z^*$

$$x'(z) = \frac{4}{DGi_{fg}} \frac{\phi}{(1 + \varepsilon)} (z - z_d) \tag{7.2}$$

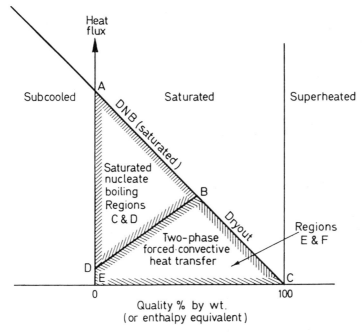

Fig. 7.1. Boundaries of saturated forced convective boiling and evaporation.

The realization that at least two different mechanisms of heat transfer could occur when liquid is evaporated in a tube came with the first measurements of local heat transfer coefficients by Dengler and Addoms (1956); many other studies have since confirmed their conclusions.

Consider the 'saturated nucleate boiling regime' (Fig. 7.1). Here, the mechanism of heat transfer is essentially identical to that in the subcooled regions. A thin layer of liquid near to the heated surface is superheated to a sufficient degree to allow nucleation. Just as with all other characterizing variables such as coolant temperature, void fraction, pressure gradient, etc., the heat transfer coefficient and heater surface temperature variation is smooth and continuous through the thermodynamic boundary AD marking the onset of saturated boiling. All the methods and equations used to correlate experimental data in the subcooled regions (see Chapter 5) remain valid in this region with the provision that $T_f(z) = T_{SAT}$. Just as the heat transfer mechanism in the subcooled region is independent of the subcooling and, to a large degree, the mass velocity, so it may be inferred that the heat transfer process in this region is independent of the 'mass quality' $x(z)$ and the mass velocity. This indeed is found to be experimentally correct (Bennett et al. 1961; Collier and Pulling 1962; Sterman and Styushin 1951) for the case of 'fully developed nucleate boiling'. Thus, because the bulk temperature is constant in this region the heat transfer coefficient is also constant since ΔT_{SAT} is fixed.

7.3 Suppression of saturated nucleate boiling

To maintain nucleate boiling on the surface, it is necessary that the wall temperature exceeds a critical value for a specified heat flux. The stability of nucleate boiling in the presence of a temperature gradient was discussed in detail in Chapter 5 and the arguments given in that section are equally valid in application to the *suppression* of nucleate boiling with decreasing wall superheat in the two-phase region. If the wall superheat is less than that given by eq. (5.23) or eq. (5.33) of Chapter 5 for the imposed surface heat flux, then nucleation does not take place (Fig. 7.2); the value of ΔT_{SAT} for

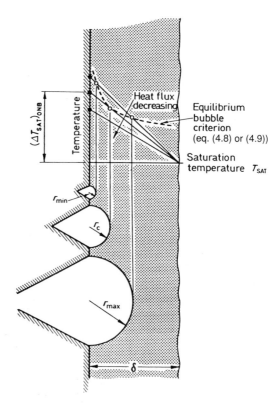

Fig. 7.2. Nucleation in a liquid film (after Hsu 1961).

comparison with these equations is calculated from the ratio (ϕ/h_{TP}), where h_{TP} is the two-phase heat transfer coefficient in the absence of nucleation. As will be seen in the following section, h_{TP} increases with increasing quality and with increasing mass velocity. Thus ΔT_{SAT} decreases, for a given flux, with increasing values of $x(z)$ or G and, when its value falls below that required to cause nucleation at the imposed flux, nucleate boiling ceases.

The complex inter-relationship between heat flux, quality, and mass velocity at the point of suppression was discussed qualitatively earlier, in Chapter 4, and can be deduced quantitatively if the equations for h_{TP} and ϕ_{ONB} are suitably combined. Early studies of this type are reported by Collier and Pulling (1962) who used the criterion derived by Hsu (1961) and assumed all the temperature drop occurred across the boundary sublayer. It is, for example, possible to combine the relationship given in eq. (5.33), Chapter 5, for ϕ_{ONB} with the Dengler and Addoms (1956) equation for h_{TP}:

$$\frac{h_{TP}}{h_{fo}} = 3.5\left(\frac{1}{X_{tt}}\right)^{0.5} \tag{7.3}$$

where h_{fo} is the heat transfer coefficient for the total flow, assumed to be liquid, and X_{tt} is the Martinelli parameter defined previously in Chapter 2 in terms of the pressure drop ratio.

$$X_{tt} = \sqrt{\left[\frac{(dp/dz)_f}{(dp/dz)_g}\right]} \approx \left(\frac{1-x}{x}\right)^{0.9}\left(\frac{\rho_g}{\rho_f}\right)^{0.5}\left(\frac{\mu_f}{\mu_g}\right)^{0.1} \tag{7.4}$$

The relationship obtained by this combination is

$$\phi_{ONB} = \frac{49Bh_{fo}^2}{k_f X_{tt}} \tag{7.5}$$

where

$$B = \left[\frac{2\sigma T_{SAT}v_{fg}}{Ji_{fg}}\right] \tag{5.27}$$

Equation (7.5) represents an equation for the boundary between the saturated nucleate boiling and the two-phase forced convection regions (Fig. 7.1). As for the case of subcooled nucleate boiling this relationship was derived on the basis of a complete range of 'active' cavities on the heating surface. The treatment by Davis and Anderson (Section 5.3) should be employed when the 'active' cavity size spectrum is limited. Methods of estimating the largest 'active' cavity size have been discussed in Chapter 4.

In subcooled boiling, a region of partial boiling occurs between the point of onset of nucleation and the region of fully developed boiling. There is a similar 'partial boiling' region between fully developed saturated nucleate boiling and the two-phase forced convective region. In this transition region, both the forced convective and nucleate boiling mechanisms are significant.

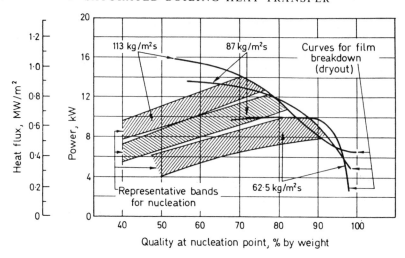

Fig. 7.3. Comparison of nucleation bands and film breakdown curves (Hewitt *et al.* 1963).

The gradual suppression of the latter as, say, quality is increased along the boiling channel can, in some cases, lead to a temporary reduction of the heat transfer coefficient with increasing quality. However, once the forced convective mechanism becomes dominant, the heat transfer coefficient increases with increasing quality. In the absence of a critical heat flux condition, the forced convective heat transfer coefficient may be regarded as a minimum value for any given quality. Results obtained by Collier and Pulling (1962) typify the variation of heat transfer coefficient with quality. The ratio h_{TP}/h_f was plotted against $1/X_{tt}$. This type of plot is preferable, for these particular results, to a straight plot of h_{TP} against x, because of the large pressure variations in these experiments both along the channel and from test to test.

Although results such as those of Collier and Pulling (1962) show qualitative agreement with the Hsu (1961) type models for the suppression of nucleation, it is difficult to identify the precise point from heat-transfer data. Direct observations of nucleation in annular flow have been reported by Hewitt *et al.* (1963). The geometry used was an internally heated annulus with a climbing film on the inner surface only. The results obtained are illustrated in Fig. 7.3 and qualitatively confirm the expected trends; ϕ_{ONB} increases with increasing G and $x(z)$. Unfortunately, it was not possible to make wall temperature measurements in these experiments, nor was it possible to infer reliable values of h_{TP} for use in the calculation of ΔT_{SAT}. As a first approximation eq. (7.5) can be used. However, the calculated values of ϕ_{ONB} are very much lower than those measured and shown in Fig. 7.3. This discrepancy is consistent with the findings of Davis and Anderson (1966) mentioned above. Equation (7.5) predicts that nucleation will persist at wall superheats down to 3°C but it is very unlikely that the large sizes of cavity required to be 'active' down to this superheat were present on the stainless steel heating surface. For this case, a maximum size of 'active' cavity present might be $\sim 1\,\mu m$ (0.4×10^{-4} in.) radius as discussed in Chapter 5.

7.4 The two-phase forced convection region

In principle, this region could be associated with any of the flow patterns described in Chapter 1. In practice, however, in boiling channels, in the absence of very high mass velocities, the two-phase forced convective region is most likely to be associated with the annular flow pattern. Heat is transferred by conduction and convection through the liquid film and vapour is generated continuously at the interface. The bulk vapour in the core region is normally assumed to be at the saturation temperature appropriate to the local pressure though this may not be true if there is a very large pressure gradient along the channel. A small temperature drop between the interface and the bulk vapour is a necessary consequence of a vapour flux away from the interface. This temperature drop is, again, normally ignored, but it may be important in some cases, particularly at low reduced pressures ($P_r < 0.001$). The calculation of this interface drop is discussed in detail in the context of condensation in Chapter 10. If the interfacial temperature drop is ignored the total temperature drop in the system ($\Delta T_{SAT} = T_W - T_{SAT}$) occurs across the liquid film and is smaller the thinner the film and the higher the degree of turbulence within it. Extremely high heat transfer coefficients are possible in this region; values can be so high as to make accurate assessment difficult. Typical figures for water of up to $200\,kW/m^2\,°C$ ($35,000$ Btu/h ft^2 °F) have been reported.

Following the suggestion of Martinelli, many workers have correlated their experimental results of heat transfer rates in the two-phase forced convective region in the form

$$\frac{h_{TP}}{h_{fo}} \left(\text{or } \frac{h_{TP}}{h_f} \right) = fn\left(\frac{1}{X_{tt}}\right) \qquad (7.6)$$

where h_{TP} is the value of the two-phase heat transfer coefficient, and h_{fo} and (h_f) are values of the single-phase liquid heat transfer coefficient based on the total (or liquid component) flow.

It is of interest to examine briefly why a relationship of this type might be successful in correlating data in this region. Consider a vertical heated tube, inside diameter D, when (a) the liquid flow completely fills the tube (as for upward flow), (b) the same liquid flow-rate occurs entirely as a liquid film flowing turbulently on the heated surface (as for downward flow). In both cases it is assumed that the mechanism of heat transfer is convection and that no nucleation occurs at the heating surface. Using the treatment of McAdams (1954) it is possible to relate, by dynamic similarity, the heat transfer coefficient obtained under the above two conditions to the physical dimensions of the tube and the liquid film. McAdams showed that for a fixed flow-rate the Reynolds number will be the same whether the fluid partially or completely fills the channel (see Section 3.5.1). Since the liquid physical properties are approximately equal for the two cases, the heat transfer coefficient should vary inversely as the equivalent diameter for each type of flow. For the case of thin film flow, it is assumed that the film is unrippled and has a thickness, δ, very much less than D. It is now possible to evaluate the equivalent diameter, D_e, for each condition.

For condition (a) $D_e = D$
For condition (b) $D_e = 4 \times$ hydraulic mean radius $= 4\delta$

Therefore the ratio of the heat transfer coefficients in the two cases may be given as

$$\frac{h_{(b)}}{h_{(a)}} = \frac{h_{TP}}{h_f} = \frac{D}{4\delta} \tag{7.7}$$

Now, when δ is very small compared with D the liquid hold-up, $(1 - \alpha)$ may be approximately given by

$$(1 - \alpha) = \frac{4\delta}{D} \tag{7.8}$$

and so

$$\frac{h_{TP}}{h_f} = \frac{1}{(1 - \alpha)} \tag{7.9}$$

It has been noted in Chapter 2 that Lockhart and Martinelli successfully correlated the liquid hold-up $(1 - \alpha)$ using the parameter, X. Thus, it might be expected that an approach such as that implied by eq. (7.6) might be successful. However, using eq. (7.9) together with actual or predicted void fraction data yields estimates of h_{TP} which are about 50 per cent higher than measured values.

   Equation (7.9) is similar in form to eq. (3.27) for the pressure gradient in annular flow and can similarly be shown to be related to the theoretical analysis outlined in Section 3.5.1. It implicitly contains, therefore, all the assumptions made in these analyses with all their deficiencies. It may sometimes be more appropriate, when there is considerable entrainment occurring, to put $h_{TP}/h_{fF} = 1/(1 - \alpha)$. h_{fF} is the full pipe heat transfer coefficient for the liquid film flow $W_f(1 - e)$ where e is the fraction of the liquid flow entrained. For further details reference should be made to Section 3.5.

The heat transfer within the liquid film can be calculated from the theoretical models for annular flow which are briefly described in Chapter 3. By analogy with the equation for turbulent momentum transfer $\tau = (\mu + \varepsilon\rho)\,du/dy$ the relationship for turbulent heat transfer can be written as

$$\phi = -(k_f + \varepsilon_H c_{pf} \rho_f)\frac{dT}{dy} \tag{7.10}$$

where ε_H is the eddy diffusivity for heat. Integrating eq. (7.10) from $y = \delta$ ($T = T_i \approx T_{SAT}$) to $y = 0$, it is possible to evaluate T_W and hence ΔT_{SAT} and h_{TP} for a given value of ϕ. It is convenient to express eq. (7.10) in a dimensionless form as follows:

$$1 = \left(\frac{1}{Pr_f} + \frac{\varepsilon_H}{\mu_f/\rho_f}\right)\frac{dT^+}{dy^+} \tag{7.11}$$

where Pr_f is the liquid Prandtl number, μ_f and ρ_f are estimated at some suitable mean temperature, and

$$T^+ = \frac{c_{pf}\rho_f u^*}{\phi}(T_W - T) \tag{7.12}$$

$$u^* = \sqrt{\frac{\tau_w}{\rho_f}} \tag{7.13}$$

$$y^+ = \frac{u^* y \rho_f}{\mu_f} \tag{7.14}$$

Assuming that the eddy diffusivities of heat and momentum are equal ($\varepsilon_H = \varepsilon$) and inserting appropriate well-known expressions for ε in the various ranges of y^+, eq. (7.11) can be integrated from $T^+ = 0$ at $y^+ = 0$ to obtain values of T_i^+ at $y^+ = y_i^+$. The two-phase heat transfer coefficient can then be computed from the relationship

$$h_{TP} = \frac{c_{pf}\rho_f u^*}{T_i^+}. \tag{7.15}$$

and the liquid film Nusselt number is given by

$$Nu_f = \frac{h_{TP}\delta}{k_f} = \frac{y_i^+ Pr_f}{T_i^+} \tag{7.16}$$

Extensive tables of T_i^+ as a function of film Reynolds number, Re_f, and Prandtl number, Pr_f, are given for various shear stress distributions by Hewitt (1961) for upflow and by Dukler (1960) for downflow. For the case

Table 7.1   Values of the liquid film Nusselt number Nu_f as a function of the liquid film Reynolds number and the Prandtl number (Hewitt and Hall-Taylor 1970)

Pr_f	Re_f					
	10	10^2	10^3	10^4	10^5	10^6
0.5	1.0	1.0	1.5	5.8	30	180
1	1.0	1.0	1.8	8.2	47	300
2	1.0	1.0	2.3	11.5	73	460
5	1.0	1.1	2.8	16	109	730
10	1.0	1.25	3.7	21.5	145	940

$$Nu_f = \left[\frac{h_{TP}\,\delta}{k_f}\right]$$

$$Re_f = \left[\frac{4\delta u_f \rho_f}{\mu_f}\right] = \left[\frac{Dj_f \rho_f}{\mu_f}\right]$$

$$Pr_f = \left[\frac{\mu c_p}{k}\right]_f$$

of a constant shear stress across the film ($\tau = \tau_W$) the friction velocity, u^*, and the film Reynolds number are known and a value of h_{TP} (or Nu_f) is readily obtained. Hewitt and Hall-Taylor (1970) have provided values of Nu_f as a function of Re_f and Pr_f for this particular case. Table 7.1 presents these values in tabular form. As might be expected at low film Reynolds numbers the value of Nu_f approaches unity—the laminar flow solution.

In general, the heat transfer coefficient predicted from eq. (7.9) or from Table 7.1 is somewhat higher than the observed one; this might be attributable to the existence of a significant interfacial temperature drop. An alternative explanation is that there is a reduction in eddy diffusivity in the region adjacent to the interface. A further explanation may be found in the transient nature of the flow. The existence of the large disturbance waves opens up the possibility that the average shear stress calculated from the mean pressure gradient is not the one which should be used in calculating u^*. In the intervals between the disturbances the shear stress may be much lower.

A number of relationships of the form of eq. (7.3) have been proposed (Dengler and Addoms 1956; Bennett et al. 1961; Guerrieri and Talty 1956; Schrock and Grossman 1959, 1962) and in some cases these have been extended to cover the saturated nucleate boiling region also.

Table 7.2   Range of conditions for data used in testing correlations

Fluid	Geometry	Flow	Pressure, (bar)	Liquid inlet velocity (m/s)	Quality (wt. %)	Heat flux ϕ (kW/m²)
Water	Tube	Up	0.55–2.76	0.06–1.45	15–71	88–630
Water	Tube	Up	2.9–34.8	0.24–4.5	3–50	205–2400
Water	Tube	Down	1.1–2.1	0.24–0.82	2–14	44–158
Water	Annulus	Up	1–2.4	0.06–0.27	1–59	100–500
Methanol	Tube	Up	1	0.3–0.76	1–4	22–54
Cyclohexane	Tube	Up	1	0.4–0.85	2–10	9.5–41
Pentane	Tube	Up	1	0.27–0.67	2–12	9.5–38
Heptane	Tube	Up	1	0.3–0.73	2–10	6.2–28
Benzene	Tube	Up	1	0.3–0.73	2–9	12.5–41

7.4.1   Chen correlation

Chen (1963a) has carried out a comparison of the correlations of Dengler and Addoms (1956), of Guerrieri and Talty (1956), of Bennett *et al.* (1961) and of Schrock and Grossman (1959, 1962) using a representative selection of 594 experimental data points. The range of experimental conditions covered by this comparison is summarized in Table 7.2 and the results are given in Table 7.3. The correlation of Dengler and Addoms represents their own data fairly well but predicts coefficients which are too high when compared with data from other sources. The equation proposed by Guerrieri and Talty correlates the data of those authors well but predicts coefficients lower than those measured by Dengler and Addoms, and by Sani (1960). It does, however, predict coefficients of the same order of magnitude as those measured by Schrock and Grossman. The correlation of Bennett *et al.* appears to predict reasonable values for the majority of the water data albeit with a wide scatter. However, the predicted coefficients are approximately 50 per cent too low when compared with the organic data. The correlation of Schrock and Grossman is some improvement on the other relationships, although it too predicts too low for the organic data by approximately 30 per cent. None of the correlations examined by Chen can be considered satisfactory. Chen (1963a), therefore, proposed a new correlation, which proved very successful in correlating all the forced convective boiling heat transfer data for water and organic systems listed in Table 7.2.

The proposed correlation covers both the 'saturated nucleate boiling region' and the 'two-phase forced convection region'. It was assumed that both mechanisms occur to some degree over the entire range of the

Table 7.3   Comparison of correlations

Data	Dengler and Addoms (1956)	Guerrieri and Talty (1956)	Bennett et al. (1969)	Schrock and Grossman (1959)	Chen (1963a)
			Average percentage deviations for correlations		
Dengler and Addoms (water)	30.5	62.3	20.0	20.3	14.7
Schrock and Grossman (water)	89.5	16.4	24.9	20.0	15.1
Sani (1960) (water)	26.9	70.3	26.5	48.6	8.5
Bennett et al. (water)	17.9	61.8	11.9	14.6	10.8
Guerrieri and Talty (methanol)	42.5	9.5	64.8	62.5	11.3
Guerrieri and Talty (cyclohexane)	39.8	11.1	65.9	50.7	13.6
Guerrieri and Talty (benzene)	65.1	8.6	56.4	40.1	6.3
Guerrieri and Talty (heptane)	61.2	12.3	58.0	31.8	11.0
Guerrieri and Talty (pentane)	66.6	9.4	59.2	35.8	11.9
Combined average for all data	38.1	42.6	32.6	31.7	11.0

correlation and that the contributions made by the two mechanisms are additive. This assumption of superposition is similar to that used by Rohsenow in the 'partial boiling' region for subcooled conditions (Chapter 5).

$$h_{TP} = h_{NcB} + h_c \qquad (7.17)$$

where h_{TP} is the local heat transfer coefficient, h_{NcB} is the contribution due to nucleate boiling and h_c is the contribution due to convection.

It was assumed that the convective component, h_c, could be represented by a Dittus–Boelter type equation,

$$h_c = 0.023 \, Re_{TP}^{0.8} \, Pr_{TP}^{0.4} \frac{k_{TP}}{D} \qquad (7.18)$$

where the thermal conductivity (k_{TP}) and the Reynolds and Prandtl numbers are effective values associated with the two-phase fluid. However, heat is transferred to a liquid film in 'annular' and 'dispersed' flow so Chen argued that it is reasonable to use the liquid thermal conductivity in eq. (7.18). Similarly, the values of the Prandtl modulus for liquid and vapour are normally of the same magnitude and it may be expected that the 'two-phase

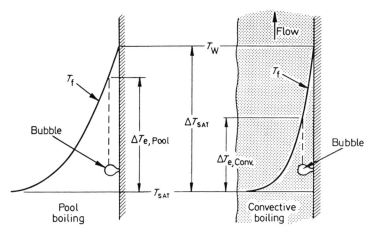

Fig. 7.4. Temperature profiles for pool boiling and for convective boiling with same superheat (Chen 1963*a*).

value' will be close to this value. A parameter F is defined, such that

$$F = \left[\frac{\text{Re}_{\text{TP}}}{\text{Re}_\text{f}}\right]^{0.8} = \left[\frac{\text{Re}_{\text{TP}}}{G(1-x)D/\mu_\text{f}}\right]^{0.8} \tag{7.19}$$

Equation (7.17) may now be rewritten as

$$h_\text{c} = 0.023\left[\frac{G(1-x)D}{\mu_\text{f}}\right]^{0.8}\left[\frac{\mu c_\text{p}}{k}\right]_\text{f}^{0.4}\left(\frac{k_\text{f}}{D}\right)(F) \tag{7.20}$$

The sole unknown function in eq. (7.20) is the expression for F. However, since this ratio is a flow parameter only, it may be expected that it can be expressed as a function of the Martinelli factor, X_{tt}. This is found to be the case. The analysis of Forster and Zuber (1955) was taken as a basis for the evaluation of the 'nucleate boiling' component, h_{NcB}. As discussed in Section 4.4 this analysis for pool boiling relates a bubble Nusselt number with a bubble Reynolds number (based on the premise that this is governed by the bubble growth rate) and the liquid Prandtl number. The bubble growth rate is described by eq. (4.26) which can be solved to show that the product of the growth rate and radius is a constant for a given superheat, ΔT_{SAT}. However, in both pool boiling and forced convective boiling the actual superheat is not constant across the boundary layer but falls (Fig. 7.4). Thus, the mean superheat of the fluid in which the bubble grows is lower than the wall superheat ΔT_{SAT}. The difference between this lower mean superheat (ΔT_e) and the wall superheat is small for the case of pool boiling and was neglected by Forster and Zuber but is significant in the forced convection case and cannot be neglected (Fig. 7.4) since the boundary layer is much thinner and the temperature gradients much steeper.

Writing the equation for h_{NcB} using the Forster and Zuber analysis with effective values of the superheat and vapour pressure difference,

$$h_{NcB} = 0.00122 \left[\frac{k_f^{0.79} c_{pf}^{0.45} \rho_f^{0.49}}{\sigma^{0.5} \mu_f^{0.29} i_{fg}^{0.24} \rho_g^{0.24}} \right] \Delta T_e^{0.24} \Delta p_e^{0.75} \tag{7.21}$$

Chen then defines a suppression factor, S, the ratio of the mean superheat (ΔT_e) to the wall superheat (ΔT_{SAT})

$$S = [\Delta T_e / \Delta T_{SAT}]^{0.99} \tag{7.22}$$

The exponent 0.99 allows S to appear to the first power in the final equation. Using the Clausius–Clapeyron equation, eq. (7.22) can be written

$$S = (\Delta T_e / \Delta T_{SAT})^{0.24} (\Delta p_e / \Delta p_{SAT})^{0.75} \tag{7.23}$$

which allows eq. (7.21) to be expressed as

$$h_{NcB} = 0.00122 \left[\frac{k_f^{0.79} c_{pf}^{0.45} \rho_f^{0.49}}{\sigma^{0.5} \mu_f^{0.29} i_{fg}^{0.24} \rho_g^{0.24}} \right] \Delta T_{SAT}^{0.24} \Delta p_{SAT}^{0.75}(S) \tag{7.24}$$

It might be expected that the suppression factor, S, would approach unity at low flows and zero at high flows. Chen suggests that S can be represented as a function of the local two-phase Reynolds number, Re_{TP}. The functions F and S were determined empirically from experimental data using an iterative procedure to obtain the best solutions. These functions are shown in Figs. 7.5 and 7.6 for F and S respectively. The shaded areas on these plots indicate the scatter of data around the two functions.

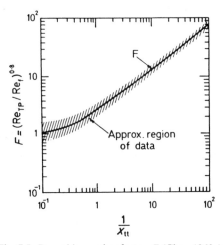

Fig. 7.5. Reynolds number factor, F (Chen 1963a).

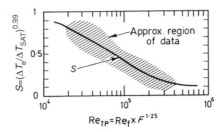

Fig. 7.6. Suppression factor, S (Chen 1963a).

This correlation developed by Chen was tested against the data listed in Table 7.2. The result of the comparison is shown in Table 7.3 and it will be seen that the correlation fits all the data remarkably well including the data of Bennett *et al.* for an internally heated annulus. In the latter case an equivalent diameter of four times the cross-sectional area divided by the *heated* perimeter was employed.

To calculate the heat transfer coefficient $(h_{TP})_1$ at a known heat flux (ϕ_1), mass velocity, and quality, the steps are as follows:

(a) calculate $1/X_{tt}$ (eq. (7.4))
(b) evaluate F from Fig. 7.5
(c) calculate h_c from eq. (7.20)
(d) calculate Re_{TP} from $Re_f(=G(1-x)D/\mu_f)$ and F (eq. (7.19), $Re_{TP} = F^{1.25} Re_f$)
(e) evaluate S from Fig. 7.6 using the calculated value of Re_{TP}
(f) calculate h_{NcB} for a range of ΔT_{SAT} values (eq. (7.24))
(g) calculate h_{TP} from eq. (7.17) for the range of ΔT_{SAT} values
(h) Plot ϕ $(=h_{TP}\,\Delta T_{SAT})$ for the ΔT_{SAT} range against h_{TP} and interpolate to obtain $(h_{TP})_1$ at ϕ_1.

Chen originally derived the functional form of the multiplier F empirically. However, using a Reynolds analogy he later derived the expression

$$F = (\phi_f^2)^{0.444} \tag{7.25}$$

If we make use of the relationship between ϕ_f^2 and X (cf. eq. (2.68)), this also agrees with the empirical relationship given by Fig. 7.5.

Bennett and Chen (1980) later extended this analysis to cover fluids of Prandtl numbers greater than unity using a modified Chilton–Colburn analogy

$$F = \left[\left(\frac{Pr_f + 1}{2}\right)\phi_f^2\right]^{0.444} \tag{7.26}$$

This approach gives good agreement with forced convective boiling data for ethylene glycol (Pr_f is about 8).

The correlation developed by Chen has been used extensively in practice for a wide variety of fluids and operating conditions, many beyond those cited in Table 7.2. The method is *not*, however, suitable for boiling in horizontal tubes.

Whilst it is assumed that the Chen correlation will be satisfactory for cryogenic liquids (Zuber and Fried 1962) modifications must be expected for the case of boiling liquid metals (Chen 1963b) although the same mechanisms of heat transfer would be expected to apply. The convective component, h_c of eq. (7.17) is now calculated from a modified form of the Lyon–Martinelli equation for convective single-phase liquid metal heat transfer. The Forster–Zuber equation is assumed to remain valid for the nucleate boiling component, h_{NcB}. Comparison with available experimental data (Hoffman and Krakoviak 1962 and Longo 1963) gives encouraging results.

The Chen correlation may be extended for use in the subcooled boiling region (Butterworth 1970). As discussed previously in Chapter 5, it can be assumed that the total surface heat flux is made up of a nucleate boiling contribution and a single-phase forced convective contribution

$$\phi = h_{NcB}(T_W - T_{SAT}) + h_c(T_W - T_f(z)) \tag{7.27}$$

For subcooled boiling the value of h_c may be obtained from eq. (7.20) with F set equal to unity. The nucleate boiling coefficient h_{NcB} is evaluated from eq. (7.24) with the value of S obtained from the single-phase liquid Reynolds number Re_F ($= GD/\mu_f$) using Fig. 7.6. Since h_{NcB} is itself a function of the wall superheat ΔT_{SAT} ($= T_W - T_{SAT}$) it is necessary to assume values of T_W in eq. (7.27) and produce curves of T_W versus ϕ for the various values of $T_f(z)$. This extension of the Chen correlation has been tested (Butterworth 1970) against experimental information for the subcooled boiling of water, n-butyl alcohol, and ammonia with satisfactory results.

Hall, Bjorge, and Rohsenow (1982) have extended the superposition approach, successful for 'partial' subcooled boiling to two-phase forced convection. In this approach eq. (5.40) is adapted to

$$\phi_{TP} = \phi_c + \phi_{SCB} - \phi_{c''} \tag{7.28}$$

where ϕ_c is the two-phase forced convective heat flux ($= h_c \Delta T_{SAT}$), ϕ_{SCB} is the fully developed nucleate boiling heat flux (given, for example, by the Forster and Zuber equation (4.46) or the Rohsenow equation (4.45) and $\phi_{c''}$ is the value of ϕ_{SCB} at the onset of nucleation (see Fig. 5.10).

This modification ensures that ϕ_{TP} is equal to ϕ_c at the onset of nucleation.

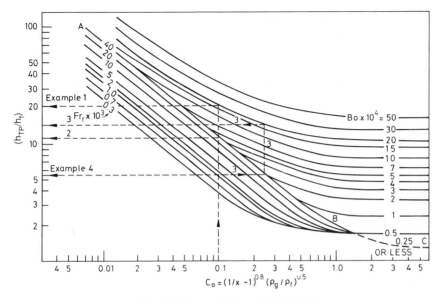

Fig. 7.7. The correlation of Shah (1976).

Since it is assumed that $\phi_{SCB} \propto \Delta T_{SAT}^{1/n}$ then eq. (7.28) may be rewritten

$$\phi_{TP} = \phi_c + \phi_{SCB}\left[1 - \left\{\frac{(\Delta T_{SAT})_{ONB}}{\Delta T_{SAT}}\right\}^{1/n}\right] \qquad (7.29)$$

The term in the brackets multiplying ϕ_{SCB} becomes the boiling suppression factor S replacing that used in the Chen correlation. Equation (5.33b) was used to estimate $(\Delta T_{SAT})_{ONB}$. This approach gave a slight improvement over the Chen correlation for experimental data with water as the fluid.

7.4.2   Shah correlations

A graphical alternative to the Chen correlation is that proposed by Shah (1976). His method is in the form of a graphical chart (Fig. 7.7) and can be used for boiling in both vertical and horizontal channels.

The ordinate is the ratio (h_{TP}/h_f) and the abscissa (C_0) is

$$\left(\frac{1-x}{x}\right)^{0.8}\left(\frac{\rho_g}{\rho_f}\right)^{0.5}$$

Two other parameters are used:

$$Bo = \text{the 'boiling number'} = \left(\frac{\phi}{Gi_{fg}}\right)$$

$$Fr_f = \text{the Froude number} = \left(\frac{G^2}{\rho_f^2 gD}\right)$$

$$(7.30)$$

Figure 7.7 also illustrates the use of the chart. Values of the abscissa C_o, of Bo and Fr_f are calculated for the specified conditions. Examples of different situations are as follows:

Example 1. Vertical tube $C_o = 0.10$, $Bo = 20 \times 10^{-4}$, $Fr_f = 0.002$. Since the tube is vertical, ignore Fr_f. Draw a vertical line at $C_o = 0.10$ to intersect the curve for $Bo = 20 \times 10^{-4}$. From this point of intersection, draw a horizontal line to the left and read (h_{TP}/h_f) at the ordinate ($=20$).

Example 2. Vertical tube $C_o = 0.10$, $Bo = 2 \times 10^{-4}$, $Fr_f = 0.002$. Since the tube is vertical, ignore Fr_f. Draw a vertical line at $C_o = 0.10$ until it intersects line AB. Curves for values of Bo less than 4.5×10^{-4} all merge into line AB; therefore the boiling number has no influence in this example. Draw a horizontal line from the point of intersection to the ordinate and read (h_{TP}/h_f) ($=11$).

Example 3. Horizontal tube $C_o = 0.10$, $Bo = 20 \times 10^{-4}$, $Fr_f = 0.002$. Draw a vertical line at $C_o = 0.10$. From the point where it intersects the line for $Fr_f = 2 \times 10^{-3}$ draw a horizontal line to the right to intersect the line AB. From this intersection draw a vertical line to intersect the curve for $Bo = 20 \times 10^{-4}$. Then draw a horizontal line from this last intersection to the left to the ordinate and read (h_{TP}/h_f) ($=14$).

Example 4. Horizontal tube $C_o = 0.10$, $Bo = 2 \times 10^{-4}$, $Fr_f = 0.002$. Draw a vertical line at $C_o = 0.10$. From the point where it intersects the line $Fr_f = 2 \times 10^{-3}$ draw a horizontal line to the right to intersect the line AB. Curves for values of Bo less than 2.0×10^{-4} all merge into the line AB, therefore the boiling number has no influence in this example. Extend the horizontal line to the left and read (h_{TP}/h_f) at the ordinate ($=5.4$).

The correlation is valid over the reduced pressure range 0.004 to 0.8. The r.m.s. error on 780 data points is 14 per cent; somewhat higher than for the Chen correlation. About 10 per cent of the data show deviations greater than ± 30 per cent. It will be noted that when the correlation is used for a horizontal tube the value of h_{TP} is the mean coefficient over the tube perimeter. The Shah correlation agrees well with the data of Chawla (1967) and Bandel and Schlünder (1974) for heat transfer coefficients to Freon 11 evaporating in 6, 14 and 25 mm horizontal tubes.

Shah (1982) has also developed equations that represent his chart method. Like Chen, he considered two distinct mechanisms: nucleate boiling and convective boiling. However, instead of adding the two contributions, the larger of his nucleate coefficient h_{NcB} or his convective coefficient h_c is used. The method is as follows:

(a) Calculate the dimensionless parameter N: For vertical tubes at all values of Fr_f and for horizontal tubes when $Fr_f > 0.04$:

$$N = C_o \qquad (7.31a)$$

For horizontal tubes when $Fr_f < 0.04$,

$$N = 0.38 Fr_f^{-0.3} C_o \qquad (7.31b)$$

(b) Then, for $N > 1.0$, calculate the values of h_{NcB} and h_c from the following expressions and choose the larger value for h_{TP}:

$$\text{For } Bo > 0.0003 \quad h_{NcB}/h_f = 230 \, Bo^{0.5} \qquad (7.32a)$$

$$\text{For } Bo < 0.0003 \quad h_{NcB}/h_f = 1 + 46 \, Bo^{0.5} \qquad (7.32b)$$

$$h_c/h_f = 1.8/N^{0.8} \qquad (7.32c)$$

For $1.0 > N > 0.1$, calculate the value h_c from eq. (7.32c) and h_{NcB} in the bubble suppression regime as

$$h_{NcB}/h_f = F \, Bo^{0.5} \exp(2.74 \, N^{-0.1}) \qquad (7.33)$$

and choose the larger value for h_{TP}.

For $N < 0.1$, calculate the value h_c from eq. (7.32c) and h_{NcB} in the bubble suppression regime as

$$h_{NcB}/h_f = F \, Bo^{0.5} \exp(2.47 \, N^{-0.15}) \qquad (7.34)$$

and choose the larger value for h_{TP}. The constant F is determined as follows:

$$Bo > 0.0011, \qquad F = 14.7$$

$$Bo < 0.0011, \qquad F = 15.43$$

The value of h_f is for the liquid flow $G(1 - x)$ and is calculated from the Dittus–Boelter correlation using liquid properties.

Shah also demonstrated that these equations apply to boiling of water in vertical annuli. For annular gaps between the inner and outer tubes greater than 4 mm, the equivalent diameter is $(D_o - D_i)$; for gaps less than 4 mm, the equivalent diameter is obtained by evaluating the hydraulic diameter using only the *heated* perimeter. One disadvantage of this approach is utilizing the boiling number Bo to model nucleate boiling since it only includes the latent heat as a fluid property; moreover, the typical exponent for nucleate boiling is about 0.7 rather than 0.5. Kandlikar (1990, 1991) has proposed a Shah-like correlation that utilizes fluid dependent constants.

7.4.3 Gungor–Winterton correlation

A new correlation of the Chen form has been developed by Gungor and Winterton (1986). A large data base (3693 points) was used covering both the nucleate and convection dominated boiling regimes for water, refrigerants (R11, R12, R22, R113 and R114) and ethylene glycol for vertical upward and downward flows and horizontal flows. The basic equation for saturated boiling is

$$h_{TP} = Eh_f + Sh_{NcB} \qquad (7.35)$$

with h_f calculated from the Dittus–Boelter correlation, using the local liquid fraction of the flow, i.e., $G(1 - x)$. The new convection enhancement factor E is given by the expression

$$E = 1 + 24{,}000 \text{Bo}^{1.16} + 1.37(1/X_{tt})^{0.86} \qquad (7.36)$$

where X_{tt} is the Martinelli parameter defined by eq. (7.4) and the new completely empirical boiling suppression factor S is

$$S = [1 + 0.000\,001\,15E^2\,\text{Re}_f^{1.17}]^{-1} \qquad (7.37)$$

with Re_f based on $G(1 - x)$. The nucleate pool boiling coefficient is calculated from the Cooper (1984a) reduced-pressure correlation:

$$h_{NcB} = 55P_r^{0.12}(-0.4343 \ln P_r)^{-0.55}M^{-0.5}\phi^{0.67} \qquad (4.50)$$

which gives the heat transfer coefficient in $W/m^2\,°C$ where M is the molecular weight, P_r is the reduced pressure and the heat flux is in watts per square metre. If the tube is horizontal and the Froude number is below 0.05, stratified flow occurs and E must be multiplied by

$$E_2 = \text{Fr}_f^{(0.1-2\,\text{Fr}_f)} \qquad (7.38)$$

and S must be multiplied by

$$S_2 = (\text{Fr}_f)^{1/2} \qquad (7.39)$$

This method correlated the saturated boiling data base to a mean deviation of 21.4 per cent, compared to 57.7 per cent for the Chen correlation and 21.9 per cent for the Shah (1982) correlation. Using the same equivalent diameter definitions as Shah, heat transfer to boiling water flowing in a vertical annulus was predicted with a mean error of 29.4 per cent. In the subcooled boiling regime, the temperature differences driving heat transfer for nucleate boiling and convection are different and the heat flux is calculated from (cf eq. (7.27))

$$\phi = h_f(T_W - T_f(z)) + Sh_{NcB}(T_W - T_{SAT}) \qquad (7.40)$$

This predicts their data base with a mean error of 25 per cent.

7.4.4 Steiner–Taborek correlation

A new general application boiling correlation has been proposed by Steiner and Taborek (1992) for boiling in vertical tubes with upward (and downward) flow. Elements of the method are drawn from many years of research by Steiner (1986, 1988) and co-workers. Rather than following the trend towards *statistical* development of boiling correlations, they have placed the main emphasis on arriving at a sound *mechanistic* model that will respect the established principles of pool and convective boiling. Even so, they have used a very extensive data base (12,607 data points) assembled at the University of Karlsruhe. It contains 10,262 data points for water and 2,345 data points for four refrigerants (R11, R12, R22 and R113), seven hydrocarbons (benzene, n-pentane, n-heptane, cyclohexane, methanol, ethanol and n-butanol), three cryogens (nitrogen, hydrogen and helium) and ammonia. The data points were closely scrutinized and some were re-evaluated from raw data using newer, more accurate physical properties. No single data reference was allowed to dominate the data base. Importantly, an attempt was made to obtain a general distribution of data over the whole range of boiling conditions (reduced pressures, quality, flow rates, etc.) and not allow easily measured data points to dominate the data bank at the expense of the fewer data points available for severe test conditions.

The Steiner–Taborek method is arranged so that one can utilize all of it or substitute one's own correlations or data for various parts of it. From initial indications, it is the most accurate vertical tube boiling correlation that is currently available and its format allows for continued improvement of its various parts without necessarily affecting other parts of the model. Its development and underlying ideas are described in detail below.

7.4.4.1 Boiling models

Boiling models can be classified by how the nucleate boiling and convective boiling coefficients are combined to obtain h_{TP}. The general power law type of model is of the form

$$h_{\text{TP}} = [(h_{\text{NcB}})^n + (h_{\text{c}})^n]^{1/n} \tag{7.41}$$

The effect of the value of n is shown in Fig. 7.8. A simple addition of the respective values is obtained by setting $n = 1$, which was originally proposed by Rohsenow (1952) and then extended by Chen (1963a) with his boiling suppression factor, indicating that the simple additive method was insufficient. Kutateladze (1961) proposed a power law method using $n = 2$, which is an asymptotic approach since the value of h_{TP} tends to the larger of the two values. Instead, n equal to infinity is the case of choosing the larger of the two coefficients.

For single-phase convection, this correlational approach has been formalized by Churchill (1974) for modelling the transition between two different

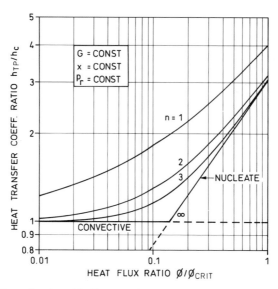

Fig. 7.8. Power law based boiling model representation (Steiner and Taborek 1992).

heat transfer regimes, namely forced and natural convection, each dominated by different limiting thermal mechanisms. Thus, this approach was chosen by Steiner and Taborek, applying it to very carefully obtained results such as those of Robertson and Wadekar (1988) covering convective dominated boiling data, nucleate boiling dominated boiling data and the transition between these two for ethanol boiling in a vertical, 10 mm internal diameter tube (Fig. 7.9). The best value of n was found to be 3 for flow boiling. Similarly, for convective condensation on the outside of tube bundles $n = 4$ has been used for combining the shear-controlled and gravity-controlled condensate drainage regimes (Honda *et al.* 1988).

7.4.4.2 *Fundamental limitations to flow boiling*

For boiling in vertical tubes, Steiner and Taborek proposed that the following limitations apply to the process and should be included in a realistic flow boiling model:

(a) At heat fluxes below that of onset of nucleate boiling (see Chapter 5), no nucleate boiling occurs and only pure convective boiling is present. Here the coefficient h_{TP} is nearly independent of heat flux over a wide range of mass velocity and vapour quality.

(b) In the fully developed nucleate boiling regime, h_{TP} is largely independent of mass velocity and flow quality.

(c) At $x = 0$, h_{TP} will correspond to the nucleate boiling coefficient when

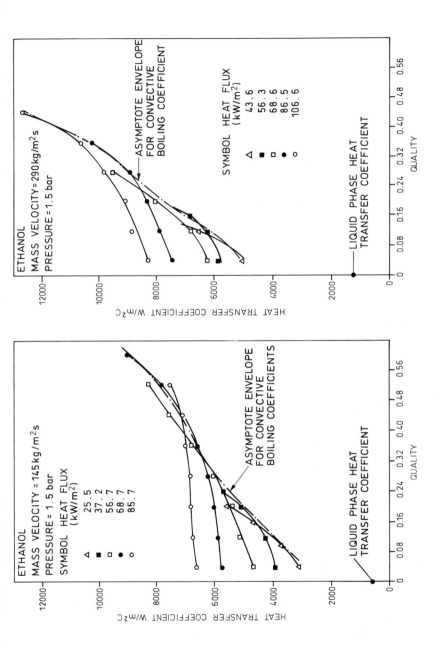

Fig. 7.9. Boiling ethanol in a vertical tube at two mass fluxes from Robertson and Wadekar (1988).

$\phi > \phi_{ONB}$ while h_{TP} will be equal to the single-phase convection coefficient when $\phi < \phi_{ONB}$ (in this case convective boiling is important right from the start).

(d) At intermediate heat fluxes, the transition from nucleate to convective boiling is more gradual while at high heat fluxes (especially at low mass velocities), h_{TP} is virtually independent of vapour quality over a wide range of x until convective boiling becomes important. This is especially true at high reduced pressure, where the nucleate boiling coefficient is very large.

(e) At $x = 1.0$, h_{TP} reverts to the single-phase coefficient for vapour flowing alone in the tube. This condition may not be reached if the temperature differences are large and deviation from thermal equilibrium is significant, such that liquid droplets will remain as a mist in the flow to relatively high superheatings of the vapour phase.

The process of boiling in vertical tubes together with its mechanistic limitations are shown schematically in Fig. 7.10 (cf. Fig. 4.22). The model can be described as follows:

Region A–B. As subcooled liquid approaches point A, only single-phase convection occurs if the heat flux is below that for onset of nucleate boiling (ϕ_{ONB}) under subcooled conditions while subcooled boiling will exist if this heat flux is surpassed. If subcooled boiling exists, vapour bubbles grow and collapse near the tube wall and the heat transfer coefficient increases substantially because of the two-phase flow effect and the disturbance of the laminar sublayer.

Region B–C–D. The flow quality is now greater than zero and the velocity of the two-phase flow rapidly increases with increasing x. For $\phi < \phi_{ONB}$, pure convective boiling will occur along the pure convective boiling curve indicated. For most industrial applications, the conditions exist for $\phi > \phi_{ONB}$ and the nucleate and convective contributions are superimposed. The *dashed* lines represent the nucleate boiling coefficient, which is only a function of heat flux but not flow velocity and vapour quality. The *full* lines represent the actual value of the combined nucleate boiling and convective boiling contributions. The flow pattern progresses from the bubbly flow regime through the churn flow regime.

Region D–E–F–G. For $\phi < \phi_{ONB}$, only convective boiling is present and the curve continues along that indicated for pure convective boiling. No dryout occurs up to very high vapour qualities, approaching 1.0. This type of boiling can exist in direct expansion evaporators in the refrigeration industry, where the flow enters with a vapour quality in the range of 0.15 to 0.25 and heat fluxes are low due to the controlling air side coefficients. For $\phi > \phi_{ONB}$, the annular flow regime is entered and characterized by a thin liquid layer on the channel wall and a higher velocity central vapour core.

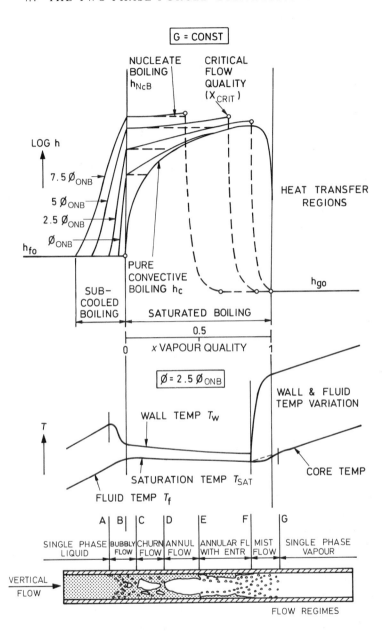

Fig. 7.10. Steiner–Taborek (1992) schematic representation of the vertical flow boiling process.

As vapour quality and hence vapour velocity continue to increase, the liquid film cannot be sustained and the dry-out point is reached at the critical flow quality point. Then either the liquid film is sheared off the wall or it dries out and the mist-flow regime is encountered. In the mist-flow regime neither nucleate boiling nor convective boiling occurs; instead heat is primarily transferred by single-phase convection to the continuous vapour phase with evaporation of the entrained liquid droplets by the superheated vapour.

7.4.4.3 *Flow boiling model*

The local flow boiling heat transfer coefficient is defined as

$$h_{TP} = \phi/(T_W - T_{SAT}) \tag{7.42}$$

where ϕ, T_W and T_{SAT} are local values for the axial flow position and T_{SAT} corresponds to the local saturation pressure. The local flow boiling coefficient is written as

$$h_{TP} = [(h_{NcB})^3 + (h_c)^3]^{1/3} \tag{7.43}$$

or

$$h_{TP} = [(h_{NcB,o} F_{NcB})^3 + (h_{fo} F_{TP})^3]^{1/3} \tag{7.44}$$

where

$h_{NcB,o}$ is the local 'normalized' nucleate pool boiling coefficient at a standard condition of heat flux and reduced pressure. It can be obtained from experimental data or a nucleate pool boiling correlation of the user's choice (or from the method recommended here.)

F_{NcB} is the nucleate boiling correction factor that accounts for the differences between pool and flow boiling, including the effects of pressure, heat flux, tube diameter, surface roughness, etc. but not boiling suppression (which does not occur according to Steiner and Taborek).

h_{fo} is the local liquid-phase forced convection coefficient based on the total flow as liquid (G). The Gnielinski (1976) correlation valid for both the transition and fully turbulent flow regimes (Reynolds number > 4000) was used in developing the correlation. While it is more complex to use than the Dittus–Boelter equation (eq. (5.7)), it is more accurate and used here.

F_{TP} is the two-phase multiplier that accounts for the enhancement of convection in a two-phase flow. It is a function of x and the liquid/vapour density ratio.

Physical properties at the local saturation pressure and temperature are used.

7.4.4.4 *Local liquid-phase forced convection coefficient h_{fo}*

The Gnielinski correlation is an adaption of the Petukhov *et al.* (1970) turbulent flow correlation that extends its coverage to the transition flow regime. It is given as

$$Nu = \frac{(f/8)(Re - 1000)Pr}{1 + 12.7(f/8)^{1/2}(Pr^{2/3} - 1)} \qquad (7.45)$$

where the Fanning friction factor is obtained from

$$f = [0.7904 \ln(Re) - 1.64]^{-2} \qquad (7.46)$$

and is valid for

$$4000 < Re < 5,000,000$$

$$0.5 < Pr < 2000$$

The total flow rate of liquid plus vapour is used for evaluating the above expression using liquid physical properties. For lower Reynolds numbers, eq. (5.6) is recommended.

7.4.4.5 *Two-phase multiplier F_{TP}* Convective boiling will occur if $x < x_{CRIT}$ and $\phi > \phi_{ONB}$ or over the entire range of x if $\phi < \phi_{ONB}$. For general industrial applications (thermosyphon reboilers and similar) where $x < x_{CRIT}$ at the tube exit, the following equation for F_{TP} was derived:

$$F_{TP} = [(1 - x)^{1.5} + 1.9x^{0.6}(\rho_f/\rho_g)^{0.35}]^{1.1} \qquad (7.47)$$

This expression has been developed for boiling data with a density ratio ranging from 3.75 to 5000. The expression correctly converges to unity as x goes to 0. Normally x_{CRIT} is in the proximity of 0.5 for these applications.

For the case of very low heat fluxes where only pure convective boiling is present and extends over the entire range from $x = 0$ up to x approaching unity, the limiting case at $x = 1.0$ applies and the value of h_c corresponds to h_{go} (the forced convection coefficient with the total flow as a gas). For this case, the following expression is given:

$$F_{TP} = \{[(1 - x)^{1.5} + 1.9x^{0.6}(1 - x)^{0.01}(\rho_f/\rho_g)^{0.35}]^{-2.2}$$
$$+ [(h_{go}/h_{fo})x^{0.01}(1 + 8(1 - x)^{0.7})(\rho_f/\rho_g)^{0.67}]^{-2}\}^{-0.5} \qquad (7.48)$$

where the terms $x^{0.01}$ and $(1 - x)^{0.01}$ were included to arrive at the proper limits at $x = 0$ and $x = 1$. This expression correlated boiling data over the density ratio range from 3.75 to 1017.

7.4.4.6 *Nucleate pool boiling coefficient $h_{NcB,o}$* In the present method, the Gorenflo (1988) type of relationship is used to predict these values. In this approach a 'normalized' nucleate flow boiling coefficient at the following

Table 7.4   Normalized heat flux ϕ_0 for various fluid classes

Fluid class	ϕ_0 (W/m²)
Inorganic fluids: water, ammonia, CO_2, etc.	150,000
Hydrocarbons, refrigerants, organics	20,000
Cryogens: H_2 to O_2 (including N_2)	10,000
Helium I	1,000

standard conditions is used: reduced pressure $P_r = 0.1$, mean surface roughness $R_{po} = 1$ μm and the heat flux equal to the value listed in Table 7.4 for
various classes of fluids. This coefficient is then modified by a generalized
expression to account for the effects of pressure, heat flux, etc. as described
in the following section.

Values of $h_{NcB,o}$ can be obtained from Table 7.5. (Note that this is not
Gorenflo's method exactly (see Section 4.4.3) but the method interpreted
at heat fluxes where most data were available for flow boiling, which is not
the same range as for nucleate pool boiling). Otherwise, the Cooper (1984b)
reduced pressure correlation (described in Section 4.4.3) can quickly be
evaluated at the standardized conditions or the user may choose any other
alternative correlation. Also, experimental nucleate pool boiling data for the
fluid of interest can be used if the value of $h_{NcB,o}$ at the actual design condition
is available (then set F_{NcB} equal to 1), or at any other condition if it is then
converted to the standard conditions using the Gorenflo nucleate pool
boiling method.

7.4.4.7   *Nucleate boiling correction factor* F_{NcB} Reviewing extensive data,
Steiner and Taborek observed *no* effect of mass velocity and vapour quality
on flow nucleate boiling over wide ranges of heat flux and pressure. Thus,
they have excluded the notion of a 'nucleate boiling suppression factor'
originally included in the Chen (1963a) correlation from their model.
Presently, F_{NcB} includes the effects of reduced pressure, heat flux, tube
diameter, surface roughness and a residual correction factor on the 'normalized' nucleate pool boiling heat transfer coefficient. The expression is as
follows:

$$F_{NcB} = F_{PF}[\phi/\phi_o]^{nf}[D/D_o]^{-0.4}[R_p/R_{po}]^{0.133}F(M) \qquad (7.49)$$

where the pressure correction factor F_{PF} valid for $P_r < 0.95$ is

$$F_{PF} = 2.816P_r^{0.45} + \left(3.4 + \frac{1.7}{(1 - P_r^7)}\right)P_r^{3.7} \qquad (7.50)$$

Table 7.5 Nucleate flow boiling coefficients at the normalized conditions $P_r = 0.1$, ϕ_0 and $R_{po} = 1$ μm

Fluid	P_{CRIT} (bar)	Molecular weight	ϕ_0 (W/m²)	$h_{NcB,0}$ (W/m² °C)
Methane	46.0	16.04	20,000	8,060
Ethane	48.8	30.07	20,000	5,210
Propane	42.4	44.10	20,000	4,000
n-Butane	38.0	58.12	20,000	3,300
n-Pentane	33.7	72.15	20,000	3,070
Isopentane	33.3	72.15	20,000	2,940
n-Hexane	29.7	86.18	20,000	2,840
n-Heptane	27.3	100.20	20,000	2,420
Cyclohexane	40.8	84.16	20,000	2,420
Benzene	48.9	78.11	20,000	2,730
Toluene	41.1	92.14	20,000	2,910
Diphenyl	38.5	154.21	20,000	2,030
Methanol	81.0	32.04	20,000	2,770
Ethanol	63.8	46.07	20,000	3,690
n-Propanol	51.7	60.10	20,000	3,170
Isopropanol	47.6	60.10	20,000	2,920
n-Butanol	49.6	74.12	20,000	2,750
Isobutanol	43.0	74.12	20,000	2,940
Acetone	47.0	58.08	20,000	3,270
R11	44.0	137.37	20,000	2,690
R12	41.6	120.91	20,000	3,290
R13	38.6	104.47	20,000	3,910
R13B1	39.8	148.93	20,000	3,380
R22	49.9	86.47	20,000	3,930
R23	48.7	70.02	20,000	4,870
R113	34.1	187.38	20,000	2,180
R114	32.6	170.92	20,000	2,460
R115	31.3	154.47	20,000	2,890
R123	36.7	152.93	20,000	2,600
R134a	40.6	102.03	20,000	3,500
R152a	45.2	66.05	20,000	4,000
R226	30.6	186.48	20,000	3,700
R227	29.3	170.03	20,000	3,800
RC318	28.0	200.03	20,000	2,710
R502	40.8	111.6	20,000	2,900
Chloromethane	66.8	50.49	20,000	4,790
Tetrachloromethane	45.6	153.82	20,000	2,320
Tetrafluoromethane	37.4	88.0	20,000	4,500

(*Continued*)

Table 7.5   (*Continued*)

Fluid	P_{CRIT} (bar)	Molecular weight	ϕ_0 (W/m^2)	$h_{NcB,0}$ (W/m^2 °C)
Helium I*	2.275	4.0	1,000	1,990
Hydrogen (para)	12.97	2.02	10,000	12,220
Neon	26.5	20.18	10,000	8,920
Nitrogen	34.0	28.02	10,000	4,380
Argon	49.0	39.95	10,000	3,870
Oxygen	50.8	32.0	10,000	4,120
Water	220.64	18.02	150,000	25,580
Ammonia	113.0	17.03	150,000	36,640
Carbon dioxide†	73.8	44.01	150,000	18,890
Sulphur hexafluoride	37.6	146.05	150,000	12,230

* Physical properties at $P_r = 0.3$ rather than 0.1 (0.68 bar).
† Calculated with properties to triple point temperature

and the exponent on the normalized heat flux term is

$$nf = 0.8 - 0.1 \exp(1.75 P_r) \tag{7.51}$$

for all fluids except cryogens, and for cryogens is

$$nf = 0.7 - 0.13 \exp(1.105 P_r) \tag{7.52}$$

Thus, the slope of the boiling curve is modified with reduced pressure, decreasing with increasing pressure. The effect of tube diameter on nucleate pool boiling has been determined from matching the flow boiling model to flow boiling data for tests with tubes ranging from 1 to 32 mm in internal diameter. The standard tube reference diameter D_0 is equal to 0.01 m. The surface roughness term covers values of R_p from 0.1 to 18 μm. The standard value of 1 μm is about average for commercial tubes and should be used if the roughness is not known. The residual correction factor was included to improve the performance of the correlation and is given as a function of liquid molecular weight M as

$$F(M) = 0.377 + 0.199 \ln M + 0.000028427 M^2 \leqslant 2.5 \tag{7.53}$$

This expression covers M from 10 to 187. Instead, higher accuracy can be attained by using values listed for $F(M)$ in Table 7.6 for the specific fluid that were obtained by statistical regression of the data for that fluid. Table 7.6 *must* be used for $M < 10$. The maximum value of $F(M)$ is limited to 2.5.

Table 7.6 Values of $F(M)$ for specific fluids

Fluid	H_2	He	NH_3	Water	N_2	R22	R12	R113
M	2	4	17	18	28	86	120	187
$F(M)$	0.35	0.86	1.24	0.72	0.80	1.20	1.21	2.20

7.4.4.8 Minimum heat flux ϕ_{ONB} The criterion for the onset of nucleate boiling is the expression

$$\phi_{ONB} = \frac{2\sigma T_{SAT} h_{fo}}{r \rho_g i_{fg}} \tag{7.54}$$

where the critical radius $r = 0.3 \times 10^{-6}$ m is recommended for commercial tubing as based on analysis of the flow boiling data. For heat fluxes above ϕ_{ONB}, nucleate boiling is predicted to occur, while below this value it is not. Other methods described in Section 5.3 can be used as appropriate.

The mean errors for the method are cited by fluid in the paper. In addition the method was compared to the other correlations described earlier for several classes of fluids, namely hydrocarbons (ethanol), refrigerants (R12) and ammonia. The Steiner–Taborek method had lower standard and maximum deviations than the best of these in each case.

7.5 Heat transfer in areas where the critical heat flux has been exceeded

7.5.1 Introduction

The processes of heat transfer in those areas where the critical heat flux has been exceeded will now be discussed. The critical heat flux condition (shown as a ridge in the 3-D representation of the 'boiling surface'—Fig. 4.23) represents a situation at which there is a more or less sudden decrease in the value of the heat transfer coefficient of one or two orders of magnitude as compared with the values obtained in the nucleate boiling and two-phase forced convective areas. This decrease is a consequence of the fact that the heat transfer surface is no longer completely wetted by the liquid phase. Vapour now completely or partially blankets the surface and heat transfer is no longer to a highly conducting liquid but to a poorly conducting gas.

Referring to Fig. 7.11 (cf. Fig. 4.24) and Fig. 4.23, it will be noted that such a condition can occur for both the subcooled and saturated liquid states. Four separate areas can be identified and it is convenient to consider the heat transfer mechanisms in terms of these four regions—*transition boiling*,

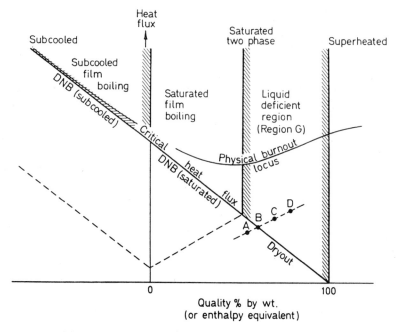

Fig. 7.11. Regions of heat transfer where the critical heat flux has been exceeded.

subcooled film boiling, saturated film boiling, and the *liquid deficient region*—
although there would appear to be no well-defined boundaries between the
regions. The *transition boiling* region occupies the reverse slope (partly
obscured) in Fig. 4.23, while the other regions—*subcooled film boiling,*
saturated film boiling, and the *liquid deficient region* all occupy parts of the
slope on the other side of the valley representing the minimum heat flux.

Knowledge of the conditions and heat transfer rates in these areas is
required in a variety of engineering fields. The design of once-through steam
generators (where complete evaporation of the water feed is carried out) and
of very-high-pressure recirculation boilers (where the critical heat flux levels
are low) requires the tube wall operating temperature to be known to ensure
against overheating. The safe design and operation of nuclear power reactors
cooled by pressurized or boiling liquid coolants often requires a knowledge
of heat transfer rates beyond the critical heat flux point for transient and
accident analysis. Other fields using such information include those of
refrigeration and cryogenic engineering and also the chemical and pharma-
ceutical industries (involving the design of high performance evaporators).

Film boiling can also occur in the heat transfer from very hot bodies
present in effectively unbounded liquid flow systems. Examples of such
systems are transverse flow over tube banks and boundary layer flows.

7.5.2 Transition boiling

The use of experimental techniques (Iloeje *et al.* 1974; Plummer *et al.* 1974; Groenveld 1974) where the surface temperature rather than the surface heat flux is the controlling variable, has established quite definitely the existence of a *transition boiling* region in the forced convection condition as well as for pool boiling. A comprehensive review of the published information on *transition boiling* under forced convection conditions has been prepared by Groeneveld and Fung (1976). Table 7.7, taken from the Groeneveld and Gardiner (1977) review, lists those studies which were carried out prior to 1976. Since that date, in response to the urgent demands of the nuclear industry, further studies have been initiated.

Various attempts have been made to produce correlations for the *transition boiling* region. Probably the most useful currently available is that by Tong and Young (1974) which is given in terms of the heat flux in the transition boiling region (ϕ_{TB})

$$\phi_{TB} = \phi_{NB} \exp\left[-0.001 \frac{x^{2/3}}{(dx/dz)} \left(\frac{\Delta T}{55.5}\right)^{(1+0.0029\,\Delta T)} \right] \qquad (7.55)$$

where ϕ_{NB} is the nucleate boiling heat flux (presumably equated with the critical heat flux) in W/m^2 and ΔT is the temperature difference between the surface and the saturation condition in °C. A transition boiling correlation having a wider range of application may be developed if the heat flux and wall temperature difference at the points of maximum and minimum in the boiling surface can be predicted with confidence. The present state of the art allows an accurate prediction of ϕ_{CRIT} and also ΔT_{CRIT} but the conditions at the minimum point are still subject to a large degree of uncertainty. The experimental data of Iloeje *et al.* (1975) show that the conditions at the minimum point are a complex function of the mass quality and flow-rate in the channel (see Figs. 7.23 and 7.24).

7.5.3 Film boiling

A schematic representation of the conditions under consideration is given in Fig. 7.12. The study of forced convective film boiling inside vertical tubes made by Dougall and Rohsenow (1963) included visual observations of the flow structure. At low qualities and mass flow-rates the flow regime would appear to be an 'inverted annular' one with liquid in the centre and a thin vapour film adjacent to the heating surface. The vapour–liquid interface is not smooth, but irregular. These irregularities occur at random locations, but appear to retain their identity to some degree as they pass up the tube with velocities of the same order as that of the liquid core. The vapour in the film adjacent to the heating surface would appear to travel at a higher

Table 7.7 Transition boiling data for water in forced convection (taken from Groeneveld and Gardiner 1977)

Geometry	Reference	Range of data				Comments
		p bar	$G \times 10^{-3}$ kg/m² s	$\phi \times 10^{-3}$ kW/m²	Subcooling (°C) or quality	
Annulus, 0.64 cm i.d. 6.35 cm o.d.	Ellion (1954)	1.10–4.13	0.33–1.49	1.47–1.96	28–56°C	ϕ-controlled system with stabilizing fluid, $z_H = 7.62$ cm
Tube $D_e = 0.386$ cm	McDonough et al. (1961)	5.51 8.27 13.78	0.27–2.04	0.32–3.78	Subcooled and low quality	NaK used as heating fluid. T_W inferred from heat transfer correlation for NaK. Data no longer available
Annulus 0.013 cm i.d. 1.21 cm o.d.	Peterson et al. (1973)	1.01	0.64–1.93	0.41–1.99	Saturated	Heat flux controlled by electronic feedback. $z_H = 5.08$ cm
Tube $D_e = 1.25$ cm	Plummer et al. (1973)	68.9	0.07–0.34	0.06–0.27	$x = 0.30$–1.00	Transient test $z_H = 10.16$ cm
Annulus 1.37 cm i.d. 2.54 cm o.d.	Ramu and Weisman (1975)	1.72–2.06	0.02–0.05	0.03–0.26	$x = 0$–0.500	Hg used as heating fluid. x not reported. Limited range in T_W
Rod bundles $D_e = 1.27$ cm	Westinghouse Flecht Cadeck et al. (1971)	1.03–6.20	0.05–0.25	0.01–0.27	0–78°C	Transient test. ΔT_{SUB} or x unknown
Tube $D_e = 1.27$ cm	Cheng and Ng (1976)	1.03	0.19	0.016–0.158	0–26°C	Transient and steady state test, high inertia. Copper block $z_H = 10.16$ cm
Tube $D_e = 1.27$ cm	Fung (1977)	1.01	0.068–1.35	0.008–1.89	0–76°C	Similar tests to Cheng and Ng
Tube $D_e = 1.27$ cm	Newbold et al. (1976)	3.03	0.016–1.25	0.016–0.948	0–80°C	Similar tests to Cheng and Ng. However, guard heaters were employed to reduce axial conduction

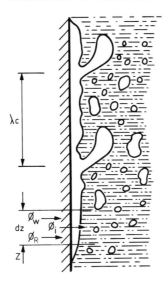

Fig. 7.12. Schematic representation of low quality post-CHF condition.

velocity. The liquid core may be in up-flow, stationary, or in down-flow. Vapour bubbles may be present in the liquid core, but will have little influence except to modify the core velocity and density which in turn determine the interfacial shear stress.

7.5.3.1 *Simple theories* Because the liquid is displaced from the heating surface by the vapour film and the uncertainties associated with bubble nucleation are removed, film boiling is amenable to analytical solution. In general, the problem is treated as an analogy of film-wise condensation (see Chapter 10) and solutions are available for horizontal and vertical flat surfaces and also inside tubes under both laminar and turbulent conditions with and without interfacial stress.

The simplest solution is obtained by assuming the vapour film is laminar and that the temperature distribution through the film is linear. For a vertical flat surface various boundary conditions may be imposed on such an analysis, viz.

(a) zero interfacial stress ($\tau_i = 0$)
(b) zero interfacial velocity ($u_i = 0$).

The local heat transfer coefficient $h(z)$ at a distance z up the surface from the start of film boiling is given by

$$h(z) = C \left[\frac{k_g^3 \rho_g (\rho_f - \rho_g) g i_{fg}'}{z \mu_g \Delta T} \right]^{1/4} \tag{7.56}$$

C depends on the assumed boundary condition; for zero interfacial stress $C = 0.707$, for zero interfacial velocity $C = 0.5$.

A similar analysis was carried out by Wallis and Collier (1968) for the assumption of turbulent flow in the vapour film. In this case

$$\left[\frac{h(z)z}{k_g}\right] = 0.056 \, \mathrm{Re}_g^{0.2}[\mathrm{PrGr}^*]^{1/3} \tag{7.57}$$

where

$$\mathrm{Gr}^* = \left[\frac{z^3 g \rho_g (\rho_f - \rho_g)}{\mu_g^2}\right]$$

Note that the coefficient is constant, independent of distance z in this equation. Fung $et\ al.$ (1979) has also developed a model which covers both the laminar and the turbulent film boiling regions.

The vapour film in reality is not smooth but breaks away from the wall in large globular voids. Various attempts have been made to produce an equation which is more consistent with this physical behaviour. Bailey (1971) suggested that the globular voids are formed by a process of 'varicose' instability of a hollow gas cylinder within a dense liquid as proposed by Chandrasekhar. The wavelength for maximum growth for a surface tension governed instability of this type is

$$\frac{\lambda_c}{2\pi} = \frac{r}{0.484} \tag{7.58}$$

Substitution of λ_c from this equation for z in eq. (7.56) leads to

$$h = \left[\frac{k_g^3 \rho_g (\rho_f - \rho_g) g i'_{fg}}{\mu_g \Delta T r}\right]^{1/4} \tag{7.59}$$

where r is the radius of the rod on which boiling is taking place. An alternative suggestion for vertical rods which predicts heat-transfer coefficients about 33 per cent higher is the use of the original Bromley equation for film boiling outside horizontal tubes:

$$h = 0.62 \left[\frac{k_g^3 \rho_g (\rho_f - \rho_g) g i'_{fg}}{D \mu_g \Delta T}\right]^{1/4} \tag{4.65}$$

Denham (1984) extended the Bromley model (eq. 7.56) to account for the heat transfer from the interface into the bulk subcooled liquid. Hence, condensation of the vapour occurs:

$$\frac{d\Gamma_g}{dz} = (\phi_w + \phi_r - \phi_i)/i'_{fg} \tag{7.60}$$

where Γ_g is the mass rate of flow of vapour per unit width of plate, ϕ_w is the heat flux from the wall by conduction to the vapour film, ϕ_r is the heat transferred by radiation and ϕ_i is the heat flux across the vapour film–liquid interface removed by conduction into the subcooled liquid core. These heat fluxes are given by the following expressions:

$$\phi_w = \frac{k_g \, \Delta T}{\delta} \tag{7.61a}$$

$$\phi_r = \frac{\sigma(T_w^4 - T_f^4(z))}{1/\varepsilon_f + (1/\varepsilon_w - 1)} \tag{7.61b}$$

$$\phi_i = -k_f \left(\frac{\partial T}{\partial r}\right)_i = k_f \frac{(T_{SAT} - T_f(z))}{(\pi\alpha_f t)} \frac{1}{1 + 3.75(\alpha_f t/r^2)} \tag{7.61c}$$

where the thickness of the vapour film used in the Bromley model (eq. 7.56) is

$$\delta = \left[\frac{4\Gamma_g \mu_g}{g\rho_g(\rho_f - \rho_g)}\right]^{1/3} \tag{7.62}$$

and where r is the radius of the liquid core (approximately $D/2$) and the time t is that taken by the liquid core to rise from the start of the film boiling region ($z = 0$), i.e. z/u_f.

These equations (7.60), (7.61a–c) and (7.62) form a first-order non-linear differential equation which Denham solved using different boundary conditions. He found good agreement with his experimental results when the constant in eq. (7.62) was modified from 4 to 5.2; he also used values of $\varepsilon_w = 0.8$ and $\varepsilon_f = 0.95$ in equation (7.61b). The model correctly predicted the effects of quality and flow rate, provided the vapour film remained stable. Instability occurred when the Weber number ($\rho_g u_g^2 D/\sigma$) exceeded 20.

7.5.3.2 *Experimental studies*

Bromley et al. (1953) have given some information for saturated film boiling of fluids other than water. In their experiments film boiling was from a horizontal graphite tube heated electrically. A variety of fluids was tested including *n*-hexane, carbon tetrachloride, benzene and ethanol.

The following relationships were proposed consisting of two terms, due to conduction and radiation, for upward flow over a horizontal tube. For the condition

$$\frac{u}{(gD)^{1/2}} < 1.0$$

the heat transfer coefficient is

$$h = h_c + 0.75h_r \tag{7.63}$$

where h_r is the radiation coefficient, h_c is the value suggested by Bromley for natural convection (eq. (4.66)), D is the outside tube diameter and u is the free stream velocity. For higher velocities where

$$\frac{u}{(gD)^{1/2}} > 2$$

the heat transfer coefficient is

$$h = h_c + 0.875h_r \tag{7.64}$$

and h_c is now given by

$$h_c = 2.7\left[\frac{uk_g\rho_g i'_{fg}}{D\,\Delta T_{SAT}}\right]^{1/2} \tag{7.65}$$

where i'_{fg} is defined by eq. (4.67). The heat transfer coefficient increases with higher velocities and smaller diameters. Typical values of h_c for n-hexane, benzene and ethanol vary from 230 W/m² °C (40 Btu/h ft² °F) for no flow to 625–800 W/m² °C (110–140 Btu/h ft² °F) at a velocity of 4 m/s (13 ft/s). The values for carbon tetrachloride were approximately one half these values.

In an extension of this work, Motte and Bromley (1957) studied film boiling from a horizontal tube for a subcooled liquid in forced convection. As we have seen already, some of the heat will be transferred from the vapour–liquid interface into the liquid. Consequently the vapour film thickness is less and the heat transfer rate across the vapour film is increased. The experimental results reported by Motte and Bromley show that at velocities (u) above about 2.5 m/s (8 ft/s), the heat transfer coefficients (h_c) are increased approximately fourfold by subcooling the liquid approximately 45°C (80°F). These values, thus, numerically approach the values expected for nucleate boiling of these organic fluids.

Liu et al. (1992) have proposed a correlation for forced convection subcooled film boiling over a horizontal cylinder which appears to represent the available experimental data over a range of liquids, subcoolings, velocities, and cylinder diameters.

To summarize, the evidence at the present time is that classical laminar film boiling occurs only over a relatively short distance (~ 5 cm) downstream of the dryout or 'rewet' front. Beyond this region vapour is released in large globular voids which ensures a time-varying but thinner vapour film. The coefficient becomes independent of distance and takes on a value considerably higher (approximately twice) than the laminar solution would indicate.

There is experimental evidence which suggests that film boiling heat transfer coefficients in downflow may be lower than for up-flow. This phenomenon was first observed by Papell (1970, 1971) with nitrogen and

hydrogen and has been confirmed by Newbold *et al.* (1976) and by Costigan
et al. (1984) for water.

Figure 7.13 shows some of the experimental results obtained by Costigan
for upflow and downflow film boiling within an 8 mm bore vertical tube
compared with the Bromley equation (7.56).

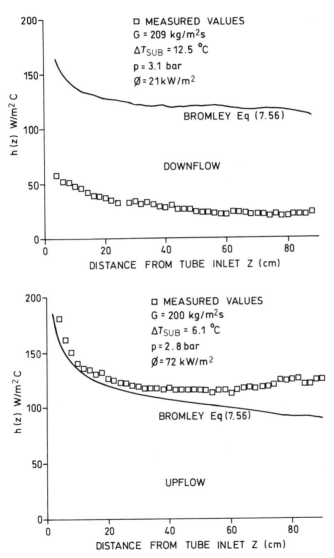

Fig. 7.13. Film boiling heat transfer for water in an 8 mm diameter vertical tube (Costigan *et al.* 1984).

The maximum effect of buoyancy in reducing film boiling heat transfer coefficients is seen under conditions where the downflow velocity is equal to the bubble drift flux velocity. This observation has allowed Bjornard and Griffith (1977) to rewrite the recommended post-dryout transfer correlations in terms of the drift flux model to account for the directional influence at low velocities.

Useful information on the forced convective boiling characteristics of various hydrocarbons beyond the critical heat flux is given in a paper by Glickstein and Whitesides (1967). Heat transfer coefficients were reported for the subcooled and saturated film boiling regions and for the liquid-deficient region. The fluids examined were methane, propane, propylene, butene-1, and certain eutectic mixtures.

7.5.4 *Heat transfer in the liquid deficient region*

In the 'saturated film boiling region' the flow pattern was that of an annular vapour film with a central liquid core. As the vapour quality is increased, the flow pattern changes to one in which liquid droplets are dispersed in a high velocity vapour core. This latter 'drop' flow pattern is characteristic of the 'liquid deficient regime'. The fact that the liquid film on the heat transfer surface in the 'two-phase forced convective region' has been partly or wholly removed results in the manifestation of the critical heat flux condition. Depending upon the particular conditions, the liquid droplets may or may not 'wet' the heat transfer surface. Heat transfer coefficients in this region are considerably higher than in the 'saturated film boiling' region. This fact coupled with the reduced critical heat flux values in the higher quality regions means that the liquid deficient region is often quite extensive and that the physical burnout heat flux may be considerably higher than the critical heat flux value.

In order to obtain some understanding of the processes which occur in the 'liquid deficient region' it is, perhaps, appropriate to consider what happens as the critical heat flux is reached and exceeded in this area. Consider Figs. 7.11, 7.14 and 7.15; Fig. 7.15 shows actual temperature recordings taken during such an experiment using a steam–water mixture at 69 bar (1000 psia). Initially, the heat flux (A) is below the critical heat flux value and the wall surface temperature is steady (Fig. 7.15) and close to the saturation temperature (Fig. 7.14). As the critical heat flux value is reached (point B), small oscillations in the surface temperature are observed (Fig. 7.15). These small temperature oscillations denote the onset of the 'dryout' condition where the liquid film covering the surface starts to break up into rivulets. A further increase in the heat flux (to point C) results in temperature oscillations of considerably increased magnitude (Figs. 7.14 and 7.15). These oscillations would appear to be the result of a given area of the heating

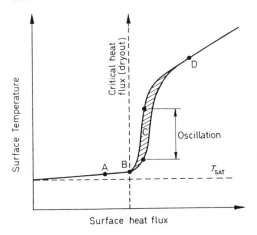

Fig. 7.14. Heating surface temperature at and above the critical heat flux (after Leong and Bray 1963).

surface being alternatively in contact with steam or rewetted by the random passage of a rivulet (Hewitt *et al.* 1963). Finally, a further increase still in heat flux (to point D) results in a temperature of the heating surface considerably above the saturation temperature but one which is steady (Figs. 7.14 and 7.15). The rivulets have dried out and the heating surface would now appear to be totally 'dry'.

Liquid deficient heat transfer rates are important in the design of very high pressure once-through and recirculation boilers. Experimental data for steam–water mixtures in the pressure range 138–250 bar (200–3500 psia) have been published by Schmidt (1959), Swenson *et al.* (1961), Bahr *et al.* (1969), and Herkenrath *et al.* (1967). Typical results from these studies are shown in Figs. 7.16 and 7.17. These diagrams show clearly the characteristic sharp increase in wall temperature at the critical point followed by decreasing wall temperatures as the vapour velocity increases with increasing vapour quality. Wall temperatures increase once again in the superheat region as a result of the increase in bulk fluid temperature. As the heat flux is increased so the critical point moves to lower vapour qualities and the wall temperatures in the liquid deficient region are increased.

Three types of correlating procedure have been adopted. They are:

(a) Correlations of an empirical nature which make no assumptions whatever about the mechanism involved in post-dryout heat transfer, but solely attempt a functional relationship between the heat transfer coefficient (assuming the coolant is at the saturation temperature) and the independent variables.

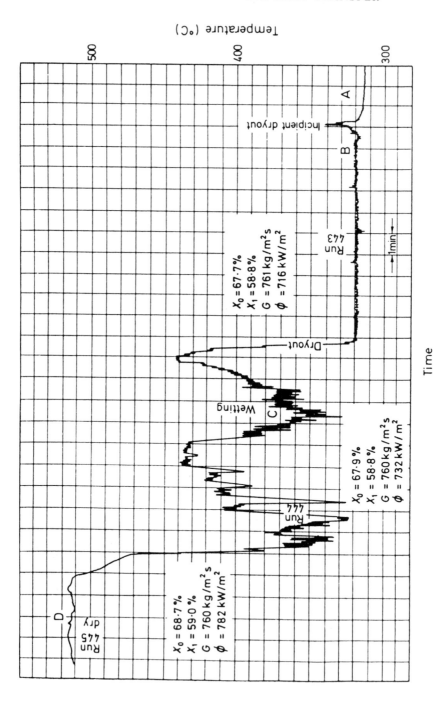

Fig. 7.15. Typical heater surface temperature plot obtained during a dryout test.

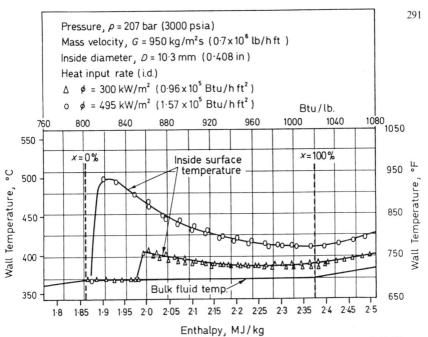

Fig. 7.16. Tube wall temperatures in the liquid deficient region (Swenson *et al.* 1961).

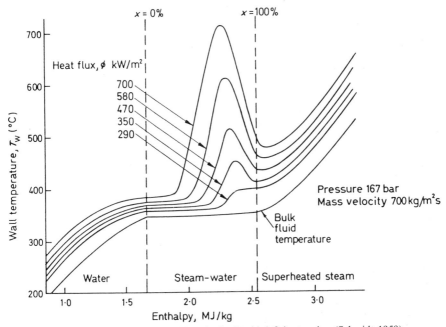

Fig. 7.17. Tube wall temperatures in the liquid deficient region (Schmidt 1959).

(b) Correlations which recognize that departure from a thermodynamic equilibrium condition can occur and attempt to calculate the 'true' vapour quality and vapour temperature. A conventional single-phase heat transfer correlation is then used to calculate the heated wall temperature.
(c) Semi-theoretical models where attempts have been made to look at and write down equations for the various individual hydrodynamic and heat transfer processes occurring in the heated channel and relate these to the heated wall temperature.

Groeneveld (1973) has compiled a bank of carefully selected data drawn from a variety of experimental post-dryout studies in tubular, annular, and rod bundle geometries for steam/water flows.

7.5.4.1 Empirical correlations

A considerable number of empirical equations have been presented by various investigators for the estimation of heat transfer rates in the post-dryout region. Almost all of these equations are modifications of the well-known Dittus–Boelter type relationship for single-phase flow and take no account of the non-equilibrium effects.

Probably the most accurate of these purely empirical correlations is that proposed by Groeneveld (1973). This has the form

$$\mathrm{Nu_g} = a\left[\mathrm{Re_g}\left\{x + \frac{\rho_g}{\rho_f}(1-x)\right\}\right]^b \mathrm{Pr_{g,w}^c}\, Y^d \tag{7.66}$$

where

$$Y = 1 - 0.1\left(\frac{\rho_f}{\rho_g} - 1\right)^{0.4}(1-x)^{0.4} \tag{7.67}$$

and the coefficients a, b, c, and d are given in Table 7.8 together with the range of independent variables on which the correlations are based. A number of improvements in the Groeneveld correlations have been made by Slaughterback et al. (1973a, b).

The use of the above correlations outside the range of variables for the respective geometries given in Table 7.8 is strongly discouraged.

7.5.4.2 Correlations allowing for departure from thermodynamic equilibrium

It has long been known that wall temperatures in the post-dryout region are bounded by two limiting situations, viz:

(i) *Complete departure from equilibrium.* The rate of heat transfer from the vapour phase to the entrained liquid droplets is so slow that their presence is simply ignored and the vapour temperature $T_g(z)$ downstream of the dryout point is calculated on the basis that all the heat added to the fluid goes into superheating the vapour. The wall temperature $T_w(z)$ is calculated using a conventional single-phase heat transfer correlation (Fig. 7.18a).

Table 7.8 Groeneveld (1973) empirical post-dryout correlations

Geometry	a	b	c	d	No. of points	RMS error %
Tubes	1.09×10^{-3}	0.989	1.41	-1.15	438	11.5
Annuli	5.20×10^{-2}	0.688	1.26	-1.06	266	6.9
Tubes and annuli	3.27×10^{-3}	0.901	1.32	-1.50	704	12.4

Range of data on which correlations are based

	Geometry	
	Tube	Annulus
Flow direction	Vertical and horizontal	Vertical
D_e, cm	0.25 to 2.5	0.15 to 0.63
p, bar	68 to 215	34 to 100
G kg/m^2 s	700 to 5300	800 to 4100
x, fraction by weight	0.10 to 0.90	0.10 to 0.90
ϕ, kW/m^2	120 to 2100	450 to 2250
Nu_g	95 to 1770	160 to 640
$Re_g(x + (1-x)\rho_g/\rho_f)$	6.6×10^4 to 1.3×10^6	1.0×10^5 to 3.9×10^5
Pr_w	0.88 to 2.21	0.91 to 1.22
Y	0.706 to 0.976	0.610 to 0.963

(ii) *Complete thermodynamic equilibrium.* The rate of heat transfer from the vapour phase to the entrained liquid droplets is so fast that the vapour temperature $T_g(z)$ remains at the saturation temperature until the energy balance indicates all the droplets have evaporated. The wall temperature $T_W(z)$ is again calculated using a conventional single-phase heat transfer correlation, this time with allowance made for the increasing vapour velocity resulting from droplet evaporation (Fig. 7.18b).

Post-dryout heat transfer behaviour tends towards situation (i) at low pressures and low velocity, while at high pressure (approaching the critical condition) and high flow-rates (>3000 kg/m^2 s) situation (ii) pertains.

It might have been expected that correlations allowing for various degrees of non-equilibrium behaviour between these two limiting situations would have been published analogous to those available for subcooled void fraction. It is, however, only in the past few years such correlations have appeared.

Consider Fig. 7.19 which represents the physical situation in the post-dryout region for a vertical tube diameter D heated uniformly with heat flux ϕ. Dryout occurs after a length z_{DO} and it is assumed that thermodynamic

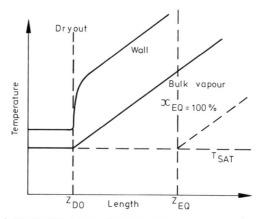

(a) COMPLETE LACK OF THERMODYNAMIC EQUILIBRIUM.

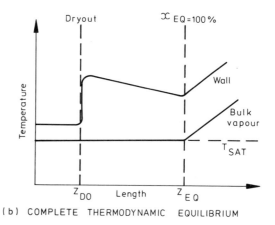

(b) COMPLETE THERMODYNAMIC EQUILIBRIUM

Fig. 7.18. Limiting conditions for post-dryout heat transfer.

equilibrium exists at the dryout point. An energy balance indicates that all the liquid is evaporated by z_{EQ} if equilibrium is maintained. However, the actual situation depicted in Fig. 7.19 is one in which, following dryout, a fraction (ε) of the surface heat flux is used in evaporating liquid, while the remainder is used to superheat the bulk vapour. The liquid is completely evaporated a distance z^* from the tube entrance (downstream from z_{EQ}).

Let the surface heat flux $\phi(z)$ be divided into two components $\phi_f(z)$, the heat flux associated with droplet evaporation, and $\phi_g(z)$, the heat flux associated with vapour superheating:

$$\phi(z) = \phi_f(z) + \phi_g(z) \tag{7.68}$$

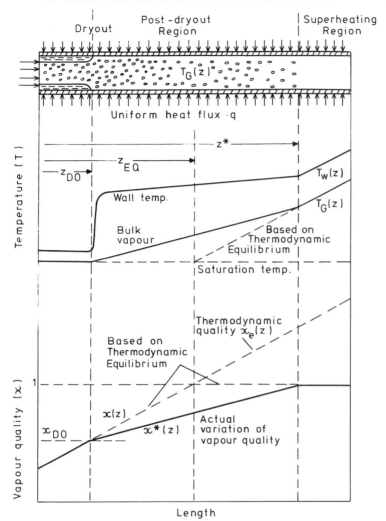

Fig. 7.19. Departure from thermodynamic equilibrium in post-dryout region.

Let $\varepsilon = \phi_f(z)/\phi(z)$; ε, in general, can be considered a function of tube length (z) (or alternatively local vapour quality x^*) but to keep the analysis simple we will initially assume ε is independent of tube length. This leads to the linear profiles of 'actual' bulk vapour temperature and vapour quality with tube length shown in Fig. 7.19, rather than the smooth curves obtained in experiment (Fig. 7.20).

The 'thermodynamic' vapour quality variation with length is given by

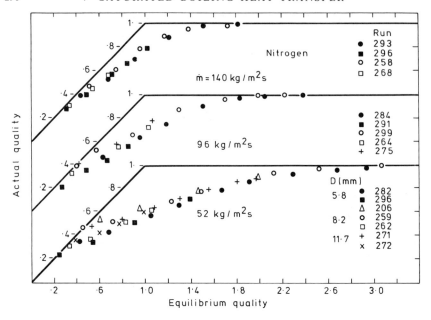

Fig. 7.20. Actual vs equilibrium quality values for nitrogen at several mass fluxes (Forslund and Rohsenow 1966).

$$(x(z) - x_{\text{DO}}) = \frac{4}{DGi_{\text{fg}}}(z - z_{\text{DO}}) \tag{7.69}$$

for $z < z_{\text{EQ}}$ and z_{EQ} is given by

$$z_{\text{EQ}} = \left[\frac{DGi_{\text{fg}}}{4\phi}(1 - x_{\text{DO}})\right] + z_{\text{DO}} \tag{7.70}$$

The 'actual' vapour quality variation with length is given by

$$(x^*(z) - x_{\text{DO}}) = \frac{4\phi}{DGi_{\text{fg}}}(z - z_{\text{DO}}) = \frac{4\varepsilon\phi}{DGi_{\text{fg}}}(z - z_{\text{DO}}) \tag{7.71}$$

for $z < z^*$ and z^* is given by

$$z^* = \left[\frac{DGi_{\text{fg}}}{4\varepsilon\phi}(1 - x_{\text{DO}})\right] + z_{\text{DO}} \tag{7.72}$$

Thus, from eqs. (7.69) and (7.72) ε is also given by

$$\varepsilon = \left[\frac{x^*(z) - x_{\text{DO}}}{x(z) - x_{\text{DO}}}\right] = \left[\frac{z^* - z_{\text{DO}}}{z_{\text{EQ}} - z_{\text{DO}}}\right] \tag{7.73}$$

Also, the 'actual' bulk vapour temperature ($T_g(z)$) is given by

$$T_g(z) = T_{SAT} + \left[\frac{4(1 - \varepsilon)\phi(z - z_{DO})}{Gc_{pg}D}\right] \tag{7.74}$$

for $z < z^*$ and

$$T_g(z) = T_{SAT} + \left[\frac{4\phi(z - z_{EQ})}{Gc_{pg}D}\right] \tag{7.75}$$

for $z > z^*$.

The two limiting conditions referred to above are clearly recognized by setting ε equal to zero and unity respectively.

Correlations of this type have been proposed by Plummer, by Groeneveld and Delorme (1976), by Jones and Zuber (1977), Jones (1977) and Chen *et al.* (1977). In considering these correlations it is necessary to distinguish clearly between the 'actual' vapour quality, $x^*(z)$, the 'thermodynamic' vapour quality, $x_E(z)$ (which can have a value greater than unity) and the 'equilibrium' vapour quality, $x(z)$ (which has a maximum value of unity and which equals $x_E(z)$ for $z < z_{EQ}$) (Fig. 7.19).

The Groeneveld–Delorme (1976) method is a significant improvement over the empirical correlation discussed earlier but does appear to suffer from the disadvantage of discontinuities in the slope of the $x^*(z)$ curve both at the dryout point and at $x(z) = 1$.

The correlation by Jones and Zuber (1977) has the attraction that their first order relaxation equation removes the restriction made earlier that ε is a constant for any particular set of flow parameters. This assumption implies that the rate of return to equilibrium is a constant value which is clearly at variance with the facts (Fig. 7.20). The correlation of Chen *et al.* (1977) is probably the most comprehensive of this set of models but is consequently difficult to describe concisely. It is, however, strongly recommended.

7.5.4.3 *Semi-theoretical models*
A comprehensive theoretical model of heat transfer in the post-dryout region must take into account the various paths by which heat is transferred from the surface to the bulk vapour phase. Six separate mechanisms can be identified.

(a) heat transfer from the surface to liquid droplets which impact with the wall ('wet' collisions)
(b) heat transfer from the surface to liquid droplets which enter the thermal boundary layer but which do not 'wet' the surface ('dry' collisions)
(c) convective heat transfer from the surface to the bulk vapour
(d) convective heat transfer from the bulk vapour to suspended droplets in the vapour core

(e) radiation heat transfer from the surface to the liquid droplets
(f) radiation heat transfer from the surface to the bulk vapour

One of the first semi-theoretical models proposed was that of Bennett *et al.* (1968) which is a one-dimensional model starting from known equilibrium conditions at the dryout point. It is assumed that there is negligible pressure drop along the channel, that the wall surface temperature increases to such an extent that the droplets no longer wet the surface and that the heat transfer coefficient between the wall and the fluid is given by one of the well-known single-phase relationships. These assumptions mean that mechanisms (a) and (b) above were not considered.

The position of the dryout point must be known or calculated. At that point it is assumed that the liquid and vapour are in equilibrium, and therefore the quality $x(z_{CR})$ is known and the vapour is at the saturation temperature. To obtain the bulk vapour temperature downstream of the dryout point, four simultaneous differential equations are solved using the Runga–Kutta method. The first of these equations is a mass balance which assumes that the *number* of droplets passing through unit cross section per unit time (N) does not change:

$$\frac{dx}{dz} = -\frac{N\pi\rho_f}{2G} \cdot d^2 \cdot \frac{dd}{dz} \tag{7.76}$$

where

$$N = \frac{6G(1 - x(z_{CR}))}{\pi d_{CR}^3 \rho_f} \tag{7.77}$$

x = vapour quality (vapour mass flow/total mass flow) and d = droplet diameter. The second equation is a heat balance giving the bulk vapour temperature, $T_g(z)$

$$\frac{dT_g}{dz} = \frac{\phi\pi D - \{AGi_{fg}\,dx/dz\}}{AGxc_{pg}} \tag{7.78}$$

The third equation describes the acceleration of the droplets in the vapour stream using Ingebo's (1956) drag coefficient:

$$\frac{du_d}{dz} = \left[K_1(u_g - u_d)^{1.16} - g\left(1 - \frac{\rho_g}{\rho_f}\right) \right]\bigg/ u_d \tag{7.79}$$

where u_d is the droplet velocity and u_g is the vapour velocity

$$K_1 = 20.25 \left[\frac{\mu_g^{0.84}\rho_g^{0.16}}{\rho_f d^{1.84}} \right] \tag{7.80}$$

The last equation describes the evaporation of the droplets using Ryley's method (1961, 1962).

$$\frac{\mathrm{d}d}{\mathrm{d}z} = \left[\frac{2}{d} \times \frac{\mathrm{d}r^2}{\mathrm{d}t}\right] \times \frac{1}{u_\mathrm{d}} \tag{7.81}$$

$$\frac{\mathrm{d}r^2}{\mathrm{d}t} = -\frac{2FMK_\mathrm{g}(p_\mathrm{g} - p_\infty)}{\rho_\mathrm{f} R T_\mathrm{g}} = -\frac{2k_\mathrm{g}F(T_\mathrm{g} - T_{\mathrm{SAT}})}{\rho_\mathrm{f} i_\mathrm{fg}} \tag{7.82}$$

where p_g = pressure in droplet, T_{SAT} = saturation temperature at pressure p_g, R = universal gas constant, K_g = self diffusion coefficient in the vapour = $Mk_\mathrm{g}(\gamma - 1)/\rho_\mathrm{g}R$, $r = d/2$ and F = 'ventilation' factor

$$= \left[1 + 0.276\left(\frac{\rho_\mathrm{g}(u_\mathrm{g} - u_\mathrm{d})d}{\mu_\mathrm{g}}\right)^{0.5}\left(\frac{\mu_\mathrm{g}}{K_\mathrm{g}\rho_\mathrm{g}}\right)^{1/3}\right]$$

Bennett *et al.* (1968) compared estimates of surface temperature made using this method with experimental data taken in the liquid deficient region for water evaporating at 69 bar (1000 psia) in a vertical 5.8 m long × 12.6 mm i.d. (19 ft long × 0.497 in. i.d.) tube. Figure 7.21 shows this comparison for different values of mass velocity. The equations proposed by Heineman (1960) were taken for the heat transfer coefficient to superheated steam. It was found that a reasonable fit was obtained with an assumed droplet size at the dryout point (d_{CR}) of about 0.3 mm.

More recently Iloeje *et al.* (1974) have proposed a three-step model taking into account mechanisms (a), (b), and (c). The physical picture postulated by Iloeje is shown in Fig. 7.22. Liquid droplets of varying sizes are entrained in the vapour core and have a random motion due to interactions with eddies. Some droplets arrive at the edge of the boundary layer with sufficient momentum to contact the wall even allowing for the fact that, as the droplet approaches the wall, differential evaporation coupled with the physical presence of the wall leads to a resultant force trying to repel the droplet. When the droplet touches the wall a contact boundary temperature is set up which depends on the initial droplet and wall temperatures and on $\sqrt{(k\rho c)_\mathrm{f}}/\sqrt{(k\rho c)_\mathrm{w}}$. If this temperature is less than some limiting superheat for the liquid then heat will be transferred, firstly by conduction until a thermal boundary layer is built up sufficient to satisfy the conditions necessary for bubble nucleation. Bubbles will grow within the droplet ejecting part of the liquid back into the vapour stream. The remaining liquid is insufficient in thickness to support nucleation and therefore remains until it is totally evaporated. The surface heat flux transferred by this mechanism can be arrived at by estimating the product of the heat transferred to a single drop

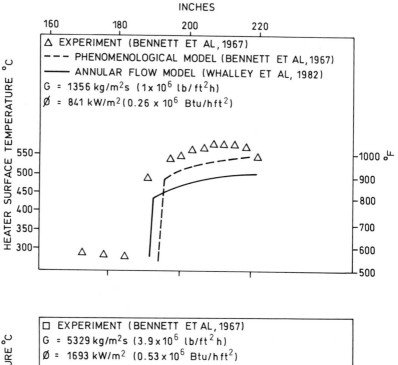

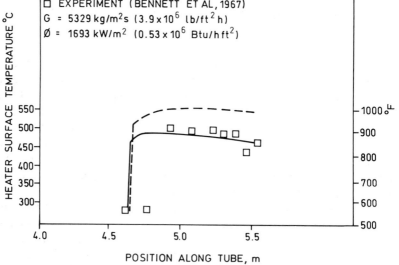

Fig. 7.21. Calculation of wall temperature profile for post-dryout heat transfer to steam–water at 70 bar in a 12.6 mm diameter tube (Bennett *et al.* 1967; Whalley *et al.* 1982).

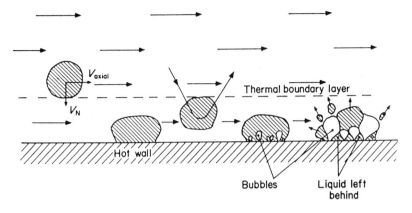

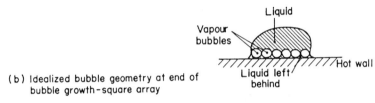

(a) Dispersed flow heat transfer process

(b) Idealized bubble geometry at end of bubble growth-square array

Fig. 7.22. Dispersed flow heat transfer model (Iloeje *et al.* 1975).

and the number of droplets per unit time and per unit area which strike the wall. Iloeje *et al.* attempt to quantify the various mechanisms identified above and finally arrive at a somewhat complex expression for the droplet-wall contact heat flux (ϕ_{dc}).

The droplet mass flux to the wall can be estimated from one of a number of turbulent deposition models (Hutchinson *et al.* 1971). It is important to appreciate that at very low values of ($T_W - T_{SAT}$) the heat flux (ϕ_{dc}) predicted from the Iloeje model must coincide with the droplet deposition flux contribution of the Hewitt film-flow model of dryout.

Indeed, Whalley *et al.* (1982) have extended this model, described in Chapters 3 and 8, to the prediction of post-dryout heat transfer (see Fig. 7.21). This extension of the model allows the treatment of simultaneous departures from both hydrodynamic and thermodynamic equilibrium.

Two basic approaches have been taken to estimate the heat flux (ϕ_{dow}) to droplets entering the thermal boundary layer but which do not touch the wall. This heat flux can be estimated as the product of the heat flux that would occur across a vapour film separating the droplet from the heating surface and the fractional area covered by such droplets. This approach has

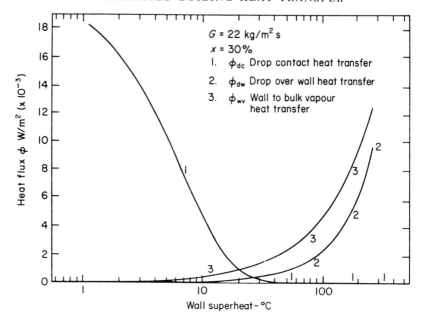

Fig. 7.23. Behaviour of components of total heat flux with quality at $22 \, \text{kg/m}^2 \, \text{s}$ fluid nitrogen (Iloeje *et al.* 1975).

been adopted with slight modifications by Iloeje (1974), Groeneveld (1972), Plummer *et al.* (1974), and, more recently, by Chen *et al.* (1977).

Alternatively, Forslund and Rohsenow (1966), Hynek *et al.* (1966), and Course and Roberts (1974) have assumed the heat transfer coefficient to a single droplet in the spheroidal state condition on a flat heated plate and then multiplied this by the number of droplets approaching the surface per unit time per unit area.

Finally, the various treatments of mechanisms (c) and (d) offered by various workers follow that proposed by Bennett *et al.* (1968) fairly closely. In some cases slightly differing assumptions are made concerning the shattering of droplets due to the droplet Weber number increasing above a critical value.

Figures 7.23 and 7.24, taken from Iloeje *et al.* (1974) show, firstly, the variation of each separate heat flux component for a particular value of vapour quality and mass velocity for nitrogen at atmospheric pressure and, secondly, the variation of the total heat flux curve at a given mass velocity but varying vapour qualities. It is important to note that the minimum in the boiling curve occurs at varying values of $(T_W - T_{SAT})$ dependent upon the values of the individual heat flux contributions.

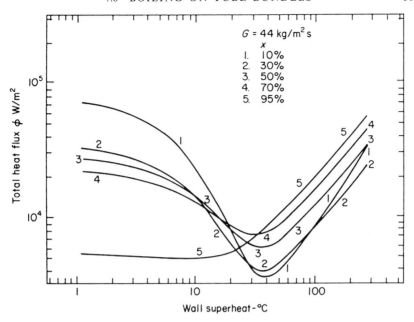

Fig. 7.24. Theoretical boiling curves at 44 kg/m² s fluid nitrogen (Iloeje *et al.* 1975).

7.6 Boiling on tube bundles

The foregoing sections have treated boiling on the inside of channels. Another important process is boiling on the outside of horizontal tube bundles. Kettle and thermosyphon reboilers, waste heat boilers, fire-tube steam generators and flooded evaporators in refrigeration systems utilize this configuration. Research has focused on this area only in recent years, and much still remains to be done. Fortunately, much of what we know about in-tube boiling can be applied, although *qualitatively*.

Consider a simplified tube bundle layout with all the tubes uniformly heated. Figure 7.25 shows in schematic form the various flow patterns encountered from the bottom to the top, together with the corresponding heat-transfer regimes. Subcooled liquid flows upwards into the bottom row of tubes from the inlet nozzle. While the liquid is being heated up to the saturation temperature and the wall temperature remains below that necessary for nucleation to occur, the heat transfer process is single-phase convective heat transfer to the liquid. When conditions are met for nucleation to occur, the first vapour is generated in the subcooled boiling regime. At some tube row above this point the fluid reaches its saturation temperature and the saturated boiling regime is entered.

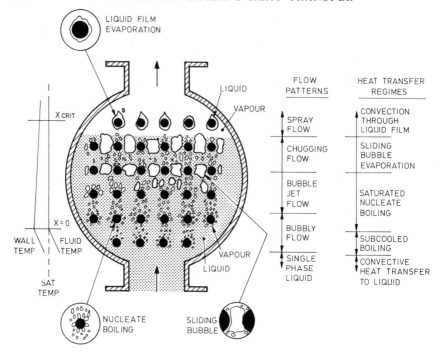

Fig. 7.25. Flow patterns for boiling on a tube bundle.

In the lower part of the bundle bubbly flow exists, similar to that which occurs for nucleate pool boiling on a single tube. The bubbly flow from the top of the tubes forms a two-phase jet and impinges on the tubes above. As the local void fraction increases, large vapour plugs are formed and pass between adjacent tubes, trapping thin layers of evaporating liquid on the sides of the tubes. This is the 'sliding bubble' phenomenon observed by Cornwell (1989). Still higher up in the bundle, the vapour becomes the continuous phase and liquid is evaporated from thin films covering the tubes, producing a frothy spray type of flow. At some critical value of quality and heat flux, dryout of the tubes (not shown) can occur with a substantial decrease in heat transfer performance. This latter process is described in Chapter 9. Figure 7.26 depicts flow regimes observed by Cornwell (1989) for boiling on a kettle reboiler tube bundle. The water flow includes very large bubbles or slugs interspersed with smaller bubbles and could be described as chugging or plug flow. The R-113 flow has only small diameter bubbles and is in the bubbly flow regime.

The following sections summarize the experimental literature on external flow boiling on plain tubes and several of the more promising bundle boiling models. For those seeking additional details, Jensen (1988) has reviewed

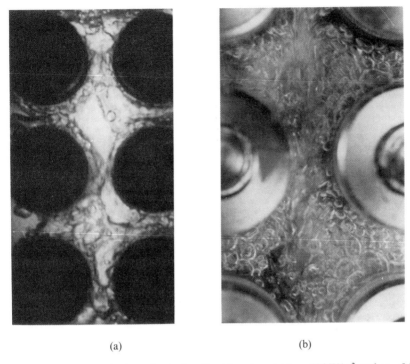

(a)   (b)

Fig. 7.26. Boiling on a horizontal tube bundle: (a) water at 1 bar, $50 \, kW/m^2$ and $x = 0.05$; (b) R-113 at 1 bar, $15 \, kW/m^2$ and $x = 0.05$ (photographs courtesy of K. Cornwell).

boiling and two-phase flow across bundles with plain tubes and Thome (1988, 1990) has surveyed the literature for boiling on low finned tube and enhanced tube bundles.

7.6.1 Boiling in crossflow

Boiling on a single tube with an externally imposed flow of the liquid across the tube is a simplified configuration relative to a bundle of tubes and hence has been studied by various researchers. Yilmaz and Westwater (1980) obtained complete boiling curves for R-113 at 1.01 bar with a steam-heated horizontal copper tube of 6.4 mm diameter and velocities from 0 to 6.8 m/s. The effect of velocity was shown to be important in all boiling regimes: nucleate, transition and film boiling. In addition, the critical heat flux was increased substantially with increasing velocity. This effect in the nucleate boiling regime has been corroborated for water and R-12 by Singh *et al.* (1983, 1985), even for liquid crossflow velocities as low as 0.013 m/s.

For R-11/R-113 mixtures, Fink *et al.* (1982) found similar trends for a

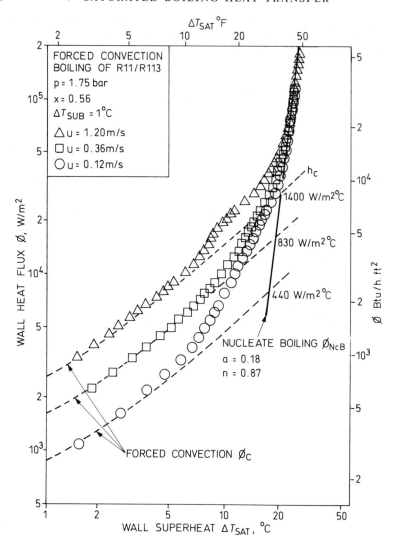

Fig. 7.27. Effect of crossflow on boiling on a single tube (Fink *et al.* 1982).

25 mm diameter, electrically heated copper cylinder. Figure 7.27 depicts their data for a 56 per cent R-11/44 per cent R-113 molar mixture. Their measured nucleate pool boiling curve with no externally imposed flow is represented by the solid line and the single-phase forced convection curves by the dashed lines. The superposition of forced convection and nucleate boiling is evident at heat fluxes of up to about 50 kW/m². Their data in the transition between the forced convection and nucleate pool boiling regimes were satisfactorily

predicted by the Bergles and Rohsenow (1964) correlation, eq. (5.48). They found that increased accuracy could be obtained using the simple additive superposition expression also proposed by Bergles and Rohsenow.

7.6.2 Experimental bundle boiling studies

Experimental bundle boiling studies can be run to obtain either (a) the overall boiling side heat transfer coefficient for the entire bundle or (b) the local values within the bundle. Overall values are typically obtained by heating a bundle with steam or a hot water source. Local boiling coefficients are obtained by instrumenting individual tubes with thermocouples and heating with cartridge heaters. As opposed to in-tube boiling studies, many bundle boiling studies are run without determining the flow rate of the evaporating fluid (and hence without determining the vapour quality). In most studies, the test bundle is immersed in a large pool of liquid with a condenser in the vapour space above the bundle. Hence, flow is due to natural recirculation generated within the pool. Some test facilities utilize a forced flow induced by a pump through a rectangular channel with numerous tube rows in the direction of flow and hence the flow rate is measured and the local vapour qualities determined from a heat balance.

If the entire bundle is treated as a single heat transfer surface, a bundle boiling curve similar to that of a single tube can be formulated. Figure 7.28,

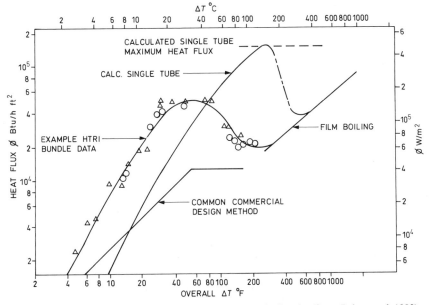

Fig. 7.28. Tube bundle boiling data compared to a single tube (from Palen *et al.* 1982).

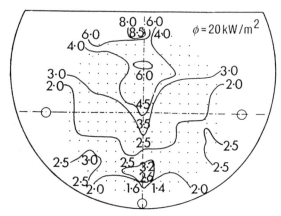

Fig. 7.29. Heat transfer results of Leong and Cornwell (1980). Lines are contours of constant heat transfer coefficient (kW/m² °C).

from Palen *et al.* (1972), shows an illustration. In the nucleate boiling regime, the bundle boiling curve is above and to the left of the single-tube curve because of the convective contribution to heat transfer, similar to that already described for in-tube boiling. However, the critical heat flux of a tube bundle is normally reached before that of a single tube. Also shown is the design curve of Kern (1950), who recommended a fixed value of 1703 W/m² K (300 Btu/h ft² °F) for the boiling heat transfer coefficient in reboilers.

Leong and Cornwell (1980) and Cornwell *et al.* (1980) tested a 241-tube bundle heated electrically. The tests were run with R-113 at atmospheric pressure with 19.05 mm diameter tubes on a 25.4 mm square pitch. Their iso-heat transfer curves are shown in Fig. 7.29 for the heat flux of 20 kW/m². The bundle was a simulated kettle reboiler with an oversized shell. The 25.4 mm long tubes represented a 'slice' of a normal tube bundle. The coefficients at the bottom of the bundle are comparable to those predicted by single-tube correlations while the two-phase circulation up through the bundle increases the boiling coefficient substantially.

The effect of tube rows for very deep bundles has been investigated by Rebrov *et al.* (1989) for boiling R-12 and R-22. They tested two bundles, one with 30 and another with 50 vertical tube rows. The local boiling curves for individual rows were measured from the bottom to the top of the bundles (22 mm diameter steel tubes on a staggered pitch of 31.9 mm). At the heat flux of 10 kW/m², the heat transfer coefficient at the 16th row was 33 per cent higher than that at the 2nd row and the 48th and 50th rows were 100 per cent higher. Thus, they concluded that the bundle effect does not level off after the first 5–6 rows but continues to increase.

Many other bundle boiling studies (Myers and Katz 1952; Heimbach

1972; Wallner 1974; Nakajima and Shiozawa 1975; Nakajima 1978; Yilmaz and Palen 1984; Nelson and Burnside 1955; Fujita *et al.* 1986; Hwang and Yao 1986a; Chan and Soukri 1987; Ivanov *et al.* 1988; Marto and Anderson 1992) have confirmed the increase in performance of a tube bundle relative to a single tube. At low heat fluxes (1 to $10 \, kW/m^2$) the increase in the bundle coefficient relative to a single tube is large because the nucleate boiling coefficient is still small and convective heat transfer is important; at heat fluxes above about $50 \, kW/m^2$ the bundle and single-tube curves coincide since nucleate boiling has become dominant, similar to forced convection in crossflow over a single tube as described earlier. Contrary to these results, Jensen and Hsu (1988) found little or no increment in the heat transfer coefficient with increasing tube row in their 27-row test bundle using forced flow rather than natural circulation, although the coefficients are larger than those predicted by single-tube correlations such as that of Cooper (1984). The uniformity in h can apparently be attributed to the forced circulation effect, i.e. the bottom tube rows are subjected to a significant fluid velocity similar to a single tube in crossflow whilst in natural circulation tests fluid velocities are very small at the lower periphery of the bundle with coefficients equal to single tube values.

The effect of pressure on bundle boiling has not been investigated extensively. Most tests have been run at test conditions between 1 and 7 bars. At large reduced pressures the nucleate boiling coefficient becomes very large and thus the bundle boiling curve should approach that of a single tube over most of the heat flux range. Under vacuum conditions the nucleate boiling coefficients are low and the effect will be the opposite, which has been confirmed for water with tests run from 0.06 to 1.0 bar by Slesarenko *et al.* (1982).

7.6.3 *Single-phase flow across tube bundles*

Similar to in-tube flow boiling, a mechanistic approach to formulation of a bundle boiling model requires a single-phase correlation for the convective contribution to heat transfer. The heat transfer for forced convective flow across a tube bank is governed primarily by the flow velocity, bundle geometry and fluid properties. Experimental results for tube bundles are correlated with the exponential expression of the form

$$Nu = c \, Re^m \, Pr^n (Pr/Pr_w)^{0.25} \tag{7.83}$$

where the empirical constant c and exponents m and n are dependent on the Reynolds number and the tube layout. Zukauskas and Ulinskas (1985) have summarized their extensive work on tube bundles, covering Reynolds numbers from 2 to 2,000,000 for staggered and in-line tube banks. Their recommended methods are summarized in Table 7.9 for the inner rows of

Table 7.9 Zukauskas and Ulinskas (1985) parameters for eq. (7.83)

Geometry	Re range	c	m	n
Staggered:				
	2–50	1.04	0.4	0.36
	50–1,000	0.71	0.5	0.36
$(s_1/s_2) < 2$	1,000–200,000	$0.35(s_1/s_2)^{0.2}$	0.6	0.36
$(s_1/s_2) > 2$	1,000–200,000	0.4	0.6	0.36
	200,000–2,000,000	$0.031(s_1/s_2)^{0.2}$	0.8	0.4
In-line:				
	2–200	0.9	0.4	0.36
	200–850	0.52	0.5	0.36
	850–200,000	0.27	0.63	0.36
	200,000–2,000,000	0.033	0.8	0.4

Table 7.10 Ratio of h for N rows deep to that for 10 rows deep

n	1	2	3	4	5	6	7	8	9	10
Staggered tubes	0.68	0.75	0.83	0.89	0.92	0.95	0.97	0.98	0.99	1.00
In-line tubes	0.64	0.80	0.87	0.90	0.92	0.94	0.96	0.98	0.99	1.00

smooth ideal tube banks, where s_1 is the transverse tube pitch between adjacent tubes in a row and s_2 is the longitudinal pitch between successive tube rows. For the first ten tube rows, the heat-transfer coefficient is less than the fully developed value and Table 7.10 of Kays and Lo (1952) can be used together with eq. (7.83). It is recommended that future bundle boiling models be formulated using these convection correlations in order to proceed with a standardized approach.

In actual heat exchangers the flow is not ideal and bypass flows can be significant. For more on these effects, one should refer to the Delaware method described by Bell (1986) and also the comprehensive experimental studies completed by Matsushima *et al.* (1986–88). For natural convection on horizontal tube bundles, Shklover and Gusev (1988) have reviewed the correlations available.

7.6.4 *Bundle boiling models*

Models in the open literature for boiling on tube bundles are available but nearly all have been developed with a very limited database. The key aspects of these models are reviewed below.

7.6.4.1 *Mean bundle boiling methods* A simple method for estimating bundle boiling coefficients has been presented by Palen (1983). The mean bundle boiling heat transfer coefficient h_b is obtained from the superposition of the contributions of boiling and natural convection as

$$h_b = h_{NcB} F_b F_c + h_{nc} \qquad (7.84)$$

where h_{NcB} is the nucleate pool boiling coefficient, F_b is the bundle boiling factor, F_c is the mixture boiling correction factor (see Chapter 12) and h_{nc} is the heat-transfer coefficient for single-phase natural convection on the tube bundle. The nucleate pool boiling coefficient is calculated using a suitable single-component correlation while F_c accounts for the degradation of h_{NcB} in mixtures and varies from 0.1 for wide boiling range mixtures to 1.0 for single-component fluids and azeotropes. The natural convection coefficient can be assumed to be 250 W/m² K. The factor F_b is an empirical multiplier that ranges between 1.0 and 3.0; at heat fluxes above 50 kW/m², F_b equals 1.0, because the bundle boiling curve intersects the single tube boiling curve near this heat flux. Most evaporators operate in the heat flux range from 5 to 30 kW/m², for which Palen recommends using $F_b = 1.5$. For wide boiling range mixtures, F_b reaches values as large as 3 because F_c approaches 0.1 and convection effects dominate. Several other simple methods have been proposed by Palen and Taborek (1962) and Palen and Small (1964).

A completely empirical method has been proposed by Rebrov et al. (1989) based on test data for several refrigerants boiling on tube bundles with 5, 6, 18, 30 and 50 vertical tube rows. Their boiling expression for the mean bundle heat transfer coefficient is determined from the expression

$$Nu_f = 0.04 \, e^{0.087N} \, Re_f^{(0.7-0.0078N)} K_p^{(0.6-0.0058N)} \, Pr_f^{0.4} \qquad (7.85)$$

where the boiling Nusselt number is defined as

$$Nu_f = \frac{h_b[\sigma/g(\rho_f - \rho_g)]^{1/2}}{k_f} \qquad (7.86)$$

and the boiling Reynolds number is given as

$$Re_f = \frac{\phi[\sigma/g(\rho_f - \rho_g)]^{1/2}}{i_{fg}\rho_f \nu_f} \qquad (7.87)$$

K_p is the pressure correction factor determined as

$$K_p = \frac{p_{SAT}}{[\sigma g(\rho_f - \rho_g)]^{1/2}} \qquad (7.88)$$

This correlation is valid for plain tubes for heat fluxes from 1 to 15 kW/m², tube pitch ratios of 1.45, N from 1 to 50 and T_{SAT} from -30 to 10°C. It is interesting to note that as the number of tube rows increases from 1 to 50, the exponent on the heat flux decreases from the single-tube value of 0.7 to 0.32.

7.6.4.2 *Local bundle boiling methods* It is desirable to calculate local heat transfer coefficients within a tube bundle, either by position or as a function of local vapour quality. Using in-tube boiling correlations directly for the tube bundles yields poor results, so various adaptions have to be made. For instance, Hwang and Yao (1986b) modified the Chen (1963a) correlation

$$h_{TP} = Sh_{NcB} + Fh_f \qquad (7.89)$$

using a new empirical expression for F and the Bennett *et al.* (1980) expression for the boiling suppression factor S:

$$S = \frac{k_f}{Fh_f Y}\left[1 - \exp\left(\frac{Fh_f Y}{k_f}\right)\right] \qquad (7.90)$$

where Y is a dimensionless variable. However, visual observations of bubbly and frothy flow made by Cornwell (1989), Leong and Cornwell (1979), and Cornwell *et al.* (1980) do *not* show evidence of a suppression of nucleate boiling in a bundle and coefficients in the first tube rows are *not* below single-tube values. In addition, crossflows over single tubes do not show effects of boiling suppression and neither does in-tube flow boiling according to recent findings of Steiner and Taborek (1992). Thus, it can be concluded that the boiling suppression factor should be set equal to 1.0 for a tube bundle and the complete effect of convection included in the model through the two-phase flow correction factor F.

Cornwell *et al.* (1986) have assumed the convective and nucleate boiling contributions to be additive without a boiling suppression factor as

$$Nu_b = cRe^m Pr^n + CRe_b^{0.67} \qquad (7.91)$$

where the Zukauskas and Ulinskas (1985) parameters are used for c, m, and n. The local bundle Nusselt number in this expression is the standard single-phase definition

$$Nu_b = \frac{h_b D}{k_f} \qquad (7.92)$$

and the bundle Reynolds number is determined based on the liquid velocity as

$$Re = \frac{\rho_f u_f D}{\mu_f} \qquad (7.93)$$

The liquid velocity is determined from the following equations, where A_{min}

is the minimum crossflow area between the tubes and α is the void fraction:

$$u_f = u_g \left(\frac{\alpha}{1-\alpha} \right) \left(\frac{\rho_g}{\rho_f} \right) \left(\frac{1-x}{x} \right) \tag{7.94}$$

$$u_g = \frac{Gx}{\rho_g A_{\min} \alpha} \tag{7.95}$$

$$\frac{1}{1-\alpha} = 1 + \left(\frac{6}{X_{tt}} \right)^{0.71} \tag{7.96}$$

$$X_{tt} = \left(\frac{\mu_f}{\mu_g} \right)^{0.1} \left(\frac{\rho_g}{\rho_f} \right)^{0.5} \left(\frac{1-x}{x} \right)^{0.9} \tag{7.97}$$

Based on their tests with R-113, they found $C = 150$ for data at $20 \, \text{kW/m}^2$. Use of this expression at other pressures or for other fluids requires a general expression for the nucleate boiling contribution.

Nakajima (1978) proposed that the bundle boiling coefficient is the summation of nucleate boiling and thin-film evaporation, the latter resulting from conduction through thin liquid films created between the tubes and rising bubbles. The general expression for the local heat transfer coefficient on the Nth tube row from the bottom of the bundle is given as

$$h_b = (1-\alpha)h_{NcB} + \alpha h_{tf} \tag{7.98}$$

where the void fraction is utilized to prorate the two contributions. At low void fractions nucleate boiling is dominant while at large values thin-film evaporation becomes controlling. The thin-film heat transfer coefficient is given by the dimensional equation

$$h_{tf} = 2326 + 1512 \exp[-(0.5556/u_g)^{1.5}] \tag{7.99}$$

where h_{tf} is in $\text{W/m}^2 \, \text{K}$ and the superficial vapour velocity is in m/s. The superficial vapour velocity is calculated from the energy input of the lower tube rows as

$$u_g = \sum_{N}^{N-1} \frac{\phi \pi D}{i_{fg} \rho_g N s_1} \tag{7.100}$$

where s_1 is the transverse tube pitch. The vapour void fraction is determined from a new expression that is difficult to evaluate. No convective heat-transfer contribution was considered; thus at the limit when the void fraction equals 1.0 the model breaks down and predicts thin-film evaporation to occur when no liquid is present.

For boiling on horizontal kettle reboilers, flow recirculation within the oversized shell has been observed by Cornwell *et al.* (1980). Thus, the flow rate through the bundle must be determined by balancing the two-phase pressure drop up through the bundle with the liquid static head of the pool. Brisbane *et al.* (1980) and Palen and Yang (1983) have described the development of circulation boiling models.

References and further reading

Bahr, A., Herkenrath, H., and Mork-Morkenstein, P. (1969). 'Anomale Druck-abhängigkeit der Wärmeübertragung im Zweiphasengebeit bei Annäherung an der kritischen Druck'. *Brennstoff-Wärme-Kraft*, **21**(12), 631–633.

Bailey, N. A. (1971). 'Film boiling on submerged vertical cylinders'. AEEW-M1051.

Bandel, J. and Schlünder, E. U. (1973). 'Frictional pressure drop and convective heat transfer of gas–liquid flow in horizontal tubes'. Paper B.5.2 presented at 5th International Heat Transfer Conference, Tokyo, Sept. (1974). See also: Bandel, J. and Schlünder, E. U. 'Druckverlust und Wärmeübengang bei der Vendampfung seidender Kältemittel im durchstromten Rohr'. *Chemie Ingenieur Technik*, **45** Jahrg Nr. 6, 345–349.

Bell, K. J. (1986). 'Delaware method for shell-side design'. In Palen, J. W. (ed.), *Heat transfer exchanger sourcebook*, Hemisphere, New York, 129–166.

Bennett, D. L. and Chen, J. C. (1980). 'Forced convection boiling in vertical tubes for saturated pure fluids and binary mixtures'. *AIChE J.*, **26**(3), 454–464.

Bennett, J. A. R., Collier, J. G., Pratt, H. R. C., and Thornton, J. D. (1961). 'Heat transfer to two-phase gas–liquid systems. Part I: Steam/water mixtures in the liquid dispersed region in an annulus'. *Trans. Inst. Chem. Engrs.*, **39**, 113. Also AERE-R 3519 (1959).

Bennett, A. W., Hewitt, G. F., Kearsey, H. A., and Keeys, R. K. F. (1968). 'Heat transfer to steam–water mixtures flowing in uniformly heated tubes in which the critical heat flux has been exceeded'. Paper 27 presented at Thermodynamics and Fluid Mechanics Convention, IMechE, Bristol, 27–29 March. See also AERE-R 5573 (1967).

Bennett, D. L., Davis, M. W., and Hertzler, B. C. (1980). 'The suppression of saturated nucleate boiling by forced convective flow'. *AIChE Symp. Ser.*, **76**, 91–103.

Bergles, A. E. and Rohsenow, W. M. (1964). 'The determination of forced convection surface boiling heat transfer'. *J. Heat Transfer*, **C86**, 365.

Bjornard, T. A. and Griffith, P. (1977). 'PWR blowdown heat transfer'. Symposium on the Thermal and Hydraulic Aspects of Nuclear Reactor Safety, Atlanta, Georgia, 27 Nov.–2 December, 1977. *Vol. 1. Light Water Reactors*, 17–41. ASME.

Brisbane, T. W. C., Grant, I. D. R., and Whalley, P. B. (1980). 'A prediction method for kettle reboiler performance'. ASME Paper 80-HT-42.

Bromley, L. A., LeRoy, N. R., and Robbers, J. A. (1953). 'Heat transfer in forced convection film boiling'. *Ind. and Engng. Chem.*, **45**(12), 2639–2646.

Butterworth, D. (1970). Private communication.

Cadeck, F. F., Dominicis, O. P., and Deyse, R. H. (1971). PWR FLECHT Final Report WCAP-7665.

Chan, A. M. C. and Shoukri, M. (1987). 'Boiling characteristics of small multitube bundles'. *J. Heat Transfer*, **109**, 753–760.

Chawla, J. M. (1967). 'Heat transfer and pressure drop for refrigerants evaporating in horizontal tubes'. *VDI Forschungschaft*, 523.

Chen, J. C. (1963a). 'A correlation for boiling heat transfer to saturated fluids in convective flow'. ASME preprint 63-HT-34 presented at 6th National Heat Transfer Conference, Boston, 11–14 August.

Chen, J. C. (1963b). 'A proposed mechanism and method of correlation for convective boiling heat transfer with liquid metals'. Paper presented at 3rd Annual Conference on High Temperature Liquid Metal Heat Transfer Technology, ORNL, September. Also BNL 7319.

Chen, J. C. (1965). 'Non-equilibrium inverse temperature profile in boiling liquid metal two phase flow'. AIChE J., 11(6), 1145–1148.

Chen, J. C., Sundaram, R. K., and Ozkaynak, F. T. (1977). 'A phenomenological correlation for post CHF Heat Transfer'. Lehigh University. NUREG-0237.

Cheng, S. C. and Ng, W. (1976). 'Transition boiling heat transfer in forced vertical flow via a high thermal capacity heating process'. Letters in Heat Transfer, 3, 333–342.

Churchill, S. W. (1974). The interpretation and use of rate data, Hemisphere, New York.

Collier, J. G. (1958). 'A review of two-phase heat transfer'. AERE CE/R 2496, HMSO.

Collier, J. G. and Pulling, D. J. (1962). 'Heat transfer to two-phase gas–liquid systems. Part II: Further data on steam/water mixtures in the liquid dispersed region in an annulus'. AERE-R 3809.

Cooper, M. G. (1984). 'Heat flow rates in saturated nucleate pool boiling—a wide-ranging examination using reduced properties'. Adv. Heat Transfer, 16, 157–239.

Cornwell, K. (1989). 'The influence of bubbly flow on boiling from a tube in a bundle'. Advances in Pool Boiling Heat Transfer, Eurotherm No. 8, Paderborn, 177–184, May 11–12.

Cornwell, K., Duffin, N. W., and Schuller, R. B. (1980). 'An experimental study of the effects of fluid flow on boiling within a kettle reboiler tube bundle'. ASME Paper 80-HT-45.

Cornwell, K., Einarsson, J. G., and Andrews, P. R. (1986). 'Studies on boiling in tube bundles'. Proc. 8th Int. Heat Transfer Conf., San Francisco, 5, 2137–2141.

Costigan, G., Holmes, A. W., and Ralph, J. C. (1984). 'Steady-state postdryout heat transfer in a vertical tube with low inlet quality'. 1st UK National Conference on Heat Transfer, 1, 1–11 (Inst. of Chemical Engineers Symposium Ser. No. 86).

Course, A. F. and Roberts, H. A. (1974). 'Progress with heat transfer to a steam film in the presence of water drops—a first evaluation of Winfrith SGHWR Cluster loop data'. AEEW-M1212 paper presented at European Two-Phase flow Group Meeting, Harwell, June.

Davis, E. J. and Anderson, G. H. (1966). 'The incipience of nucleate boiling in forced convection flow'. AIChE J., 12(4), 774–780.

Dengler, C. E. and Addoms, J. N. (1956). 'Heat transfer mechanism for vaporization of water in a vertical tube'. Chem. Engng. Prog. Symp. Series No. 18, 52, 95–103.

Denham, M. K. (1984). 'Inverted annular flow film boiling and the Bromley model'. 1st UK National Conference on Heat Transfer, 1, 13–23. (Inst. of Chemical Engineers Symposium Ser. No. 86).

Dougall, R. S. and Rohsenow, W. M. (1963). 'Film boiling on the inside of vertical tubes with upward flow of the fluid at low qualities'. Mech. Engng. Dept., Engineering Project Laboratory, MIT report No. 9079-26, September.

Dukler, A. E. (1960). 'Fluid mechanics and heat transfer in vertical falling film systems'. *Chem. Engng. Prog. Symp. Series* No. 30, **56**, 1–10.

Ellion, M. E. (1954). 'Study of mechanism of boiling heat transfer'. *Jet Prop. Lab. Memo.*, 20–88.

Fink, J., Gaddis, E. S., and Vogelpohl, A. (1982). 'Forced convection boiling of a mixture of Freon-11 and Freon-113 flowing normal to a cylinder'. *Proc. 7th Int. Heat Transfer Conf.*, Munich, **4**, 207–212.

Forslund, R. P. and Rohsenow, W. M. (1966). 'Thermal non-equilibrium in dispersed flow film boiling in a vertical tube'. MIT report 75312-44, November.

Forster, H. K. and Zuber, N. (1955). 'Dynamics of vapour bubbles and boiling heat transfer'. *AIChE Journal*, **1**(4), 531–535.

Fujita, Y., Ohta, H., Hidaka, S., and Nishikawa, K. (1986). 'Nucleate boiling heat transfer on horizontal tubes in bundles'. *Proc. 8th Int. Heat Transfer Conf.*, San Francisco, **5**, 2131–2136.

Fung, K. K. (1977). 'Forced convection transition boiling'. MSc Thesis, University of Toronto.

Fung, K. K., Gardiner, S. R. M., and Groeneveld, D. C. (1979). 'Subcooled and low quality flow boiling of water at atmospheric pressure'. *Nucl. Engng. Des.*, **55**, 51–57.

Glickstein, M. R. and Whitesides, R. H. (1967). 'Forced convection nucleate and film boiling of several aliphatic hydrocarbons'. Paper 67-HT-7 presented at 9th ASME–AIChE Heat Transfer Conference, Seattle, August.

Gnielinski, V. (1976). *Int. Chem. Eng.*, **6**, 359–368.

Gorenflo, D. (1988). 'Behältersieden', *VDI Wärmeatlas*, Sect. Ha, VDI Verlag, Düsseldorf, 5.

Gouse, S. W. and Dickson, A. J. (1966). 'Heat transfer and fluid flow inside a horizontal tube evaporator, Phase II'. Paper presented as ASHRAE Conference, Houston, January.

Groeneveld, D. C. (1972). 'The thermal behaviour of a heated surface at and beyond dryout'. AECL-4309.

Groeneveld, D. C. (1973). 'Post-dryout heat transfer at reactor operating conditions'. Paper AECL-4513 presented at the National Topical Meeting on Water Reactor Safety, ANS, Salt Lake City, Utah, 26–28 March.

Groeneveld, D. C. (1974). 'Effect of a heat flux spike on the downstream dryout behaviour'. *J. of Heat Transfer*, 121–125, May.

Groeneveld, D. C. and Delorme, G. G. J. (1976). 'Prediction of the thermal non-equilibrium in the post-dryout regime'. *Nucl. Engng. and Design*, **36**, 17–26.

Groeneveld, D. C. and Fung, K. K. (1976). 'Forced convection transition boiling—a review of literature and comparison of prediction methods'. AECL-5543.

Groeneveld, D. C. and Gardiner, S. R. M. (1977). 'Post CHF heat transfer under forced convection conditions'. Symposium on the Thermal and Hydraulic Aspects of Nuclear Reactor Safety, Atlanta, Georgia, 27 Nov.–2 Dec. 1977, **1**, *Light Water Reactors*, 43–73, ASME.

Guerrieri, S. A. and Talty, R. D. (1956). 'A study of heat transfer to organic liquids in single tube, natural circulation vertical tube boilers'. *Chem. Engng. Prog. Symp. Series: Heat Transfer*, Louisville, No. 18, **52**, 69–77.

Gungor, K. E. and Winterton, R. H. S. (1986). 'A general correlation for flow boiling in tubes and annuli'. *Int. J. Heat Mass Transfer*, **29**, 351–358.

Hall, G. R., Bjorge, R. W., and Rohsenow, W. M. (1982). 'Correlation for forced convection boiling heat transfer data'. *Int. J. Heat Mass Transfer*, **20**, 763.

Heimbach, P. (1972). 'Boiling coefficients on refrigerant–oil mixtures outside a finned-tube bundle'. In *Heat and mass transfer in refrigeration systems and in air conditioning*, Inst. of Refrigeration, Paris, pp. 117–125.

Heineman, J. B. (1960). 'An experimental investigation of heat transfer to superheated steam in round and rectangular channels'. ANL-6213.

Herkenrath, H., Mork-Morkenstein, P., Jung, U., and Weckermann, F. J. (1967). 'Heat transfer in water with forced circulation in 140–250 bar pressure range'. EUR 3658d.

Hewitt, G. F. (1961). 'Analysis of annular two-phase flow: application of the Dukler analysis to vertical upwards flow in a tube'. AERE-R 3680.

Hewitt, G. F. and Hall-Taylor, N. (1970). *Annular two-phase flow*, Pergamon Press.

Hewitt, G. F., Kearsey, H. A., Lacey, P. M. C., and Pulling, D. J. (1963). 'Burnout and nucleation in climbing film flow'. AERE-R 4374.

Hoffman, H. W. and Krakoviak, A. I. (1962). 'Forced convection boiling of potassium at near atmospheric pressure'. *Proceedings of 1962 High Temperature Liquid Metal Heat Transfer Technology meeting* BNL 756 (c-35).

Honda, H., Fujü, T., Uchima, B., Nozu, S., and Nakata, H. (1988). 'Condensation of Downward Flowing R-113 Vapour on Bundles of Smooth Tubes', *Trans. Jpn. Soc. Mech. Engrs.*, *54–502*, B, 1453–1460.

Hsu, Y. Y. (1961). 'On the size range of active nucleation cavities on a heating surface'. Paper 61-WA-177 presented at ASME Winter Annual Meeting, New York, December.

Hutchinson, P., Hewitt, G. F., and Dukler, A. E. (1971). *Chem. Engng. Sci.*, **26**, 419.

Hwang, T. H. and Yao, S. C. (1986). 'Crossflow boiling heat transfer in tube bundles'. *Int. Comm. Heat Mass Transfer*, **13**, 493–502.

Hwang, T. H. and Yao, S. C. (1986). 'Forced convective boiling in horizontal tube bundles'. *Int. J. Heat Mass Transfer*, **29**, 785–795.

Hynek, S. J., Rohsenow, W. M., and Bergles, A. E. (1966). 'Forced convection dispersed vertical flow film boiling'. MIT report 70586-03, November.

Iloeje, O. C., Plummer, D. N., Rohsenow, W. M., and Griffith, P. (1974). 'A study of wall rewet and heat transfer in dispersed vertical flow'. Mech. Engng. Dept., MIT, Report 72718–92, September.

Iloeje, O. C., Plummer, D. N., Rohsenow, W. M., and Griffith, P. (1975). 'An investigation of the collapse and surface rewet in film boiling in forced vertical flow'. *J. Heat Transfer*, 166–172, May.

Ingebo, R. D. (1956). 'Drag coefficients for droplets and solid spheres in clouds accelerating in air streams'. NACA-TN 3762.

Ivanov, O. P., Mamchenko, V. O., and Yemel'yanov, A. L. (1988). 'Boiling and condensation of refrigerants on bundles of smooth horizontal tubes of a two-phase thermosiphon'. *Heat Transfer-Sov. Res.*, *20*(3), 289–293.

Jensen, M. K. (1988). 'Boiling on the shellside of horizontal tube bundles'. In *Two-Phase Flow Heat Exchangers*, Kluwer, Dordrecht, pp. 707–746.

Jensen, M. K. and Hsu, J. T. (1988). 'A parametric study of boiling heat transfer in a tube bundle'. *J. Heat Transfer*, **110**, 976–981.

Jones, O. C. (1977). 'Liquid deficient cooling in dispersed flows. A non-equilibrium relaxation model'. BNL-NUREG-50639.

Jones, O. C. and Zuber, N. (1977). 'Post-CHF heat transfer—a non-equilibrium relaxation model'. ASME paper 77-HT-79 presented at the 17th Nat. Heat Transfer Conference, Salt Lake City, Utah, 14–17 August.

Kandlikar, S. G. (1990). 'A General Correlation for Saturated Two-Phase Flow Boiling Heat Transfer Inside Horizontal and Vertical Tubes'. *J. Heat Transfer*, **112**, 219–228.

Kandlikar, S. G. (1991). 'Correlating Flow Boiling Heat Transfer Data in Binary Systems'. *Phase Change Heat Transfer*, ASME HTD 159.

Kays, W. M. and Lo, R. K. (1952). 'Basic heat transfer and flow friction data for gas flow normal to banks of staggered tubes: Use of a transient technique'. *Stanford Univ. Tech. Rep.* 15, Navy Contract N6-OHR251 T.O. 6.

Kern, D. Q. (1950). *Process Heat Transfer*, McGraw-Hill, New York, p. 474.

Kutateladze, S. S. (1961). 'Boiling Heat Transfer'. *Int. J. Heat Mass Transfer*, **4**, 3–45.

Leong, L. S. and Cornwell, K. (1979). 'Heat transfer coefficients in a reboiler tube bundle'. *Chemical Engineer*, **343**, 219–221.

Levy, S. and Bray, A. P. (1963). 'Reliability of burnout calculations in nuclear reactors'. *Nuclear News*, ANS, 3–6 February.

Liu, Q. S., Shiotsu, M., and Sakurai, A. (1992). 'A correlation for forced convection film boiling heat transfer from a hot cylinder under subcooled conditions'. Presented in *Fundamentals of Subcooled Flow Boiling*, HTD-Vol. 217, pp. 21–32 (1992). Paper presented at ASME Winter Ann. Mtg, Anaheim, CA.

Longo, J. (ed.) (1963). *Alkali metal boiling and condensing investigations.* Quarterly progress reports 2 and 3. Space, Power and Propulsion Section, General Electric Co., April.

McAdams, W. H. (1954). *Heat transmission*, 244, McGraw-Hill.

McDonough, J. B., Milich, W., and King, E. C. (1961). 'An experimental study of partial film boiling region with water at elevated pressures in a round vertical tube'. *Chem. Engng. Prog. Symp. Series* No. 32, **57**, 197–208.

Marto, P. J. and Anderson, C. L. (1992). 'Nucleate boiling characteristics of R-113 in a small tube bundle'. *J. Heat Transfer*, **114**, 425–433.

Matsushima, H., Nakayama, W., and Daikoku, T. (1986). 'Shell-side single-phase flows and heat transfer in shell-and-tube heat exchangers (Part 1: Experimental study on local heat transfer coefficients)'. *Heat Transfer-Jap. Res.*, **15**(2), 60–74.

Matsushima, H., Nakayama, W., Yanagida, T., and Kudo, A. (1987). 'Shell-side single-phase flows and heat transfer in shell-and-tube heat exchangers (Part 2: Effects of inlet nozzle diameter on flow patterns and local heat transfer coefficients)'. *Heat Transfer-Jap. Res.*, **16**(2), 1–14.

Matsushima, H., Nakayama, W., Yanagida, T., and Kudo, A. (1988). 'Shell-side single-phase flows and heat transfer in shell-and-tube heat exchangers (Part 3: Experimental study on the distribution of heat transfer coefficients in a tube bundle)'. *Heat Transfer-Jap. Res.*, **17**(6), 60–73.

Motte, E. I. and Bromley, L. A. (1957). 'Film boiling of flowing subcooled liquids'. *Ind. and Engng. Chem.*, **49**(11), 1921–1928.

Myers, J. W. and Katz, D. L. (1952). *Refrig. Eng.*, **60**, 56.

Nakajima, K. (1978). 'Boiling heat transfer outside horizontal multitube bundles'. *Heat Transfer-Jap. Res.*, **7**(1), 1–24.

Nakajima, K. and Shizowa, A. (1975). 'An experimental study of the performance of a flooded type evaporator, *Heat Transfer-Jap. Res.*, **4**(3), 49–66.

Nelson, P. J. and Burnside, B. M. (1985). 'Boiling the immiscible water/n-nonane system from a tube bundle'. *Int. J. Heat Mass Transfer*, **28**, 1257–1267.

Newbold, F. J., Ralph, J. C., and Ward, J. A. (1976). 'Post dryout heat transfer under low flow and low quality conditions'. AERE-R 8390.

Palen, J. W. (1983). 'Shell-and-tube reboilers'. *Heat Exch. Des. Handbook*, 3:3.6.1–1 to 3.6.5–6.

Palen, J. W. and Small, W. M. (1964). 'Kettle and internal reboilers'. *Hydrocarbon Processing*, **43**(11), 199–208.

Palen, J. W. and Taborek, J. J. (1962). 'Refinery kettle reboilers—proposed method for design and optimization'. *Chem. Eng. Prog.*, **58**(7), 37–46.

Palen, J. W. and Yang, C. C. (1983). 'Circulation boiling model for analysis of kettle and internal reboiler performance'. In *Heat Exchangers for Two-Phase Applications*, ASME HTD Vol. 27, 55–61.

Palen, J. W., Yarden, A., and Taborek, J. (1972). 'Characteristics of boiling outside large-scale horizontal multitube bundles'. *AIChE Symp. Ser.*, **68**(118), 50–61.

Papell, S. S. (1970). 'Buoyancy effects on liquid nitrogen film boiling in vertical flow'. *Advances in Cryogenic Engng.*, **16**, 435–444.

Papell, S. S. (1971). 'Film boiling of cryogenic hydrogen during upward and downward flow'. Paper NASA-TMX-67855 presented at 13th International Congress on Refrigeration, Washington.

Peterson, W. C., Aboul Fetouh, M. M., and Zaalouk, M. G. (1973). 'Boiling curve measurements from a controlled forced convection process'. *Proceedings BNES Conference on Boiler Dynamics and Control in Nuclear Power Stations, London.*

Petukhov, B. S. (1970). In *Advances in Heat Transfer*, Vol. 6, Academic Press, New York.

Plummer, D. N., Iloeje, O. C., Griffith, P., and Rohsenow, W. M. (1973). 'A study of post critical heat flux transfer in a forced convection system'. MIT Technical Report No. 73645-80.

Plummer, D. N., Iloeje, O. C., Rohsenow, W. M., Griffith, P., and Ganic, E. (1974). 'Post critical heat transfer to flowing liquid in a vertical tube'. Mech. Engng. Dept., MIT, Report 72718-91, September.

Quinn, E. P. (1966). 'Forced flow heat transfer to high pressure water beyond the critical heat flux'. Paper presented at the ASME Winter Annual Meeting, 66WA/HT-36.

Ramu, K. and Weisman, J. (1975). 'Transition boiling heat transfer to water in a vertical annulus'. Paper presented at US National Heat Transfer Conference, San Francisco.

Rebrov, P. N., Bukin, V. G., and Danilova, G. N. (1989). 'A correlation for local coefficients of heat transfer in boiling of R12 and R22 refrigerants on multirow bundles of smooth tubes. *Heat Transfer-Sov. Res.*, **21**(4), 543–548.

Robertson, J. M. and Wadekar, V. V. (1988). 'Vertical upflow boiling of ethanol in a 10 mm tube'. *Trans. 2nd UK Nat. Heat Transfer Conf.*, **1**, 67–77, Inst. Mech. Eng., London.

Rohsenow, W. M. (1952). 'A method of correlating heat transfer data for surface boiling of liquids'. *ASME Trans.*, **74**, 969–976.

Ryley, D. J. (1961/62). 'The evaporation of small liquid drops with special reference to water drops in steam'. *J. Liverpool Engng. Soc.*, **7**(1), 1.

Sani, R. L. (1960). 'Downward boiling and non-boiling heat transfer in a uniformly heated tube'. University of California, UCRL-9023, January.

Schmidt, K. R. (1959). 'Wärmetechnische Untersuchungen an hoch belasteten Kesselheizflächen'. *Mitteilungen der Vereinigung der Grosskessel-bezitzer*, December, 391–401.

Schrock, V. E. and Grossman, L. M. (1959). 'Forced convection boiling studies'. *Forced Convection Vaporization Project—Final Report 73308*—UCX 2182, 1 November, University of California, Berkeley, USA.

Schrock, V. E. and Grossman, L. M. (1962). 'Forced convection boiling in tubes'. *Nuclear Science & Engineering*, **12**, 474–480.

Shah, M. M. (1976). 'A new correlation for heat transfer during boiling flow through pipes'. *ASHRAE Trans.*, **82**(2), 66–86.

Shah, M. M. (1982). 'Chart correlation for saturated boiling heat transfer: equations and further study'. *ASHRAE Trans.*, **88**(1), 185–196.

Shklover, G. G. and Gusev, S. E. (1988). 'Natural convection heat transfer in horizontal tube bundles'. *Heat Transfer-Sov. Res.*, **20**(6), 746–758.

Singh, R. L., Saini, J. S., and Varma, H. K. (1983). 'Effect of cross-flow on boiling heat transfer in water'. *Int. J. Heat Mass Transfer*, **26**, 1882–1885.

Singh, R. L., Saini, J. S., and Varma, H. K. (1985). 'Effect of cross-flow on boiling heat transfer of refrigerant-12'. *Int. J. Heat Mass Transfer*, **28**, 512–514.

Slaughterback, D. C., Ybarrondo, L. J., and Obenchain, C. F. (1973). 'Flow film boiling heat transfer correlations—parametric study with data comparisons. Paper presented at ASME–AIChE Heat Transfer Conference, Atlanta, August.

Slaughterback, D. C., Vesely, W. E., Ybarrondo, L. J., Condie, K. G., and Mattson, R. J. (1973). 'Statistical regression analyses of experimental data for flow film boiling heat transfer'. Paper presented at ASME–AIChE Heat Transfer Conference, Atlanta, August.

Slesarenko, V. N., Rudakova, A. Y., and Zakharov, G. A. (1982). 'Effect of operating conditions and tube bundle geometry on boiling heat transfer'. *Heat Transfer-Sov. Res.*, **14**(2), 119–123.

Steiner, D. (1986). 'Heat transfer during flow boiling of cryogenic fluids in vertical and horizontal tubes'. *Cryogenics*, **26**, 309–318.

Steiner, D. (1988). 'Wärmeübertragung beim sieden gesättigter Flüssigkeiten'. *VDI Wärmeatlas*, VDI Verlag, Düsseldorf.

Steiner, D. and Taborek, J. (1992). 'Flow boiling heat transfer in vertical tubes correlated by an asymptotic model'. *Heat Transfer Engng.*, **13**(2), 43–69.

Sterman, L. S. and Styushin, N. G. (1951). 'Heat transfer and aerodynamics'. *Teplopenedacha I. Aerodirakika, Trudy UKTI*, **21**, Mashiga.

Swenson, H. S., Carver, J. R., and Szoeke, G. (1961). 'The effects of nucleate boiling versus film boiling on heat transfer in power boiler tubes'. Paper 61-W-201 presented at ASME Winter Annual Meeting, New York, 26 November–1 December.

Thome, J. R. (1988). 'Reboilers with enhanced boiling tubes'. *Heat Transfer Engng.*, **9**(4), 45–62.

Thome, J. R. (1990). *Enhanced boiling heat transfer*, Hemisphere, New York.

Tong, L. S. and Young, J. D. (1974). 'A phenomenological transition and film boiling heat transfer correlation'. *Proc. 5th Int. Heat Transfer Conf., Tokyo.* Paper B.3.9, **4**, September.

Wallner, R. (1974). 'Heat transfer in flooded shell and tube evaporators'. *Proc. 5th Int. Heat Transfer Conf., Tokyo*, **5**, 214–217.

Whalley, P. B., Azzopardi, B. J., Hewsett, G. F. and Owen, R. G. (1982). 'A physical model for two-phase flows with thermodynamic and hydrodynamic

EXAMPLE 321

non-equilibrium'. *Heat Transfer*, **5**, 181–188. *7th Int. Heat Transfer Conf., Munich FRG.*

Yilmaz, S. and Palen, J. W. (1984). 'Performance of finned tube reboilers in hydrocarbon service'. ASME Paper 84-HT-91 (1984).

Yilmaz, S. and Westwater, J. W. (1980). 'Effect of velocity on heat transfer to boiling Freon-113'. *J. Heat Transfer*, **102**, 26–31.

Zuber, N. and Fried, E. (1962). 'Two-phase flow and boiling heat transfer to cryogenic liquids'. *Am. Rocket Society Jn.*, 1332–1341, September.

Zukauskas, A. and Ulinskas, R. (1985). 'Efficiency parameters for heat transfer in tube banks'. *Heat Transfer Engng.*, **6**(1), 19–25.

Example

Example 1
Water at a pressure of 1.186 bar is to be evaporated in a 12.7 mm i.d. tube. The water flow-rate is 136 kg/h. Prepare a table of values of heat flux against temperature difference for steam qualities of 1, 5, and 20 per cent by weight using Chen's correlation.

Solution

$$\text{Tube area} = \frac{\pi \times (12.7)^2 \times 10^{-6}}{4} = 1.267 \times 10^{-4} \text{ m}^2$$

$$\text{Mass velocity } G = \frac{136}{1.267 \times 10^{-4} \times 3600} = 298 \text{ kg/m}^2 \text{ s.}$$

$$h_c = 0.023 \text{ Re}_f^{0.8} \text{ Pr}_f^{0.4}\left(\frac{k_f}{D}\right) F$$

$$\text{Re}_f = \frac{DG(1-x)}{\mu_f} \qquad \mu_f = 2.725 \times 10^{-4} \text{ N s/m}^2$$

$$= \frac{12.7 \times 10^{-3} \times 298(1-x)}{2.725 \times 10^{-4}} = 1.389 \times 10^4(1-x)$$

For $x = 0.05$, $\quad \text{Re}_f = 1.319 \times 10^{-4}$, $\text{Re}_f^{0.8} = 1.97 \times 10^3$,

$$\text{Pr}_f = \frac{c_{pf}\mu_f}{k_f} = 1.69 \text{ and } \text{Pr}_f^{0.4} = 1.234$$

$$\frac{k_f}{D} = \frac{0.683}{12.7 \times 10^{-3}} = 53.8$$

F is a function of the term $1/X_{tt}$

$$1/X_{tt} = \left(\frac{x}{1-x}\right)^{0.9}\left(\frac{v_g}{v_f}\right)^{0.5}\left(\frac{\mu_g}{\mu_f}\right)^{0.1} = \left(\frac{x}{1-x}\right)^{0.9}\left(\frac{1.448}{1.048 \times 10^{-3}}\right)^{0.5}\left(\frac{1.365 \times 10^{-5}}{2.725 \times 10^{-4}}\right)^{0.1}$$

$$= 27.54\left(\frac{x}{1-x}\right)^{0.9}$$

For $x = 0.05$

$$\frac{1}{X_{tt}} = 27.54\left(\frac{0.05}{0.95}\right)^{0.9} = 1.946$$

From Fig. 7.5

$F = 4.25$

$h_c = 0.023 \times 1.978 \times 10^3 \times 1.234 \times 53.8 \times 4.25 = \underline{12.85\ \text{kW/m}^2\,^\circ\text{C}}$

For $x = 0.01$, $h_c = \underline{5.46\ \text{kW/m}^2\,^\circ\text{C}}$

$x = 0.20$, $h_c = \underline{29.0\ \text{kW/m}^2\,^\circ\text{C}}$

$$h_{\text{NcB}} = 0.00122\left[\frac{k_f^{0.79}c_{pf}^{0.45}\rho_f^{0.49}}{\sigma^{0.5}\mu_f^{0.29}i_{fg}^{0.24}\rho_g^{0.24}}\right]S(\Delta T_{\text{SAT}})^{0.24}(\Delta p_{\text{SAT}})^{0.75}$$

For $k_f = 0.683\ \text{W/m}\,^\circ\text{C}$, $k_f^{0.79} = 0.740$

$c_{pf} = 4.21 \times 10^3\ \text{J/kg}\,^\circ\text{C}$, $c_{pf}^{0.45} = 42.65$

$\rho_f = 955\ \text{kg/m}^3$, $\rho_f^{0.49} = 28.84$

$\sigma = 0.058\ \text{N/m}$, $\sigma^{0.5} = 0.241$

$\mu_f = 2.725 \times 10^{-4}\ \text{N s/m}^2$, $\mu_f^{0.29} = 9.259 \times 10^{-2}$

$i_{fg} = 2.245 \times 10^6\ \text{J/kg}$, $i_{fg}^{0.24} = 33.35$

$\rho_g = 0.691\ \text{kg/m}^3$, $\rho_g^{0.24} = 0.9152$

$$h_{\text{NcB}} = \left(\frac{0.00122 \times 0.740 \times 42.65 \times 28.84}{0.241 \times 9.259 \times 10^{-2} \times 33.35 \times 0.9152}\right)S(\Delta T_{\text{SAT}})^{0.24}(\Delta p_{\text{SAT}})^{0.75}$$

$$= 1.63 \times (\Delta T)^{0.24} \times (\Delta p)^{0.75}S$$

S is a function of Re_{TP}, $\text{Re}_{\text{TP}} = \text{Re}_f \times F^{1.25}$

For $x = 0.05$

$\text{Re}_{\text{TP}} = 1.319 \times 10^{-4} \times (4.25)^{1.25} = 1.319 \times 10^4 \times 6.102 = \underline{8.048 \times 10^4}$

From Fig. 7.6 $S = 0.435$

Evaluate h_{NcB} and therefore values of ϕ at temperature differences ΔT_{SAT} of 2.78°C, 11.1°C, and 22.2°C.

For $\Delta T_{\text{SAT}} = 2.78$°C

p_{SAT} at 104.5°C $= 1.186 \times 10^5\ \text{N/m}^2$

p_{SAT} at 107.28°C $= 1.304 \times 10^5\ \text{N/m}^2$

$\Delta p_{\text{SAT}} = 1.184 \times 10^4\ \text{N/m}^2$

Therefore

$h_{\text{NcB}} = 1.63 \times 0.435 \times (2.78)^{0.24} \times (1.184 \times 10^4)^{0.75}$

$= 1.03\ \text{kW/m}^2\,^\circ\text{C}$

$h_{\text{TP}} = 12.85 + 1.03 = 13.88\ \text{kW/m}^2\,^\circ\text{C}$

$\phi = 2.78 \times 13.88 = 38.6\ \text{kW/m}^2$

For $\Delta T_{SAT} = 11.1°C$

$$p_{SAT} \text{ at } 104.5°C = 1.186 \times 10^5 \text{ N/m}^2$$
$$p_{SAT} \text{ at } 115.6°C = 1.721 \times 10^5 \text{ N/m}^2$$
$$\Delta p_{SAT} = 5.355 \times 10^4 \text{ N/m}^2$$

Therefore

$$h_{NcB} = 1.63 \times 0.435 \times (11.1)^{0.24} \times (5.355 \times 10^4)^{0.75}$$
$$= 4.46 \text{ kW/m}^2 \, °C$$
$$h_{TP} = 12.85 + 4.46 = 17.31 \text{ kW/m}^2 \, °C$$
$$\phi = 17.31 \times 11.1 = 192 \text{ kW/m}^2$$

For $\Delta T_{SAT} = 22.2°C$

$$p_{SAT} \text{ at } 104.5°C = 1.186 \times 10^5 \text{ N/m}^2$$
$$p_{SAT} \text{ at } 126.7°C = 2.444 \times 10^5 \text{ N/m}^2$$
$$\Delta p_{SAT} = 1.258 \times 10^5 \text{ N/m}^2$$

Therefore

$$h_{NcB} = 1.63 \times 0.435 \times (22.2)^{0.24} \times (1.258 \times 10^5)^{0.75}$$
$$= 9.94 \text{ kW/m}^2 \, °C$$
$$h_{TP} = 12.85 + 9.94 = 22.79 \text{ kW/m}^2 \, °C$$
$$\phi = 22.79 \times 22.2 = 506 \text{ kW/m}^2.$$

Table 7.11 gives the values of the surface heat flux $\phi(\text{kW/m}^2)$ for superheats ΔT of 2.78, 11.1, and 22.2°C and steam qualities of 1, 5, and 20 per cent by weight.

Table 7.11

Superheat °C	Steam quality percent by weight		
	1	5	20
2.78	19.9	38.6	81.5
11.1	142.5	192	342
22.2	486.5	506	735

Problems

1. Repeat the calculation illustrated in Example 1 of Chapter 5 using the Chen correlation (eq. (7.27)) to estimate the surface temperature distribution.

2. Examine the influence of liquid entrainment upon the value of the two-phase heat transfer coefficient for the case where a steam–water mixture at a pressure of 6.9 bar flows through a 5 mm tube. The mass velocity of the flow is 136 kg/m² s and the steam quality is 20 per cent. Compute the heat transfer coefficient for values of e of 0, 0.5, and 0.95.

3. In an electrically heated cryogenic vaporizer, liquid nitrogen passes through the once-through vaporizer tube 0.5 in i.d. at a flow of 1 lb/s. The pressure at the exit of the tube is 60 psia. Dryout occurs at a mass quality of 20 per cent. The heat flux supplied through the tube wall is 27,000 Btu/h ft². Calculate the wall temperature distribution downstream of the dryout point assuming (a) complete thermodynamic equilibrium and (b) complete lack of equilibrium between the vapour and liquid phases in the liquid-deficient region.

4. Using the same situation as in Example 1, calculate the heat transfer coefficient and boiling temperature difference for a steam quality of 5 per cent and a heat flux of 192 kW/m² using the Steiner–Taborek method. Assume a tube roughness of 2 μm.

5. Repeat Problem 4 using the Gungor–Winterton correlation.

6. For refrigerant R134a boiling inside an 8.0 mm horizontal tube, determine the local heat transfer coefficient for a flow rate of 0.01257 kg/s, a local vapour quality of 40 per cent and a heat flux of 10 kW/m² at a saturation temperature of 3°C (3.27 bar) using the Gunger–Winterton correlation:

(Properties: liquid density = 1282 kg/m³, vapour density = 16.0 kg/m³, liquid viscosity = 0.211 × 10⁻³ N s/m², vapour viscosity = 0.0121 × 10⁻³ N s/m², liquid thermal conductivity = 0.0929 W/m °C, latent heat = 196.4 kJ/kg, liquid specific heat = 1.334 kJ/kg °C, critical pressure = 40.56 bar, molecular weight = 102.03).

7. Use the Groeneveld correlation to estimate the heat transfer coefficient in the post-dryout region for water flowing inside a 30 mm bore tube. The water is flowing at 0.55 kg/s at 86 bar and the local vapour quality is 80 per cent. How does this value compare with those assuming single-phase flow of (i) all vapour and (ii) all liquid?

8. Determine the bundle boiling coefficient for R134a boiling on a horizontal tube bundle with 30 vertical tube rows for the same conditions as Problem 6 (surface tension = 0.0114 N/m) using the Rebrov et al. method. Calculate the Palen bundle boiling factor using this bundle coefficient.

8

CRITICAL HEAT FLUX IN FORCED CONVECTIVE FLOW—1. VERTICAL UNIFORMLY HEATED TUBES

8.1 Introduction

The nature of the critical heat flux condition in forced convection was briefly discussed in Chapter 4 in the context of the boiling surface. Critical heat flux in pool boiling was also discussed in Chapter 4. This limiting condition forms a most important boundary when considering the performance of heat exchange equipment in which evaporation is occurring. The critical heat flux condition is characterized by a sharp reduction of the local heat transfer coefficient which results from the replacement of liquid by vapour adjacent to the heat transfer surface. For the case where the surface heat flux is the independent variable, the condition manifests itself as a sharp increase in surface temperature as the critical heat flux value is reached. Likewise, a considerably reduced heat flux will result when the condition is reached using a temperature-controlled heating surface.

There is considerable disparity in nomenclature for the critical heat flux condition. The most common name is 'burnout' but this implies a physical destruction of the heated surface. Most measurements of the 'burnout heat flux' are, in fact, measurements of the heat flux at which a sharp rise in surface temperature takes place and, as has been noted earlier, melting of the heated surface would not necessarily be expected to occur at the same flux level. The alternative forms 'DNB' (departure from nucleate boiling) and 'dryout' are equally unsatisfactory for a *general* description of the phenomenon since they imply definite mechanisms. The term 'critical heat flux condition' has therefore been chosen to denote the *state* of the system when the characteristic reduction in heat transfer coefficient has just occurred and the term 'critical heat flux' (CHF) to describe the value of heat flux at which and local to the point at which, this state of the system first occurs. Some authors have used the term 'boiling crisis' to describe the critical heat flux condition, and in some ways this term is preferable, though not as commonly acceptable. The main difficulty in using the chosen nomenclature is that it implies an approach to the critical condition by increasing the heat flux, whereas, in fact, the critical condition may also be approached for a given system by varying any of the other independent variables; pressure, inlet temperature (or mass quality), and mass velocity. However, it is hoped that the reader will appreciate this difficulty and will be sympathetic to the choice of terms.

In Chapter 4 the various regimes of heat transfer were discussed and it has been noted that the critical heat flux condition can occur both for subcooled liquids and for saturated liquid–vapour mixtures. Indeed, for a uniformly heated tube the critical heat flux (CHF) may be shown as a continuous function of the fluid enthalpy (Fig. 4.24). On examining this diagram it will be seen that the critical heat flux curve is divided into three regions depending mainly on the enthalpy and on the mechanism of heat transfer prior to the critical condition. When the bulk fluid is subcooled at the location where the critical heat flux is exceeded, the critical condition is termed 'DNB' (subcooled)—departure from subcooled nucleate boiling. Similarly, where saturated nucleate boiling occurs just prior to the critical condition, the term is 'DNB' (saturated)—departure from saturated nucleate boiling. For higher fluid enthalpies where nucleation is supposed and where the flow pattern is likely to be annular, the term 'dryout' is used to imply that drying out of the liquid film is the cause of the critical heat flux condition. The corresponding regions of heat transfer where the critical heat flux has been exceeded are discussed in Chapter 7.

Practically all of the experimental studies of the onset of the critical heat flux condition have been carried out using heated surfaces for which the heat flux rather than the surface temperature is independently controlled. In this instance the severity of the temperature transient, as the critical heat flux value is exceeded, differs markedly in the three regions, decreasing considerably at the higher fluid enthalpies. Indeed, whereas in the subcooled region the transient is often sufficient to melt or rupture the heating surface (physical burnout), under particular conditions it may become very difficult or even impossible to clearly distinguish a temperature discontinuity as a result of dryout of the liquid film.

Considerable differences are also found to exist in the practical methods used to define and measure the critical condition. Thus, before attempting to compare data from different sources it is always necessary to ensure that the definition and methods used in the different works are the same or, at least, are likely to give similar results.

Initially, the literature on this subject was confined to the reporting of experiment data for different geometrical systems and coolant conditions, including those specific to high-performance heat-exchanger equipment such as nuclear power reactors and to empirical correlations of these data which are invariably very limited in their areas of satisfactory application. More recently the various mechanisms by which the critical heat flux condition is initiated in forced convection have become better understood and a number of analytical treatments have become available.

The following sections attempt a review of the current status of knowledge of the critical heat flux for forced convective flow of the coolant inside conduits of various shapes and orientation.

8.2 Critical heat flux for forced convective flow of water in circular conduits with uniform heat flux

8.2.1 *Introduction*

In many fields of engineering heat is removed from a piece of equipment by means of an evaporating liquid fed by natural or forced convection through tubular channels. Despite this fact, it is only during the past forty or so years that studies of the critical heat flux condition have been made.

This section will cover the case for water evaporating in vertical, uniformly heated tubes. For the present the discussion is restricted to conditions where the fluid fed to the channel is subcooled and contains no entrained vapour or non-condensable gases. The critical heat flux condition with vapour–liquid mixtures at the tube inlet is discussed in Chapter 9. The discussion is also restricted to the case where the flow is stable and oscillations do not occur. The presence of flow oscillations may considerably reduce the critical heat flux relative to its value for steady flow.

In general, for uniformly heated channels, it can be assumed that the onset of the critical heat flux condition occurs first at the exit end of the channel. For non-uniform axial heat flux distributions, overheating may occur first either at the exit or upstream of the exit. There have been reports (Waters *et al.* 1963) of upstream overheating for uniformly heated tubes but, whilst there is no reason to doubt the validity of such observations, it may be stated that they are unusual and, for the vast majority of cases, overheating occurs initially at the channel exit. Under unstable hydrodynamic conditions, overheating may occur at the tube exit or entrance or any location between.

8.2.2 *Some initial considerations*

Consider first the possible independent variables which can influence the critical heat flux. There are certainly five major variables (Fig. 8.1); the inlet flow-rate, the inlet temperature, the system pressure, the tube internal diameter, and the tube length. For the present discussion it will be assumed that the channel pressure drop is small compared with the imposed static pressure and that a single pressure, p, can be ascribed to the entire system. Other variables which might have been listed are such items as the material and surface finish of the tube and perhaps the method of heating. The influence of these variables is obscure and they are therefore omitted for the present.

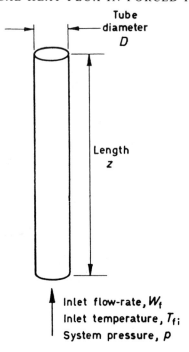

Tube
diameter
D

Length
z

Inlet flow-rate, W_f
Inlet temperature, T_{fi}
System pressure, P

Fig. 8.1. Independent variables affecting the critical heat flux for uniformly heated vertical round ducts.

It is instructive to consider very briefly the limits inside which the critical heat flux must lie. It has been stated earlier that the critical heat flux condition cannot exit if the heater surface temperature lies below the saturation temperature. Thus the minimum possible critical heat flux is given by

$$(\phi_{CRIT})_{MIN} = \frac{(\Delta T_{SUB})_i}{[4z/(Gc_{pf}D) + 1/h_{fo}]} \tag{8.1}$$

The critical heat flux condition must occur at or before all the liquid fed to the channel is evaporated ($x(z) = 1$) leading to

$$(\phi_{CRIT})_{MAX} = \frac{GDi_{fg}}{4z}\left[1 + \frac{c_{pf}(\Delta T_{SUB})_i}{i_{fg}}\right] \tag{8.2}$$

The possible range of conditions for the critical heat flux can be shown diagrammatically (Fig. 8.2). These limits are derived from very simple self-evident considerations but when correlations are applied to any particular problem it is prudent to check that the estimated critical heat flux does,

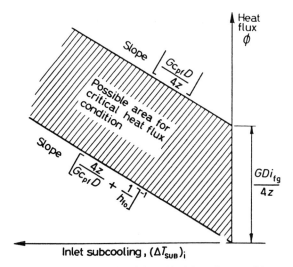

Fig. 8.2. Defining the limits of the critical heat flux condition.

in fact, lie within the specified limits particularly if the values of the independent variables are outside the range covered by the original correlation.

8.2.3 *Availability of experimental information*

Experiments to determine the critical heat flux for water in vertical uniformly heated round tubes have been carried out in many countries over the past forty years or so. It would be very tedious to review, or even to list, the very many publications and papers which contain relevant experimental data. Fortunately a compilation of many of these data was published by Thompson and Macbeth (1964). A total of 4389 separate experimental results are recorded covering a very wide range of independent variables—Table 8.1 lists the ranges of the independent variables covered by the compilation. This listing of data supersedes that issued by De Bortoli *et al.* (1958). None of the wide range of experimental data from the USSR is included in this compilation. The main reason is the lack of detailed information in the published sources; however, the USSR data (for 'non-pulsating' flow) are generally in good agreement with the data in the compilation.

8.2.4 *General relationships between independent variables and methods of display of experimental data*

Earlier it was postulated that the critical heat flux for water in a vertical uniformly heated tube is a function of five independent variables,

Table 8.1 Experimental range of independent variables for round tubes uniformly heated with liquid water at inlet (Thompson and Macbeth 1964)

Actual or nominal pressure		Actual pressure range (p/p_{NOM})	Range of tube diameters (D)		Range of tube lengths (z)		Range of mass velocity (G)		Range of inlet subcooling ($\Delta i_{SUB})_i$		No. of Expts.
bar	psia		mm	in.	m	in.	kg/m² s	lb/h ft² × 10⁻⁶	MJ/kg	Btu/lb	
1	15	—	1–23.9	0.04–0.94	0.025–0.86	1–33.84	10–5750	0.0073–4.24	1–0.35	0–152	107
6.9	100	—	4.6	0.18	0.239	9.4	13–92	0.0096–0.0677	0.093–0.58	40–249	20
19	280	0.89–1.11	1.1–3	0.045–0.12	0.114–0.152	4.5–6.0	54–15800	0.04–11.6	0.0074–0.775	3.2–334.5	71
38.5	560	0.89–1.30	3–10.8	0.12–0.426	0.076–1.73	3–68	38–10600	0.028–7.82	0–0.94	0–404.9	534
49	720	0.945–1.05	3–5.6	0.12–0.22	0.152–1.73	6–68	38–8150	0.028–5.99	0–0.364	0–156.6	37
69	1000	0.96–1.065	3–37.5	0.12–1.475	0.076–3.66	3–144	27–18600	0.02–13.7	0–1.1	0–473.1	934
89	1300	0.984–1.015	5.7–11.5	0.226–0.451	0.625–1.52	24.6–60	1020–4080	0.75–3.0	0.075–0.426	32.3–183.4	19
107	1550	0.955–1.05	1.9–20	0.075–0.78	0.152–1.52	6–60	28–9900	0.0207–7.28	0.080–1.32	34.6–567.1	269
124	1800	0.972–1.028	1.9–20	0.075–0.78	0.152–0.915	6–36	79–4120	0.058–3.03	0.117–1.40	50.4–599.3	63
138	2000	—	1.9–11.1	0.075–0.436	0.076–1.83	3–72	31–10600	0.023–7.79	0–1.45	0–626.1	649
163	2250	—	1.9	0.075	0.152–0.695	6–27.4	2040–3890	1.5–2.86	0.42–1.51	180.8–650.7	30
173	2500	—	1.9	0.075	0.152–0.695	6–27.4	1950–3670	1.43–2.70	0.46–1.58	198.4–681.6	30
183	2650	—	3	0.118	0.035–0.15	1.38–5.91	860–2900	0.635–2.14	0.049–0.39	20.9–170.7	10
190	2750	—	1.9	0.075	0.152–0.695	6–27.4	1850–3790	1.36–2.79	0.50–1.66	215.2–713.2	30
Misc. Swedish		2.34–41 bar	3.9–10	0.155–0.395	0.6–3.12	23.6–122.8	100–2510	0.074–1.85	0.08–0.92	34.7–396.2	1576
Misc. Others		1.38–207 bar	3–5.7	0.12–0.226	0.076–0.625	3–24.6	50–6860	0.037–5.05	0.042–0.995	18–427.6	10
											4389

thus,

$$\phi_{CRIT} = fn(G, (\Delta T_{SUB})_i, p, D, z) \tag{8.3}$$

Since overheating of the tube surface at the critical condition almost invariably begins at the exit of the heated section, it can be argued that display and correlation should be terms of the exit conditions. The thermodynamic state at the tube exit may be such that the fluid is either subcooled or saturated. Either the fluid enthalpy $i(z)$ or the thermodynamic mass quality, $x(z)$ may be chosen to characterize the exit conditions. Each of these dependent variables is related to the inlet subcooling via the heat balance equation thus,

$$i(z) = i_f + \frac{4\phi z}{DG} - (\Delta T_{SUB})_i c_{pf} \tag{8.4}$$

$$x(z) = \frac{1}{i_{fg}} \left[\frac{4\phi z}{DG} - (\Delta T_{SUB})_i c_{pf} \right] \tag{8.5}$$

Thus, as alternatives to eq. (8.3) it is possible to write

$$\phi_{CRIT} = fn(G, i(z), p, D, z) \tag{8.6}$$

or

$$\phi_{CRIT} = fn(G, x(z), p, D, z) \tag{8.7}$$

It is instructive to examine briefly the manner in which the critical heat flux varies with some of the independent variables. Examples of such relationships are as follows:

(a) *The relationship between* ϕ_{CRIT}, $(\Delta T_{SUB})_i$ *and G.* As an example of the variation of critical heat flux with inlet subcooling for fixed tube length and diameter, the data of Weatherhead (1963) for an inside tube diameter of 7.7 mm (0.304 in), tube length 45.7 cm (18 in) and a system pressure of 138 bar (2000 psia) may be taken. Weatherhead's critical heat flux values are plotted against the inlet subcooling (expressed in terms of enthalpy, $(\Delta i_{SUB})_i$ in Fig. 8.3 with mass velocity as parameter. It is noted that the critical heat flux appears to increase linearly with inlet subcooling and this linear relationship is often, indeed usually, found, though by no means universally. It is used as a basic factor in a number of empirical correlations. It is also noted that increases in mass velocity cause an increase in critical heat flux for a given inlet subcooling. Here again this effect is quite general but not universal. It is instructive to replot the data of Fig. 8.3 in terms of the exit conditions (Fig. 8.4). The linear relationship between critical heat flux and exit mass quality is to be expected as a consequence of eq. (8.5). The relationship between critical heat flux and mass velocity is, however, considerably complicated by the method of display given in Fig. 8.4. In the subcooled region (negative values of $x(z)$), the critical heat flux increases

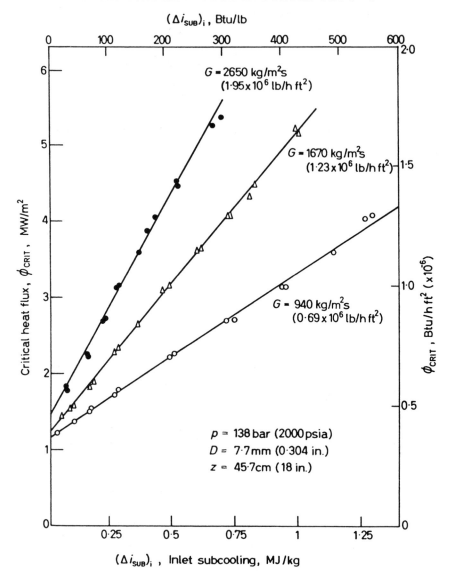

Fig. 8.3. The effect of inlet subcooling on critical heat flux.

with increased mass velocity at constant exit conditions. In the saturated region (positive values of $x(z)$) there is a cross-over and at high values of exit quality the critical heat flux decreases with increased mass velocity.

This example demonstrates a very important point. When the influence of one particular variable on the critical heat flux is considered it is essential

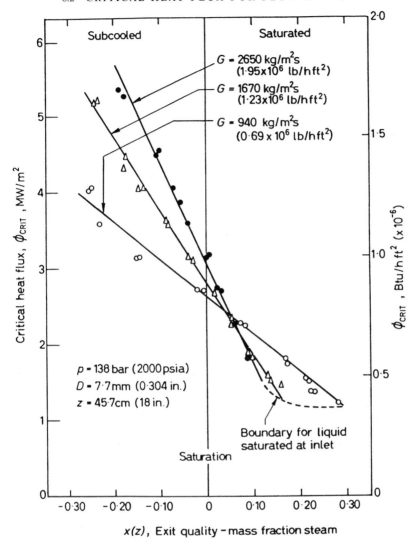

Fig. 8.4. The influence of exit quality on critical heat flux.

to specify which of the other independent (or dependent) variables is held constant. A further interesting feature of Fig. 8.4 is the continuity of the straight lines through the saturation ($x(z) = 0$) condition. This implies that the cause of the critical heat flux condition is similar in the regions either side of the saturation condition though the same reasoning would not necessarily extend to the high quality region.

(b) *The relationship between* ϕ_{CRIT}, $(\Delta T_{SUB})_i$ *and* z. Experimental data obtained by Lee and Obertelli (1963) and Lee (1965) for a pressure of 69 bar (1000 psia), a mass velocity of 2000 kg/m² s (1.5 × 10⁶ lb/h ft²) and a tube diameter in the range of 10.75–10.85 mm (0.424–0.426 in) serve as an example of the variation of the critical heat flux with length and subcooling for fixed mass velocity and tube diameter. Figure 8.5 shows data for six tube lengths between 0.216 m and 3.66 m (8.5 in and 144 in) plotted as critical heat flux against inlet subcooling. It will be seen that the critical heat flux increases as tube length is decreased for constant inlet subcooling. For the longer lengths ($z > 1$ m) the relationship between heat flux and subcooling is linear. This linearity breaks down for short tubes.

If these data are replotted as critical heat flux versus outlet quality as shown in Fig. 8.6 it is seen that the data points for different lengths all fall on one curve, though admittedly with some scatter. It is concluded from this observation that the effect of tube length on the critical heat flux for fixed exit quality is small.

The observation that the critical heat flux is independent of tube length for a given exit quality has interesting implications. The tube may be divided into two lengths; over the first length (z_{SC}) the liquid is raised to the saturation condition ($x(z_{SC}) = 0$) and over the second length (z_{SAT}) the quality is raised from zero to $x(z)$, the outlet quality corresponding to the critical condition. The following relationships hold:

$$z = z_{SC} + z_{SAT} \tag{8.8}$$

$$z_{SC} = \frac{DG(\Delta i_{SUB})_i}{4\phi_{CRIT}} \tag{8.9}$$

$$z_{SAT} = \frac{x(z)DGi_{fg}}{4\phi_{CRIT}} \tag{8.10}$$

For a fixed $x(z)$ and ϕ_{CRIT}, the value of z_{SAT} is also fixed, though z_{SC} will vary with subcooling. This implies that the only function of z_{SC} is to heat the fluid up to the saturation condition at the appropriate flux. With the assumed relationship

$$\phi_{CRIT} = fn(x(z), G, p, D) \tag{8.11}$$

it follows from the above equation that these further equations hold:

$$\phi_{CRIT} = fn(z_{SAT}, G, p, D) \tag{8.12}$$

$$x_{CRIT} = fn(z_{SAT}, G, p, D) \tag{8.13}$$

Though eq. (8.13) represents a relationship between two independent variables it is a particularly useful one since it provides an alternative way of looking at the data. Equation (8.11) suggests there is a local relationship between critical heat flux and quality whereas eq. (8.13) suggests that the

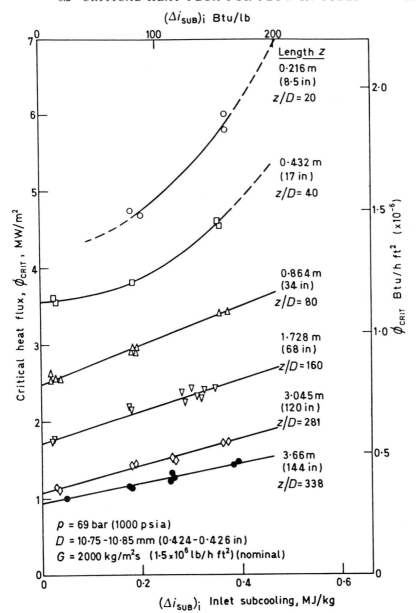

Fig. 8.5. The effect of tube length on critical heat flux at fixed inlet conditions.

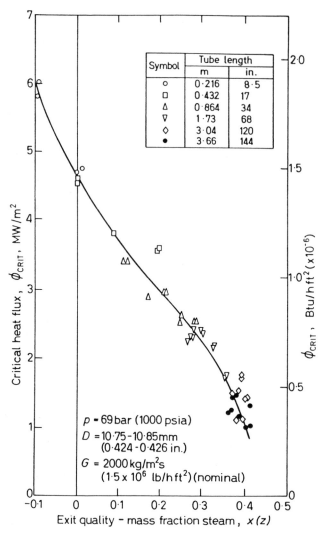

Fig. 8.6. The effect of tube length on critical heat flux at fixed exit conditions.

mass fraction of the liquid which can be evaporated (x_{CRIT}) in the channel before the onset of the critical condition is a function of the length over which the evaporation takes place, (z_{SAT}). Neither of these two views of the critical heat flux phenomenon is completely correct and neither, as it turns out, can be used as a general basis for the prediction of more complex cases. However, for the critical condition in a uniformly heated channel it is impossible to distinguish between the two hypotheses. Figure 8.7 shows

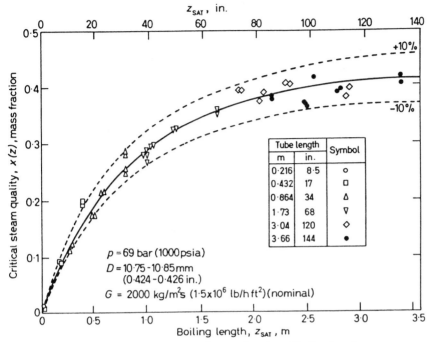

Fig. 8.7. Critical steam quality as a function of boiling length.

the data of Lee and Obertelli (1963) plotted in the form suggested by eq. (8.13).

(c) *The relationship between* ϕ_{CRIT}, $(\Delta T_{\text{SUB}})_i$ *and D.* The effect of tube diameter on the relationship between ϕ_{CRIT} and $(\Delta i_{\text{SUB}})_i$ and between ϕ_{CRIT} and $x(z)$ for fixed values of G, z, and p can be well illustrated by plotting the data of Lee and Obertelli (1963) and Matzner (1963). These data were obtained for a pressure of 69 bar (1000 psia), a mass velocity of 2000 kg/m² s (1.5 × 10⁶ lb/h ft²) and for tube lengths in the range 1.93–2 m (76–79 in).

Figure 8.8 shows the critical heat flux plotted against inlet subcooling for six tube diameters ranging from 5.58 mm to 37.45 mm (0.220 in to 1.475 in). As the tube diameter is increased, so the critical heat flux increases at constant inlet subcooling. The relationship between critical heat flux and inlet subcooling is linear for small tube diameters ($D \leqslant 12.8$ mm) but the larger diameters show a marked curvature. Figure 8.9 shows the same data plotted against exit quality. Where the data for different tube diameters overlap, it can be seen that the critical heat flux *decreases* with increasing tube diameter for a given exit quality.

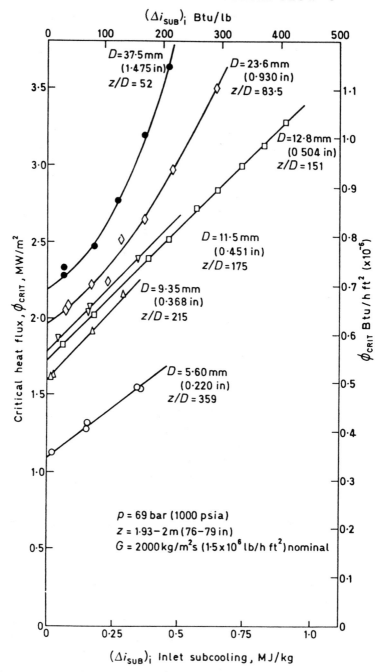

Fig. 8.8. The influence of tube diameter on critical heat flux at fixed inlet conditions.

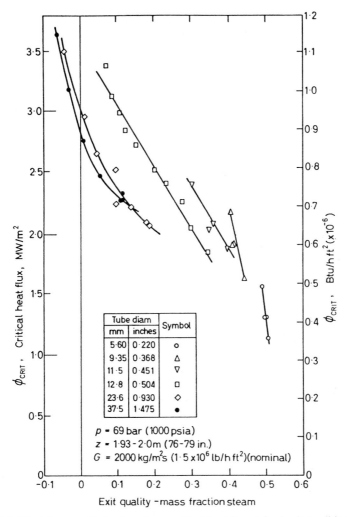

Fig. 8.9. The influence of tube diameter on critical heat flux at fixed exit conditions.

8.2.5 *Empirical correlation of experimental data*

The technical importance of the critical heat flux condition has led to the development of a bewildering variety of correlations. Milioti (1964) catalogues fifty-nine correlations and detailed comparisons of a wide range of correlations reveal considerable differences. Many of the more recent correlations do, however, show fairly reasonable agreement despite wide variations in their functional form. Two types of correlating procedure have been adopted. They are:

(a) Correlations of an empirical nature which make no assumptions what-ever about the mechanisms involved in the critical heat flux condition, but solely attempt a functional relationship between the critical heat flux and the independent variables.
(b) Correlations where attempts have been made to look at and write down equations for the hydrodynamic and heat transfer processes occurring in the heated channel and to relate these to the critical heat flux condition.

The USSR Academy of Sciences (Kirillov et al. 1991) has produced a series of standard tables of critical heat flux as a function of the *local* bulk mean water condition and for various pressures and mass velocities for a fixed tube diameter of 0.315 in (8 mm). These tables are reproduced in Table 8.2 and are valid for $z/D \geq 20$. For tube diameters other than 8 mm the critical heat flux is given by the approximate relationship

$$\phi_{\text{CRIT}} = \phi_{\text{CRIT, 8mm}}(D/0.008)^k \tag{8.14}$$

where k is a complex function of both the pressure and flow pattern (Kirillov et al. 1991).

8.2.5.1   *The Macbeth–Barnett hypothesis*   Probably the most general of the empirical correlations suggested to date that fall into the first category mentioned above is that by Macbeth (1963a, b) and Thompson and Macbeth (1964) based on a hypothesis proposed and tested by Barnett (1963). This hypothesis, the 'local conditions hypothesis' suggests that the critical heat flux is solely a function of the mass quality at the point of overheating, as suggested by eq. (8.11). Figure 8.6 is a proof of the approximate validity of the hypothesis—for uniformly heated round tubes, at least. Barnett and Macbeth make the simplifying assumption that the critical heat flux is a linear function of inlet subcooling, thus

$$\phi_{\text{CRIT}} = A + B(\Delta i_{\text{SUB}})_i \tag{8.15}$$

where A and B are assumed to be functions of G, p, z, and D. The heat balance equation may be written from eq. (8.5) (remembering that $(\Delta i_{\text{SUB}})_i = (\Delta T_{\text{SUB}})_i c_{\text{pf}}$

$$x(z)i_{\text{fg}} + (\Delta i_{\text{SUB}})_i = \frac{4\phi z}{DG} \tag{8.16}$$

Eliminating $(\Delta i_{\text{SUB}})_i$ between eq. (8.15) and eq. (8.16)

$$\phi_{\text{CRIT}} = \frac{A - Bx(z)i_{\text{fg}}}{\left[1 - \dfrac{4Bz}{DG}\right]} \tag{8.17}$$

Table 8.2 Critical heat flux in pipes of 8 mm diameter with uniform heating, MW/m². Look-up table for quality values from −0.5 to 0.15. (Kirillov et al.) 1991)

Pressure (MPa)	Mass flux. (kg/m²s)	Quality													
		−0.50	−0.45	−0.40	−0.35	−0.30	−0.25	−0.20	−0.15	−0.10	−0.05	0.00	0.05	0.10	0.15
1.0	500														
	750														
	1000														
	1500														
	2000											8.97	8.17	7.11	6.41
	2500											8.97	7.93	6.89	6.22
	3000											8.95	7.73	6.72	6.07
	4000									9.59	9.24	8.90	7.43	6.47	5.83
	5000							11.73	10.85	10.00	9.44	8.85	7.21	6.27	5.66
	6000							12.14	11.23	10.35	9.61	8.80	7.04	6.12	5.52
1.5	500														
	750														
	1000														
	1500														
	2000											8.59	7.95	6.92	6.24
	2500											8.60	7.71	6.71	6.05
	3000											8.60	7.52	6.54	5.90
	4000									9.90	9.26	8.57	7.24	6.29	5.68
	5000									10.32	9.48	8.53	7.02	6.11	5.51
	6000									10.68	9.67	8.50	6.85	5.96	5.37

(*Continued*)

Table 8.2 (*Continued*)

Pressure (MPa)	Mass flux (kg/m² s)	−0.50	−0.45	−0.40	−0.35	−0.30	−0.25	−0.20	−0.15	−0.10	−0.05	0.00	0.05	0.10	0.15
2.0	500														
	750														
	1000														6.64
	1500														6.29
	2000											8.27	7.71	6.70	6.05
	2500											8.30	7.48	6.50	5.87
	3000									9.50	8.91	8.30	7.29	6.34	5.72
	4000									10.03	9.22	8.29	7.01	6.10	5.50
	5000							12.27	11.34	10.45	9.46	8.27	6.80	5.92	5.34
	6000							12.70	11.74	10.82	9.67	8.24	6.64	5.77	5.21
3.0	500													7.16	6.46
	750													6.89	6.21
	1000										7.60	7.60	7.60	6.52	5.88
	1500								9.04	8.33	7.92	7.67	7.47	6.27	5.65
	2000							10.32	9.55	8.80	8.25	7.75	7.21	6.08	5.49
	2500							10.76	9.95	9.17	8.51	7.80	6.99	5.93	5.35
	3000							11.14	10.30	9.49	8.73	7.83	6.82	5.70	5.15
	4000							11.76	10.87	10.02	9.08	7.84	6.56	5.53	4.99
	5000							12.26	11.34	10.45	9.36	7.84	6.36	5.40	4.87
	6000							12.69	11.74	10.81	9.60	7.82	6.21		

5.0	500								
	750					6.38	6.33	6.21	5.50
	1000		7.98	7.35	6.82	6.57	6.36	5.97	5.39
	1500	9.31	8.61	7.93	7.35	6.80	6.32	5.65	5.10
	2000	9.83	9.09	8.37	7.74	6.94	6.20	5.44	4.90
	2500	10.25	9.48	8.73	8.07	7.03	6.06	5.27	4.76
	3000	10.60	9.81	9.04	8.34	7.09	5.92	5.14	4.64
	4000	11.19	10.35	9.54	8.79	7.16	5.69	4.95	4.46
	5000	11.67	10.80	9.95	9.15	7.19	5.52	4.80	4.33
	6000	12.08	11.17	10.30	9.46	7.21	5.38	4.68	4.22
7.0	500						5.86	5.86	5.59
	750		7.11	6.55	6.02	5.73	5.64	5.59	4.89
	1000	8.29	7.67	7.07	6.49	6.07	5.79	5.55	4.45
	1500	9.22	8.53	7.86	7.22	6.57	5.95	5.38	3.90
	2000	9.95	9.20	8.48	7.79	6.94	6.02	5.16	3.55
	2500	10.55	9.76	8.99	8.26	7.24	6.06	4.95	3.30
	3000	11.07	10.24	9.44	8.67	7.49	6.07	4.73	3.10
	4000	11.94	11.04	10.18	9.35	7.90	6.07	4.33	2.83
	5000	12.66	11.71	10.79	9.91	8.23	6.05	4.03	2.63
	6000	13.28	12.28	11.32	10.40	8.52	6.02	3.79	2.47
	7500	14.08	13.03	12.01	11.03	8.88	5.96	3.52	2.30

(Continued)

Table 8.2 (*Continued*)

Pressure (MPa)	Mass flux (kg/m² s)	−0.50	−0.45	−0.40	−0.35	−0.30	−0.25	−0.20	−0.15	−0.10	−0.05	0.00	0.05	0.10	0.15
10.0	500									4.08	3.73	3.62	3.61	3.61	3.61
	750						5.80	5.36	4.94	4.54	4.15	3.89	3.70	3.53	3.24
	1000				7.24	6.74	6.25	5.78	5.33	4.90	4.48	4.08	3.72	3.40	2.94
	1500				8.06	7.50	6.96	6.43	5.93	5.45	4.98	4.34	3.67	3.05	2.58
	2000				8.69	8.09	7.50	6.94	6.40	5.87	5.37	4.50	3.56	2.77	2.35
	2500				9.22	8.58	7.96	7.36	6.78	6.23	5.70	4.62	3.42	2.58	2.18
	3000				9.67	9.00	8.35	7.72	7.12	6.54	5.98	4.71	3.27	2.43	2.05
	4000				10.43	9.70	9.00	8.33	7.68	7.05	6.45	4.23	2.97	2.21	1.87
	5000				11.06	10.29	9.55	8.83	8.14	7.48	6.84	6.03	5.02	4.42	3.11
	6000				11.60	10.80	10.02	9.26	8.54	7.84	7.18	6.24	5.22	4.39	2.87
	7500				12.31	11.45	10.62	9.82	9.06	8.32	7.61	6.53	5.54	4.30	2.46
12.0	500			5.31	4.95	4.60	4.27	3.95	3.64	3.35	3.06	2.81	2.72	2.66	2.62
	750			5.90	5.51	5.12	4.75	4.40	4.05	3.72	3.40	3.11	2.86	2.65	2.42
	1000			6.37	5.94	5.53	5.13	4.74	4.37	4.01	3.67	3.33	2.93	2.56	2.20
	1500			7.08	6.61	6.15	5.70	5.27	4.86	4.47	4.09	3.68	2.95	2.28	1.93
	2000			7.64	7.13	6.63	6.15	5.69	5.24	4.82	4.41	3.94	2.90	2.07	1.75
	2500			8.10	7.56	7.03	6.52	6.03	5.56	5.11	4.67	4.15	2.82	1.93	1.63
	3000			8.50	7.93	7.38	6.84	6.33	5.83	5.36	4.90	4.33	2.71	1.82	1.54
	4000			9.17	8.55	7.96	7.38	6.83	6.29	5.78	5.29	4.63	2.44	1.65	1.40
	5000			9.72	9.07	8.44	7.83	7.24	6.67	6.13	5.61	4.86	4.15	3.08	2.41
	6000			10.20	9.51	8.85	8.21	7.60	7.00	6.43	5.88	5.06	4.91	3.55	2.19
	7500			10.82	10.09	9.39	8.71	8.05	7.42	6.82	6.24	5.31	5.08	3.45	1.82

14.0	500	4.87	4.56	4.27	3.98	3.70	3.44	3.18	2.93	2.69	2.46	2.24	2.06	1.96	1.89
	750	5.42	5.08	4.75	4.43	4.12	3.82	3.54	3.26	2.99	2.74	2.49	2.25	1.99	1.76
	1000	5.84	5.48	5.12	4.78	4.44	4.12	3.81	3.52	3.23	2.95	2.69	2.37	1.94	1.60
	1500	6.50	6.09	5.70	5.31	4.94	4.59	4.24	3.91	3.59	3.29	2.99	2.53	1.70	1.40
	2000	7.01	6.57	6.15	5.73	5.33	4.95	4.58	4.22	3.87	3.54	3.23	2.60	1.51	1.28
	2500	7.44	6.97	6.52	6.08	5.66	5.25	4.85	4.47	4.11	3.76	3.42	2.62	1.40	1.19
	3000	7.80	7.31	6.84	6.38	5.93	5.50	5.09	4.69	4.31	3.94	3.59	2.60	1.32	1.12
	4000	8.41	7.89	7.37	6.88	6.40	5.94	5.49	5.06	4.65	4.25	3.87	2.44	1.20	1.02
	5000	8.92	8.36	7.82	7.29	6.79	6.30	5.82	5.37	4.93	4.51	4.11	3.44	2.85	1.83
	6000	9.36	8.77	8.20	7.65	7.12	6.60	6.11	5.63	5.17	4.73	4.31	3.67	2.82	1.64
	7500	9.93	9.30	8.70	8.12	7.55	7.00	6.48	5.97	5.48	5.02	4.57	3.73	2.74	1.43
16.0	500	3.03	2.82	2.63	2.44	2.26	2.08	1.92	1.76	1.60	1.45	1.35	1.25	1.16	1.06
	750	3.59	3.35	3.11	2.89	2.67	2.47	2.27	2.08	1.90	1.69	1.56	1.42	1.29	1.16
	1000	4.04	3.77	3.51	3.26	3.02	2.78	2.56	2.35	2.14	1.89	1.72	1.55	1.39	1.22
	1500	4.79	4.47	4.16	3.86	3.57	3.30	3.03	2.78	2.54	2.21	1.98	1.75	1.52	1.29
	2000	5.40	5.04	4.69	4.35	4.03	3.72	3.42	3.14	2.86	2.48	2.21	1.94	1.66	1.39
	2500	5.93	5.53	5.15	4.78	4.42	4.08	3.76	3.44	3.14	2.73	2.43	2.14	1.84	1.55
	3000	6.40	5.97	5.56	5.16	4.77	4.41	4.05	3.71	3.39	2.95	2.63	2.32	2.01	1.70
	4000	7.22	6.73	6.27	5.82	5.38	4.97	4.57	4.19	3.82	3.33	2.99	2.64	2.30	1.95
	5000	7.92	7.39	6.88	6.39	5.91	5.45	5.02	4.60	4.20	3.66	3.29	2.92	2.55	2.18
	6000	8.55	7.98	7.42	6.89	6.38	5.89	5.41	4.96	4.53	3.96	3.57	3.17	2.78	2.39
	7500	9.39	8.76	8.15	7.57	7.00	6.46	5.94	5.45	4.97	4.36	3.94	3.51	3.09	2.67

(Continued)

Table 8.2 (*Continued*)

Pressure (MPa)	Mass flux. (kg/m² s)	Quality													
		−0.50	−0.45	−0.40	−0.35	−0.30	−0.25	−0.20	−0.15	−0.10	−0.05	0.00	0.05	0.10	0.15
18.0	500	2.66	2.49	2.31	2.15	1.99	1.83	1.69	1.55	1.41	1.26	1.17	1.08	0.98	0.89
	750	3.16	2.95	2.74	2.54	2.35	2.17	2.00	1.83	1.67	1.48	1.35	1.22	1.09	0.96
	1000	3.56	3.32	3.09	2.87	2.66	2.45	2.25	2.07	1.89	1.65	1.49	1.33	1.17	1.02
	1500	4.22	3.94	3.66	3.40	3.15	2.90	2.67	2.45	2.23	1.95	1.74	1.54	1.34	1.14
	2000	4.76	4.44	4.13	3.83	3.55	3.27	3.01	2.76	2.52	2.20	1.98	1.76	1.54	1.35
	2500	5.22	4.87	4.53	4.21	3.90	3.59	3.31	3.03	2.77	2.42	2.18	1.95	1.71	1.47
	3000	5.64	5.26	4.89	4.54	4.20	3.88	3.57	3.27	2.99	2.62	2.37	2.11	1.86	1.61
	4000	6.36	5.93	5.52	5.12	4.74	4.38	4.02	3.69	3.37	2.96	2.69	2.41	2.14	1.86
	5000	6.98	6.51	6.06	5.62	5.20	4.80	4.42	4.05	3.70	3.26	2.97	2.67	2.38	2.09
	6000	7.53	7.02	6.54	6.07	5.62	5.18	4.77	4.37	3.99	3.53	3.22	2.91	2.60	2.29
	7500	8.27	7.71	7.18	6.66	6.17	5.69	5.23	4.80	4.38	3.88	3.55	3.23	2.90	2.57
20.0	500	2.03	1.89	1.76	1.64	1.51	1.40	1.28	1.18	1.07	0.95	0.87	0.79	0.71	0.64
	750	2.40	2.24	2.09	1.94	1.79	1.65	1.52	1.39	1.27	1.12	1.01	0.91	0.80	0.70
	1000	2.71	2.53	2.35	2.18	2.02	1.87	1.72	1.57	1.44	1.26	1.15	1.03	0.92	0.80
	1500	3.21	3.00	2.79	2.59	2.40	2.21	2.03	1.86	1.70	1.50	1.37	1.24	1.11	0.98
	2000	3.62	3.38	3.14	2.92	2.70	2.49	2.29	2.10	1.92	1.70	1.56	1.42	1.28	1.14
	2500	3.98	3.71	3.45	3.20	2.97	2.74	2.52	2.31	2.11	1.88	1.73	1.58	1.43	1.28
	3000	4.29	4.00	3.73	3.46	3.20	2.95	2.72	2.49	2.27	2.03	1.88	1.72	1.57	1.41
	4000	4.84	4.51	4.20	3.90	3.61	3.33	3.06	2.81	2.56	2.30	2.14	1.97	1.81	1.64
	5000	5.31	4.96	4.61	4.28	3.96	3.66	3.36	3.08	2.81	2.53	2.36	2.19	2.02	1.85
	6000	5.73	5.35	4.98	4.62	4.28	3.95	3.63	3.33	3.04	2.74	2.57	2.39	2.22	2.04
	7500	6.29	5.87	5.46	5.07	4.69	4.33	3.98	3.65	3.33	3.02	2.84	2.66	2.48	2.30

Table 8.3 Critical heat flux in pipes of 8 mm diameter with uniform heating, MW/m². Look-up table for quality values from 0.2 to 0.9. *Note:* * declares the region where the data are of huge scattering; here the interpolation procedure can be used

Pressure MPa	Mass (kg/m² s)	Quality														
		0.20	0.25	0.30	0.35	0.40	0.45	0.50	0.55	0.60	0.65	0.70	0.75	0.80	0.85	0.90
1.0	500							3.98	*	*	*				0.06	0.04
	750		6.15	5.66	5.21	4.78	4.38					0.11	0.09	0.08	0.07	0.05
	1000	6.44	5.91	5.45	5.01	4.60	4.21	*	*	*	0.13	0.12	0.11	0.09	0.08	0.06
	1500	6.09	5.60	5.15	4.74	*	*	0.21	0.20	0.18	0.16	0.15	0.13	0.11	0.09	0.07
	2000	5.86	5.38	4.96	4.56	*	0.27	0.25	0.23	0.21	0.19	0.17	0.15	0.13	0.11	0.08
	2500	5.68	5.22	4.81	*	0.32	0.30	0.27	0.25	0.23	0.21	0.19	0.16	0.14	0.12	0.09
	3000	5.54	5.09	4.69	*	0.35	0.32	0.30	0.27	0.25	0.23	0.20	0.18	0.15	0.13	0.10
	4000	5.33	4.90	*	0.43	0.40	0.37	0.34	0.31	0.28	0.26	0.23	0.20	0.18	0.15	0.11
	5000	5.17	*	0.52	0.48	0.44	0.41	0.38	0.35	0.32	0.29	0.26	0.23	0.20	0.16	0.13
	6000	5.04	1.31	0.56	0.53	0.48	0.45	0.41	0.38	0.34	0.31	0.28	0.25	0.21	0.18	0.14
1.5	500									3.31	2.93	*	*	*	*	0.06
	750					4.66	4.26	3.88	3.50	*	*	*	0.13	0.11	0.09	0.07
	1000	6.26	5.76	5.30	4.88	4.48	4.10	3.73	*	*	*	0.16	0.14	0.12	0.10	0.08
	1500	5.93	5.45	5.01	4.61	4.24	*	*	0.26	0.24	0.22	0.20	0.17	0.15	0.12	0.10
	2000	5.70	5.24	4.82	4.44	*	*	0.33	0.30	0.28	0.25	0.22	0.20	0.17	0.14	0.11
	2500	5.53	5.08	4.68	*	*	0.40	0.37	0.34	0.31	0.28	0.25	0.22	0.19	0.16	0.12
	3000	5.39	4.96	4.56	*	0.47	0.43	0.40	0.37	0.33	0.30	0.27	0.24	0.21	0.17	0.13
	4000	5.19	4.77	*	0.58	0.54	0.50	0.46	0.42	0.38	0.35	0.31	0.27	0.24	0.20	0.15
	5000	5.03	4.62	0.70	0.65	0.60	0.55	0.51	0.47	0.42	0.39	0.35	0.30	0.26	0.22	0.17
	6000	4.91	3.76	0.77	0.71	0.65	0.60	0.55	0.51	0.46	0.42	0.38	0.33	0.29	0.24	0.19

(*Continued*)

Table 8.3 (*Continued*)

Pressure MPa	Mass (kg/m² s)	Quality														
		0.20	0.25	0.30	0.35	0.40	0.45	0.50	0.55	0.60	0.65	0.70	0.75	0.80	0.85	0.90
2.0	500						4.36	3.97	3.59	3.21	2.84	*	*	*	*	*
	750				4.91	4.51	4.13	3.76	3.39	*	*	*	*	0.13	0.11	0.09
	1000	6.07	5.58	5.14	4.73	4.34	3.97	3.61	*	*	*	0.20	0.18	0.15	0.13	0.10
	1500	5.75	5.28	4.86	4.47	4.11	*	*	0.32	0.30	0.27	0.24	0.21	0.18	0.15	0.12
	2000	5.53	5.08	4.67	4.30	*	*	0.40	0.37	0.34	0.31	0.28	0.24	0.21	0.17	0.14
	2500	5.36	4.93	4.53	*	*	0.49	0.45	0.41	0.38	0.34	0.31	0.27	0.23	0.19	0.15
	3000	5.23	4.80	4.42	*	0.58	0.53	0.49	0.45	0.41	0.37	0.33	0.29	0.25	0.21	0.16
	4000	5.03	4.62	*	0.71	0.66	0.61	0.56	0.51	0.47	0.42	0.38	0.34	0.29	0.24	0.19
	5000	4.88	4.48	*	0.79	0.73	0.67	0.62	0.57	0.52	0.47	0.42	0.37	0.32	0.27	0.21
	6000	4.76	4.37	0.94	0.86	0.80	0.73	0.68	0.62	0.57	0.51	0.46	0.41	0.35	0.29	0.23
3.0	500		5.73	5.28	4.86	4.46	4.08	3.71	3.35	3.00	2.65	2.31	*	*	*	*
	750	5.90	5.43	5.00	4.60	4.22	3.86	3.51	3.17	2.84	*	*	*	0.17	0.14	0.11
	1000	5.68	5.22	4.80	4.42	4.06	3.71	3.38	3.05	*	*	0.26	0.23	0.20	0.17	0.13
	1500	5.37	4.94	4.55	4.18	3.84	3.51	*	*	0.39	0.35	0.32	0.28	0.24	0.20	0.16
	2000	5.17	4.75	4.37	4.02	3.69	*	0.53	0.49	0.44	0.40	0.36	0.32	0.28	0.23	0.18
	2500	5.01	4.61	4.24	3.90	*	0.64	0.59	0.54	0.49	0.45	0.40	0.35	0.31	0.25	0.20
	3000	4.89	4.49	4.14	3.81	*	0.70	0.64	0.59	0.54	0.49	0.44	0.39	0.33	0.28	0.22
	4000	4.70	4.32	3.98	*	0.86	0.80	0.73	0.67	0.61	0.56	0.50	0.44	0.38	0.32	0.25
	5000	4.56	4.19	*	1.04	0.96	0.88	0.81	0.75	0.68	0.62	0.55	0.49	0.42	0.35	0.27
	6000	4.45	4.09	1.23	1.13	1.04	0.96	0.89	0.81	0.74	0.67	0.60	0.53	0.46	0.38	0.30

5.0	500	*	4.97	4.58	4.21	3.87	3.54	3.22	2.91	2.60	2.30	2.00	1.70	*	*	*
	750	5.12	4.71	4.33	3.99	3.66	3.35	3.05	2.75	2.46	2.18	*	*	*	0.20	0.15
	1000	4.92	4.53	4.17	3.83	3.52	3.22	2.93	2.65	*	*	*	0.32	0.27	0.23	0.18
	1500	4.66	4.28	3.94	3.63	3.33	3.05	*	*	0.53	0.48	0.43	0.38	0.33	0.27	0.21
	2000	4.48	4.12	3.79	3.49	3.20	*	*	0.67	0.61	0.55	0.49	0.44	0.38	0.31	0.24
	2500	4.35	4.00	3.68	3.38	3.11	*	0.80	0.74	0.68	0.61	0.55	0.49	0.42	0.35	0.27
	3000	4.24	3.90	3.59	3.30	*	0.95	0.88	0.80	0.74	0.67	0.60	0.53	0.46	0.38	0.30
	4000	4.08	3.75	3.45	*	1.18	1.09	1.00	0.92	0.84	0.76	0.68	0.60	0.52	0.44	0.34
	5000	3.96	3.64	*	1.41	1.31	1.21	1.11	1.02	0.93	0.85	0.76	0.67	0.58	0.48	0.38
	6000	3.86	3.55	1.93	1.54	1.42	1.32	1.21	1.11	1.02	0.92	0.83	0.73	0.63	0.53	0.41
7.0	500	4.90	4.38	3.96	3.60	3.28	3.00	2.73	2.48	2.25	2.02	1.80	1.57	1.35	*	*
	750	4.29	3.84	3.47	3.15	2.87	2.62	2.39	2.17	1.97	1.77	*	*	*	*	0.03
	1000	3.91	3.49	3.15	2.87	2.61	2.39	2.18	1.98	*	*	*	0.08	0.07	0.06	0.04
	1500	3.42	3.06	2.76	2.51	2.29	2.09	*	*	0.15	0.14	0.12	0.11	0.09	0.07	0.05
	2000	3.11	2.78	2.51	2.28	*	*	0.22	0.20	0.18	0.16	0.15	0.13	0.11	0.08	0.06
	2500	2.89	2.58	2.34	*	*	0.28	0.26	0.23	0.21	0.19	0.17	0.14	0.12	0.10	0.07
	3000	2.72	2.43	2.20	*	0.34	0.31	0.29	0.26	0.24	0.21	0.19	0.16	0.13	0.11	0.08
	4000	2.48	2.22	*	0.44	0.40	0.37	0.34	0.31	0.28	0.25	0.22	0.19	0.16	0.13	0.09
	5000	2.30	*	0.54	0.50	0.46	0.43	0.39	0.35	0.32	0.29	0.25	0.22	0.18	0.15	0.11
	6000	2.17	0.72	0.61	0.56	0.52	0.48	0.44	0.40	0.36	0.32	0.28	0.24	0.20	0.16	0.12
	7500	1.56	0.75	0.70	0.64	0.59	0.54	0.50	0.45	0.41	0.37	0.32	0.28	0.23	0.19	0.14

(*Continued*)

Table 8.3 (*Continued*)

Pressure MPa	Mass (kg/m² s)	Quality														
		0.20	0.25	0.30	0.35	0.40	0.45	0.50	0.55	0.60	0.65	0.70	0.75	0.80	0.85	0.90
10.0	500	3.24	2.90	2.62	2.38	2.17	1.98	1.81	1.64	1.49	1.34	1.19	*	*	*	*
	750	2.84	2.54	2.29	2.08	1.90	1.73	1.58	1.44	*	*	*	*	0.08	0.06	0.05
	1000	2.58	2.31	2.09	1.90	1.73	1.58	1.44	*	*	0.14	0.13	0.11	0.09	0.07	0.05
	1500	2.26	2.02	1.83	1.66	*	*	0.25	0.22	0.20	0.18	0.16	0.14	0.12	0.09	0.07
	2000	2.06	1.84	1.66	*	*	0.32	0.29	0.27	0.24	0.22	0.19	0.16	0.14	0.11	0.08
	2500	1.91	1.71	*	*	0.40	0.37	0.34	0.31	0.28	0.25	0.22	0.19	0.16	0.13	0.09
	3000	1.80	1.61	*	0.48	0.44	0.41	0.37	0.34	0.31	0.28	0.24	0.21	0.18	0.14	0.10
	4000	1.64	1.15	0.62	0.57	0.53	0.49	0.45	0.41	0.37	0.33	0.29	0.25	0.21	0.17	0.12
	5000	1.80	0.77	0.71	0.66	0.61	0.56	0.51	0.47	0.42	0.38	0.33	0.29	0.24	0.19	0.14
	6000	1.34	0.85	0.79	0.73	0.68	0.62	0.57	0.52	0.47	0.42	0.37	0.32	0.27	0.22	0.16
	7500	1.06	0.98	0.91	0.84	0.77	0.71	0.65	0.60	0.54	0.48	0.42	0.37	0.31	0.25	0.18
12.0	500	2.42	2.17	1.96	1.78	1.62	1.48	1.35	1.23	1.11	1.00	*	*	*	*	0.04
	750	2.12	1.90	1.71	1.56	1.42	1.30	1.18	*	*	*	0.12	0.10	0.09	0.07	0.05
	1000	1.93	1.73	1.56	1.42	1.29	1.18	*	*	0.18	0.16	0.14	0.12	0.10	0.08	0.06
	1500	1.69	1.51	1.37	1.24	*	0.31	0.28	0.26	0.23	0.21	0.18	0.16	0.13	0.11	0.08
	2000	1.54	1.38	1.24	*	0.40	0.37	0.34	0.31	0.28	0.25	0.22	0.19	0.16	0.13	0.09
	2500	1.43	1.28	*	0.49	0.45	0.42	0.38	0.35	0.32	0.28	0.25	0.22	0.18	0.15	0.11
	3000	1.35	1.20	0.59	0.55	0.51	0.47	0.43	0.39	0.35	0.32	0.28	0.24	0.20	0.16	0.12
	4000	1.23	0.76	0.71	0.65	0.60	0.56	0.51	0.47	0.42	0.38	0.33	0.29	0.24	0.19	0.14
	5000	1.23	0.87	0.81	0.75	0.69	0.64	0.59	0.53	0.48	0.43	0.38	0.33	0.28	0.22	0.16
	6000	1.05	0.97	0.90	0.84	0.77	0.71	0.65	0.60	0.54	0.48	0.43	0.37	0.31	0.25	0.18
	7500	1.20	1.11	1.03	0.96	0.88	0.82	0.75	0.68	0.62	0.55	0.49	0.42	0.35	0.28	0.21

14.0	500	1.76	1.58	1.42	1.29	1.18	1.08	0.98	0.89	0.81	*	*	*	0.08	0.06	0.05
	750	1.54	1.38	1.25	1.13	1.03	0.94	*	*	0.17	0.15	0.14	0.12	0.10	0.08	0.06
	1000	1.40	1.26	1.13	1.03	0.94	*	0.25	0.23	0.20	0.18	0.16	0.14	0.12	0.09	0.07
	1500	1.23	1.10	0.99	*	0.37	0.34	0.32	0.29	0.26	0.23	0.21	0.18	0.15	0.12	0.09
	2000	1.12	1.00	*	0.48	0.44	0.41	0.38	0.34	0.31	0.28	0.25	0.21	0.18	0.14	0.11
	2500	1.04	0.92	0.59	0.55	0.51	0.47	0.43	0.39	0.36	0.32	0.28	0.24	0.20	0.16	0.12
	3000	0.98	0.74	0.66	0.61	0.57	0.52	0.48	0.44	0.40	0.37	0.31	0.27	0.23	0.18	0.14
	4000	0.90	0.84	0.78	0.73	0.67	0.62	0.57	0.52	0.47	0.42	0.37	0.32	0.27	0.22	0.16
	5000	1.04	0.97	0.90	0.83	0.77	0.71	0.66	0.60	0.54	0.49	0.43	0.37	0.31	0.25	0.18
	6000	1.16	1.08	1.00	0.93	0.86	0.80	0.73	0.67	0.60	0.54	0.48	0.41	0.35	0.28	0.21
	7500	1.33	1.24	1.15	1.07	0.99	0.91	0.84	0.76	0.69	0.62	0.55	0.47	0.40	0.32	0.24
16.0	500	0.97	0.87	0.78	0.68	0.58	0.49	0.39	0.30	0.20	0.12	0.10	0.09	0.07	0.06	0.04
	750	1.02	0.89	0.76	0.62	0.49	0.36	0.22	0.20	0.18	0.16	0.14	0.12	0.10	0.08	0.05
	1000	1.05	0.88	0.71	0.55	0.38	0.29	0.26	0.24	0.21	0.19	0.17	0.14	0.12	0.09	0.07
	1500	1.06	0.83	0.60	0.44	0.41	0.38	0.34	0.31	0.28	0.25	0.22	0.19	0.15	0.12	0.09
	2000	1.12	0.85	0.58	0.53	0.49	0.46	0.42	0.38	0.34	0.30	0.27	0.23	0.19	0.15	0.11
	2500	1.26	0.96	0.67	0.62	0.57	0.53	0.48	0.44	0.40	0.35	0.31	0.26	0.22	0.17	0.12
	3000	1.38	1.07	0.76	0.70	0.65	0.60	0.55	0.50	0.45	0.40	0.35	0.30	0.25	0.19	0.14
	4000	1.61	1.26	0.92	0.85	0.79	0.73	0.66	0.60	0.54	0.48	0.42	0.36	0.30	0.24	0.17
	5000	1.81	1.44	1.07	0.99	0.92	0.84	0.77	0.70	0.63	0.56	0.49	0.42	0.35	0.27	0.20
	6000	1.99	1.60	1.21	1.12	1.04	0.95	0.87	0.79	0.71	0.64	0.56	0.48	0.39	0.31	0.22
	7500	2.25	1.83	1.40	1.30	1.21	1.11	1.02	0.92	0.83	0.74	0.65	0.55	0.46	0.36	0.26

(Continued)

Table 8.3 (*Continued*)

Pressure MPa	Mass (kg/m² s)	Quality														
		0.20	0.25	0.30	0.35	0.40	0.45	0.50	0.55	0.60	0.65	0.70	0.75	0.80	0.85	0.90
18.0	500	0.79	0.70	0.61	0.51	0.42	0.32	0.23	0.17	0.15	0.14	0.12	0.10	0.08	0.07	0.05
	750	0.84	0.71	0.58	0.45	0.32	0.27	0.24	0.22	0.20	0.18	0.16	0.13	0.11	0.09	0.06
	1000	0.86	0.70	0.54	0.38	0.35	0.32	0.30	0.27	0.24	0.22	0.19	0.16	0.13	0.11	0.08
	1500	0.94	0.74	0.54	0.50	0.46	0.43	0.39	0.36	0.32	0.29	0.25	0.21	0.18	0.14	0.10
	2000	1.09	0.87	0.65	0.61	0.56	0.52	0.46	0.43	0.39	0.35	0.30	0.26	0.22	0.17	0.12
	2500	1.23	0.99	0.76	0.70	0.65	0.60	0.55	0.50	0.45	0.40	0.35	0.30	0.25	0.20	0.14
	3000	1.36	1.11	0.86	0.80	0.74	0.68	0.62	0.57	0.51	0.46	0.40	0.34	0.28	0.22	0.16
	4000	1.59	1.31	1.04	0.97	0.90	0.83	0.76	0.69	0.62	0.55	0.48	0.41	0.34	0.27	0.19
	5000	1.79	1.50	1.21	1.12	1.04	0.96	0.88	0.80	0.72	0.64	0.56	0.48	0.40	0.31	0.22
	6000	1.98	1.67	1.37	1.27	1.18	1.09	0.99	0.91	0.82	0.73	0.64	0.54	0.45	0.36	0.25
	7500	2.24	1.92	1.59	1.48	1.36	1.26	1.16	1.05	0.95	0.84	0.74	0.63	0.52	0.41	0.30
20.0	500	0.56	0.48	0.40	0.32	0.25	0.23	0.21	0.19	0.17	0.15	0.13	0.11	0.10	0.08	0.05
	750	0.59	0.49	0.39	0.35	0.32	0.30	0.26	0.25	0.23	0.20	0.18	0.15	0.13	0.10	0.07
	1000	0.68	0.57	0.45	0.42	0.39	0.36	0.33	0.30	0.26	0.24	0.21	0.18	0.15	0.12	0.09
	1500	0.85	0.72	0.59	0.55	0.51	0.48	0.44	0.40	0.36	0.32	0.28	0.24	0.20	0.16	0.11
	2000	1.00	0.86	0.72	0.67	0.62	0.58	0.53	0.48	0.44	0.39	0.34	0.29	0.24	0.19	0.14
	2500	1.13	0.99	0.84	0.78	0.73	0.67	0.62	0.56	0.51	0.45	0.40	0.34	0.28	0.22	0.16
	3000	1.26	1.10	0.95	0.88	0.82	0.76	0.70	0.64	0.57	0.51	0.45	0.38	0.32	0.25	0.18
	4000	1.48	1.31	1.15	1.07	1.00	0.92	0.85	0.77	0.70	0.62	0.54	0.47	0.39	0.31	0.22
	5000	1.68	1.51	1.34	1.25	1.16	1.07	0.98	0.90	0.81	0.72	0.63	0.54	0.45	0.35	0.25
	6000	1.86	1.69	1.51	1.41	1.31	1.21	1.11	1.01	0.92	0.82	0.72	0.61	0.51	0.40	0.29
	7500	2.12	1.94	1.76	1.64	1.52	1.41	1.29	1.18	1.06	0.95	0.83	0.71	0.59	0.47	0.33

Equation (8.17) indicates a linear relation between ϕ_{CRIT} and $x(z)$ and implies that eq. (8.11) may be written as:

$$\phi_{CRIT} = M + Nx(z) \tag{8.18}$$

where M and N are assumed to be functions of G, p, and D only.
  Comparing eq. (8.17) and eq. (8.18)

$$B = \frac{-N}{i_{fg} - (4Nz/DG)} \tag{8.19}$$

and

$$A = \frac{M}{1 - (4Nz/DGi_{fg})} \tag{8.20}$$

Substituting these expressions for A and B into eq. (8.15) gives

$$\phi_{CRIT} = \frac{M - (N(\Delta i_{SUB})_i/i_{fg})}{1 - (4Nz/DGi_{fg})} \tag{8.21}$$

and finally, writing $A' = MC'$ and $C' = -DGi_{fg}/4N$ gives the general equations

$$\phi_{CRIT} = \frac{A' + DG(\Delta i_{SUB})_i/4}{C' + z} \tag{8.22}$$

and

$$\phi_{CRIT} = \frac{A' - DGi_{fg}x(z)/4}{C'} \tag{8.23}$$

Barnett and Macbeth developed their correlation in terms of eq. (8.22) rather than eq. (8.23) since the former has the advantage of being in terms of the independent variables. The values of A' and C' can be correlated in terms of G, D, and p as indicated below.

8.2.5.2 *The Bowring correlation* A more recent correlation which retains the basis and accuracy of the Macbeth correlation but which is more convenient to use has been provided by Bowring (1972). The correlation makes use of eq. (8.22) where ϕ_{CRIT} is expressed in W/m^2; $(\Delta i_{SUB})_i$ is the inlet subcooling expressed in J/kg; z is the tube length expressed in m and A' and C' are functions given by

$$\left. \begin{array}{l} A' = 2.317 \left[\dfrac{DGi_{fg}}{4} \right] F_1/[1.0 + 0.0143F_2D^{1/2}G] \\[12pt] C' = 0.077F_3DG/[1.0 + 0.347F_4(G/1356)^n] \end{array} \right\} \tag{8.24}$$

Table 8.4 The Bowring critical heat flux
correlation. Tabulation of functions F_1, F_2, F_3,
and F_4 for water only

Pressure bar	F_1	F_2	F_3	F_4
1	0.478	1.782	0.400	0.0004
5	0.478	1.019	0.400	0.0053
10	0.478	0.662	0.400	0.0166
15	0.478	0.514	0.400	0.0324
20	0.478	0.441	0.400	0.0521
25	0.480	0.403	0.401	0.0753
30	0.488	0.390	0.405	0.1029
35	0.519	0.406	0.422	0.1380
40	0.590	0.462	0.462	0.1885
45	0.707	0.564	0.538	0.2663
50	0.848	0.698	0.647	0.3812
60	1.043	0.934	0.890	0.7084
68.9	1.000	1.000	1.000	1.000
70	0.984	0.995	1.003	1.030
80	0.853	0.948	1.033	1.322
90	0.743	0.903	1.060	1.647
100	0.651	0.859	1.085	2.005
110	0.572	0.816	1.108	2.396
120	0.504	0.775	1.129	2.819
130	0.446	0.736	1.149	3.274
140	0.395	0.698	1.168	3.760
150	0.350	0.662	1.186	4.227
160	0.311	0.628	1.203	4.825
170	0.277	0.595	1.219	5.404
180	0.247	0.564	1.234	6.013
190	0.220	0.534	1.249	6.651
200	0.197	0.506	1.263	7.320

where D is the internal tube diameter in m, G is the total mass velocity in
kg/m^2 s, and i_{fg} is the latent heat of vaporization in J/kg. The exponent n is
given by

$$n = 2.0 - 0.00725p \qquad (8.25)$$

where p is the system pressure in bar. F_1, F_2, F_3, and F_4 are functions of
system pressure and are tabulated in Table 8.4. The correlation was derived

from data covering the following parameter ranges:

pressure (p)   2–190 bar
tube diameter (D)   0.002–0.045 m
tube length (z)   0.15–3.7 m
mass velocity (G)   136–18,600 kg/m^2 s

The RMS error of the correlation is 7 per cent and the 95 per cent confidence level is ± 14 per cent. Extrapolation outside the above ranges of parameters is not recommended.

No attempt will be made to review the wide range of alternative correlations available for vertical uniformly heated round tubes. Many of the hypotheses upon which these correlations are based have been shown to be invalid by Barnett (1963).

8.2.5.3   *The Biasi correlation*   One alternative which has the advantage of being continuous with respect to the variable 'system pressure' with apparently little or no loss of accuracy is the correlation proposed by Biasi et al. (1967).

$$\phi_{\text{CRIT}} = \frac{1.883 \times 10^3}{D^n G^{1/6}} \left[\frac{f(p)}{G^{1/6}} - x(z) \right] \qquad (8.26)$$

for the low quality region

$$\phi_{\text{CRIT}} = \frac{3.78 \times 10^3 h(p)}{D^n G^{0.6}} [1 - x(z)] \qquad (8.27)$$

for the high quality region

where $n = 0.4$ for $D \geqslant 1$ cm
$\quad\quad n = 0.6$ for $D < 1$ cm

and $f(p) = 0.7249 + 0.099p \exp(-0.032p)$

$$h(p) = -1.159 + 0.149p \exp(-0.019p) + \frac{8.99p}{10 + p^2}$$

The correlation is evaluated in c.g.s. units, i.e., ϕ_{CRIT} in W/cm^2 and is valid over the following range of variables:

0.3 cm $< D <$ 3.75 cm
20 cm $< z <$ 600 cm
2.7 bar $< p <$ 140 bar
10 g/cm^2 s $< G <$ 600 g/cm^2 s
$1/(1 + \rho_f/\rho_g) < x(z) < 1$

The critical heat flux is given as the higher of the two values obtained by the intersection of eqs. (8.26) and (8.27) with the heat balance eq. (8.5). For values of G below 30 g/cm^2 s eq. (8.27) is always used. The RMS error for over 4500 data points examined was 7.26 per cent and 85.5 per cent of all points were correlated within ± 10 per cent.

8.2.5.4   The Becker correlation

The reader interested in lower pressure applications may wish to note that Becker (1965) has proposed a correlation of the form:

$$\phi_{\text{CRIT}} = G^{-0.5}f_1[x(z), p] \cdot f_2(D) \tag{8.28}$$

where $f_1[x(z), p]$ and $f_2(D)$ are functions presented graphically. This correlation has a functional form suggested by a physical mechanism and is based on a considerable amount of data taken at pressures below 69 bar (1000 psia). It may be considered as an alternative correlation for these low pressures.

8.2.5.5   The Katto correlation

Katto (1980, 1981, 1982) and Katto and Ohne (1984) have published a general correlation applicable to a range of fluids which they have progressively improved over the years. The formulation of Katto and Ohne (1984) used the following equation:

$$\phi_{\text{CRIT}} = XG(i_{\text{fg}} + K(\Delta i_{\text{SUB}})_i) \tag{8.29}$$

The terms X and K are functions of three dimensionless groupings

$$Z' = z/D \tag{8.30}$$

$$R' = \rho_{\text{g}}/\rho_{\text{f}} \tag{8.31}$$

$$W' = \left[\frac{\sigma\rho_{\text{f}}}{G^2 z}\right] \tag{8.32}$$

Five values of X need then to be determined:

$$X_1 = \frac{CW'^{0.043}}{Z'} \tag{8.33}$$

$$X_2 = \frac{0.1R'^{0.133}W'^{0.333}}{1 + 0.0031Z'} \tag{8.34}$$

$$X_3 = \frac{0.098R'^{0.133}W'^{0.433}Z'^{0.27}}{1 + 0.0031Z'} \tag{8.35}$$

$$X_4 = \frac{0.0384R'^{0.6}W'^{0.173}}{1 + 0.28W'^{0.233}Z'} \tag{8.36}$$

$$X_5 = \frac{0.234R'^{0.513}W'^{0.433}Z'^{0.27}}{1 + 0.0031Z'} \tag{8.37}$$

Table 8.5   Choice of X and K values for Katto and Ohne (1984) correlation

For	$R' < 0.15$		$R' > 0.15$	
If	$X_1 < X_2$	$X = X_1$	$X_1 < X_5$	$X = X_1$
	$\left.\begin{array}{l} X_1 > X_2 \\ \text{and} \\ X_2 < X_3 \end{array}\right\}$	$X = X_2$	$\left.\begin{array}{l} X_1 > X_5 \\ \text{and} \\ X_5 > X_4 \end{array}\right\}$	$X = X_5$
	$\left.\begin{array}{l} X_1 > X_2 \\ \text{and} \\ X_2 > X_3 \end{array}\right\}$	$X = X_3$	$\left.\begin{array}{l} X_1 > X_5 \\ \text{and} \\ X_5 < X_4 \end{array}\right\}$	$X = X_4$
	$K_1 > K_2$	$K = K_1$	$K_1 > K_2$	$K = K_1$
	$K_1 < K_2$	$K = K_2$	$\left.\begin{array}{l} K_1 < K_2 \\ \text{and} \\ K_2 < K_3 \end{array}\right\}$	$K = K_2$
			$\left.\begin{array}{l} K_1 < K_2 \\ \text{and} \\ K_2 > K_3 \end{array}\right\}$	$K = K_3$

The value of C in eq. (8.33) is given by

$$C = 0.25 \text{ for } Z' < 50$$
$$C = 0.25 + 0.0009(Z' - 50) \text{ for } 50 < Z' < 150$$
$$C = 0.34 \text{ for } Z' > 150$$

Three values of K also need to be established:

$$K_1 = \frac{0.261}{CW'^{0.043}} \tag{8.38}$$

$$K_2 = \frac{0.833[0.0124 + (1/Z')]}{R'^{0.133}W'^{0.333}} \tag{8.39}$$

$$K_3 = \frac{1.12[1.52W'^{0.233} + (1/Z')]}{R'^{0.6}W'^{0.173}} \tag{8.40}$$

The appropriate values of X and K are then chosen according to Table 8.5. The correlation is valid over the following ranges of data:

$$0.01 < z < 8.8 \text{ m}$$
$$0.001 < D < 0.038 \text{ m}$$
$$5 < Z' < 880$$
$$0.0003 < R' < 0.41$$
$$3 \times 10^{-9} < W' < 2 \times 10^{-2}$$

8.2.5.6 *The Shah correlation* Over the years Shah (1979, 1980, 1987) has also proposed a series of progressively improved methods. Shah's most recent one (1987) is really two separate correlations. The first covers the situation where the critical heat flux depends on the upstream conditions, namely the inlet subcooling and the distance along the tube, that is '*the upstream conditions correlation*' (UCC). The second relates the critical heat flux only to the local quality, and is referred to as '*the local condition correlation*' (LCC)—except for very short tubes.

The UCC correlation is

$$\text{Bo} = 0.124 \left(\frac{D}{z_{eq}} \right)^{0.89} \left(\frac{10^4}{Y} \right)^n (1 - x_{i\,eq}) \tag{8.41}$$

where $\text{Bo} = (\phi_{\text{CRIT}}/Gi_{fg})$ 'the boiling number'

$$Y = \left[\frac{GDc_{pf}}{k_f} \right] \left[\frac{G^2}{\rho_f^2 gD} \right]^{0.4} \left[\frac{\mu_f}{\mu_g} \right]^{0.6}$$

z_{eq} is the effective length of the tube, $x_{i\,eq}$ is the effective inlet quality and when

$$x_i \leqslant 0 \qquad z_{eq} = z_{\text{CRIT}} \quad \text{and} \quad x_{i\,eq} = x_i$$
$$x_i > 0 \qquad z_{eq} = z_{\text{SAT}} \quad \text{and} \quad x_{i\,eq} = 0.$$

When $Y \leqslant 10^4$, $n = 0$ for all fluids. When $Y > 10^4$, n is given by the following relations. For helium (all values of Y) n is given by

$$n = \left(\frac{D}{z_{eq}} \right)^{0.33} \tag{8.42}$$

For all fluids other than helium

$$n = \left(\frac{D}{z_{eq}} \right)^{0.54} \qquad \text{for } Y \leqslant 10^6 \tag{8.43a}$$

and

$$n = \frac{0.12}{(1 - x_{i\,eq})^{0.5}} \qquad \text{for } Y > 10^6 \tag{8.43b}$$

Figure 8.10 shows the relationship between Bo and x_i for $Y < 10^4$ and $x_i < 0$.

The LCC correlation is given by the expression

$$\text{Bo} = F_E F_x \text{Bo}_0 \tag{8.44}$$

where F_E is an entrance factor given by

$$F_E = 1.54 - 0.032(z_{\text{CRIT}}/D) \tag{8.45}$$

If eq. (8.45) yields values of $F_E < 1$, then F_E is set equal to one.

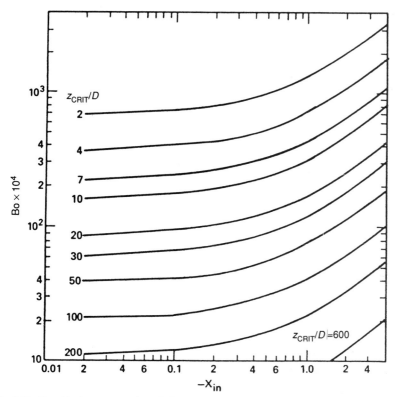

Fig. 8.10. Graphical representation of the present upstream condition correlation (UCC) at $Y \leqslant 10^4$ (Shah 1987).

Bo_0 is the boiling number for $x_{CRIT} = 0$. This is a function of Y and reduced pressure P_r as shown in graphical form in Fig. 8.11. F_x is the ratio of the boiling number for $x_{CRIT} = x$ (Bo_x) to the value of the boiling number for $x_{CRIT} = 0$ (Bo_0). It is represented in graphical form as a function of Y, x_{CRIT} and P_r in Fig. 8.12. The original Shah reference gives the function Bo_0 and F_x also in the form of equations.

The choice between the UCC and LCC correlations is made on the following basis:

(a) for helium always use UCC
(b) for other fluids for $Y \leqslant 10^6$ use UCC and for $Y > 10^6$ use the correlation that gives the lower value of Bo.

The only exception to this rule is that UCC is used when

$$z_{eq} > 160/P_r^{1.14}$$

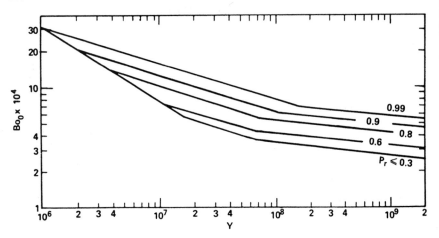

Fig. 8.11. Graphical representation of the present local condition correlation (LCC), part 1: value of boiling number at $x_{CRIT} = 0$ (Shah 1987).

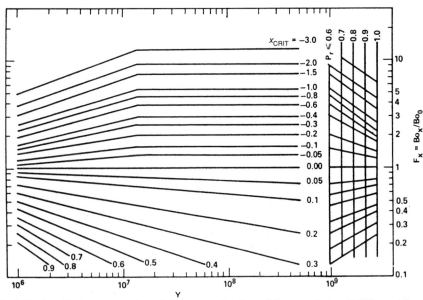

Fig. 8.12. Graphical representation of the present local condition correlation (LCC), part 2: ratio of boiling numbers at $x_{CRIT} = x$ and $x_{CRIT} = 0$ (Shah 1987).

Table 8.6 Comparison of Bowring, Katto and Shah correlations

Fluid	Data range	No. of data	Correlation of	Deviation (%)		No. of data with deviation > 30%
				Mean	Average	
Water	All data	427	Shah	14.4	−0.9	40
			Katto	16.1	+3.1	51
			Bowring	18.6	−9.6	86
Water	Verified range of Bowring correlation	251	Shah	14.4	−0.3	20
			Katto	14.1	−3.2	17
			Bowring	11.7	−3.9	19
All fluids	$x_{CRIT} < 0$	397	Shah	16.9	−2.1	56
			Katto	30.7	+21.8	141
All fluids	$x_{CRIT} \geqslant 0$	1046	Shah	15.4	−3.2	129
			Katto	18.8	+11.8	187
All fluids	All data	1443	Shah	16.0	−2.9	185
			Katto	22.3	+14.6	328

The Shah correlation was tested with 23 different fluids from 62 sources that cover the following conditions:

$$0.315 < D < 37.5 \text{ mm}$$
$$1.3 < z/D < 940$$
$$4 < G < 29,051 \text{ kg/m}^2 \text{ s}$$
$$0.0014 < P_r < 0.96$$
$$-4 < x_i < +0.85$$
$$-2.6 < x_{CRIT} < +1$$

Shah also compared his correlation with those of Bowring and Katto; the results are shown in Table 8.6.

8.2.6 *The influence of system pressure on the critical heat flux*

Consideration of the influence of system pressure as an independent variable on the critical heat flux has been deferred until now. To illustrate the major effect, the Bowring correlation has been evaluated in the pressure range up to 190 bar for a condition in which the tube geometry and mass velocity are held constant at fixed values ($z = 0.76$ m (30 in), $D = 10.15$ mm (0.4 in), $G = 2720$ kg/m^2 s (2×10^6 lb/h ft^2)). Other experimental data have been assembled to supplement the correlation above 138 bar (2000 psia) (Herkenrath *et al.* 1967) and to validate certain trends.

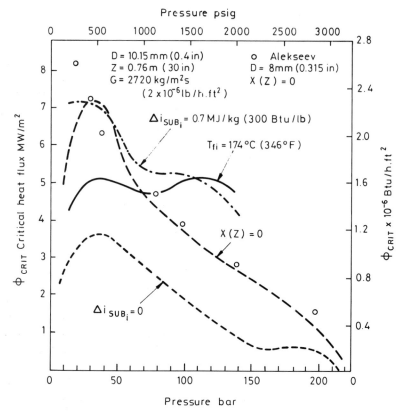

Fig. 8.13. The influence of system pressure on the critical heat flux.

Figure 8.13 shows the critical heat flux plotted against pressure for four fixed inlet or exit conditions. In the case of the two upper curves the exit condition for pressures above 138 bar is highly subcooled and no data were available to supplement these curves.

If the exit mass quality is held fixed at zero, $x(z) = 0$ then the critical heat flux falls sharply as the pressure is increased above 30 bar up to the critical pressure. The experimental data of Alekseev *et al.* (1965) for 8–9 mm i.d. tubes are shown for comparison and confirm the predicted trend remarkably well. At lower mass velocities the value of critical heat flux at $x(z) = 0$ will be increased for pressures below 100 bar and decreased at higher pressures. For higher mass velocities the converse is true.

For the condition where the inlet subcooling is held constant, the critical heat flux passes through a maximum at low pressures (below 30 bar) and then falls as the pressure is increased. In a number of cases a secondary maximum is found in the pressure range 100–200 bar. This maximum is enhanced by increasing the subcooling and the mass velocity and reducing the tube length-to-diameter ratio.

If the inlet water temperature T_{fi} is held constant then the inlet subcooling increases as the pressure is raised. The influence of system pressure on the critical heat flux is much reduced over the complete pressure range. Holding the inlet water temperature constant tends to enhance the secondary maximum which occurs at high pressures. At high mass velocities this effect becomes dominant and the critical heat flux increases as the pressure is increased (Lee 1966) for a fixed inlet temperature.

8.3 Important second order effects on the critical heat flux

In the previous sections the influence of the major independent variables on the critical heat flux has been discussed. In this section some important second order effects will be noted and these comments will, in many ways, qualify the more general conclusions arrived at earlier.

(a) *The relationship between critical heat flux and inlet subcooling.* Earlier it was seen that the critical heat flux is generally a linear function of inlet subcooling; Macbeth used this observation as a key factor in his correlation. Inspection of Figs. 8.5 and 8.8 suggests that above a z/D ratio of 100 the linearity is preserved. These findings were confirmed by experiments carried out using R-12 (Stevens *et al.* 1964). Non-linearity may also be associated with a large pressure drop across the channel or with data taken under hydrodynamically unstable conditions.

(b) *The relationship between critical heat flux and tube length.* Experiments carried out by Stevens *et al.* (1964) using refrigerants have shown that the 'local conditions' hypothesis, whilst being very nearly true, is not universally valid. Figure 8.14 shows the critical heat flux for Refrigerant 12 plotted against vapour mass quality for various tube lengths with the tube diameter and system pressure held constant. It will be observed that for tube lengths greater than 0.61 m (25 in), $z/D > 75$, all data points are collinear, confirming the 'local conditions' hypothesis. However, at shorter lengths the data points for each separate length lie on individual curves. In the case of the higher mass velocity value this effect is combined with a departure from a linear relationship between ϕ_{CRIT} and $x(z)$. Close examination of the experimental data for water taken by Lee and Obertelli (1963) and Lee (1965) reveals a similar effect for channels with a z/D ratio less than 80.

(c) *The relationship between critical heat flux and mass velocity.* Normally the critical heat flux increases with mass velocity at constant inlet subcooling. However, there are exceptions to this rule which are also associated with high mass velocity values, small inlet subcooling values, and low z/D ratios.

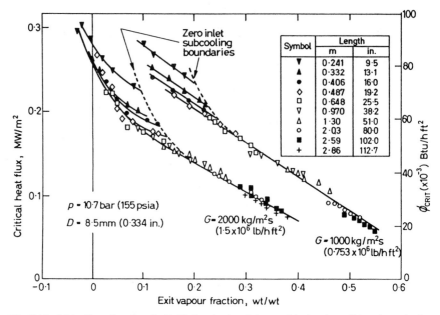

Fig. 8.14. Critical heat flux data for R-12 showing breakdown of the local conditions hypothesis.

(d) *The effect of channel orientation.* An investigation into buoyancy effects on the critical heat flux in forced convection vertical flow has been reported by Papell *et al.* (1966). Liquid nitrogen was the test fluid and critical heat flux data were obtained over a wide range of test conditions, for a tubular section 12.8 mm (0.505 in) i.d. and 30.5 cm (12 in) long with the flow direction both upwards and downwards. Under certain conditions—associated with low flows and high exit qualities in the annular flow region—the critical heat flux for downflow was significantly lower than for upflow. The buoyancy effect increased as the pressure and subcooling decreased.

Similar tests for water have been reported by Kirby *et al.* (1967) and Remizov *et al.* (1983). Changing the flow direction from upflow to downflow reduced the critical heat flux by 10–30 per cent, the greatest reductions being at the lowest flows. It would appear that provided the liquid downflow velocity is significantly above the bubble rise velocity given by eq. (3.13) then the influence of flow direction on the critical heat flux is small.

The critical heat flux condition in horizontal and inclined tubes has not been extensively investigated. Most practical designs of evaporator using horizontal elements include such items as 180° return bends. It has been found that such features have a profound influence on the critical heat flux condition in horizontal tubes and this aspect is discussed in Chapter 11.

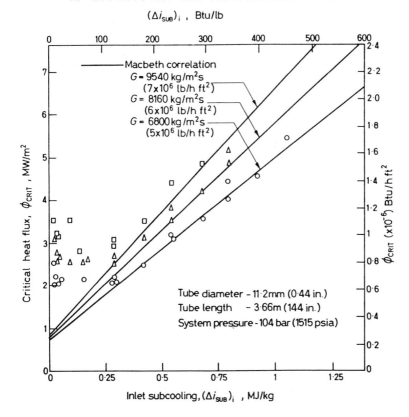

Fig. 8.15. Anomalous behaviour at high mass velocity (data of Waters *et al.* 1963).

Experimental data in the absence of such features are sparse. Macbeth (1963) quotes an example which suggests that flow stratification is important in the 'low mass velocity' region. The difference in critical heat flux between horizontal and vertical channels will decrease as the mass velocity and/or the system pressure is increased and the tube diameter is decreased.

(e) *Anomalous effects due to ultra-high mass velocities and small tube diameters.* A number of studies have been published where experiments have been carried out at very high mass velocities or with very small diameter tubes. A certain type of anomalous behaviour has been found under these conditions. This behaviour can be illustrated by the results obtained by Waters *et al.* (1963). Figure 8.15 shows ϕ_{CRIT} plotted against $(\Delta i_{SUB})_i$ for three mass velocity values at a system pressure of 104 bar (1515 psia). It is observed that the data are well represented by the Macbeth (1963) correlation provided the inlet subcooling is greater than 0.35 MJ/kg (150 Btu/lb). At values below this, the points depart from the linear relationship and lie considerably above the correlation. In addition, Waters *et al.* (1963) report that, at low subcooling, serious overheating first occurred not at the exit end of the tube but at various locations upstream.

Studies in which short tubes of small bore have been employed also reveal this anomaly.

(f) *Compressibility effects.* Fauske (1966) has pointed out that choking of the two-phase flow (maximum, critical, or sonic flow) can occur and pass unrecognized in critical heat flux experiments at low pressures and high liquid inlet velocities. In particular he cites the work of Lowdermilk *et al.* (1958) in which the steam–water mixture after leaving the heated test section passed onto a downstream plenum where the exit pressure gauge was located. The conditions at the exit from the heated section were reported on the basis of this measured pressure which was always close to atmospheric. Examination of these quoted conditions show that the flow is required to be supersonic (or supercritical). In actual fact, the true exit condition from the heated section was probably more likely to be 'DNB' (subcooled) at 14 bar (200 psia).

(g) *Axial heat conduction effects.* Passos and Gentile (1991) experimentally investigated transition boiling in subcooled R-113 under forced flow conditions. Regulating the electric current with an analogue electronic feedback control device, they were able to obtain the coexistence of the nucleate, transition and film boiling regimes in their 50 mm long, 8 mm i.d. test section. From these tests, they observed that the high tube wall temperature in the film boiling region resulted in significant axial heat conduction, which in turn moved the location of the transition from nucleate to film boiling farther upstream from its original point of occurrence where the local heat flux was much lower than that of the original critical value. Thus, once the critical condition was reached, its location propagated upstream in the flow.

(h) *The influence of surface finish and tube wall thickness on the critical heat flux.* It is possible to list four secondary variables associated with the test section which, so far, have not been considered. These are:

(i) method of heating the tube
(ii) material of the tube wall
(iii) tube wall thickness
(iv) internal bore surface finish.

To this list might be added the method used for detection of the critical heat flux. Since an accurate correlation of the available world data has been achieved without the inclusion of the above factors as independent variables, it seems reasonable to conclude that they do not play a significant role in the determination of the critical heat flux value. However, few systematic investigations of the importance of these variables have been undertaken.

The experimental critical heat flux data listed by Macbeth and Thompson were taken using Joule heating of the test channel. Most experiments used direct current heating. It has been suggested (Bertoletti *et al.* 1959; Tippets 1962) that the use of alternating current with test elements having a low thermal capacity and short time constant causes a reduced critical heat flux, all other independent variables remaining fixed.

Aladyev *et al.* (1961) report that the thickness of the tube wall did not affect the critical heat flux in the range 0.016 to 0.079 in. Barnett (1963) suggests that some influence of tube wall thickness might explain an apparent slight inconsistency in the data of Lee and Obertelli (1963). Data taken at AERE Harwell and AEE Winfrith (Lee 1965) suggest that there is a small reduction in the critical heat flux (~ 5 per cent) at low inlet subcoolings as the tube wall thickness is decreased from 0.084 to 0.034 in, all other independent variables remaining constant. Tippets (1962) reports a decrease in critical heat flux of approximately 20 per cent when a 0.010 in thick ribbon heater was replaced by a 0.006 in thick ribbon. Some influence of tube wall thickness might be expected by comparison with the influence of this variable in pool boiling experiments.

There appears to be some difference of opinion on the influence of tube wall material and internal bore surface finish. It should be observed that for the purposes of this discussion 'roughness' refers to the 'micro-roughness' of normal machined engineering surfaces. The influence of certain types of macro-roughness will be considered in Chapter 11. De Bortoli *et al.* (1963) concluded that the heat transfer surface material or roughness of the surface did not influence the critical heat flux. The materials investigated were nickel, Zircaloy II, and stainless steel. With Zircaloy II two surface roughness values were investigated, namely 29 microinches and 120 microinches. Aladyev *et al.* (1961) report no influence of surface finish on the critical heat flux. In these experiments eleven tubes with various internal finishes were tested, including polished, etched, and machined surfaces. The evidence for some effect of surface finish comes from the data of Weatherhead (1963) in the USA and Chirkin and Iukin (1956) in the USSR. Both of these works conclude that the critical heat flux is lowered for very smooth surfaces upon which bubble nucleation might be difficult.

It has been reported that porous deposits on heat transfer surfaces can influence the critical heat flux both favourably and adversely. In tests reported by Goldstein *et al.* (1967) the addition of contaminants to a test circuit caused reductions in the critical heat flux. The effect was temporary and the critical heat flux value returned to the clean tube figure after a period of several hours.

References and further reading

Aladyev, I. T., Miropolsky, Z. L., Doroshchuk, V. E., and Styrikovich, M. A. (1961). 'Boiling crisis in tubes'. *Int. Develop. in Heat Transfer*, Vol. II, Paper 28, University of Colorado, Boulder.

Alekseev, G. V., Zenkevitch, B. A., Peskov, O. L., Sergeev, N. D., and Subbotin, V. I. (1965). 'Burnout heat fluxes under forced water flow'. Paper A/CONF 28/p/327a presented at 3rd UN Conference on Peaceful Uses of Atomic Energy, 1964; see also *Teploenergetika*, **12**(3), 47–51.

Barnett, P. G. (1963). 'An investigation into the validity of certain hypotheses implied by various burnout correlations'. AEEW-R 214.

Becker, K. M. (1965). 'An analytical and experimental study of burnout conditions in vertical round ducts'. A. B. Atomenergi Report AE 178, Sweden.

Bertoletti, G. *et al.* (1959). 'A facility used for wet steam cooling experiments: pressure drop, heat transfer and burnout'. *Energia Nucleare*, **6**(7), 458–471.

Biasi, L., Clerici, G. C., Garribba, S., and Sara, R. (1967). 'Studies on burnout. Part 3'. *Energia Nucleare*, **14**(9), 530–536.

Bowring, R. W. (1972). 'A simple but accurate round tube uniform heat flux, dryout correlation over the pressure range 0.7–17 MN/m² (100–2500 psia)'. AEEW-R 789.

Chirkin, V. S. and Iukin, V. P. (1956). 'Critical point in heat removal from boiling water flowing through an annular gap'. *Soviet Physics (Technical Physics)*, **1**(7), 1503–1515.

De Bortoli, R. A., Green, S. I., Letourneau, B. W., Troy, M., and Weiss, A. (1958). 'Forced convection heat transfer burnout studies for water in rectangular channels and round tubes at pressures above 500 psia'. WAPD-188.

Fauske, H. (1966). 'Compressibility affects flow instability and burnout'. *Power Reactor Technology*, **9**(2), 65–68.

Goldstein, P., Dick, I. B., and Rice, J. K. (1967). 'Internal corrosion of high pressure boilers'. *J. Engineering for Power, Trans. ASME*, Series A, **89**(3), 378–394, July.

Herkenrath, H., Mörk-Mörkenstein, P., Jung, U., and Weckermann, F. J. (1967). 'Wärmeübergang an Wasser bei Erzwungener Strömung im Druckbereich von 140 bis 250 bar'. EUR 3658d.

Katto, Y. (1980). 'General features of CHF of forced convection boiling in uniformly heated vertical tubes with zero inlet subcooling'. *Int. J. Heat Mass Transfer*, **23**, 493–504.

Katto, Y. (1981). 'General features of CHF of forced convection boiling in uniformly heated rectangular channel'. *Int. J. Heat Mass Transfer*, **24**(8), 1413–1419.

Katto, Y. (1982). 'An analytical investigation on CHF of flow boiling in uniformly heated vertical tubes with special reference to governing dimensionless groups'. *Int. J. Heat Mass Transfer*, **25**(9), 1353–1361.

Katto, Y. and Ohne, H. (1984). 'An improved version of the generalized correlation of critical heat flux for forced convection boiling in uniformly heated vertical tubes'. *Int. J. Heat Mass Transfer*, **27**(9), 1641–1648.

Kirby, G. J., Staniforth, R., and Kinneir, J. H. (1967). 'A visual study of forced convection boiling. Part 2. Flow patterns and burnout for a round tube test section'. AEEW-R 506.

Kirillov, P. L., Bobkov, V. P., Boltenko, E. A., Katan, I. B., Smogalev, I. P., and

Vinogradov, V. N. (1991). 'New CHF Table for Water in Round Tubes'. *Atomic Energy*, **70**, 18–28 (in Russian), Obninsk.

Lee, D. H. (1965). 'An experimental investigation of forced convection boiling in high pressure water. Pt. III'. AEEW-R 355.

Lee, D. H. (1966). 'An experimental investigation of forced convection burnout in high pressure water. Part 4. Large diameter tubes at about 1600 psia'. AEEW-R 479.

Lee, D. H. and Obertelli, J. D. (1963). 'An experimental investigation of forced convection boiling in high pressure water. Pt. I'. AEEW-R 213.

Lowdermilk, W. H., Lanzo, C. D., and Siegel, B. L. (1958). 'Investigation of boiling burnout and flow stability for water flowing in tubes'. NACA-TN-4382.

Macbeth, R. V. (1963). 'Burn-out analysis. Part 3. The low velocity burnout regime'. AEEW-R 222.

Macbeth, R. V. (1963). 'Burn-out analysis. Part 4. Application of a local condition hypothesis to world data for uniformly heated round tubes and rectangular channels'. AEEW-R 267.

Matzner, B. (1963). 'Basic experimental studies of boiling fluid flow and heat transfer at elevated pressures'. TID 18978.

Milioti, S. (1964). 'A survey of burnout correlations as applied to water cooled nuclear reactors'. MEng. Thesis, Pennsylvania State University, September.

Papell, S. S., Simoneau, R. J., and Brown, D. D. (1966). 'Buoyancy effects on critical heat flux of forced convective boiling in vertical flow'. NASA-TND-3672.

Passons, J. C. and Gentile, D. (1991). 'An experimental investigation of transition boiling in subcooled Freon-113 forced flow'. *J. Heat Transfer*, **113**, 459–462.

Remizov, O. V., Sergeev, V. V., and Yurkov, Yu. I. (1983). 'Experimental investigation of deterioration in heat transfer with up-flow and down-flow of water in a tube'. *Thermal Engng.*, **30**(9), 549–551.

Shah, M. M. (1979). 'A generalized graphical method for predicting CHF in uniformly heated vertical tubes'. *Int. J. Heat Mass Transfer*, **22**, 557–568.

Shah, M. M. (1980). 'A general correlation for critical heat flux in annuli'. *Int. J. Heat Mass Transfer*, **23**, 225–234.

Shah, M. M. (1987). 'Improved general correlation for critical heat flux during upflow in uniformly heated vertical tubes'. *Heat and Fluid Flow*, **8**(4), 326–335.

Stevens, G. F., Elliott, D. F., and Wood, R. F. (1964). 'An experimental investigation into forced convection burnout in Freon, with reference to burnout in water'. AEEW-R 321.

Thompson, B. and Macbeth, R. V. (1964). 'Boiling water heat transfer—burnout in uniformly heated round tubes: A compilation of world data with accurate correlations'. AEEW-R 356; available price 80p from HMSO.

Tippets, F. E. (1962). 'Critical heat fluxes and flow patterns in high pressure boiling water flows'. Paper 62-WA-162 presented at the ASME Winter Annual Meeting, New York, 25–30 November.

Waters, E. D., Anderson, J. K., Thorne, W. L., and Batch, J. M. (1963). 'Experimental observations of upstream boiling burnout'. AIChE reprint 7 presented at 6th National Heat Transfer Conference, Boston, Mass., August; *see also* HW-73902 (Rev.).

Weatherhead, R. J. (1963). 'Nucleate boiling characteristics and the critical heat flux occurrence in sub-cooled axial flow water systems'. ANL 6675.

Example

Calculate the critical heat flux (ϕ_{CRIT}) and the critical steam quality $x(z)$ for a uniformly heated vertical tube cooled by water under the following conditions:

tube diameter (D) = 8 mm
tube length (z) = 0.5 m
pressure (p) = 69 bar
mass velocity (G) = 1500 kg/m² s
inlet temperature (T_{fi}) = 204°C

Compare the results obtained using the following methods:

(a) the revised Russian look-up tables (Tables 8.2, 8.3);
(b) the Bowring correlation;
(c) the Biasi correlation;
(d) the Katto correlation;
(e) the Shah correlation.

Solution
(a) Russian look-up tables (Tables 8.2 and 8.3). The heat balance equation (eq. 8.5) states

$$x(z) = \frac{1}{i_{fg}}\left[\frac{4\phi z}{DG} - (\Delta i_{SUB})_i\right]$$

Now for a pressure of 68.9 bar

$$i_{fg} = 1.51 \text{ MJ/kg}$$

$$\left.\begin{array}{l} i_f = 1.26 \text{ mJ/kg} \\ i_{fi} = 0.872 \text{ MJ/kg} \end{array}\right\}(\Delta i_{SUB})_i = 0.388 \text{ MJ/kg}$$

So $$x(z) = \frac{1}{1.51 \times 10^6}\left[\frac{4\phi \times 0.5}{0.008 \times 1500} - 0.388 \times 10^6\right]$$

or rearranging,

$$\phi \text{ (in MW/m}^2) = \frac{x(z) + 0.2569}{0.1103}$$

$x(z)$	ϕ from heat balance eq. MW/m²	ϕ_{CRIT} from Tables 8.2, 8.3 MW/m²
0	2.33	5.95
0.1	3.23	4.61
0.15	3.69	3.90
0.20	4.14	3.42

EXAMPLE   371

Thus by interpolation

$$x(z) = 0.16 \qquad \phi_{CRIT} = \underline{3.78 \text{ MW/m}^2}.$$

(b) The Bowring correlation

$$\phi_{CRIT} = \frac{A' + DG(\Delta i_{SUB})_i/4}{C' + z}$$

$$A' = 2.317 \left[\frac{DGi_{fg}}{4} \right] F_1 \bigg/ [1.0 + 0.0143 F_2 D^{1/2} G]$$

$$C' = 0.077 F_3 DG / [1.0 + 0.347 F_4 (G/1356)^n]$$

Because $p = 68.9$ bar

$$F_1 = F_2 = F_3 = F_4 = 1$$

and

$$n = 2.0 - 0.00725p = 0.5$$

Evaluating A' and C' we have

$$A' = 3.596 \times 10^6 \qquad C' = 0.677$$

so

$$\phi_{CRIT} = \frac{3.596 \times 10^6 + 1.164 \times 10^6}{0.677 + 0.5}$$

$$= \underline{4.04 \text{ MW/m}^2}$$

(c) The Biasi correlation (c.g.s. units). The low-quality correlation (eq. 8.26) applies:

$$\phi_{CRIT} = \frac{1.883 \times 10^3}{D^n G^{1/6}} \left[\frac{f(p)}{G^{1/6}} - x(z) \right]$$

Since $D < 1$ cm, $n = 0.6$

$$f(p) = 0.7249 + 0.099p \exp(-0.032p)$$

$$= 1.481$$

$$\phi_{CRIT} = 0.933 \times 10^3 [0.6425 - x(z)]$$

or

$$\phi_{CRIT}(\text{in MW/m}^2) = 9.33[0.6425 - x(z)]$$

This equation is now solved with the heat balance equation

$x(z)$	ϕ from heat balance eq. MW/m^2	ϕ_{CRIT} BIASI MW/m^2
0	2.33	5.99
0.1	3.23	5.06
0.15	3.69	4.59
0.20	4.14	4.13

Thus by interpolation $x(z) = 0.20$,

$$\phi_{CRIT} = \underline{4.13 \text{ MW/m}^2}$$

(d) The Katto correlation

$$\phi_{CRIT} = XG(i_{fg} + K(\Delta i_{SUB})_i)$$

we need the following physical properties for water at 68.9 bar

$$\rho_f = 742.3 \text{ kg/m}^3$$
$$\rho_g = 35.71 \text{ kg/m}^3$$
$$\sigma = 0.01796 \text{ N/m}$$

X and K are functions of the groupings

$$Z' = z/D = 0.5/0.008 = 62.5$$

$$R' = \rho_s/\rho_f = 35.71/742.3 = 0.0481$$

$$W' = \frac{\sigma\rho_f}{G^2z} = \left[\frac{0.01796 \times 742.3}{1500^2 \times 0.5}\right] = 11.85 \times 10^{-6}$$

Since $50 < z' < 150$

$$C = 0.25 + 0.0009(z' - 50)$$
$$= 0.261$$

Evaluating $X_1 \rightarrow X_5$, $K_1 \rightarrow K_3$ from the relevant equations

X_1	X_2	X_3	X_4	X_5
0.00256	0.00128	0.00123	0.000385	0.000929

	K_1	K_2	K_3	
	1.628	1.548	6.113	

Now since $R' < 0.15$ and $X_1 > X_2$, $X_2 > X_3$,

$$X = X_3 = 0.00123$$

and $K_1 > K_2$ so $K = K_1 = 1.628$.

So $\phi_{CRIT} = 0.00123 \times 1500(1.51 \times 10^6 + 1.628 \times 0.388 \times 10^6)$
$$= \underline{3.95 \text{ MW/m}^2}$$

(e) the Shah correlation. This correlation depends on the parameters

$$Y = \left[\frac{GDc_{pf}}{k_f}\right]\left[\frac{G^2}{\rho_f^2 gD}\right]^{0.4}\left[\frac{\mu_f}{\mu_g}\right]^{0.6}$$

EXAMPLE 373

Further physical property information is needed:

$$c_{pf} = 5407 \text{ J/kg K}$$
$$k_f = 0.5637 \text{ W/m K}$$
$$\mu_f = 0.972 \times 10^{-4} \text{ N s/m}$$
$$\mu_g = 1.89 \times 10^{-5} \text{ N s/m}$$

Evaluating Y we have $Y = 1.493 \times 10^6$.
The Shah correlation is in two parts:

(i) the upstream correlation:

Since $Y > 10^6$ $n = 0.12/(1 - x_{i\,eq})^{0.5}$

$$x_{i\,eq} = x_i = -0.388/1.51 = -0.257$$

$$n = 0.107$$

$$\text{Bo} = 0.124 \left(\frac{0.008}{0.5}\right)^{0.89} \left(\frac{10^4}{1.493 \times 10^6}\right)^{0.107} \qquad (1.257)$$

$$= 0.002297$$

$$\phi_{\text{CRIT}} = \text{Bo}\, Gi_{fg} = 0.002297 \times 1500 \times 1.51 \times 10^6$$
$$= 5.20 \text{ MW/m}^2 \quad \text{(from UCC)}$$

(ii) the local conditions correlation:

$$\text{Bo} = F_E \times F_x \times \text{Bo}_0$$

$$F_E = 1.54 - 0.032 \left(\frac{z}{D}\right)$$

since this equation gives $F_E < 1$, F_E is set equal to 1.

F_x is established from Fig. 8.12 $\left.\begin{array}{l} \\ \end{array}\right\}$ $P_r = \dfrac{68.9}{221.2} = 0.311.$
Bo_0 is established from Fig. 8.11 $\left.\begin{array}{l} \end{array}\right.$

For $Y = 1.493$, $\text{Bo}_0 = 0.002498$.

For

$x_c = 0$	$F_x = 1$	$\text{Bo} = 0.002498$
$x_c = 0.05$	$F_x = 0.9$	$\text{Bo} = 0.002248$
$x_c = 0.1$	$F_x = 0.8$	$\text{Bo} = 0.001998$
$x_c = 0.2$	$F_x = 0.7$	$\text{Bo} = 0.001749$

This is now solved with the heat balance equation

x(z)	ϕ from heat balance eq. MW/m^2	ϕ_{CRIT} (Shah) MW/m^2
0	2.33	5.66
0.05	2.78	5.09
0.1	3.23	4.52
0.2	4.14	3.96

$$x_{CRIT} = 0.185 \qquad \phi_{CRIT} = 4.0 \text{ MW/m}^2 \text{ (from LCC)}$$

This value is correct since $Y > 10^6$. This gives lowest Bo value.

Problems

1. Using the Bowring correlation, determine the critical heat flux for water at 10 bar flowing up a vertical tube of 45 mm bore and 3 m length. The water enters at 160°C with a flow rate of 6000 kg/h. What is the maximum vapour exit quality that can be attained?
2. For the same conditions as Problem 1, repeat the exercise using the Biasi method for $x = 0.378$.

9

CRITICAL HEAT FLUX IN FORCED CONVECTIVE FLOW—2. MORE COMPLEX SITUATIONS

9.1 Introduction

In the previous chapter the critical heat flux conditions for water flowing in vertical uniformly heated tubes were discussed. In this chapter the effects on the critical heat flux of (a) non-uniform heating, (b) non-circular geometry, (c) the use of fluids other than water, (d) liquid–vapour mixtures at the channel entrance, and (e) unstable hydrodynamic conditions will be reviewed. In addition the present knowledge regarding the mechanism of the critical heat flux will be discussed. Finally the question or critical heat flux on the outside of horizontal tube bundles in crossflow is addressed.

9.2 The critical heat flux in round tubes with non-uniform heating

Three types of non-uniform heating will be considered: (a) the heat generation in the tube wall is a smooth continuous function of the tube length only, (b) the heat generation in the tube wall is uniform except for a short length at a higher or lower uniform value, and (c) the heat generation in the tube wall in the axial direction is uniform but there is a variation in heat generation around the tube circumference.

9.2.1 Non-uniform axial heat flux profile

Table 9.1 lists experiments carried out with vertical round tubes having a non-uniform heat flux distribution in the axial direction. Since many of these studies are concerned with the limitations of the critical heat flux in nuclear reactor design, the form of the heat flux axial profile often resembles a sine wave

$$\phi(z) = \phi_{MAX} \sin\left(\frac{\pi z}{L}\right) \tag{9.1}$$

In some of the studies, the heat flux profile is only part of a full sine wave—i.e., 'chopped'—and, in others, the heat flux profile has been built up in a series of steps to ease fabrication problems—i.e., 'stepped'. The form factor 'ff' is the ratio of the maximum to average heat flux ($\phi_{MAX}/\bar{\phi}$).

Table 9.1 Range of variables for critical heat flux in tubes (vertically upwards, axial flow, non-uniform heating)

Source and reference	Axial heat flux profile	Tube dimensions					Range of experimental variables				
		Inside diameter		Heated length		Inlet subcooling	Pressure		Mass velocity		Outlet quality % by wt.
		mm	in	cm	in		bar	psia	kg/m² s	10^6 lg/h ft²	
Columbia Univ. Casterline (1964)	Chopped, smooth sine ff = 1.541	10.15	0.400	487	192	Subcooled	69	1000	1360–9650	1.0–7.1	16–98
AEEW, Winfrith, Lee, and Obertelli (1963)	Chopped, smooth, and stepped sine ff = 1.27	9.73	0.383	183	72	Subcooled	38.6/69	560/1000	850–4080	0.7–3.0	4–76
	= 1.42	9.85	0.388	183	72		38.6/69	560/1000			
	= 1.45	9.73	0.383	183	72		69/110	1000/1600			
AEEW, Winfrith Lee (1965)	Chopped, smooth sine ff = 1.40	9.46	0.373	368	144	Subcooled 10–200 Btu/lb 23–465 kJ/kg	69	1000	2040–4080	1.5–3.0	17–46
AEEW, Winfrith Lee (1966a)	Inlet peak ff = 1.66	22.1	0.869	119	47	Subcooled	110	1600	340–1360	0.25–1.0	0–50
	Inlet peak ff = 1.63	28.2	1.11	119	47		110/124	1600/1800	340–1360	0.25–1.0	
	Central peak ff = 1.27	15.9	0.625	100	39.4	40–200 Btu/lb	86/124	1250/1800	540–3400	0.4–2.5	
	Central peak ff = 1.17	15.9	0.625	100	39.4		86/124	1250/1800	1630–2720	1.2–2.0	
	Outlet peak ff = 1.63	28.2	1.11	119	47	93–465 kJ/kg	110/124	1600/1800	340/1360	0.25–1.0	

(Continued)

Source	Axial flux shape									
Babcock and Wilcox Swenson et al. (1962)	Inlet peak Outlet peak Central peak	10.72	0.422	183	72	Subcooled 0–240 Btu/lb 0–557 kJ/kg	138	2000	680–1760	0.5–1.3	9–53
MIT Todreas and Rohsenow (1965)	Sine Linear increasing Linear decreasing Inlet peak Outlet peak	5.43	0.214	76/122	30/48	Subcooled	4.14/13.8	60/200	680–2720	0.5–2.0	0–68
Babcock and Wilcox. Swenson et al. (1964)	Inlet peak $ff = 1.94$ Outlet peak $ff = 1.39$ Central peak $ff = 1.94$	11.32	0.446	183	72	Subcooled	69 103.5 138	1000 1500 2000	—	—	—
SORIN, Italy, Biancone et al. (1965)	Chopped, smooth Sine $ff = 1.4$ Outlet peak Inlet peak	17.1 11.6	0.673 0.456	132 118	51.8 46.6	Subcooled 0–500°F 0–278°C	82.6/138	1200/2000	435–3400	0.32–2.5	0–63
AERE, Harwell. Hewitt et al. (1966)	Hyperbolic Decreasing	12.6	0.495	368	144	Subcooled	69	1000	1360–2720	1.0–2.0	—

9.2.1.1 *The 'local conditions' hypothesis*   If it is assumed that there is a unique relationship between critical heat flux and local mass quality, then the case of a non-uniformly heated tube can be dealt with in a straightforward manner. The level of critical mean heat flux (or power) is that which, for any locality within the channel, causes the unique critical heat flux/mass quality relationship to be first satisfied. For the case of a linear form of the relationship (as in eq. (8.15)) the operation is much simplified and may be formulated as follows.

The distance along the tube is denoted by z, the functional form of the heat flux profile by $f(z)$, the local heat flux by $\phi(z)$, and the peak heat flux by ϕ_{MAX}

$$\phi(z) = \phi_{\text{MAX}} f(z) \tag{9.2}$$

The mean enthalpy of the steam–water mixture at a distance z, is given by

$$i(z) - i_{\text{f}} = \frac{4\phi_{\text{MAX}}}{DG}\left(\int_0^z f(z)\,\mathrm{d}z\right) - (\Delta i_{\text{SUB}})_{\text{i}} \tag{9.3}$$

The linear critical heat flux/quality relationship (eq. (8.23)) can be written in terms of the local enthalpy as follows:

$$\phi_{\text{CRIT}}(z) = \frac{A' - DG(i(z) - i_{\text{f}})/4}{C'} \tag{9.4}$$

The critical heat flux will occur at z when $\phi(z)$ and $i(z)$ related by eq. (9.3) also satisfy eq. (9.4). If $(\phi_{\text{MAX}})_{\text{CRIT}}(z)$ is the value of the peak heat flux when the critical heat flux is exceeded at z, then

$$(\phi_{\text{MAX}})_{\text{CRIT}}(z) = \frac{A' + DG(\Delta i_{\text{SUB}})_{\text{i}}/4}{C'f(z) + \int_0^z f(z)\,\mathrm{d}z} \tag{9.5}$$

Equation (9.5) gives the predicted value of the peak heat flux when the critical heat flux first occurs at a distance z from the inlet, regardless of whether it has occurred elsewhere in the system. Clearly, the predicted first occurrence of the critical condition, as the total power applied to the tube is increased, will correspond to the lowest value of $(\phi_{\text{MAX}})_{\text{CRIT}}(z)$ given by eq. (9.5). This will be when

$$\frac{\mathrm{d}}{\mathrm{d}z}\left[C'f(z) + \int_0^z f(z)\,\mathrm{d}z\right] = 0 \tag{9.6}$$

Using values of A' and C' derived from data taken from uniformly heated tubes, Barnett (1964) compared the predictions of eq. (9.5) with experimental data from two sources (Lee and Obertelli 1963; Swenson *et al.* 1962). Predicted values of the peak heat flux were consistently higher than the experimental values. The accuracy of the prediction varied considerably with pressure. The method did, however, appear satisfactorily to predict the area of the tube over which the critical condition would occur. This comparison showed that the 'local conditions' hypothesis is not generally valid.

Kirby (1966) has shown that critical heat flux data for non-uniformly heated tubes can be correlated on a 'local conditions' basis but with the actual critical heat flux/mass quality relationship (assumed linear as in eq. (8.18)) re-optimized using the non-uniform heating data alone. Table 9.2 shows the resultant re-optimized values of M and N.

The basis of Kirby's reasoning was that the new 'local conditions' equation

Table 9.2 Kirby (1966) correlation for non-uniformly heated round tubes

Correlating equation

$$\phi_{\text{CRIT}} = M + Nx(z) \text{ (eq. 8.18)}$$

$$M = Y_1 G^{Y_2} D^{Y_3}$$

$$-N = Y_4 G^{Y_5} D^{Y_6}$$

with ϕ in Btu/h ft^2($\times 10^{-6}$), G in lb/h ft^2($\times 10^{-6}$), D in inches

Range of pressures psia (bar)	Range of data used in correlation and optimized constants						
	560 (38.6 bar)	1000 (69 bar)	1250 (86 bar)	1500 (103.5 bar)	1600 (110 bar)	1800 (124 bar)	2000 (138 bar)
D (in)	0.383	0.373 0.446	0.625	0.446 0.625	0.383 1.110	0.625 1.110	0.373 0.446
z (in)	72.0	60.0 192.0	39.4	39.4 72.0	47.0 72.0	39.4 47.0	72.0 144.0
$G \times 10^{-6}$ lb/h ft^2	0.73 3.00	0.50 7.01	1.20 2.00	0.50 3.50	0.27 3.01	0.25 2.50	0.50 3.50
Form factor ff	1.27 1.42	1.27 1.92	1.16 1.26	1.16 1.92	1.30 1.63	1.16 1.63	1.37 1.92
Y_1	2.411	0.7468	0.988	0.7423	0.776	0.616	6.18
Y_2	−0.271	−0.1285	−0.191	−0.1057	−0.118	−0.155	0.231
Y_3	0.0*	−0.544	0.0*	−0.2335	−0.452	−0.293	2.907
Y_4	3.511	1.344	2.918	2.384	2.130	2.086	50.8
Y_5	0.345	0.526	0.287	0.6910	0.739	0.743	0.827
Y_6	0.0*	−0.2278	0.0*	0.0903	−0.296	−0.695	4.228
No. of expts.	34	269	22	79	98	80	69
RMS error %	3.66	5.73	3.37	4.82	3.42	5.95	6.60

* No diameter variation
Overall RMS deviation = 5.30 per cent

represented the point of equality of droplet deposition and local evaporation. However, the location of the critical condition was predicted only approximately by Kirby's analysis and in further work Kirby (1966) attempted to modify the analysis to take account of liquid film flow. The relationship between droplet deposition, liquid film flow, and film evaporation will be returned to in more detail when discussing the mechanisms of the critical heat flux condition.

9.2.1.2 *The 'overall power' hypothesis* Lee and Obertelli (1963) have suggested alternative methods of estimating the total power which could be fed to a tube with non-uniform axial flux distribution before the critical condition is initiated. One such method, the overall power hypothesis, suggests that the total power which can be fed to the tube with non-uniform heating will be the same as for a uniformly heated tube of the same bore and heated length, with the same inlet conditions. No prediction of the location of the critical conditions is made with this method. Lee and Obertelli (1963) and Lee (1965, 1966a) found the method predicted the critical power for a number of sine wave profile tubes reasonably well. Barnett (1964), however, showed the hypothesis was not generally valid for different heat flux profiles when he compared the data of Swenson et al. (1962).

Of these two methods, the 'local conditions' hypothesis and the 'overall power' hypothesis, it is found that the latter is slightly more accurate for symmetrical flux profiles. Lee and Obertelli have recommended calculation of the 'overall power' with a deduction of 10 per cent for the non-uniform heat flux. This procedure was later confirmed by Lee (1966a). For steeper flux gradients associated with skewed flux profiles, the 'local conditions' hypothesis is recommended (Lee 1966a) with a deduction of 15 per cent.

Studies (Todreas and Rohsenow 1965; Swenson et al. 1964; Biancone et al. 1965; Lee 1966a) on heat flux distributions other than the sine wave profile have confirmed that the critical power which can be applied to a tube is seldom greater than that for a uniform heat flux profile and may be up to 30 per cent lower if the maximum heat flux occurs near the tube exit. The one exception to this finding is the case of a linear or hyperbolic decreasing heat flux profile for which improvements of up to 10 per cent in critical power above that for the uniform profile case have been observed in two studies (Todreas and Rohsenow 1965; Biancone et al. 1965).

9.2.1.3 *The 'F-factor' method* A consequence of the failure of the 'local conditions' hypothesis for non-uniform heat flux distributions is that the value of the local critical heat flux must depend, to some degree, on the heat flux profile upstream of the point considered.

An empirical method has been derived by Tong et al. (1965), Smith et al. (1965), and Tong (1966), which enables the effect of the upstream flux profile on the local critical heat flux value to be taken into account. A similar approach has been put forward independently by Silvestri (1966). A factor, F, is defined such that

$$F = \frac{[\phi_{\text{CRIT}}(z)]_u}{[\phi_{\text{CRIT}}(z)]_{\text{nu}}} \tag{9.7}$$

where $[\phi_{\text{CRIT}}(z)]_u$ is the value of the critical heat flux at any given local enthalpy for the uniform heat flux profile case and $[\phi_{\text{CRIT}}(z)]_{\text{nu}}$ is the value of the critical heat flux at the same given local enthalpy for the particular non-uniform heat flux profile case.

The form of the correction factor, F, was obtained by considering an energy balance on the superheated boundary layer in the bubbly flow region (Tong *et al.* 1965) or a mass balance on the liquid film in the annular flow region (Smith *et al* 1965) and was found to be

$$F = \frac{\Omega}{1 - \exp(-\Omega z_{\text{CRIT}})} \int_0^{z_{\text{CRIT}}} \frac{\phi(z)}{\phi(z_{\text{CRIT}})} \exp[-\Omega(z_{\text{CRIT}} - z)] \, dz \tag{9.8}$$

By combining eqs. (9.3) and (9.4) the following expression for the critical heat flux for a uniform heat flux profile is obtained:

$$[\phi_{\text{CRIT}}(z)]_u = \frac{A' - [\phi_{\text{MAX}} \int_0^z f(z) \, dz - (DG(\Delta i_{\text{SUB}})_i/4)]}{C'} \tag{9.9}$$

from the definition of F

$$[\phi_{\text{CRIT}}(z)]_{\text{nu}} = \frac{A' + DG(\Delta i_{\text{SUB}})_i/4}{FC' + \int_0^{z_{\text{CRIT}}} \phi(z)/\phi(z_{\text{CRIT}}) \, dz} \tag{9.10}$$

The ratio $[\phi(z)/\phi(z_{\text{CRIT}})]$ in eqs. (9.8) and (9.10) may be replaced by $[f(z)/f(z_{\text{CRIT}})]$ and thus both integrals are functions of the flux profile only and can be evaluated directly as a function of z_{CRIT}.

The value of the factor Ω has been determined by fitting the method to experimental data. Tong *et al.* (1965) found that Ω was a function of both local steam quality $x(z_{\text{CRIT}})$ and mass velocity G, thus

$$\Omega = 0.44 \frac{[1 - x(z_{\text{CRIT}})]^{7.9}}{(G/10^6)^{1.72}} \, (\text{in}^{-1}) \tag{9.11}$$

where the mass velocity is given in lb/h ft^2. Smith *et al.* (1965) suggested that, for the limited range of variables investigated in their tests ($D = 0.2285$ in (5.8 mm), $p = 515$ psia (35 bar), $G = 0.55$–1.10×10^6 lb/h ft^2 (750–1500 kg/m^2 s)), Ω was constant and equal to 0.59 in^{-1}. Lee (1966a) found, when correlating a large number of non-uniform critical heat flux data, that Ω was

Table 9.3   Optimized values of K for eq. (9.12) and the corresponding RMS
error on the predicted total critical power (Lee 1966a)

Pressure psia	560	1000	1250	1550	1800	2000
bar	38.6	69	86	107	124	138
K	3.67	5.00	3.98	3.59	2.48	1.03
RMS error %	3.96	7.18	8.8	5.78	5.27	8.30

well represented by

$$\Omega = \frac{K\delta}{D(\phi_{MAX}/\bar{\phi})} \ (\text{in}^{-1}) \tag{9.12}$$

where K is a constant dependent on pressure (see Table 9.3), δ is the
dimensionless axial flux gradient taken at the point of average flux in the
upper part of the tube

$$\delta = \left[\frac{d(\phi(z)/\bar{\phi})}{d(z/D)}\right] \ \text{at} \ \left[\frac{\phi(z)}{\bar{\phi}}\right] = 1 \tag{9.13}$$

$\bar{\phi}$ is the average heat flux over the tube length, L.

To use the 'F-factor' method to predict the critical heat flux for any given
heat flux profile and inlet condition the following procedure is recommended
(Lee 1966a).

(a) Define the heat flux profile in terms of the ratio $[\phi(z)/\bar{\phi}]$ with the
tube divided into zones about 2 in (5 cm) long. Evaluate the integrals in eqs.
(9.8) and (9.10) as a function of a series of likely values of z_{CRIT}. If eq. (9.12)
is being used for Ω evaluate δ.

(b) Select values of constants for the uniform heat flux correlation
(Bowring). Obtain values of A' and C' for the chosen values of pressure, mass
velocity, and tube diameter. Select appropriate value of K if eq. (9.12) is used
and evaluate Ω.

(c) Use eqs. (9.8), (9.9), and (9.10) to obtain the predicted local critical
heat flux $[\phi_{CRIT}(z)]_{nu}$ at a series of values of z_{CRIT} along the tube.

(d) Evaluate the ratio of the local predicted heat flux to the local operation
heat flux at each data of z_{CRIT} and obtain the minimum value of this ratio.
This will provide the margin of safety between the operating and critical
heat flux conditions and will also provide the location of the critical
condition. Lee (1966a) recommends that a safety factor of twice the RMS
error given in Table 9.3 should be applied to the results to allow for
uncertainties in predicted critical power.

To summarize, no completely general analytical method is available for the prediction of the critical heat flux condition in tubes with a non-uniform axial heat flux profile. The 'F-factor' method as proposed by Tong et al. (1965), Smith et al. (1965), and Tong (1966) and as applied by Lee (1966a) is recommended as the best purely empirical method for smooth non-uniform axial heat flux profiles at the present time.

9.2.2 Step changes in the heat flux level over short lengths

The second type of non-uniform heating occurs where there is a step change in the level of heating over a short length or short lengths of the tube. This type of test is commonly known as a 'hot patch' or 'cold patch' test. De Bortoli et al. (1958) report some interesting experiments of this type carried out with a rectangular channel, $0.097 \times 1.0 \times 27$ in ($0.248 \times 2.54 \times 68.5$ cm) with water at 2000 psia (138 bar). The initial 26.625 in (67.55 cm) of the channel were at a uniform heat flux ϕ_1 and the last 0.375 in (0.95 cm) was operated at a heat flux ϕ_2 where the ratio (ϕ_2/ϕ_1) was maintained at 1.98 through the tests. The results are shown in diagrammatic form in Fig. 9.1 where the upper curve is ϕ_2 at the critical condition and the lower curve is the corresponding magnitude of ϕ_1 ($=\phi_2/1.98$). Also shown on the diagram is the curve for the critical heat flux against exit quality for a uniformly heated channel of the same total length at the same exit conditions—ϕ_3. At high exit subcoolings $\phi_2 = \phi_3$ and at high exit qualities

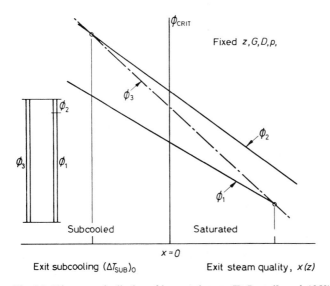

Fig. 9.1. Diagrammtic display of hot patch tests (DeBortoli et al. 1958).

$\phi_1 = \phi_3$. Clearly, for the 'local conditions' hypothesis to be valid, the curves for ϕ_2 and ϕ_3 would need to be co-linear. Thus, it would appear from these tests that the 'local conditions' hypothesis is only likely to be approximately true in the highly subcooled regions. It is also of interest to compare the 'overall power' from each arrangement. For the uniformly heated channel the critical power S_3 is simply

$$W_3 = P_h \phi_3 L \tag{9.14}$$

where P_h is the heated perimeter. For the channel with the 'hot patch' the critical power, $W_1 + W_2$ is

$$W_1 + W_2 = P_h \phi_1 (L - 0.014L) + 1.98 \phi_1 P_h (0.0141L) = 1.0138 P \phi_1 L \tag{9.15}$$

If the 'overall power' concept discussed previously is applied then $(W_1 + W_2)$ should equal W_3 or comparing eq. (9.14) and eq. (9.15), $\phi_1 \approx \phi_3$. Figure 9.1 shows this to be approximately true only at high exit steam qualities. Experiments carried out in Italy (Bertoletti *et al.* 1965; Silvestri 1966) with round tubes having a similar 'hot patch' arrangement have shown that this concept of a constant critical channel power independent of heat flux distribution, although approximately valid at high exit qualities for moderate value of the ratio (ϕ_2/ϕ_1), becomes increasingly incorrect as (ϕ_2/ϕ_1) is increased. Experiments (De Bortoli *et al.* 1958; Bertoletti *et al.* 1965; Smith *et al.* 1965) such as these have clearly demonstrated the inadequacy of the 'local conditions' and 'overall power' hypotheses as general statements. The results shown in Fig. 9.1 also suggest the existence of at least two different mechanisms by which the critical condition is reached—one for subcooled conditions and another at high qualities. The need to consider the influence of the upstream heat flux profile in these 'hot patch' experiments has led to the development of the '*F*-factor' methods (Smith *et al.* 1965; Silvestri 1966) of dealing with non-uniform heat flux profiles.

For the situation shown diagrammatically in Fig. 9.1 the factor F is given by the ratio (ϕ_3/ϕ_2) which varies from a value of unity at high subcooling to a value of approximately 0.5 at high exit qualities. This type of variation for F can be calculated (Tong *et al.* 1965) using the method outlined in Section 9.2.1.3 with the value of Ω given by eq. (9.11).

It should be borne in mind that the value of F calculated from this method can be either greater or less than unity depending on the particular heat flux profile. For a symmetrical sine wave profile, F may typically take up values between 1 and 3.

Tests in which part of the channel was unheated are reported by Hewitt *et al.* (1966). These authors report experiments at both 3.5 and 69 bar (50 and 1000 psia). For given inlet conditions the unheated zone ('cold patch') was moved successively downstream with respect to the inlet. Figure 9.2 illustrates some of the results obtained at 69 bar (1000 psia). With the cold

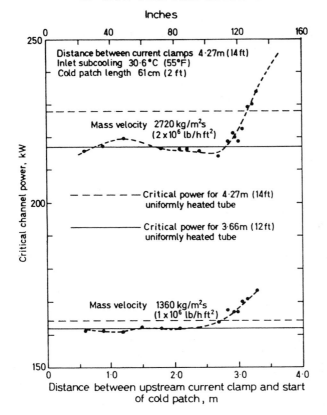

Fig. 9.2. Variation of critical power with position of cold patch for a 12.7 mm (0.497 in) i.d. tube at 69 bar (1000 psia) (data of Hewitt *et al.* 1966).

patch near the inlet end of the tube, the power to the critical condition was close to that obtained for the same *heated* length with uniform heating. However, as the cold patch was moved towards the downstream end of the tube the power increased sharply and could in fact be greater than that obtained for the same *total* length with uniform heating.

9.2.3   *Non-uniform circumferential heat flux profile*

The third type of non-uniform heating is that in which there is a variation in heat generation around the tube circumference—the axial heat flux profile is maintained uniform. Experiments of this type have been reviewed by Lee (1966*b*) and by Alekseev *et al.* (1964). The non-uniform circumferential heat generation was produced by passing a heavy electrical current through the tube wall, the inside and outside diameters of which were eccentric. In the

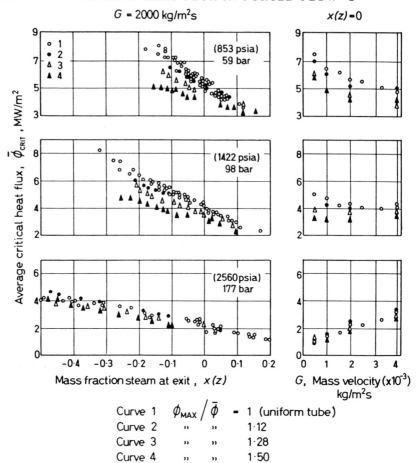

Fig. 9.3. The influence of non-uniform heating around the tube circumference.

experiments reported by Alekseev *et al.* (1964) three tubes 40 cm (15.75 in) long and 10 mm (0.394 in) i.d. having different eccentricities were used, producing values of the ratio $(\phi_{MAX}/\bar{\phi})$ of 1.12, 1.28, and 1.50. Allowance was made for the circumferential conduction in the tube wall when calculating the heat flux profile. The experimental results for each tube are shown in Fig. 9.3 for three different pressures together with the results for a uniform tube of the same dimensions. It is seen that, as the circumferential heat flux profile becomes more asymmetric, so the average critical heat flux $\bar{\phi}_{CRIT}$ (evaluated from the total power to the tube) falls. At high subcoolings the value of $(\phi_{MAX})_{CRIT}$ for each of the eccentric tubes approximately corresponds to the uniform tube value, ϕ_{CRIT}. As the exit steam mass fraction increases so the average heat flux approaches the value for a uniform tube. These experiments confirm the 'local' nature of the critical condition at high subcoolings and the tendency to an 'overall power' condition in the higher steam quality regions. It is recommended that the 'overall power' hypothesis be used to calculate the critical power which can be applied to a tubular channel with a non-uniform circumferential heat flux profile. The power calculated should be reduced by 10 per cent to cover the worst condition (see Section 9.2.1.2) for cases with $(\phi_{MAX}/\bar{\phi})$ between 1.0 and 1.5 and by 20 per cent for cases with $(\phi_{MAX}/\bar{\phi})$ between 1.5 and 2.0.

The combined effects of non-uniform circumferential heating and tube inclination on the critical heat flux has been investigated by Kitto and Weiner (1982) and Kitto and Albrecht (1987). To produce the non-uniform heat flux, they welded Inconel bars longitudinally to one side of stainless steel tubes. Heating the tube electrically, the non-uniform electrical resistance of the tube allowed them to attain nominal peak-to-average heat flux ratios $(\phi_{MAX}/\bar{\phi})$ up to 2.05. Steam/water flows at 186 bar were studied for 15°, 30° and 90° inclinations from the horizontal. The following effects were observed:

(a) With respect to the peak non-uniform heat flux for vertical tubes, the critical heat flux $(\phi_{MAX})_{CRIT}$ is higher than ϕ_{CRIT} for uniformly heated tubes.
(b) With respect to the average heat flux $\bar{\phi}$ of the non-uniform heat flux for vertical tubes, the critical heat flux is less than that of a uniformly heated tube at vapour qualities below 30 per cent while essentially the same above this vapour quality. The difference was explained by the passage from the bubbly flow regime to the annular flow regime at this quality.
(c) For inclined tubes, the effect of non-uniform heating depended on where the peak heat flux was applied. For the peak applied to the top of the tube, the peak heat flux $(\phi_{MAX})_{CRIT}$ was the same as for uniformly heated inclined tubes while $\bar{\phi}$ was substantially lower; for the peak applied to the side of the tube, the critical heat flux phenomenon still occurred at the top of the tube first because of stratification effects.

The vertical tube results agreed with other previous experimental studies by Styrikovich and Mostinski (1959) and an analytical study by Butterworth (1972). The physical phenomenon proposed in these two studies to explain the increase in $(\phi_{MAX})_{CRIT}$ relative to a uniformly heated tube was the formation of a secondary circumferential flow of liquid around the inner tube wall. This circumferential flow superimposed on the axial flow makes it more difficult for a stable vapour film to form, resulting in a higher $(\phi_{MAX})_{CRIT}$ value before reaching the critical condition.

Butterworth (1972) analysed a circumferential film spreading model for the annular flow regime, arriving at

$$\frac{\bar{\phi}}{\phi} = \frac{1}{1 + (D_i^2/Kz)\alpha}$$

$$\alpha = \frac{(\phi_{MAX})_{CRIT}}{\bar{\phi}} - 1$$

(9.16)

with $\bar{\phi}$ the average heat flux for non-uniform heating and ϕ the heat flux for uniform heating. D_i and z are the internal diameter and length of the tube, respectively. The constant K represents the spreading coefficient for the secondary circumferential flow, assumed to be 0.9 mm by Butterworth. Comparing this method to their data, Kitto and Weiner (1982) found qualitative agreement but noted that K should be made a function of vapour quality.

   More recently, Kuzma-Kichta et al. (1988) and Komendantov et al. (1991) have developed a new method for studying the effect of non-uniform circumferential heat fluxes on the critical heat flux condition. Their test section was made from a very thick-walled copper cylinder that had very high longitudinal fins machined in the wall. The tube was non-uniformly heated on the outside by electron bombardment, produced by a focused electron beam. Thermocouples installed in the fins provided the temperature profiles and local heat fluxes around the periphery of the tube. These workers confirmed earlier tests showing that the critical heat flux condition develops under the portion of the tube subject to the peak non-uniform heat flux and observed that its occurrence is accompanied by large wall temperature fluctuations. The liquid film was destroyed only on the side with the peak heat flux and remained intact on the other side. The fluctuation thus indicates the existence of an unstable vapour liquid film boundary.

9.3   The critical heat flux in non-circular geometries

A considerable amount of work has been carried out to determine critical heat flux values for a variety of non-circular ducts. Many of these geometries are identical or similar to the fuel channels in particular types of nuclear reactor. The three most common geometries, rectangular ducts, annular ducts, and multi-rod cluster assemblies will be briefly considered.

9.3.1   Rectangular channels

Table 9.4 lists sources of available experimental data for uniformly heated vertical rectangular channels. In many of the studies only the two longer sides of the rectangle were heated. The critical heat flux appears to vary with the individual independent variables in the same general manner as with round tubes. The linear relation between the critical heat flux and the inlet subcooling is again evident.

Macbeth (1963) has correlated the data available from De Bortoli *et al.* (1958) for channels 2.54 cm (1 in) in width and with spacings 1.27 mm (0.050 in) and 2.54 mm (0.100 in). These channels have three characteristic dimensions, viz., the flow width, an equivalent heated width which measures 2.29 cm (0.9 in), and the channel spacing, S. From these figures the corresponding heat balance equation is

$$x(z) = \frac{1}{i_{fg}}\left[\frac{1.8\phi z}{SG} - (\Delta i_{SUB})_i\right] \tag{9.17}$$

Comparing eq. (9.17) with the same equation for round tubes, it will be seen that the tube diameter is replaced by $2.22S$, the heated equivalent diameter. The Macbeth correlation then becomes

$$\phi_{CRIT} \times 10^{-6} = \frac{A^1 + 0.555C^1S(G \times 10^{-6})(\Delta i_{SUB})_i}{1 + C^1 z} \tag{9.18a}$$

or

$$\phi_{CRIT} \times 10^{-6} = A^1 - 0.555C^1S(G \times 10^{-6})i_{fg}x(z) \tag{9.18b}$$

where A^1 and C^1 are now unknown functions of S, G, and P. The units of ϕ are Btu/h ft^2, of G are lb/h ft^2, and of z and S are inches. For the low mass velocity region,

$$\phi_{CRIT} \times 10^{-6} = \frac{(G \times 10^{-6})(i_{fg} + (\Delta i_{SUB})_i)}{3.78S^{-1.73}(G \times 10^{-6})^{1.1} + 1.8z/S} \tag{9.19}$$

To see how well his round tube equation would fit, Macbeth applied it directly to the rectangular channel data, putting $D = 2.22S$. He concluded that the round tube equation cannot be applied to rectangular channels without significant loss of accuracy.

The rectangular channel data in the high mass velocity region were examined pressure by pressure. The parameters A^1 and C^1 in eqs. (9.18a) and (9.18b) were assumed to be simple power functions of mass velocity and channel spacing

$$A^1 = y_0 S^{y_1}(G \times 10^{-6})^{y_2} \tag{9.20}$$

$$C^1 = y_3 S^{y_4}(G \times 10^{-6})^{y_5} \tag{9.21}$$

The results of optimizing eq. (9.18a) using eqs. (9.20) and (9.21) are given in Table 9.5.

Because of the small number of data the optimized values of 'y' do not

Table 9.4 Available data for critical heat flux in rectangular channels

Reference	Range of pressures		Duct cross-section dimensions	Duct length		Range of mass velocity		Range of inlet subcooling		Comments
	bar	psia		cm	in	kg/m² s	lb/h ft² × 10⁻⁶	kJ/kg	Btu/lb	
De Bortoli et al. (1958)	41.5–138	600–2000	1 in × 0.050 in 1 in × 0.100 in 0.9 in heated width 2.54 cm × 1.27 mm 2.54 cm × 2.54 mm 2.28 cm heated width	15.2–68.5	6.0–27.0	27.2–6500	0.02–4.78	27.9–1522	12–654	511 data points
Finat and Saunier (1963)	78.5 108 138	1140 1570 2000	Similar to above; exact dimensions not given	—	—	500–3500	0.368–2.57	126–879	54–378	Some data at the lower pressures were taken under unstable flow conditions
Tippets (1962)	69	1000	2.10 in × 0.50 in 2.10 in × 0.25 in 5.33 cm × 1.27 cm 5.33 cm × 0.685 cm	94	37.0	245–1950	0.18–1.44	35–600	15–259	Thin ribbons used for heating surface—possible effect of ribbon thickness

Reference								$(\Delta T_{SUB})_0$		
Levy et al. (1959)	4.5–9.5	65–138	2.5 in × 0.1 in 6.35 cm × 2.54 mm	45.7 and 91.5	18 and 36	1630–14,400	1.2–10.6	50–111°C	90–200°F	Heating occurred over whole perimeter
Gambill and Bundy (1961)	11–39.3	160–570	1.060 in × 0.043 in (heated 1.0 in) 1 in × 0.057 in 2.69 cm × 1.09 mm (heated 2.54 cm) 2.54 cm × 1.45 mm	30.5 and 47.7	12 and 18.8	2720–23,100	2.0–17.0	14.5–36°C	26–64°F	Nickel and aluminium surface
Troy (1958)	41.5–138	600–2000	2.25 in × 0.070 in 5.72 cm × 1.78 mm	183	72	516–5440	0.38–4.0	0–8°C	0–15°F	
Kafengauz and Bauarov (1959)	40	590	Spacing between parallel plates 0.2–2 mm (0.008–0.079 in)	5.08	2	815–23,100	0.6–17.0	—	—	
Zenkevich et al. (1962)	61.6–152	880–2200	0.5 in × 0.25 in 1.27 cm × 0.685 cm	20	7.9	815–4900	0.6–3.6	2–50°C	4–90°C	Also only available data for isosceles triangle cross-section duct

Table 9.5   Macbeth correlation—optimized value of 'y' for 1 in. wide rectangular channel data in the high mass velocity region

Pressure psia (bar)	y_0	y_1	y_2	y_3	y_4	y_5	RMS error %	No. of expts.
600 (41.5)	23.4	−0.472	−3.29	0.123	−1.4	−3.93	6.1	22
800 (55)	0.445	−1.01	0.384	0.0096	−1.4	−0.0067	12.9	28
1200 (83)	1.88	−0.081	−0.526	0.0035	−1.4	−1.29	4.9	42
2000 (138)	0.546	−0.315	−0.056	0.0027	−1.4	−0.725	9.4	359

show a smooth variation with pressure. Extrapolation of the correlation outside the limits of the experimental data is unwise. The data at 138 bar (2000 psia) are the most extensive, and, to see how well it would fit, Macbeth applied his corresponding round tube correlation, again substituting $2.22S$ for D. This exercise produced a similar conclusion to that arrived at for the low-mass velocity region, namely, a significant loss of accuracy.

In the experiments discussed so far both long sides of the rectangular duct have been heated. If the use of a heated equivalent diameter as the cross-sectional dimension is correct then there will be some influence on the critical heat flux if only one long side is heated. Tippets (1962) carried out experiments in a rectangular channel 5.33 × 1.27 cm (2.10 × 0.5 in) in which first both long sides were heated and then just one side. The results are shown in Fig. 9.4. It is noted that the critical heat flux increases as the heated equivalent diameter increases at constant inlet subcooling whereas the critical flux decreases as the heated equivalent diameter increases at constant exit quality.

Apart from the Macbeth correlation, Katto (1981) has given an empirical correlation for uniformly heated rectangular channels applicable to fluids other than water. In many cases it has been stated that correlations developed for round tubes are applicable to rectangular channels. In cases where the whole perimeter of the rectangular channel is heated, such as in the studies of Levy *et al.* (1959) and Zenkevich *et al.* (1962), overheating occurs in the corner near the exit of the channel at a considerably reduced average heat flux.

9.3.2   *Annular channels*

Several extensive compilations (Becker and Hernborg 1963; Janssen *et al.* 1963; Barnett 1966) of data exist for the critical heat flux in vertical heated annular channels with uniform heating on the internal surface. The critical heat flux appears to vary qualitatively in the same general manner as for

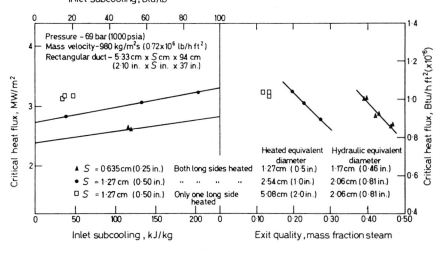

Fig. 9.4. Influence of heated equivalent diameter.

round tubes. The critical heat flux may be assumed to be a function of six independent variables thus:

$$\phi_{\text{CRIT}} = f(G, (\Delta i_{\text{SUB}})_i, p, D_i, D_o, z) \tag{9.22}$$

where D_i and D_o are the internal and external diameters of the annulus respectively.

In a similar manner to that used for round tubes, and providing the annulus is reasonably long, the critical heat flux in uniformly heated annuli can be correlated in terms of the exit mass quality only (Janssen *et al.* 1963) and the variables of length and inlet subcooling are eliminated except in their influence in determining the exit mass quality from the heat balance equation. Thus

$$\phi_{\text{CRIT}} = f(G, x(z), p, D_i, D_o) \tag{9.23}$$

The data for annuli can be correlated by means of a Macbeth-type correlation and this has been done by Barnett (1966). The correlation is as follows:

$$\phi_{\text{CRIT}} \times 10^{-6} = \frac{A + B(\Delta i_{\text{SUB}})_i}{C + z} \tag{9.24}$$

where for a pressure of 69 bar (1000 psia)

$$A = 67.45 D_h^{0.68}(G \times 10^{-6})^{0.192}[1 - 0.744 \exp(-6.512 D_e(G \times 10^{-6}))] \tag{9.25}$$

$$B = 0.2587 D_h^{1.261}(G \times 10^{-6})^{0.817} \tag{9.26}$$

$$C = 185.0 D_e^{1.415}(G \times 10^{-6})^{0.212} \tag{9.27}$$

D_e is the hydraulic equivalent diameter given by $(D_o - D_i)$; D_h is the heated equivalent diameter given by $(D_o^2 - D_i^2)/D_i$. The correlation is in the British System of Units and covers the following range of parameters:

$$
\left.
\begin{array}{lll}
D_i: & 0.375 \text{ to } 3.798 \text{ in } (0.952 \text{ to } 9.65 \text{ cm}) \\[4pt]
D_o: & 0.551 \text{ to } 4.006 \text{ in } (1.4 \text{ to } 10.16 \text{ cm}) \\[4pt]
z: & 24.0 \text{ to } 108.0 \text{ in } (0.61 \text{ to } 2.74 \text{ m}) \\[4pt]
G: & 0.140 \text{ to } 6.20 \text{ lb/h ft}^2 \ (\times 10^{-6})(190 \text{ to } 8430 \text{ kg/m}^2 \text{ s}) \\[4pt]
(\Delta i_{SUB})_i: & 0 \text{ to } 412 \text{ Btu/lb } (0 \text{ to } 0.958 \text{ MJ/kg})
\end{array}
\right\} \quad (9.28)
$$

Equation (9.24) represents a slight departure from the 'local conditions' hypothesis (in which B is fixed as $0.25D_h$ $(G \times 10^{-6})$). This modification resulted in a small but significant improvement in the RMS error from 7.3 per cent to 5.9 per cent. As with round tubes the local conditions hypothesis is not obeyed for annuli of short length: this is illustrated by the results of Alekseev et al. (1964) in experiments carried out at 98 bar (1422 psia) in an internally heated annulus ($D_i = 12$ mm (0.472 in), $D_o = 15$ mm (0.590 in)). The critical heat flux increased with decreasing length at constant exit conditions. As the heated length was decreased from 40 cm to 20 cm (15.75 in to 7.88 in) the critical heat flux at zero exit quality rose by a factor 1.16. When the heated length was reduced to 10 cm (3.94 in) the factor becomes 1.68. The effect appears to start at about the same length to heated equivalent diameter ratio as for round tubes (~ 70).

It is interesting to compare the data for internally heated annuli with similar data for round tubes. Barnett (1964a) has carried out such an exercise and one example of his results is shown in Fig. 9.5, which shows some data from Janssen et al. (1963) for an internally heated annulus plotted as critical heat flux against inlet subcooling at a fixed mass velocity and pressure. Also shown are data from Columbia University for round tubes having diameters close to the value of the heated equivalent diameter and the hydraulic diameter of the annulus. It is not possible to relate round tube and annulus data using either of these two equivalent diameter concepts (Barnett 1964a).

Shah (1980) has presented a generalized correlation for CHF in vertical annuli for various fluids and with combinations of internal and external heating. Over 800 data points from four fluids were correlated with a mean deviation of 14 per cent. The ranges of parameters include reduced pressures from 0.017 to 0.9, mass flow rate (G) from 100 to 16,000 kg/m² s, inlet quality from -3.1 to 0, critical quality from -2.8 to 0.74, annular gap size from 0.5 to 11.1 mm and D_i from 1.5 to 96.5 mm.

One complicating factor with the internally heated annulus is the need to support the heated rod centrally in the unheated duct. It has been shown (Bertoletti et al. 1965; Alekseev et al. 1964; Barnett 1966) that these supports

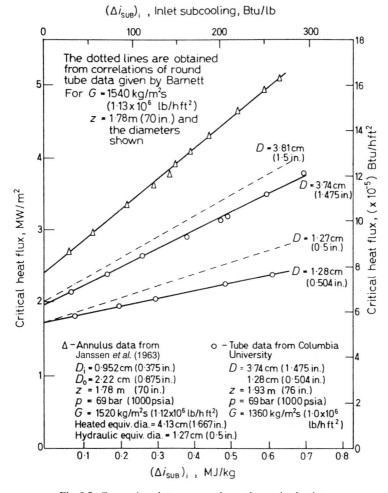

Fig. 9.5. Comparison between annulus and round tube data.

can transport liquid from the unheated outer wall to the inner heated wall. A wire wrapped around the central heater rod can also influence the critical heat flux. If the pitch of the wrap is small then the critical heat flux is reduced relative to the unwrapped case. As the pitch is increased so the reduction becomes less until for the case where the wires are lying on the heated surface parallel to the direction of flow there may be a net increase in critical heat flux over the unwrapped situation. It is uncertain whether this improvement would occur if the wrap is not in contact with the outer heated surface.

The effect of eccentricity of the heated rod in the unheated duct is to reduce the critical heat flux. The magnitude of the effect appears to increase as the ratio of the minimum to average gap and the ratio of average gap to rod diameter are decreased. If both of these ratios are greater than 0.25 the effect is probably small (Barnett 1966).

Experiments to determine the critical heat flux for the internal surface of an annulus in which the outer wall is also heated produce interesting results (Bertoletti *et al.* 1965; Alekseev *et al.* 1964; Becker and Hernborg 1963; Collier *et al.* 1964). At a fixed mass velocity and pressure the critical quality (x_{CRIT}) passes through a maximum depending upon the fraction of the total power input which is applied to the outer tube (Fig. 9.12).

9.3.3 Rod clusters

In water-cooled nuclear-power reactors the fuel element is often in the form of a cluster of small diameter rods spaced one from another by spirally wrapped wires or by grid type spacers. A considerable number of electrically heated models of such fuel elements have been tested in various laboratories and the detailed results have been compiled, correlated and discussed by Macbeth (1964) and Barnett (1966, 1968). Analysis of thermo-hydraulic conditions within a multi-rod cluster can be handled at various levels of detail.

9.3.3.1 *Mixed flow analyses* In this approach the entire channel is assumed to behave in a one-dimensional manner with 'average' properties being ascribed to the variables such as mass flow-rate, quality and heat flux. No attempt is made to consider any two-dimensional 'fine structure' within the channel. An early attempt at such a correlation was that of Macbeth (1964) who applied the 'local conditions' hypothesis to a selection of data from uniformly heated vertical rod clusters and obtained the following relationships for A, B, and C for substitution into eq. (9.24) for 69 bar (1000 psia) only

$$\left. \begin{aligned} A &= 67.6D_{\mathrm{h}}^{0.83}(G \times 10^{-6})^{0.57} \\ B &= 0.25D_{\mathrm{h}}(G \times 10^{-6}) \\ C &= 47.3D_{\mathrm{h}}^{0.57}(G \times 10^{-6})^{0.27} \end{aligned} \right\} \qquad (9.29)$$

Barnett (1966, 1968) showed that his annulus correlation, eq. (9.24), predicts a wide range of rod bundle data with remarkable accuracy. For rod bundles the values of D_{i} and D_{o} are defined as follows for use in eqs. (9.24)–(9.27).

'equivalent' $D_{\mathrm{i}} = D_{\mathrm{R}}$, the rod diameter (inches)

'equivalent' $D_{\mathrm{o}} = [D_{\mathrm{R}}(D_{\mathrm{R}} + D_{\mathrm{h}}^{*})]^{1/2}$ (inches)

where

$$D_h^* = \frac{4 \times (\text{flow area})}{S \times (\text{heated rod perimeter})}, \quad \text{and} \quad S = \sum_{\text{rods}} \xi.$$

ξ is the ratio of heat flux on a rod to the maximum heat flux in the cluster. If all the rods carry the same heat flux, $D_h^* = D_h$; the definition of D_h^* ensures that the rod cluster and the 'equivalent' annulus have the same exit quality for the same critical heat flux. One of the more recent mixed flow correlations is that of Bowring (1977) which can be applied to all types of nuclear fuel elements and has corrections for both non-uniform axial and radial (rod-to-rod) heating.

9.3.3.2 *Sub-channel analyses* Under this general heading, an attempt is made to sub-divide the rod cluster into a number of parallel interacting flow sub-channels between rods. The equations of mass, momentum and energy conservation are solved to give radial and axial variations in quality (or fluid enthalpy) and mass flow-rate. Interchange of mass, heat and momentum is allowed between neighbouring sub-channels, the amounts being governed by the conservation equations. Once the *local* enthalpy and flow conditions at each sub-channel node have been established, suitable constitutive relations in the form of empirical correlations are used to define the heat transfer regime, including the CHF, so as to provide the local temperature of the rod surface. Details of such calculations have been given by Weisman and Bowring (1975). Bowring (1979) has published a subchannel dryout correlation (WSC-2) for use with sub-channel analysis computer codes.

9.3.3.3 *Phenomenological analyses* Under this general heading, an attempt is made to model the kinetics of the various physical processes which occur. Clearly such analyses must relate to the actual flow patterns present and, more important, they must be able to handle departures from hydrodynamic equilibrium (i.e., not 'fully developed' flows; see Chapter 3). To illustrate these points an example of the phenomenological model developed by Hewitt and co-workers (Whalley *et al.* 1974; Whalley 1978) for two-phase flow in a vertical cluster is given in 9.7.2.

9.4 Critical heat flux in forced flow of fluids other than water

9.4.1 *Forced flow in single tubes*

Table 9.6 gives a selection of the availble sources of experimental data on the critical heat flux taken under forced convection conditions for various organic liquids, cryogenic fluids (Richards *et al.* 1961; Seader *et al.* 1965) and

Table 9.6 Available experimental data for critical heat flux in forced flow of fluids other than water

| Source date | Fluid | Geometry | Pressure | | Subcooling | | Velocity | | Critical heat flux | | No. of tests |
			bar	psia	°C	°F	m/s	ft/s	kW/m²	Btu/h ft² (×10⁻⁶)	
Noel (1961a)	Hydrazine	Tube	6.9–83	100–1200	55–281	98–506	0.3–28.4	1.0–93.0	2330–44,500	0.74–14.1	62
Hines (1959)	Hydrazine	Tube	13.2–71	191–1025	57–263	102–473	4.02–36.8	13.2–120.5	4450–43,300	1.41–13.72	22
Noel (1961)	Ammonia	Tube	11.7–126	170–1820	1.6–115	3–208	0.9–47.6	3–15.6	2840–23,400	0.90–7.42	45
Dimmock (1957)	Ammonia	Tube	11.4–88	165–1275	0.5–112	1–201	3.79–25	12.4–81.8	2460–16,350	0.78–5.19	111
Core and Sato (1958)	Diphenyl	Annulus	1.6–28	23–406	0–183	0–328	0.15–5.2	0.5–17	220–2800	0.07–0.89	48
	Santowax R		6.9	100	—	—	0.15–4.6	5–15	—	—	10
	Polyphenyl mixt.		4.4–22	64–318	5–213	9–383	0.15–4.6	5–15	—	—	25
	Monoisopropylbiphenyl (MIPB)		26.6–27.6	386–400	88–179	159–322	0.15–4.6	5.1–15	1230–3000	0.39–0.95	6
Robinson and Lurie (1962)	Santowax R and other polyphenyls	Tube	1.52–10.35	22–150	19–157	34–282	1.25–7.7	4.1–25.2	1005–3720	0.319–1.180	94
Birdseye (1960)	Nitrogen tetroxide	Tube	10.35–42.5	150–615	0–111	0–200	3.11–18.2	10.2–59.8	1730–11,300	0.55–3.59	73

Reference	Fluid	Geometry									
Gambill and Bundy (1962)	Ethylene glycol	Tube	1.38–6.1	20–88	86–147	155–264	5.3–27	17.3–88.4	6500–19,700	2.06–6.25	5
Povarnin (1964)	Ethyl alcohol	Tube	2.07–62	30–900	0–210	0–396	0.49–50	1.6–164	567–21,600	0.18–6.86	163
Lewis et al. (1962)	Nitrogen	Tube	3.45	50	1.6	3	0.027	0.09	50–177	0.016–0.056	11
	Hydrogen	Tube	3.45	50	0–3.3	0–6	0.10	0.33	19–66	0.0059–0.021	44
Stevens et al. (1964)	Refrigerant 12	Tube	10.7	155	0–28 kJ/kg	0–12 Btu/lb	—	—	22–284	0.007–0.090	Large number
Stevens et al. (1965)	Refrigerant 12	Tube (non-uniform heating)	10.7	155	0–28 kJ/kg	0–12 Btu/lb	—	—	93.5–315	0.030–0.10	81
Barnett and Wood (1965)	Refrigerant 12	Tube	7.3–13.8	106–200	−60.5 kJ/kg	0–26 Btu/lb	—	—	79–440	0.025–0.140	Large number
Hoffman and Krakoviak (1964)	Potassium	Tube	—	—	—	—	—	—	101–1100	0.032–0.350	6
Lurie (1964)	Sodium	Annulus	0.35–0.55	5–8	0–34	0–61	—	—	945–2520	0.300–0.800	11
Aladyev et al. (1966)	Potassium	Tube uniform and non-uniform heating	1.1–1.3	16–19	100–390	180–700	low	low	252–1890	0.08–0.6	—
Tippets et al. (1963)	Potassium	Tube	2.1–3.8	30–55	—	—	low	low	252–1320	0.08–0.42	—

liquid metals. Apart from the work of Stevens *et al.* (1964) the studies have been somewhat limited in scope. The very comprehensive set of experiments on Refrigerant 12 (Stevens *et al.* 1964) have shown that this fluid, at least, behaves qualitatively in an identical way to water. The data for both the cryogenic fluids (Lewis *et al.* 1962) and the liquid metals (Hoffman and Krakoviak 1964) clearly show that as the mass velocity and diameter are reduced or the tube length increased so the critical heat flux approaches that needed for complete evaporation of the fluid.

An attempt to determine the relevant physical properties and a set of scaling laws for the critical heat flux condition was made at AEE, Winfrith (Barnett 1963, 1964b), Barnett (1963) discusses which of the numerous physical property terms are likely to be important and goes on to develop appropriate dimensionless property groupings. A number of hypothetical relationships between these groupings have been examined using the experimental data for water and Refrigerant 12 (Barnett 1964b; Barnett and Wood 1965). None of the relationships tested was entirely satisfactory, although one set of laws, eq. (9.30), does appear to be near the truth.

$$\left(\frac{\phi_{CRIT}\gamma^{0.5}}{i_{fg}\rho_f^{0.5}}\right) = f\left[\left(\frac{L}{D}\right), \left(\frac{Dc_{pf}\rho_f^{0.5}}{k_f\gamma^{0.5}}\right), \left(\frac{G\gamma^{0.5}}{\rho_f^{0.5}}\right), \left(\frac{\rho_f}{\rho_g}\right), \left(\frac{(\Delta i_{SUB})_i}{i_{fg}}\right)\right] \quad (9.30)$$

where

$$\gamma = -\frac{d(\rho_f/\rho_g)_{SAT}}{dp}$$

Using scaling factors obtained empirically (Stevens and Kirby 1964) from round tube critical heat flux data for water and R-12, Stevens and Wood (1966) were able directly to relate critical heat flux data for the two fluids taken in other more complex geometries including a rod cluster. One of the more successful approaches to the problem of scaling is that of Ahmad (1973). Using classical dimensional analysis he deduced that the critical heat flux (expressed in terms of the 'boiling number' ϕ_{CRIT}/Gi_{fg}) would depend on twelve dimensionless groups of which six were eliminated on inductive arguments. Three dimensionless groups are commonly matched in scaling of the critical heat flux; $(\Delta i_{SUB})_i/i_{fg}$, ρ_f/ρ_g, L/D (see eq. (9.30)). This left three dimensionless groups which were combined into a single modelling parameter ψ_{CRIT} as follows

$$\psi_{CRIT} = \left[\frac{GD}{\mu_f}\right]\left[\frac{\mu_f^2}{\sigma D\rho_f}\right]^{2/3}\left[\frac{\mu_f}{\mu_g}\right]^{-1/5} \quad (9.31a)$$

or alternatively as

$$\psi_{CRIT} = \left[\frac{GD}{\mu_f}\right]\left[\frac{\gamma^{0.5}\mu_f}{D\rho_f^{0.5}}\right]^{2/3}\left[\frac{\mu_f}{\mu_g}\right]^{1/8} \quad (9.31b)$$

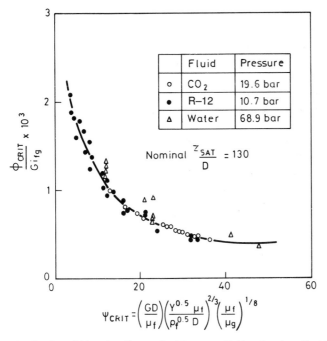

Fig. 9.6. Application of Ahmad scaling method for water, R-12 and carbon dioxide (Hauptmann *et al.* 1973).

Equation (9.31*a*) is the more basic version but eq. (9.31*b*) gives nearly identical results over wide ranges of pressure and temperature. The usefulness of this scaling law may be judged from Fig. 9.6 where data for water, R-12 and carbon dioxide are compared (Hauptmann *et al.* 1973).

A number of attempts have been made to produce empirical correlations applicable to a range of fluids. Only those generalized correlations of Katto and of Shah (see Chapter 8) can be recommended for general use. In particular Shah compares his function Y (see eq. (8.41)) with Ahmed's parameter ψ_{CHF}. Alternately the phenomenological method described in Section 9.7.2 can be used to generate critical vapour quality (x_{CRIT}) versus saturated boiling length (z_{SAT}) curves for desired values of tube diameter, mass velocity and system pressure. Such curves have been reproduced in handbooks (Collier 1983).

9.4.2 *Vertical thermosyphon evaporators*

There are very few published studies on the critical heat flux for natural circulation flows inside vertical bundles of tubes, the geometry typical of

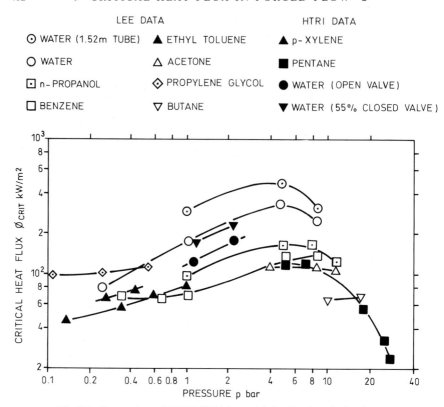

LEE DATA   HTRI DATA

⊙ WATER (1.52m TUBE)   ▲ ETHYL TOLUENE   ▲ p-XYLENE

○ WATER   △ ACETONE   ■ PENTANE

⊡ n-PROPANOL   ◇ PROPYLENE GLYCOL   ● WATER (OPEN VALVE)

□ BENZENE   ▽ BUTANE   ▼ WATER (55% CLOSED VALVE)

Fig. 9.7. Comparison of HTRI CHF data with Lee 7-tube reboiler data.

vertical thermosyphon reboilers widely used in the petrochemical industries. The design of these units must involve the calculation of the critical heat flux to avoid operation in the film boiling regime.

Lee *et al.* (1956) measured the critical heat flux for a model vertical thermosyphon reboiler with seven tubes. The study obtained data for water, *n*-propanol, benzene, acetone, propylene glycol and butane over a wide range of pressures. Shellene *et al.* (1968) later obtained more data for a 73-tube industrial reboiler, using a valve in the liquid feed line to prevent two-phase instabilities. Palen *et al.* (1974) completed a comprehensive study for a 43-tube vertical thermosyphon reboiler bundle. Figure 9.7 depicts their data compared to those of Lee. There appears to be a definite maximum in the critical heat flux at a pressure of about 5 bar for some of the fluids.

Palen *et al.* went on to identify three types of limitations to the heat flux: film boiling, mist flow and two-phase flow instabilities. The first limitation corresponds to the departure from nucleate boiling (DNB) in the bubbly flow regime, similar to that for pool boiling outside a single tube. A stable vapour film is established if the Leidenfrost point is exceeded. The mist flow limitation corresponds to the dryout of the liquid film in the annular flow regime, which occurs at relatively large exit vapour qualities. Two-phase flow instabilities instead result from choking of the flow and density wave oscillations; the latter is particularly important at low reduced pressure. Based on the data shown in Figure 9.7, they developed the following dimensional empirical expression for the critical heat flux in vertical thermo-syphon reboilers:

$$\phi_{CRIT} = 350\left(\frac{D_i^2}{L}\right)^{0.35}(p_{CRIT})^{0.61}\left(\frac{p}{p_{CRIT}}\right)^{0.25}\left(1 - \frac{p}{p_{CRIT}}\right) \qquad (9.32)$$

with D_i and L in meters and ϕ_{CRIT} in watts per square meter. The diameter D_i is the internal tube diameter and L is the length of the tube bundle. Nominal O.D. tube sizes covered by the correlation range from 19.05 mm ($\frac{3}{4}$ in) to 31.75 mm (1.25 in). Tube lengths varied from 1.52 to 3.66 m. The method predicted the data to within about -30 and $+50$ per cent. Additional data at high vacuum for mixtures and other tube diameters and tube lengths are still needed. Also, more study on the stabilizing effect of a control valve in the inlet piping is required.

9.5 Critical heat fluxes for the case where there is a liquid–vapour mixture at the channel inlet

A considerable number of experimental data for the critical heat flux exist in the published literature for the case where there is a liquid–vapour mixture at the channel inlet. Up until about 1965 it was common to consider these data together with data obtained by evaporation of subcooled liquids. However, it is now clear that a distinction must be made between these two types of experiments. Just as with the data for the evaporation of subcooled liquids, it is important to establish whether the conditions in the heated channel are hydrodynamically stable or not.

In general, the critical heat flux results obtained by mixing the two phases at the channel inlet under stable flow conditions are higher at a given exit quality, $x(z)$, than those obtained using a subcooled inlet. This type of behaviour has been observed by Styrikovich et al. (1960) who measured the critical heat flux in a uniformly heated tube for a wide range of conditions with both subcooled water and steam–water mixtures at the tube inlet. The

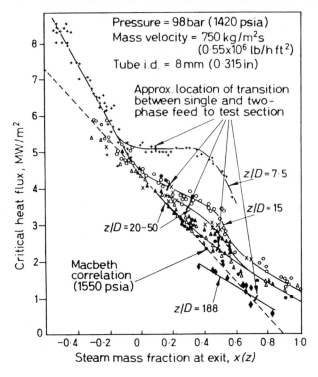

Fig. 9.8. Russian data for critical heat flux vs exit quality (Styrikovich *et al.* 1960).

type of result obtained for the *stable* system is illustrated in Fig. 9.8 and it will be seen that there are considerable increases in critical heat flux at short lengths when there is a steam–water mixture at the tube inlet. For these short lengths the lines appear to be continuous from the subcooled inlet case. A similar effect has been noted at AERE Harwell (Howell and Siegel 1966). It is obvious from these comparisons that a mere statement of the local thermodynamic quality is insufficient to define the critical condition and the 'local conditions' hypothesis breaks down completely.

It seems highly unlikely that a satisfactory correlation of data taken with vapour–liquid mixtures at the channel inlet will be possible since the results most certainly will also be a function of the method of injection of the phases (see Fig. 3.9). The possible exception to this conclusion is the case where the steam–water mixture at entry to the test section is generated in a low heat flux boiler/preheater as in the tests at CISE (Silvestri 1966). In this instance it may be possible to consider the test section as a 'hot-patch' on a very much longer heated assembly.

9.6 Critical heat flux under dynamic conditions

9.6.1 *Unstable hydrodynamic conditions*

It has been found that under certain circumstances the hydraulic characteristics of the experimental equipment in which the critical heat flux data are obtained can have a marked effect on the value of the critical heat flux. This was first emphasized by Styrikovich *et al.* (1960), who found it necessary to divide their results into two classes depending on whether they were taken under 'pulsating' or 'non-pulsating' conditions. The 'pulsating' region manifested itself in either stable or divergent oscillations of pressure and flow-rate. It is believed that these disturbances have their origin in the heated section but that they are amplified by the presence of upstream compressible volumes. The presence of these oscillations drastically reduces the critical heat flux. A detailed account of the phenomena associated with the critical heat flux under these unstable conditions is given in by Bertoletti (1965). Maulbetsch and Griffith (1966) have shown that it is possible to predict the initiation of oscillatory instability if the pump and circuit characteristics and the amount of compressibility upstream of the heated section are known. A critical slope of the overall pressure drop versus mass flow-rate curve can be found and the associated frequency of the instability computed. Agreement between the proposed analytical model and the available experimental data including that of Styrikovich *et al.* (1960) was very good.

9.6.2 *Transient conditions*

In many instances it is necessary to change the operating conditions of an oil-fired boiler or a nuclear reactor while the unit is at power. Consideration must also be given to possible fault conditions resulting from, for example, a loss of electrical supplies to the pumps or, in the case of a nuclear reactor, a loss-of-coolant accident (LOCA). In each case it is necessary to assess whether the critical heat flux is exceeded and if so at what time during the ensuing flow, pressure or power transient. It is common to assume that, during the transient, the critical heat flux will be exceeded when the 'local instantaneous conditions' are equivalent to those causing its occurrence under steady state conditions.

The literature on this subject particularly, in relation to a nuclear reactor loss-of-coolant accident (LOCA), is extensive and the reader interested in this particular subject should consult the excellent review of the available literature by Leung (1978). One study, by Moxon and Edwards (1967), involved tests on both uniformly heated and non-uniformly heated round tubes up to 3.66 m (12 ft) long and on a 37 rod cluster simulating a nuclear reactor fuel element. Flow transients were initiated by tripping the circulator, the flow being halved in less than 0.5 second. Instantaneous increases in power to the test section were brought about by switching from a dummy load. This resulted, in one particular test, in the power to the coolant rising to 1.3 times its initial value in 0.2 s.

Theoretical predictions of the time taken to exceed the critical heat flux (CHF) after the start of the transient were made. These predictions were based on the assumption that the critical heat flux during a transient occurs at the instant the local conditions are equivalent to those causing its occurrence under steady state conditions. The local hydraulic conditions at positions along the heated length were evaluated from the measured inlet flow and the non-steady equations of conservation of mass and energy. The local heat flux to the coolant was determined using the transient heat conduction equations. The results of the comparison between the predicted times and experimental times show that in all cases the critical heat flux occurs later than predicted. The delay between the predicted and experimental times varies between 0 and 0.4 s in Moxon and Edwards' tests.

It is therefore recommended that consideration of the likelihood of the CHF being exceeded during a flow or power transient be based on the 'local instantaneous conditions' approach outlined above.

A more refined approach to transients will only come from a consideration of the equations describing the mechanism of the critical heat flux. For example, Whalley *et al.* (1974*a*) have applied the annular flow model described in 9.7.2 to the Moxon and Edwards' tests with excellent results.

9.7   The mechanism of the critical heat flux condition

There is, at present, no complete understanding of the mechanisms leading to the critical heat flux condition despite the very large amount of research work that has been done. Relatively little is known about the mechanism of the critical heat flux condition corresponding to *departure from nucleate boiling* (DNB); considerably more is known about the mechanism of *dryout* in the higher quality regions.

9.7.1   *Subcooled and low quality regions*

Tong and Hewitt (1972) have identified at least three separate mechanisms for initiating the CHF condition (Fig. 9.9).

(a) *Dryout under a vapour clot.* As a result of evaporation of the micro-layer a dry patch tends to form on the heating surface under a growing vapour bubble. When the bubble departs from the surface this dry patch is rewetted. A stable situation results from an alternate heating and quenching

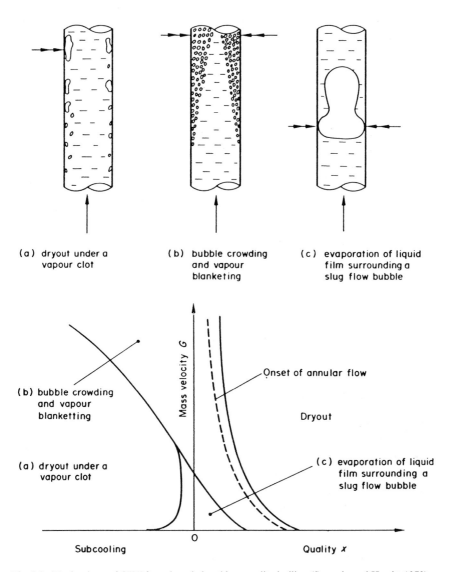

Fig. 9.9. Mechanisms of CHF for subcooled and low quality boiling (Semeria and Hewitt 1972).

of the surface at the dry spot. Fiori and Bergles (1970) and Kirby et al. (1967) found that if the heat flux is high the temperature rise of the dry patch is such that it cannot easily be rewetted following bubble departure and there is a progressive increase in the surface temperature leading to the CHF condition.

(b) *Bubble crowding and vapour blanketing.* At moderate subcoolings, a boundary layer of bubbles may grow to the point where it restricts the access of liquid to the heated surface. At some point, if access is seriously affected, then overheating occurs with the formation of a continuous vapour layer adjacent to the wall. Kutateladze and Leontev (1966) and later Tong (1968) suggest that the critical phenomenon is separation of the hydro-dynamic boundary layer. An analogy is suggested between the phenomenon of boundary layer separation induced by the injection of gas into a flowing liquid through a porous wall and the CHF condition in sub-cooled boiling. The flow stagnation which occurs in the separated boundary layer could result in vapour blanketing.

(c) *Evaporating of liquid surrounding a slug flow bubble.* At low mass velocities the slug flow pattern may occur with a liquid film initially remaining between the vapour bubble and the heated wall. If, however, the heat flux is high, this film may be completely evaporated and a form of 'dryout' with consequent overheating of the tube wall may occur.

A tentative map with mass velocity and subcooling as ordinates, showing where these various mechanisms might be expected to occur, has been published by Semeria and Hewitt (1972) (Fig. 9.9).

9.7.2   The critical heat flux condition in the saturated region

Considerably more is known about the mechanism of 'dryout' in the higher quality regions. By making measurements of the annular liquid film flow-rate at the exit of a heated tube as a function of the tube length and applied power, Hewitt et al. (1963, 1965), and Hewitt (1964) were able to build up a complete picture of the conditions occurring along the heated tube and were able to demonstrate that the CHF condition occurred when the flow-rate of the liquid film on the heated surface fell to zero; hence the name *dryout*. These experiments also revealed that the distribution of liquid phase between entrained droplets and the liquid film at a particular quality and mass velocity in a heated tube is significantly different from that observed at the same quality and mass velocity at the end of a long unheated tube, i.e., under *fully developed equilibrium conditions*.

The departures from hydrodynamic equilibrium in an evaporating flow are illustrated in Fig. 9.10 which shows the flow-rate of the liquid phase entrained as droplets versus the vapour quality. The *equilibrium* flow-rate of liquid entrained in the vapour core is plotted (as curve ABC). Also shown

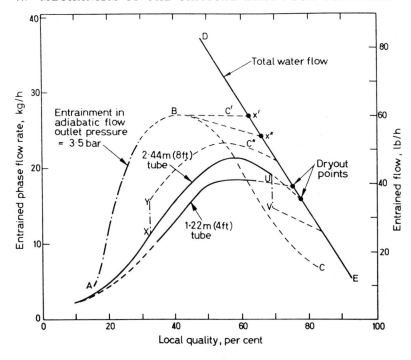

Fig. 9.10. Entrained phase flow-rate as a function of local quality (Hewitt *et al.*, 1965).

is the total liquid flow-rate in the tube (DE); the difference between curves ABC and DE is the flow-rate in the liquid film.

Now consider an evaporating flow in a heated tube. If the conditions at every axial location in the heated tube correspond to the *fully developed equilibrium state* at the corresponding quality in the unheated tube, then the amount of liquid entrained in the vapour core will increase up to an axial location corresponding to the maximum (point B). Deposition of droplets on to the film would take place in the downstream parts of the tube (region BC) and *dryout* would occur at approximately 100 per cent quality. Clearly this is not the true picture. Hewitt *et al.* (1963) initially postulated that entrainment was fairly rapid but that deposition was a rather slow process. In this instance the entrained liquid flow in the lower parts of the tube would follow curve AB closely until the maximum at B. Since deposition was regarded as a slow process, the entrained liquid flow-rate tended to remain constant (curve ABC′) or decrease only slowly (curve ABC″) with further increase in quality; *dryout* (i.e. liquid film flow-rate → 0) would now occur at x′ or x″. It was found in the later series of measurements (Hewitt *et al.*, 1965) that the measured entrained liquid flow-rates in the lower part of the tube were considerably lower than the equilibrium value. Clearly neither entrainment nor deposition occurs rapidly. The measured relationships for the tube length 2.44 m and 1.22 m are shown in Fig. 9.10. The observed approximately constant exit quality at dryout is the result of two effects—the longer tube having the higher entrained flow-rate but also the higher deposition rate.

Many of the differing effects observed with non-uniform heating and mentioned earlier can be qualitatively explained (Hewitt *et al.* 1966) using Fig. 9.10. For example, the introduction of a short unheated length into the heated section will produce different effects depending on whether it is placed in a location before or after that corresponding to point B in the tube. If it is placed before B then, although the steam quality remains constant across the cold patch, liquid entrainment will continue as the flow attempts to become 'fully developed'. Thus, the effect of the cold patch can be shown on Fig. 9.10 by the vertical line XY. Likewise, if it is placed after B then deposition occurs at constant quality and the effect is illustrated by the vertical line UV. Another example is the linear decreasing heat flux profile (Hewitt *et al.* 1966) which has a short length over which entrainment may take place and a long length for deposition to occur giving a higher steam quality at dryout than a uniform heat flux profile.

This qualitative description has been developed into a quantitative closed form theory using the methods described in Section 3.5 for the prediction of non-equilibrium annular flows. Figure 9.11 shows the mass balance on an elemental length of tube δz. A mass balance on the liquid film gives

$$W_{\text{Ff}} + \pi D(D_{\text{d}})\,\delta z = \underbrace{\left(W_{\text{Ff}} + \frac{\partial W_{\text{Ff}}}{\delta z}\,\delta z\right) + \pi DE\,\delta z + \pi D\,\frac{\phi}{i_{\text{fg}}}\,\delta z}_{\text{OUTFLOW}} + \underbrace{\frac{\partial M_{\text{Ff}}}{\partial t}\,\delta z}_{\text{ACCUMULATION}}$$

$$\underbrace{\hphantom{W_{\text{Ff}} + \pi D(D_{\text{d}})\,\delta z}}_{\text{INFLOW}}$$

$$(9.33)$$

where W_{Ff} is the liquid film flow-rate; D is the tube diameter; D_{d} is the deposition rate of droplets on to the film; E is the entrainment rate of droplets from the film; ϕ is the local heat flux; i_{fg} is the latent heat of vaporization of the liquid; and M_{Ff} is the mass of liquid contained in the film per unit length $(=\pi D\delta\rho_{\text{f}})$ where δ is the average film thickness and ρ_{f} is the liquid density. Hence

$$\frac{\partial W_{\text{Ff}}}{\partial z} = \pi D\left[D_{\text{d}} - E - \frac{\phi}{i_{\text{fg}}} - \rho_{\text{f}}\frac{\partial\delta}{\partial t}\right] \qquad (9.34)$$

Similarly, a mass balance over the liquid entrained in the vapour core gives

$$\frac{\partial W_{\text{Ef}}}{\partial z} = \pi D[E - D_{\text{d}}] - \frac{\pi D^2}{4}\frac{\partial C}{\partial t} \qquad (9.35)$$

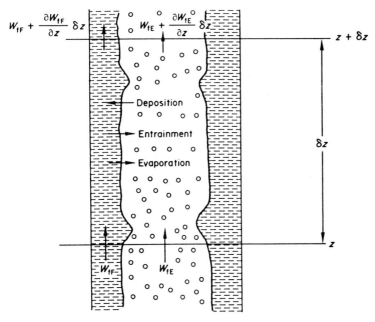

Fig. 9.11. Mass balance in annular flow on elemental length of channel (Whalley *et al.*, 1974a).

where C is the homogeneous concentration of droplets in the vapour core. Methods of calculating the entrainment and deposition rates and also the film thickness are given in Section 3.5.

Using these relationships, eq. (9.33) (with the time derivatives set equal to zero) can be integrated from the point of onset of annular flow (assumed arbitrarily to start at 1 per cent quality with 99 per cent of the liquid entrained at this point) and the integration is continued along the channel until W_{Ff} becomes zero, indicating *dryout*. Some examples (Whalley *et al.* 1974) of calculated critical heat flux values for uniformly heated tubes for steam–water mixtures at 1.5 bar, 69 bar and 180 bar pressure are given in Fig. 9.12 for comparison with experimental data. A further comparison with experimental data for R-12 is also shown.

More recently this model has been updated by Govan *et al.* (1988) and Hewitt and Govan (1989). Comparison with critical heat flux data for a wide range of fluids shows a mean error of -9.7 per cent with a standard deviation of 16 per cent for 5300 data points, provided the CHF mechanism is dryout, i.e. $G < G_{CRIT}$, where the latter is given as

$$G_{CRIT} = 1.42(z/D)^{0.85}(\sigma\rho_f/D)^{0.5} \qquad (9.36)$$

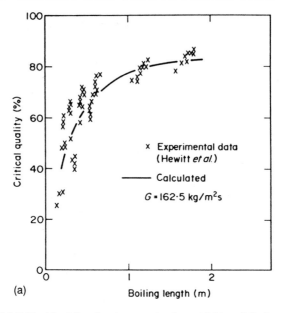

(a)

Fig. 9.12. (a) Critical heat flux for steam–water flow at 1.5 bar (Whalley *et al.* 1974*b*).

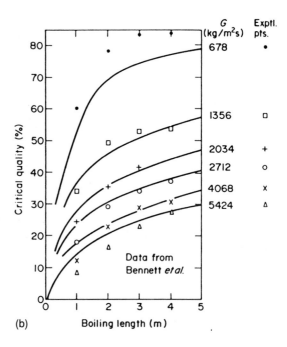

(b)

Fig. 9.12. (b) Critical heat flux for steam–water flow at 69 bar (Whalley *et al.* 1974*b*).

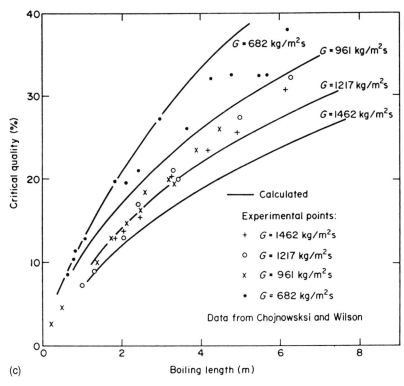

(c)

Fig. 9.12. (c) Critical heat flux for steam–water at 180 bar (Whalley *et al.* 1974*b*).

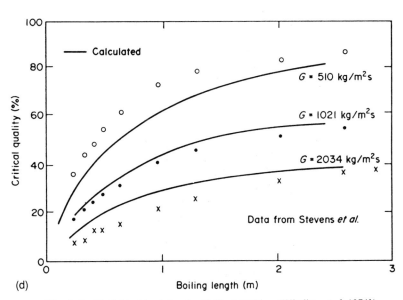

(d)

Fig. 9.12. (d) Critical heat flux for R-12 at 10.5 bar (Whalley *et al.* 1974*b*).

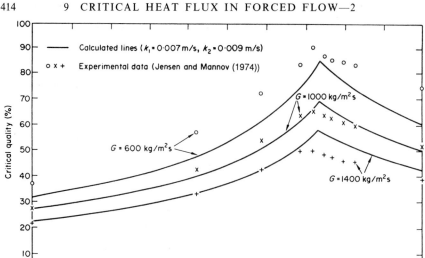

Fig. 9.13. Variation of critical quality with power distribution within annular channels (Whalley *et al.* 1974*a*).

The theoretical model has been extended to an annular geometry in which both the inside and outside surfaces may be heated (Whalley *et al.* 1974*a*) and to multi-rod clusters (Whalley 1976, 1978). Figure 9.13 shows a comparison of the predicted critical heat flux for comparison with the experimental critical heat flux value of Jensen and Mannov (1974) for an annulus with independent heating of the inner and outer surfaces. Also shown are predictions (Whalley 1976) for a seven-rod cluster cooled by R-12 at 10.7 bar (Kinneir *et al.* 1969) The geometry of the bundle was varied so that the gap between the rods was altered, the outer six rods always being at the vertices of a regular hexagon with the seventh rod at the centre of the hexagon. In one set of tests the outer rods touched the centre rod; in another they touched the outer containing tube—positions intermediate between these extremes were also studied. Experimental and predicted dryout powers as a function of the gap between the rods are shown in Fig. 9.14.

9.7.3   Critical heat flux in counter-current flow

It is well known that for the case of a vertical channel with a reservoir of fluid above the channel the critical heat flux does not approach zero as the flow to the bottom of the channel ceases. Liquid can run back from the reservoir into the channel to provide cooling. However, there will be some

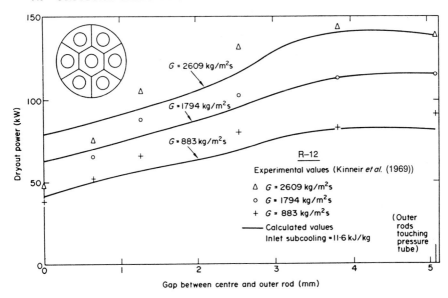

Fig. 9.14. Comparison between experimental and predicted value of CHF for annulus and multi-rod cluster (prediction using Hewitt–Whalley–Hutchinson (Whalley 1976) annular flow model).

limiting vapour flow-rate at the tube upper exit above which no liquid can run back. The heat flux corresponding to this condition can be thought of as the critical heat flux. Nejat (1978) has carried out experiments for a range of fluids (carbon-tetrachloride, n-hexane, and water) and tube diameters (8, 10, and 14 mm i.d.) and has produced the following correlation in terms of dimensionless groups:

$$\phi^*_{\text{CRIT}}(z/D)^{-0.1} = 1 - 1.5D^{*-1/2} \qquad (9.37)$$

where

$$\phi^*_{\text{CRIT}} = \frac{1}{3.2} \frac{\phi_{\text{CRIT}}}{\rho_g^{1/2} i_{fg}} [g\sigma(\rho_f - \rho_g)]^{-1/4} \qquad (9.38)$$

and

$$D^* = D \left[\frac{\sigma}{g(\rho_f - \rho_g)} \right]^{-1/2} \qquad (9.39)$$

9.8 Critical heat flux for boiling on horizontal tube bundles

Leroux and Jensen (1990) have recently surveyed the problem of critical heat flux on the outside of tube bundles with cross flow. This geometry is common to small 'kettle' type steam generators, flooded refrigerant evaporators and horizontal reboilers in petrochemical services. Kern (1950) as an initial guideline recommended a maximum bundle heat flux from 38 to 47 kW/m² based on conservative engineering practice. Palen and Small (1963) found this range to be too low, with many reboilers operating successfully at heat fluxes above 78 kW/m². Analysing a variety of industrial data for reboilers, they observed that the critical heat flux was a strong function of bundle geometry. Also, they noted that the critical heat flux is lower than that of a single tube for the same process conditions because of two physical limitations:

(a) At large heat fluxes and large bundle diameters, the quantity of vapour generated is so voluminous that the upper tubes in the bundle can become blanketed with vapour, rendering these tubes ineffective for boiling;

(b) For a poor tube bundle layout or geometric limitations (such as a small tube pitch), the liquid circulation can be restricted so as to prevent liquid from reaching and wetting the upper or center tube rows.

Palen and Small (1963) developed the following expression to relate the critical heat flux of a tube bundle to that of a single tube:

$$\phi_{\text{CRIT, B}} = \phi_{\text{CRIT}} F_{\text{tb}} \qquad (9.40)$$

where the bundle correction factor is

$$F_{\text{tb}} = 2.2\left(\frac{\pi D_B L}{A}\right) \qquad (9.41)$$

that ranges between 0.1 and 1.0. D_B is the outside diameter of the tube bundle, L is the length of the tube bundle and A is the total heat transfer surface area of the tube bundle. The ratio of these dimensions in eq. (9.41) is equivalent to the external envelope area of the tube bundle divided by its heat transfer surface area. For small bundles, F_{tb} tends towards a value of 2.2 rather than 1.0 for a single tube. Thus, it is recommended to use 1.0 as the maximum value for small bundles and they recommend 0.1 as the minimum value for large bundles. The single-tube critical heat flux ϕ_{CRIT} in eq. (9.40) is given by the Kutateladze (1948) correlation:

$$\phi_{\text{CRIT}} = 0.131 \rho_g^{1/2} i_{\text{fg}} [g(\rho_f - \rho_g)\sigma]^{1/4} \qquad (9.42)$$

Palen *et al.* (1972) later confirmed the validity of this approach based on laboratory tests on a kettle reboiler. A typical tube bundle boiling curve they obtained is shown in Fig. 7.28 for an unspecified hydrocarbon fluid. As can be seen, the critical heat flux for the bundle is significantly below that of the single-tube prediction.

Most other laboratory research studies have centred on measuring the

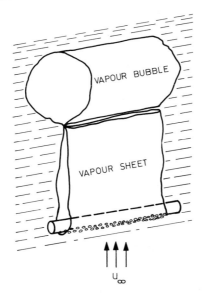

Fig. 9.15. Sheet-like vapour escape pattern in flow boiling.

critical heat flux for crossflow over a single tube or a single heated tube in a non-heated tube bundle. Vliet and Leppert (1964) noted that for a single tube in crossflow a sheet-like vapour formation was created behind the tube. McKee and Bell (1969) also observed the critical heat flux condition to initiate at the back of a heated tube in crossflow and Lienhard and Eichhorn (1976) observed the formation of a pulsating vapour sheet. Figure 9.15 from Kheyrandish and Lienhard (1985) depicts this vapour sheet schematically. They observed that during nucleate boiling on a tube in crossflow, the vapour bubbles leaving the tube formed a vapour sheet with travelling waves in the wake behind the tube. The critical heat flux condition was hypothesized to be reached when the wake became Helmholtz unstable, confining the vapour near the wall and causing dryout to occur.

Instead, Katto and Haramura (1983) concluded that the critical heat flux condition resulted from a Helmholtz instability occurring at the individual vapour jets leaving the tube wall, rather than in the vapour sheet trailing the tube. Thus, the phenomenon was thought to be similar to the Zuber (1959) model for critical heat flux on a flat plate (see Chapter 4). Inside a tube bundle, however, the vapour sheet-like wake cannot form because of the proximity of other downstream tubes. This fact was verified visually by Hasan et al. (1982).

For a single heated tube in a horizontal tube bundle, Leroux and Jensen (1991) studied the effect of mass velocity and vapour quality on the critical

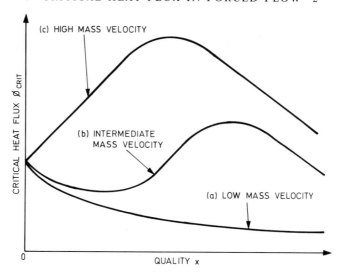

Fig. 9.16. CHF for a single heated tube within a tube bundle (Leroux and Jensen 1991).

heat flux. Figure 9.16 shows the effect of these parameters. At low mass velocities, the critical heat flux on the heated tube decreased with increasing vapour quality; at high mass velocities, a maximum in critical heat flux was found to occur; at intermediate mass fluxes the effect was more complex.

References and further reading

Ahmad, S. Y. (1973). 'Fluid to fluid modelling of critical heat flux: a compensated distortion model'. *Int. J. Heat Mass Transfer*, **16**, 641–661.

Aladyev, I. T., Gorlow, J. G., Dodonow, R. J., Sevastyanov, R. J., and Fedynsky, O. S. (1966). 'Heat transfer to boiling potassium in tubes'. Paper presented to 3rd International Heat Transfer Conference, Chicago, 1966, **III**, 93, 123–128.

Alekseev, G. V., Zenkevich, B. A., Peskov, O. L., Sergeev, N. D., and Subbotin, V. I. (1964). 'Burnout heat fluxes under forced flow'. Paper A/CONF.28/P/327a presented by USSR at the Third International Conference on Peaceful Uses of Atomic Energy, Geneva, August.

Barnett, P. G. (1963). 'The scaling of forced convection boiling heat transfer'. AEEW-R 134.

Barnett, P. G. (1964a). 'The prediction of burnout in non-uniformly heated rod clusters from burnout data for uniformly heated round tubes'. AEEW-R 362.

Barnett, P. G. (1964b). 'An experimental investigation to determine the scaling laws of forced convection boiling heat transfer Pt. I'. AEEW-R 363.

Barnett, P. G. (1966). 'A correlation of burnout data for uniformly heated annuli and its use for predicting burnout in uniformly heated rod bundles'. AEEW-R 463.

Barnett, P. G. (1968). 'A comparison of the accuracy of some correlations for burnout in annuli and rod bundles'. AEEW-R 558.

Barnett, P. G. and Wood, R. W. (1965). 'An experimental investigation to determine the scaling laws of forced convection boiling heat transfer. Part 2'. AEEW-R 443.

Becker, K. M. and Hernborg, G. (1963). 'Measurements of burnout conditions for flow of boiling water in an annulus'. Paper 63-HT-25 presented at 6th National ASME–AIChE Heat Transfer Conference, Boston, August.

Bennett, A. W., Collier, J. G., and Kearsey, H. A. (1964). 'Heat transfer to mixtures of high pressure steam and water in an annulus. Pt. IV'. AERE-R 3961.

Bertoletti, S., Gaspari, G. P., Lombardi, G., Peterlongo, G., Silvestri, M., and Tacconi, E. A. (1965). 'Heat transfer crisis with steam–water mixtures'. *Energia Nucleare*, **12**(3), 121–172.

Biancone, F., Campanile, A., Galimi, G., and Goffi, M. (1965). 'Forced convection burnout and hydrodynamic instability experiments for water at high pressure. Pt. I'. EUR 2490e.

Birdseye, D. E. (1960). 'Experimental investigation of heat transfer characteristics of liquid nitrogen tetroxide'. JPL Tech. Report No. 32–37, October.

Bowring, R. W. (1977). 'A new mixed flow cluster dryout correlation for pressures in the range 0.6–15.5 MN/m^2 (90–2250 psia)—for use in a transient blowdown code'. Paper C217/77 presented at Conference on Heat and Fluid Flow in Water Reactor Safety, IMechE, Manchester, 13–15 September.

Bowring, R. W. (1979). 'WSC-2 a subchannel dryout correlation for water-cooled clusters over the pressure range 3.4–15.9 MPa (500–2300 psia)'. AEEW-R983, UKAEA Winfrith.

Butterworth, D. (1972). 'A model for predicting dryout in a tube with circumferential variation on heat flux'. UKAEA Report No. AERE M2436.

Casterline, J. E. (1964). 'Burnout in long vertical tubes with uniform and cosine heating using water at 1000 psia'. Topical Rep. No. 1. Task XVI. Columbia University (TID 21031). Also 'Second experimental study of dryout in a long vertical tube with a cosine heat flux'. Topical Rep. No. 5.

Collier, J. G. (1983). 'Boiling and evaporation'. In *Heat exchanger design handbook*, Vol. 2, Chapter 2.7.3, pp. 28–29, Hemisphere Pub. Corp.

Collier, J. G., Lacey, P. M. C., and Pulling, D. J. (1964). 'Heat transfer to two-phase gas–liquid systems—Pt. III'. AERE-R 3960.

Core, T. C. and Sato, K. (1958). 'Determination of burnout limits of polyphenyl coolants'. IDO-28007.

De Bortoli, R. A., Green, S. J., Letourneau, B. W., Troy, M., and Weiss, A. (1958). 'Forced convection heat transfer burnout studies for water in rectangular channels and round tubes at pressures above 500 p.s.i.a.'. WAPD-188.

Dimmock, T. H. (1957). 'Heat transfer properties of anhydrous ammonia'. *Reaction Motors*, No. RMI-124-S1, June.

Mm. Finat and Saunier (1963). 'Determination of critical heat fluxes at different pressures between 80 and 140 atmospheres in a rectangular channel heated either uniformly or non-uniformly'. EAEA Symposium on Two-phase Flow, Studsvik, Sweden, 1–3 October.

Fiori, M. P. and Bergles, A. E. (1970). 'Model of critical heat flux in subcooled flow boiling'. In *Heat Transfer*, **7**, Paper B6.3 Elsevier Pub. Co., Amsterdam.

Gambill, W. R. and Bundy, R. D. (1961). 'HFIR heat transfer studies of turbulent water flow in thin regular channels'. ORNL 3079.

Gambill, W. R. and Bundy, R. D. (1962). 'High flux heat transfer characteristics of pure ethylene glycol'. Paper no. 2, AIChE National Meeting, Los Angeles, Calif., February.

Govan, A. H., Hewitt, G. F., Owen, D. G., and Bott, T. R. (1988). 'An improved CHF modelling code'. 2nd UK National Conference on Heat Transfer, 1, 33–48, Inst. of Mechanical Engineers, 14–16 September.

Hasan, M. M., Eichhorn, R., and Lienhard, J. H. (1982). 'Burnout during flow across a small cylinder influenced by parallel cylinders'. Presented at 7th Int. Heat Transfer Conf., 4, 285–290.

Hauptmann, E. G., Lee, V., and McAdam, D. (1973). 'Two-phase fluid modelling of the critical heat flux'. Proceedings of Institute Meeting, Reactor Heat Transfer, Karlsruhe, 557–576, 9–11 October.

Hewitt, G. F. (1964). 'A method of representing burnout data in two-phase heat transfer for uniformly heated round tubes'. AERE-R 4613.

Hewitt, G. F. and Govan, A. H. (1989). 'Phenomenological Modelling of Non-Equilibrium Flows with Phase Change'. Paper presented at Eurotherm Seminar No. 7, Rome.

Hewitt, G. F., Kearsey, H. A., Lacey, P. M. C., and Pulling, D. J. (1963). 'Burnout and nucleation in climbing film flow'. AERE-R 4374.

Hewitt, G. F., Kearsey, H. A., Lacey, P. M. C., and Pulling, D. J. (1965). 'Burnout and film flow in the evaporation of water in tubes'. AERE-R 4864.

Hewitt, G. F., Bennett, A. W., Kearsey, H. A., Keeys, R. K. F., and Pulling, D. J. (1966). 'Studies of burnout in boiling heat transfer to water in round tubes with non-uniform heating'. AERE-R 5076.

Hines, W. S. (1959). 'Forced convection and peak nucleate boiling heat transfer characteristics for hydrazine flowing turbulently in a round tube at pressures to 1000 psia'. Rocketdyne Report—R.2059, August.

Hoffman, H. W. and Krakoviak, A. I. (1964). 'Convective boiling with liquid potassium'. Paper presented at 1965 Heat Transfer and Fluid Mechanics Institute, Stanford University.

Janssen, E., Levy, S., and Kervinen, J. A. (1963). 'Investigations of burnout in an internally heated annulus cooled by water at 600 to 1400 p.s.i.a.'. Paper 63-WA-149 presented at ASME Winter Annual Meeting, Philadelphia, November. See also GEAP 3899 (1963).

Jensen, A. and Mannov, G. (1974). 'Measurements of burnout, film flow, film thickness and pressure drop in a concentric annulus 3500 × 26 × 17 mm with heated rod and tube'. Paper A.5 presented at European Two-phase Flow Group Meeting, Harwell, June.

Kafengauz, N. L. and Bauarov, I. D. (1959). 'The effect of the height of a flat aperture on the transformation of heat into water'. Teploenergetika (3), 76–78. Quoted in Gambill and Bundy (1961).

Katto, Y. (1981). 'Generalized features of CHF of forced convection boiling in uniformly heated rectangular channels'. Int. J. Heat Mass Transfer, 24, 1413–1419.

Katto, Y. and Haramura, Y. (1983). 'Critical heat flux on a uniformly heated horizontal cylinder in an upward cross flow of saturated liquid'. Int. J. Heat Mass Transfer, 26, 1199–1205.

Kern, D. Q. (1950). Process heat transfer, McGraw-Hill, New York.

Kheyrandish, K. and Lienhard, J. H. (1985). 'Mechanisms of burnout in saturated and subcooled flow boiling over a horizontal cylinder'. National Heat Transfer Conference, ASME Paper 85-F-41.

Kinneir, J. H., Heron, R. A., Stevens, G. F., and Wood, R. W. (1969). 'Burnout power and pressure drop measurements on 12 ft. 7 rod clusters cooled by Freon 12 at 155 psia'. European Two-phase Flow Group Meeting, Karlsruhe.

Kirby, G. J. (1966). 'A new correlation of non-uniformly heated round tube burnout data'. AEEW-R 500.

Kirby, G. J., Staniforth, R., and Kinneir, J. H. (1967). 'A visual study of forced convection boiling. Pt. I'. AEEW-R 281 (1965) and Pt. II. AEEW-R 506.

Kitto, J. B., Jr. and Albrecht (1987). 'Elements of two-phase flow in fossil boilers'. Paper Presented at the NATO Advanced Study Institute on Thermal-Hydraulic Fundamentals and Design of Two-Phase Flow Heat Exchangers, Povoa di Varzim, Portugal, July 6–16.

Kitto, J. B., Jr. and Weiner, M. (1982). 'Effects of nouniform circumferential heating and inclination on critical heat flux in smooth and ribbed bore tubes'. Paper Presented at 7th Int. Heat Transfer Conf., Munich, 4, 297–302.

Komendantov, A. S., Kuzma-Kichta, Y. A., Vasil'eva, L. T., and Ovodkov, A. A. (1991). 'Experimental investigation of heat transfer and burnout in condition of nonuniform megawatt heat fluxes'. Experimental Heat Transfer, 4, 281–288.

Kutateladze, S. S. (1948). 'On the transition to film boiling under natural convection'. Kotloturbostroenie, 3, 10.

Kutateladze, S. S. and Leontev, A. I. (1966). 'Some applications of the asymptotic theory of the turbulent boundary layer'. Proceedings of 3rd International Heat Transfer Conference, Chicago, III.

Kuzma-Kichta, Y. A., Komandantov, A. S., and Vasil'eva, L. Y. (1988). 'Experimental investigation of enhanced heat transfer with nonuniform megawatt heat flux'. Paper Presented at 1st Experimental Heat Transfer, Fluid Mechanics and Thermodynamics Conf., Dubrovnik, 252–257, September.

Lee, D. H. (1965). 'An experimental investigation of forced convection burnout in high pressure water—Pt. III'. AEEW-R 355.

Lee, D. H. (1966). 'An experimental investigation of forced convection burnout in high pressure water—Pt. IV'. AEEW-R 479.

Lee, D. H. (1966). 'Burnout in a channel with non-uniform circumferential heat flux'. AEEW-R 477.

Lee, D. H. and Obertelli, J. D. (1963). 'An experimental investigation of forced convection burnout in high pressure water—Pt. II'. AEEW-R 309.

Lee, D. C., Dorsey, J. W., Moore, G. Z., and Mayfield, F. D. (1956). CEP, 52(4), 160.

Leroux, K. M. and Jensen, M. K. (1990). 'Critical heat flux in shell side boiling on horizontal tube bundles in vertical crossflow'. Report HTL-7, Rensselaer Polytechnic Institute, Troy, New York.

Leroux, K. M. and Jensen, M. K. (1991). 'Critical heat flux in horizontal tube bundles in vertical crossflow of R113'. Phase Change Heat Transfer, ASME HTD, 159.

Leung, J. C. M. (1978). 'Critical heat flux under transient conditions: a literature survey'. NUREG/CR-0056 Argonne Nat. Lab. Rept. ANL-78-39, June.

Levy, S., Fuller, R. A., and Niemi, R. O. (1959). 'Heat transfer to water in thin rectangular channels'. J. Heat Transfer, 81(2), 129.

Lewis, J. P., Goodykoontz, J. H., and Kline, J. F. (1962). 'Boiling heat transfer to liquid hydrogen and nitrogen in forced flow'. NASA-TN-D-1314.

Lienhard, J. H. and Eichhorn, R. (1976). 'Peak boiling heat flux on cylinders in a cross flow'. *Int. J. Heat Mass Transfer*, **19**, 1135–1141.

Lurie, H. (1964). Atomics International Progress Reports NAA-SR-10501 and NAA-SR-10850. See also Paper 160 presented at 1966 International Heat Transfer Conference, Chicago.

Macbeth, R. V. (1963). 'Burnout analysis. Part 4. Application of a local condition hypothesis to world data for uniformly heated round tubes and rectangular channels'. AEEW-R267.

Macbeth, R. V. (1964). 'Burnout analysis—Part 5. Examination of published world data for rod bundles'. AEEW-R 358.

McKee, H. R. and Bell, K. J. (1969). 'Forced convection boiling from a cylinder normal to the flow'. *AIChE Symp. Ser.*, **92**(65), 222–230.

Maulbetsch, J. S. and Griffith, P. (1966). 'A study of system-induced instabilities in forced convection flows with subcooled boiling'. MIT TR 5382-35 (1965); see also paper to 3rd International Heat Transfer Conference, Chicago, **III**, **4**, 247.

Moxon, D. and Edwards, P. A. (1967). 'Dryout during flow and power transients'. AEEW-R 553.

Nejat, Z. (1978). 'Maximum heat flux for counter-current two-phase flow in a closed end vertical tube'. Paper FB-29 presented at 6th International Heat Transfer Conference, Toronto, August, **1**, 441–444.

Noel, M. B. (1961*a*). 'Experimental investigation of heat-transfer characteristics of hydrazine and a mixture of 90% hydrazine and 10% ethylenediamine'. JPL Tech. Report No. 32-109, June.

Noel, M. B. (1961*b*). 'Experimental investigation of the forced convection and nucleate boiling heat transfer characteristics of liquid ammonia'. JPL Tech. Report No. 32-125, July.

Palen, J. W. and Small, W. M. (1963). 'A new way to design kettle and internal reboilers'. *Hydrocarbon Processing*, **43**(11), 199–208.

Palen, J. W., Yarden, A., and Taborek, J. (1972). 'Characteristics of boiling outside large scale horizontal multitube bundles'. *AIChE Symp. Ser.*, **69**(118), 50–60.

Palen, J. W., Shih, C. C., Yarden, A., and Taborek, J. (1974). 'Performance limitations in a large scale thermosyphon reboiler'. Paper presented at 5th Int. Heat Transfer Conference, Tokyo, **5**, 204–208.

Povarnin, P. I. (1964). 'Critical boiling point in the flow of 96% ethyl alcohol under low heating conditions'. *Teploenergetika*, **9**(12), 57–63 (1962). Report ORNL-tr-108 (1964).

Richards, R. J., Steward, W. G., and Jacobs, R. B. (1961). 'A survey of the literature on heat transfer from solid surfaces to cryogenic fluids'. National Bureau of Standards. Tech. Note NBS-TN-122.

Robinson, J. M. and Lurie, H. (1962). 'Critical heat flux for some polyphenyl coolants'. AIChE preprint 156 presented at 55th Annual Meeting, Chicago, December.

Seader, J. D., Miller, W. S., and Kalvinskas, L. A. (1965). 'Boiling heat transfer for cryogenics'. NASA-CR-243, June.

Semeria, R. and Hewitt, G. F. (1972). 'Aspects of heat transfer in two-phase one-component flows'. Seminar Recent Developments in Heat Exchangers, International Centre for Heat and Mass Transfer, Trogir, Yugoslavia, Lecture Session H53, September.

Shah, M. M. (1980). 'A general correlation for critical heat flux in annuli'. *Int. J. Heat Mass Transfer*, **23**, 225–234.

Shellene, K. R., Sternling, C. V., Church, D. M., and Snyder, N. H. (1968). 'Experimental study of a vertical thermosyphon reboiler'. *CEP Symp. Ser.*, **64**, 102.

Silvestri, M. (1966). 'On the burnout equation and on location of burnout points'. *Energia Nucleare*, **13**(9), 469–479.

Smith, O. G., Tong, L. S., and Rohner, W. M. (1965). 'Burnout in steam–water flows with axially non-uniform heat flux'. ASME paper no. 65-WA/HT-33.

Stevens, G. F. and Kirby, G. J. (1964). 'A quantitative comparison between burnout data for water at 1000 lb/in² and Freon 12 at 155 lb/in² (abs) in uniformly heated round tubes in vertical upflow'. AEEW-R 327.

Stevens, G. F. and Wood, R. W. (1966). 'A comparison between burnout data for 19 rod-cluster test sections cooled by Freon 12 at 155 lb/in² (abs) and by water at 1000 lb/in² in vertical upflow'. AEEW-R 468.

Stevens, G. F., Elliott, D. F., and Wood, R. W. (1964). 'An experimental investigation into forced convection burnout in Freon, with reference to burnout in water'. AEEW-R 321.

Stevens, G. F., Elliott, D. F., and Wood, R. W. (1965). 'An experimental comparison between forced convection burnout in Freon 12 flowing vertically upwards through uniformly and non-uniformly heated round tubes'. AEEW-R 426.

Styrikovich, M. A. and Mostinski, I. L. (1959). 'Effect of non-uniform heat distribution on steam generating tubes'. *Doklady Akad. Nauk,. SSSR*, **127**(2), 316.

Styrikovich, M. A., Miropolsky, Z. L., Shitsman, M. Ye., Mostinski, I. L., Stavrovski, A. A., and Factorovich, L. E. (1960). 'The effect of prefixed units on the occurrence of critical boiling in steam generating tubes'. *Teploenergetika*, **6**, 81.

Swenson, H. S., Carver, J. R., and Kakarala, C. R. (1962). 'The influence of axial heat flux distribution on the departure from nucleate boiling in a water-cooled tube'. Paper 62-WA-297 presented to ASME Winter Annual Meeting, New York, November.

Swenson, H. S. (1964). 'Non-uniform heat generation experimental programme'. QPR Nos. 4, 5, 6, 7, and 8 (BAW-3238).

Tippets, F. E. (1962). 'Critical heat fluxes and flow patterns in high pressure boiling water flows'. Paper 62-WA-162 presented at ASME Winter Annual Meeting, New York, November.

Tippets, F. (1963). 'Alkali metals boiling and condensing investigations'. QPRs 2-12, Contract NAS 3-2528, General Electric Co., April.

Todreas, N. E. and Rohsenow, W. M. (1966). 'The effect of non-uniform axial heat flux distribution'. MIT-9843-37 (1965); see also paper 89 presented at 1966 International Heat Transfer Conference, Chicago.

Tong, L. S. (1966). 'Prediction of departure from nucleate boiling for an axially non-uniform heat flux distribution', WCAP-5584 Rev. 1; see also *J. Nuc. Energy*, **21**(3), 241–248 (1967).

Tong, L. S. (1968). 'Boundary layer analysis of the flow boiling crisis'. *Int. J. Heat Mass Transfer*, **11**, 1208–1211.

Tong, L. S. and Hewitt, G. F. (1972). 'Overall viewpoint of flow boiling CHF mechanisms'. ASME Paper 72-HT-54.

Tong, L. S., Currin, H. B., Larsen, P. S., and Smith, O. G. (1965). 'Influence of axially non-uniform heat flux on DNB', in 'Heat Transfer—Los Angeles', *Chem. Engng. Prog. Symp. Series* No. 64, **62**, 35–40 (1966); also WCAP-2767 (1965).

Troy, M. (1958). 'Upflow burnout data for water at 2000, 1200, 800 and 600 psia in vertical 0.070 in. × 2.25 in. × 72 in. long stainless steel rectangular channels'. WAPD-TH-408.

Vliet, G. C. and Leppert, G. (1964). 'Critical heat flux for nearly saturated water flowing normal to a cylinder'. *J. Heat Transfer*, **86**, 59–67.

Weisman, J. and Bowring, R. W. (1975). 'Methods for detailed thermal and hydraulic analysis of water-cooled reactors'. *Nuc. Sci. & Engng.*, **57**, 255–276.

Whalley, P. B. (1976). 'The calculation of dryout in a rod bundle'. AERE-R8319.

Whalley, P. B. (1978). 'The calculation of dryout in a rod bundle—a comparison of experimental and calculated results'. AERE-R 8977.

Whalley, P. B., Hutchinson, P., and Hewitt, G. F. (1974). 'Prediction of annular flow parameters for transient and for complex geometries'. European Two-phase Flow Group Meeting, Haifa, Israel (1975); see also AERE-M 2661 (1974).

Whalley, P. B., Hutchinson, P., and Hewitt, G. F. (1974). 'The calculation of critical heat flux in forced convection boiling'. Paper B6.11, 5th International Heat Transfer Conference, Tokyo, Japan.

Zenkevich, B. A. *et al.* (1962). 'The effect of channel geometry on the critical heat loads in forced flow of water'. *Teploenergetika*, **9**(7), 81–83. AEC-tr-5538.

Zuber, N. (1959). 'Hydrodynamic aspects of boiling heat transfer'. AEC Report AECU-4439, Physics and Mathematics.

Examples

Example 1
A vertical tubular test section is to be installed in an experimental high-pressure water loop. The tube is 10.16 mm i.d. and 3.66 m long heated uniformly over its length. The water enters the test section at 204°C and 68.9 bar. Calculate the minimum flow-rates to prevent occurrence of the critical heat flux condition with powers of 100 kW and 200 kW applied to the channel.

Solution
The calculation will be carried out in SI units using the Bowring correlation. The given conditions are

Pressure (p)	68.9 bar
Tube bore (D)	10.16 mm
Tube length (z)	3.66 m
Power applied	100 kW (or 200 kW)
Inlet temperature (T_i)	204°C

It is necessary to produce a graph of power at the critical condition against flow-rate

$$\phi_{\text{CRIT}} = \frac{A' + DG(\Delta i_{\text{SUB}})_i/4}{C' + z} \tag{8.22}$$

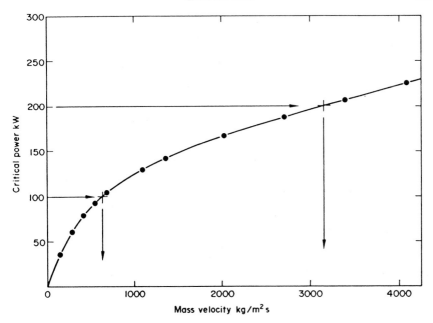

Fig. 9.17.

Table 9.7.

Mass velocity kg/m² s	$A' \times 10^{-6}$	C'	Critical heat flux (ϕ_{CRIT}) MW/m²	Critical power kW
0	—	—	0	0
136	1.010	0.1052	0.3039	35.5
272	1.736	0.2064	0.5185	60.6
407	2.280	0.3012	0.6770	79.1
542	2.703	0.3898	0.7997	93.4
678	3.047	0.4724	0.8994	105.1
1085	3.760	0.6799	1.1134	130.0
1356	4.078	0.7875	1.2182	142.3
2034	4.597	0.9717	1.4264	166.6
2712	4.909	1.0707	1.6041	187.4
3390	5.118	1.1182	1.7721	207.0
4068	5.267	1.1353	1.9365	226.2

$$A' = 2.317 \left[\frac{DGi_{fg}}{4} \right] F_1 / [1.0 + 0.0143 F_2 D^{1/2} G] \Bigg\}$$

(8.24)

$$C' = 0.077 F_3 DG / [1.0 + 0.347 F_4 (G/1356)^n] \Bigg)$$

$$n = 2.0 - 0.00725p$$

(8.25)

(for $p = 68.9$, $n = 1.5$)

$$i_{fg} = 1.51 \times 10^6 \text{ J/kg}$$

$$(\Delta i_{SUB})_i = 0.389 \times 10^6 \text{ J/kg}$$

$$F_1 = F_2 = F_3 = F_4 = 1.0$$

Figure 9.17 shows critical power versus flow-rate. It will be seen that the minimum flow-rate at which 100 kW can be sustained on the test section without exceeding the critical heat flux is 610 kg/m² s. For 200 kW the figure is 3150 kg/m² s.

Example 2
A nuclear reactor designer wishes to optimize the shape of a reactor core to give minimum capital cost. The core is to produce 120 MW(t) and may be up to 20 ft in length. The fuel element is in the form of a 19-rod cluster, each rod 0.612 in o.d. located in a flow channel 3.25 in i.d. The coolant is boiling water at 1000 psia entering the fuel channel with subcooling of 8°F and a mass velocity of 2×10^6 lb/h ft². Each fuel channel costs £1000/ft and the end fittings, instrumentation, etc. for each channel cost £10,000. On the basis that the critical heat flux is the only limitation, what is the optimum length of the core?

Solution
The critical channel power can be estimated as a function of length using Macbeth's rod cluster correlation,

$$(\phi_{CRIT} \times 10^{-6}) = \frac{A + B(\Delta i_{SUB})_i}{C + z}$$

(9.24)

where

$$A = 67.6 D_h^{0.83} (G \times 10^{-6})^{0.57} \Bigg\}$$

$$B = 0.25 D_h (G \times 10^{-6})$$

(9.29)

$$C = 47.3 D_h^{0.57} (G \times 10^{-6})^{0.27} \Bigg)$$

$$G = 2 \times 10^6 \text{ lb/h ft}^2 \quad T_i = 544.6 - 8 = 536.6°F \quad (\Delta i_{SUB})_i = 11.6 \text{ Btu/lb}$$

$$D_h = \frac{4A}{P_h} = 4\pi [3.25^2 - (19 \times 0.612^2)]/19 \times \pi \times 0.612 \times 4$$

$$= 0.2964 \text{ in.}$$

Hence

$$(\phi_{CRIT} \times 10^{-6}) = \frac{67.6 \times 0.3645 \times 1.485 + 0.25 \times 0.2964 \times 2 \times 11.6}{47.3 \times 0.500 \times 1.206 + z}$$

$$= \frac{36.591 + 1.719}{28.522 + z} = \frac{38.310}{28.522 + z} \text{ (Btu/h ft}^2\text{)}.$$

z is the channel length in inches.

Table 9.8.

Channel length ft	Critical heat flux $(\phi_{CRIT} \times 10^{-6})$ Btu/h ft^2	Critical channel power kW
0	1.343	0
2	0.7294	1301
4	0.5006	1786
6	0.3811	2039*
8	0.3076	2195
10	0.2579	2300
12	0.2220	2376
14	0.1949	2436
16	0.1737	2478
18	0.1567	2516
20	0.1427	2545

* Hesson *et al.* (BNWL-206) tested a rod cluster of this geometry with a heated length of 6.3 ft and obtained a critical channel power for the conditions assumed of 1910 kW

Table 9.9.

Channel length ft	No. of channels for 120 MW	Cost of channels £K	Cost of fittings, instrumentation, etc. £K	Total capital cost £K
2	93	186	930	1116
4	68	272	680	952
6	59	354	590	944
8	55	440	550	990
10	53	530	530	1060
12	51	612	510	1122
14	50	700	500	1200
16	49	784	490	1274
18	48	864	480	1344
20	48	960	480	1440

Assuming that the non-uniform heat flux profile likely in the reactor core does not affect the critical channel power ('overall power' hypothesis) the capital costs can be readily calculated as laid out in Table 9.9.

The optimum channel length for the core on this basis is then 6 ft, the core having 59 channels. This example illustrates the important fact that the critical power that can be extracted from a heated channel approaches a constant value at long lengths rather than increasing monotonically.

Problems

1. A vertical tubular test section 2.83 cm i.d. and 120 cm long is installed in a high-pressure water rig. The axial heat flux profile is non-uniform as shown in the adjacent figure. A critical heat flux test is carried out at a pressure of 110 bar, an inlet subcooling of 0.23 MJ/kg and a mass velocity of 344 kg/m^2 s. Use the 'F factor method' to predict the average critical heat flux and the position of first overheating.

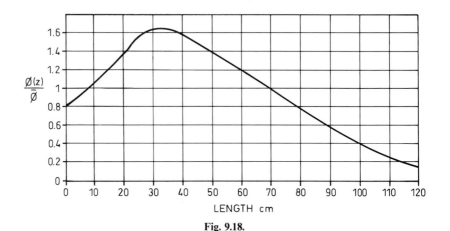

Fig. 9.18.

2. A waste-heat boiler consists of 78 vertical tubes 44.5 mm o.d., 35.6 mm i.d. each 10 m long. Saturated water taken from a steam drum is fed to the base of the tubes at a pressure of 132.5 bar ($T_{SAT} = 332$°C). The water mass velocity in the tubes is 912 kg/m^2 s. The tubes are heated by a gas stream on the shell side passing co-current with the evaporating water. The local gas temperature and overall heat transfer coefficient along the tube are as given in Table 9.10.

Table 9.10.

	Inlet $z = 0$	$z = 4$ m	Outlet $z = 10$ m
Local overall value of heat transfer coefficient U (kW/m^2 °C) (referred to gas side area)	0.96	0.875	0.85
Local gas temperature °C	975	571	390

Decide whether a critical heat flux (dryout) condition will occur on the water side for these conditions. Use the 'overall power' hypothesis.

3. As an exercise prepare a data bank of all available published information on the critical heat flux for liquid nitrogen in round vertical tubes. Attempt a correlation of this data using the critical vapour quality (x_{CRIT}) versus saturated boiling length (z_{SAT}) plot.
4. Estimate the critical heat flux in a vertical thermosyphon reboiler with 3.66 m long tubes with a 19.86 mm internal diameter. The fluid being boiled is a mixture of 50% benzene and 50 per cent toluene at 5 bar.
5. Determine the critical heat flux for a small steam generator constructed with 450 25.4 mm diameter tubes that are 2.0 m long. The tube bundle diameter is 750 mm and the operating pressure is 19 bar.

10

CONDENSATION

10.1 Introduction

Condensation is defined as the removal of heat from a system in such a manner that vapour is converted into liquid. This may happen when vapour is cooled sufficiently below the saturation temperature to induce the nucleation of droplets. Such nucleation may occur homogeneously within the vapour or heterogeneously on entrained particulate matter, for example, within the low-pressure stages of a large steam turbine (Gardner 1963–64). Heterogeneous nucleation may also occur on the walls of the system, particularly if these are cooled as in the case of a surface condenser. In this latter case there are two forms of heterogeneous condensation, drop-wise and film-wise, corresponding to the analogous cases in evaporation, of nucleate boiling and film boiling. Film-wise condensation occurs on a cooled surface which is easily wetted. On non-wetted surfaces the vapour condenses in drops which grow by further condensation and coalescence and then roll over the surface. New drops then form to take their place. This chapter will mainly deal with film-wise condensation since reliable theories of drop-wise condensation have not yet been established.

For other reviews of condensation heat transfer, the reader is referred to Marto (1984), Webb (1984a,b) and Butterworth *et al.* (1983).

10.2 The basic processes of condensation

10.2.1 *Liquid formation*

By analogy with the case of evaporation, liquid may form in one of three ways corresponding to the departure from a stable, metastable, or unstable equilibrium state. The formation of liquid at a planar interface occurs when the vapour temperature is decreased fractionally below the corresponding saturation temperature. This process is treated in Section 10.3 below. The relationships governing the formation of liquid from a metastable vapour or an unstable equilibrium state start from the equation defining the mechanical equilibrium of a spherical liquid drop.

Droplet nucleation. Consider a drop of radius r^* in equilibrium with the surrounding vapour at a system temperature T_g and pressure p_g. The vapour pressure over the drop (p_g) is higher than the vapour pressure p_∞ for a planar interface at the system temperature T_g and is given by the expression (cf. eq. (4.3)).

$$p_g = p_\infty \exp(2\sigma v_f M/r^* RT)$$

$$\approx p_\infty \left(1 + \frac{2\sigma v_f}{p_\infty r^* v_g}\right) \tag{10.1}$$

To calculate the temperature for equilibrium T_g the Clausius–Clapeyron equation can be used:

$$\frac{p_g - p_\infty}{T_{SAT} - T_g} \approx \frac{2\sigma v_f/r^* v_g}{T_{SAT} - T_g} \approx \frac{Ji_{fg}Mp_\infty}{RT_g^2} \tag{10.2}$$

Thus

$$T_{SAT} - T_g \approx \frac{(2\sigma v_f/r^* v_g)RT_{SAT}^2}{Ji_{fg}Mp_\infty} \approx \frac{2\sigma v_f T_g}{Ji_{fg}r^*} \tag{10.3}$$

where T_{SAT} is the saturation temperature corresponding to the vapour pressure p_g. For large differences between p_g and p_∞, the more accurate form of eq. (10.1) should be used to calculate r^*, thus:

$$r^* = \frac{2\sigma v_f M}{RT \ln(p_g/p_\infty)} \tag{10.4}$$

The rate of formation of embryo droplets can be obtained in an analogous manner to that used for embryo bubble nuclei. Hill et al. (1963) have applied the classical nucleation theory to droplet nucleation and their expression for the rate of nucleation (dn/dt) is as follows:

$$\frac{dn}{dt} = \left(\frac{p_g}{kT_g}\right)^2 \frac{Mv_f}{N_0} \sqrt{\left(\frac{2\sigma}{\pi m}\right)} \exp\left(-\frac{\Delta G(r^*)}{kT_g}\right) \tag{10.5}$$

where N_0 is the number of molecules in one kmol (lb mol) and where $\Delta G(r^*)$ is given by eq. (4.12). Equation (10.5) reduces to

$$\frac{dn}{dt} = N\frac{v_f}{v_g} \lambda \exp(-4\pi\sigma r^{*2}/3kT_g) \tag{10.6}$$

with $\lambda = (2\sigma/\pi m)^{1/2}$ and N is the number of vapour molecules per unit volume. Equation (10.6) is very similar to the expression for bubble nucleation rate (eq. (4.13)). The more accurate expression for r^* (eq. (10.4)) should be used for calculating (dn/dt). The nucleation rate increases rapidly with small increases in (p_g/p_∞) and the nucleation rate for 'significant' droplet growth is usually taken as 10^{17}–10^{22} m^{-3} s^{-1}.

The critical size of droplets in steam is of the order of 0.01 μm at atmospheric pressure (Gardner 1963–64, Stever 1958). For very small droplets, the surface tension itself may vary and the expression due to Tolman (1949):

$$\sigma(r) = \frac{\sigma_\infty}{1 + \delta/r} \tag{10.7}$$

may be applicable where σ_∞ is the surface tension for a flat interface and δ is a length lying between 0.25 and 0.6 of the molecular or atomic radius in the liquid state. Surface tension reductions of the order of 10 per cent are reported (Hill *et al.* 1963) for some homogeneous nucleation cases.

Nucleation of drops at solid surfaces. For nucleation on solid surfaces in the absence of pre-existing liquid in cavities, the free energy of formation of an embryo droplet $\Delta G(r^*)$ should be reduced by multiplying by the factor ϕ (see Chapter 4)

$$\phi = \frac{2 - 2\cos\theta - \cos\theta\sin^2\theta}{4} \tag{10.8}$$

For a contact angle $\theta = 0°$ (complete wetting), $\phi = 0$, and for $\theta = 180°$ (no wetting), $\phi = 1$.

If liquid is already in cracks in the surface, then nucleation can proceed at much lower subcoolings; this is analogous to the case where bubbles grow from pre-existing gas in vapour-filled cavities.

10.2.2   Droplet growth

For the critical embryo to grow, it is necessary (Ryley 1961) that the latent heat liberated by the condensation be removed by conduction and convection to the surrounding fluid. This heat transfer is countercurrent to the mass transfer towards the drop. For pure conduction and diffusion control of the growth, the following equation has been proposed by Mason (1951)

$$\left[\frac{p_g}{p_\infty} - 1\right] = \left[\frac{\rho_f R T_g}{2K_g p_g M} + \frac{i_{fg}^2 \rho_f M}{2k_g R T_g^2}\right]\frac{dr^2}{dt} \tag{10.9}$$

where K_g is the mass diffusivity or self diffusion coefficient. Equation (10.9) can be used for values of r greater than 5 µm. For smaller droplets the values of K_g and k_g must be made smaller to allow for the fact that the dimensions of the drop are of the same order as the mean free path. For droplets smaller than 2 µm the following equation based on kinetic theory and derived by Oswatitsch (1942) is applicable

$$\frac{dr}{dt} = \frac{3p_\infty}{8i_{fg}\rho_f}\sqrt{\left(\frac{3R}{T_g M}\right)}(T_\infty - T_g) \tag{10.10}$$

When droplets grow on cooled surfaces the rate of growth is controlled by the conduction of heat through the drop (McCormick and Baer 1963).

10.3 The mechanism of evaporation and condensation at a plane liquid–vapour interface

10.3.1 *Introduction*

Consider a pure saturated vapour at a pressure p_g and a temperature T_g condensing on its own liquid phase whose surface temperature is T_f. The phenomenon of such an interface mass transfer can be viewed from the standpoint of kinetic theory as a difference between two quantities—a rate of arrival of molecules from the vapour space towards the interface and a rate of departure of molecules from the surface of the liquid into the vapour space. When condensation takes place the arrival rate exceeds the departure rate. During evaporation the reverse occurs, and during an equilibrium state the two rates are equal (Knacke and Stranski 1956).

From kinetic theory it can be shown that, in a stationary container of molecules, the mass rate of flow (of molecules) passing in either direction (to right or left) through an imagined plane is given by

$$|j| = \left(\frac{M}{2\pi R}\right)^{1/2} \frac{p}{T^{1/2}} \tag{10.11}$$

where $|j|$ = flux of molecules (mass per unit time per unit area), M = molecular weight, R = universal gas constant, and p and T = pressure and temperature (absolute).

Equation (10.11) is the starting point for many theories of interfacial phase change.

In general it can be stated that the net molecular flux through an interface is the difference between the vector fluxes in the directions from gas to liquid and vice versa. Thus, referring to Fig. 10.1,

$$j = j_+ - j_- \tag{10.12}$$

Since the condition close to the surface is not one of static thermal equilibrium, for any significant rate of evaporation or condensation, it is really not meaningful to make use of the thermostatic pressure and temperature on each side of the interface. Strictly speaking one should solve the Boltzmann transport equation with appropriate boundary conditions and asymptotes which are conditions of thermal equilibrium at several mean free path distances from the interface. However, some considerable success for

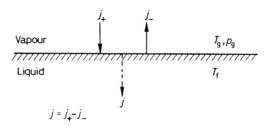

Fig. 10.1. Interfacial phase change.

engineering purposes has been achieved by using simplified kinetic theory techniques and applying correction factors to the resulting predictions.

10.3.2 The crude theory

The simplest approach considers that the fluxes in eq. (10.12) are given by eq. (10.11) for appropriate liquid and vapour 'temperatures' and 'pressures', thus,

$$j = \left(\frac{M}{2\pi R}\right)^{1/2}\left[\frac{p_g}{T_g^{1/2}} - \frac{p_f}{T_f^{1/2}}\right] \tag{10.13}$$

For small differences in pressure and temperature this may be rewritten by differentiation as follows:

$$j = \left(\frac{M}{2\pi R T}\right)^{1/2} p \left[\frac{\Delta p}{p} - \frac{\Delta T}{2T}\right] \tag{10.14}$$

The pressure which is assigned to the liquid is usually taken to be the saturation pressure corresponding to T_f. This is based on the argument that in thermostatic equilibrium the net value of j is zero and the number of molecules leaving the liquid surface is equal to the number arriving.

Equation (10.14) contains both the pressure and temperature differences and it would seem that these can in principle be varied independently since there is no *a priori* guarantee that the vapour is saturated and not superheated.

A simplification which is often made is to assume that there is no temperature jump across the interface. Equation (10.14) then simplifies to give

$$j = \left(\frac{M}{2\pi R T}\right)^{1/2} \Delta p \tag{10.15}$$

In order for condensation to occur the vapour pressure must then exceed the liquid pressure. Figure 10.2 shows the variation of the fluxes j_+ and j_-

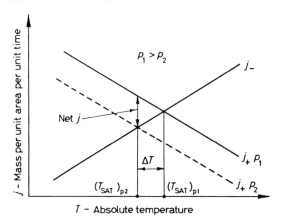

Fig. 10.2. Condensation at an interface.

with temperature. The evaporation flux, j_-, increases with temperature according to a unique curve for a given fluid but the condensation flux j_+, decreases with temperature at constant pressure according to eq. (10.11). The thermostatic saturation temperature T_{SAT} corresponding to the pressure p is then given by the intersection between the j_- line and the j_+ line. If the pressure of the vapour is increased to $p + \Delta p$, while the temperature remains at T, there will be a net condensation flux given by eq. (10.15).

The saturation temperature corresponding to the pressure $p + \Delta p$ is $T + \Delta T$, as shown in Fig. 10.2. Therefore, another way of interpreting the result is that the liquid must be at a temperature ΔT below the static saturation temperature corresponding to the prevailing vapour pressure. Thus, although no temperature jump is assumed to exist at the interface, the net effect is equivalent to a subcooling of the liquid, if one assumes that the driving force for condensation is the saturation temperature corresponding to the vapour pressure. An interfacial heat transfer coefficient can then be defined as

$$h_i = \phi/\Delta T \tag{10.16}$$

where ϕ is the condensation energy ('heat') flux. This energy flux is simply the latent heat times the mass flux, therefore

$$\phi = i_{fg} j \tag{10.17}$$

From eqs. (10.15), (10.16), and (10.17)

$$h_i = i_{fg} \left[\frac{M}{2\pi RT} \right]^{1/2} \frac{\Delta p}{\Delta T} \tag{10.18}$$

Using the Clausius–Clapeyron relationship (eq. (4.5)), eq. (10.18) becomes

$$h_i = \left[\frac{M}{2\pi RT}\right]^{1/2} \frac{Ji_{fg}^2}{Tv_{fg}} \tag{10.19}$$

If the temperature difference term in eq. (10.14) is retained and the Clausius–Clapeyron relationship is assumed to apply, eq. (10.19) would have been multiplied by the factor

$$\left[1 - \frac{pv_{fg}}{2i_{fg}}\right]$$

which is usually close to unity (Labuntsov and Smirnov 1966).

10.3.3   Modifications to the crude theory

The easiest way to modify the simple theory outlined above is by means of correction factors. For example, Silver (1964) multiplies eq. (10.19) by a factor of f, called the 'molecular exchange condensation fraction'. The value of f for water is found empirically under vacuum condenser conditions to be about 0.036.

Another technique is to apply correction factors to both the j_+ and j_- terms. Following Schrage (1953), eq. (10.13) is rewritten in the form

$$j = \left(\frac{M}{2\pi R}\right)^{1/2} \left[\Gamma\sigma_c \frac{p_g}{T_g^{1/2}} - \sigma_e \frac{p_f}{T_f^{1/2}}\right] \tag{10.20}$$

σ_c and σ_e are condensation and evaporation 'coefficients' which are usually taken to be the same, as a result of thermostatic equilibrium arguments which are not strictly valid in the dynamic case, and are written generally as σ without a subscript. The coefficient Γ results from the net motion of the vapour towards the surface which is superimposed on the assumed Maxwellian distribution. For a gas which moves with an overall speed u towards a transparent plane it is readily shown from kinetic theory that the fluxes of molecules which cross the plane are

$$\left.\begin{aligned} j_+ &= \Gamma(a)\left(\frac{M}{2\pi R}\right)^{1/2} \frac{p}{T^{1/2}} \\ j_- &= \Gamma(-a)\left(\frac{M}{2\pi R}\right)^{1/2} \frac{p}{T^{1/2}} \end{aligned}\right\} \tag{10.21}$$

where

$$\left.\begin{aligned} \Gamma(a) &= \exp(-a^2) + a\pi^{1/2}[1 + \text{erf } a] \\ \Gamma(-a) &= \exp(-a^2) - a\pi^{1/2}[1 - \text{erf } a] \end{aligned}\right\} \tag{10.22}$$

Table 10.1 The correction factor
Γ as a function of a

a	$\Gamma(-a)$	$\Gamma(a)$
1	0.089	3.634
0.8	0.162	2.998
0.6	0.276	2.403
0.4	0.447	1.865
0.2	0.685	1.394
0.1	0.833	1.187
0.08	0.8646	1.1482
0.06	0.8972	1.1100
0.04	0.9307	1.0725
0.02	0.9649	1.0359
0.01	0.9824	1.0178
0.001	0.998229	1.001773

and a represents the ratio of the overall speed, u, to a characteristic molecular velocity, thus,

$$a = \frac{u}{(2RT/M)^{1/2}} \tag{10.23}$$

In terms of the condensation rate, u can be rewritten as

$$u = \frac{j}{\rho_g} \tag{10.24}$$

Utilizing the perfect gas law (10.23) and (10.24) can then be combined to give

$$a = \frac{j}{p_g(2M/RT_g)^{1/2}} \tag{10.25}$$

The values of $\Gamma(a)$ and $\Gamma(-a)$ are tabulated in Table 10.1 for convenience.
For low values of a, Γ may be estimated from the formula, $\Gamma = 1 + a\pi^{1/2}$. The error is slightly over 1 per cent for $a = 0.1$ and increases rapidly for larger values.

Note that the effect of the net velocity towards the interface in condensation is to increase the molecular flux. In the limit of very large a, eq. (10.22) reduces to

$$\Gamma(a) \rightarrow 2a\pi^{1/2} \tag{10.26}$$

Equation (10.20) then becomes

$$j = \left(\frac{M}{2\pi R}\right)^{1/2}\left[2a\pi^{1/2}\sigma\frac{p_g}{T_g^{1/2}} - \sigma\frac{p_f}{T_f^{1/2}}\right] \tag{10.27}$$

If σ is unity, that is all the molecules hitting the surface are condensed and evaporation is negligible compared with condensation, eqs. (10.25) and (10.27) are seen to be compatible. In this limiting condition of maximum possible condensation rate on a very cold surface the kinetic theory terms are completely dominated by the net flow terms. The surface simply behaves as a 'black body' which absorbs a flow of molecules which can be indefinitely high.

In the other limit of a very small value of a, eq. (10.22) becomes

$$\Gamma(a) \approx 1 + a\pi^{1/2} \tag{10.28}$$

Substituting this value into eq. (10.20) and making use of eq. (10.25)

$$j = \left(\frac{M}{2\pi R}\right)^{1/2} \sigma \left[\frac{p_g}{T_g^{1/2}} - \frac{p_f}{T_f^{1/2}}\right] + \frac{j\sigma}{2} \tag{10.29}$$

whence, on rearrangement,

$$j = \left(\frac{2\sigma}{2 - \sigma}\right)\left(\frac{M}{2\pi R}\right)^{1/2} \left[\frac{p_g}{T_g^{1/2}} - \frac{p_f}{T_f^{1/2}}\right] \tag{10.30}$$

Equation (10.30) is in the form suggested by Silver and Simpson (1961) and referred to by Russian authors as the Kucherov–Rikenglaz (1960) equation. It is only valid for low values of a, in other words low condensation rates.

The above arguments ignore the non-equilibrium interactions between the cold molecules leaving the interface and the hot molecules approaching it. It is really assumed that the molecular fluxes of the evaporating or condensing streams can be derived from kinetic theory for each 'component' alone and the results superimposed to obtain the net flux. This is plainly incorrect. However, the difficulties which arise if attempts are made to solve the more complex problem are considerable. The interested reader is referred to the review by Wilhelm (1964).

Further background material, which treats the various experimental methods which have been used to measure the condensation coefficient, is contained in Danon (1962), Sukhatme and Rohsenow (1964), Subbotin et al. (1964), Labuntsov and Smirnov (1966), Misra and Bonilla (1956), and Aladyev et al. (1966).

Whether liquids evaporate and condense at the maximum rate calculated from eq. (10.15) has been hotly debated for many years. Various experimental studies seem to lead to the conclusion that polar liquids vaporize and condense at rates significantly less than the maximum rate while non-polar liquids vaporize and condense at the maximum rate. For water, for example, it is commonly reported that the condensation rate and the interfacial heat transfer coefficient is 0.03–0.05 of the maximum value given by eq. (10.15) and eq. (10.19) respectively. However, more recent experiments have indicated that for both non-polar and polar liquids even water (Bonacci *et al.* 1976, Maa 1970) the maximum rate is achieved and that experimental uncertainty about the liquid surface temperature is responsible for the reported very low values in the literature. In the case of evaporation, it has been established (Johnstone and Smith 1966) that the rate of evaporation falls rapidly as the time of exposure of the evaporating interface is increased. Temperature gradients are set up in the liquid phase adjacent to the interface in order to supply the latent heat of the net vapour flux. A simple transient conduction analysis (Saha 1977) based on the Plesset–Zwick approach (see eq. (4.26)) will yield the following equation for the 'effective' interface heat transfer coefficient after an exposure time (*t*).

$$h_i = k_f \left(\frac{\pi}{3} \alpha_f t \right)^{-1/2} \tag{10.31}$$

For water, the interfacial heat transfer coefficient falls from the maximum value given by eq. (10.19) to the much lower values given by eq. (10.31) within a few milliseconds.

10.3.4 *The influence of non-condensables on interfacial resistance*

The presence of even a small quantity of non-condensable gas in the condensing vapour has a profound influence on the resistance to heat transfer in the region of the liquid–vapour interface. The non-condensable gas is carried with the vapour towards the interface where it accumulates. The partial pressure of gas at the interface increases above that in the bulk of the mixture, producing a driving force for gas diffusion away from the surface. This motion is exactly counterbalanced by the motion of the vapour–gas mixture towards the surface. Since the total pressure remains constant the partial pressure of vapour at the interface is lower than that in the bulk mixture providing the driving force for vapour diffusion towards the interface. This situation is illustrated in Fig. 10.3 which also shows the variation of temperature in the region of the interface. It is usual to assume that the temperature at the interface (T_{gi}) corresponds to the saturation temperature equivalent to the partial pressure of vapour (p_{gi}) at the interface.

The molar flux of non-condensable gas (J_a) passing through a plane parallel to and at a distance y from the interface is

$$J_a = J\tilde{y}_a + D_{AG}\tilde{c}\frac{d\tilde{y}_a}{dy} = 0 \tag{10.32}$$

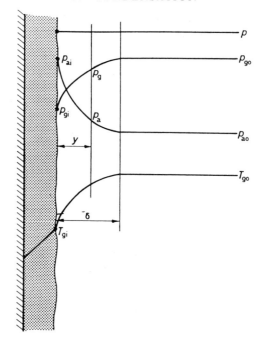

Fig. 10.3. The influence of non-condensables on interfacial resistance.

Likewise, the molar flux of vapour (J_g) is

$$J_g = J\tilde{y}_g + D_{AG}\tilde{c}\,\frac{d\tilde{y}_g}{dy} \tag{10.33}$$

where D_{AG} is the binary diffusion coefficient, $\tilde{c}$ is the total molar concentration, J is the mixture molar ('drift') flux towards the interface, $\tilde{y}_a$, $\tilde{y}_g$ are the mol fractions of non-condensable gas and condensable vapour respectively ($\tilde{y}_a = p_a/p$, etc.)

$$\frac{d\tilde{y}_g}{dy} = -\frac{d\tilde{y}_a}{dy}; \qquad \tilde{y}_g = 1 - \tilde{y}_a \tag{10.34}$$

Thus

$$J_g = J\tilde{y}_g - D_{AG}\tilde{c}\,\frac{d\tilde{y}_a}{dy} \tag{10.35}$$

and eliminating J between eqs. (10.35) and (10.32)

$$J_g = -D_{AG}\tilde{c}\,\frac{d\tilde{y}_a}{dy}\left[\frac{\tilde{y}_g}{\tilde{y}_a} + 1\right] \tag{10.36}$$

Integrating the above equation between the interface ($y = 0$) and the edge of the diffusion layer ($y = \delta_M$)

$$J_g = \frac{D_{AG}\tilde{c}}{\delta_M} \ln\left[\frac{1 - \tilde{y}_{gi}}{1 - \tilde{y}_{go}}\right] \tag{10.37}$$

where $\tilde{y}_{gi}$ is the mol fraction of vapour at the interface and $\tilde{y}_{go}$ is the mol fraction of the vapour in the bulk mixture.

Equation (10.37) can be rewritten in terms of the *mass* condensation flux j_g:

$$j_g = K_g\rho_g \ln\left[\frac{1 - \tilde{y}_{gi}}{1 - \tilde{y}_{go}}\right]$$

or

$$j_g = \frac{K_g\rho_g}{p_{am}}(p_{go} - p_{gi}) \tag{10.38}$$

where p_{go}, p_{gi} are the respective partial vapour pressures in the bulk and at the interface; p_{am} is the log mean of the partial pressure of non-condensable gas at the interface and in the bulk mixture given by

$$p_{am} = \left[\frac{p_{ai} - p_{ao}}{\ln(p_{ai}/p_{ao})}\right] \tag{10.39}$$

K_g is a mass transfer coefficient, defined by

$$K_g = \frac{D_{AG}}{\delta_M} \tag{10.40}$$

The heat transfer rate is also affected by the occurrence of mass transfer. The total heat flux at a distance y from the interface is made up of two components; a sensible heat transfer (conductive) term and a sensible heat transport (convective) term:

$$\phi_g = k_g\frac{dT}{dy} + j_g c_{pg}(T - T_{gi}) \tag{10.41}$$

Integrating from $y = 0$ to $y = \delta_T$ and using $h_g = k_g/\delta_T$

$$\phi = \frac{a}{1 - e^{-a}}[h_g(T_{go} - T_{gi})] = h'_g(T_{go} - T_{gi}) \tag{10.42}$$

where

$$a = \left[\frac{j_g c_{pg}}{h_g}\right]$$

h'_g and h_g are the sensible heat transfer coefficients with and without mass transfer respectively. This correction (Ackermann 1937) is often small and may then be ignored.

The transfer of heat from the bulk mixture to the interface is made up of two components; firstly, the sensible heat transferred through the diffusion layer to the interface and secondly, the latent heat released due to condensation of the vapour reaching the interface. This model of heat transfer in the vapour phase was first expressed by Colburn and Hougen (1934) as:

$$\phi = \phi_v + \phi_g \tag{10.43}$$

$$= \frac{K_g i_{fg} \rho_g}{p_{am}} (p_{go} - p_{gi}) + h'_g(T_{go} - T_{gi}) \tag{10.44}$$

For the case where the vapour–gas mixture is stagnant only the first term in eq. (10.44) need be considered. However, the use of the simple diffusion model for the mass transfer coefficient, K_g results in an underestimate of the heat flow for a given temperature gradient. This discrepancy has been attributed by Akers et al. (1960) and Sparrow and Eckert (1961) to the presence of natural convection effects.

Akers et al. (1960) concluded that the ratio of the mass transfer boundary layer thickness, δ_M, to some characteristic length, L, could be expressed in terms of the product of the Grashof and Schmidt numbers (Gr.Sc) thus

$$\frac{L}{\delta_M} = \left\{ \frac{K_g L}{D_{AG}} \right\} = \mathrm{fn} \left[\frac{gL^3\rho^2}{\mu^2} \left(\frac{\rho_o}{\rho_i} - 1 \right) \frac{\mu}{\rho D} \right] \tag{10.45}$$

Expressing the function as a power law and using experimental data to produce empirical relationships led to

$$\left\{ \frac{K_g L}{D_{AG}} \right\} = C_1 \left[\frac{gL^3\rho}{\mu D} \left(\frac{\rho_o}{\rho_i} - 1 \right) \right]^{0.373} \tag{10.46}$$

over the range $10^3 < (\mathrm{GrSc}) < 10^7$, where C_1 is 1.02 for vertical surfaces and 0.62 for horizontal tubes; ρ_i and ρ_o are the densities of the fluid adjacent to the interface and in the bulk mixture respectively.

For the case where the vapour–gas mixture flows parallel to the interface the heat transfer coefficient h_g in eq. (10.42) is evaluated from the correlation appropriate to the flow regime (i.e. laminar, transition, and turbulent flow) and flow geometry (i.e. flow within a tube, over a flat plate or a tube bank in crossflow). The mass transfer coefficient K_g in the first term of eq. (10.44) may be evaluated from the heat transfer coefficient h_g by making use of the analogy between heat and mass transfer. Thus,

$$K_g = \left(\frac{h_g}{\rho_g c_{pg}} \right) \left(\frac{\mathrm{Pr}}{\mathrm{Sc}} \right)^{2/3} \tag{10.47}$$

For the turbulent flow of a condensing vapour–gas mixture within a tube the well-known Dittus–Boelter eq. (5.7) could be used with the physical properties evaluated for the bulk vapour–gas mixture. However, actual heat and mass transfer coefficients will be higher than predicted by this method mainly as a result of the disturbed nature of the film interface. This effect may be allowed for in an approximate manner by multiplying the smooth tube value of the coefficient by the ratio of the interfacial friction factor, f_i, to the smooth tube value, f_g (viz., eq. (3.35)).

The application of eq. (10.44) requires some comment since the conditions at the interface are usually unknown. Consider the situation where condensation occurs from a vapour–gas mixture to a vertical wall under conditions where the wall surface temperature, T_W, the bulk temperature of the vapour–gas mixture, T_{go}, and the partial pressure of gas in the bulk vapour, p_{ao}, are known. The various stages to arrive at the overall heat transfer rate are as follows:

(a) Guess a condensate film interface temperature T_{gi} intermediate between T_{go} and T_W. The heat transfer across the liquid condensate film is given by

$$\phi_1 = h_f(T_{gi} - T_W) \tag{10.48}$$

The heat transfer coefficient, h_f, across the condensate film is evaluated using the methods described in Section 10.4.

(b) Since the interface is assumed to be at saturation condition for the vapour, the choice of T_{gi} allows the partial pressure of vapour at the interface (p_{gi}) to be evaluated. The partial pressure of vapour in the bulk mixture, $p_{go} = (p - p_{ao})$ is known. Values of h_g and K_g, the heat and mass transfer coefficients respectively, are evaluated from the appropriate equations. Use is made, where necessary, of the analogy between heat and mass transfer (eq. (10.47)) and the correction for sensible heat transfer (eq. (10.42)).

(c) The rate of heat transfer through the diffusion layer may now be calculated from eq. (10.44). In general this heat flux, ϕ_2, will differ from that found from eq. (10.48), ϕ_1. The correct solution occurs when a value of interface temperature T_{gi} is reached which allows ϕ_1 to equal ϕ_2. This may be achieved by an iterative procedure in which successively better guesses for the interfacial temperature are made. Alternatively ϕ_1 and ϕ_2 may be evaluated for three or more values of interface temperature and plotted against T_{gi}. The intersection of the two curves gives the correct solution for ϕ and T_{gi}.

Such a calculation is only valid at one point in the condenser and the computation must be repeated for other locations.

More sophisticated boundary layer treatments of the influence of noncondensable gas on forced convection condensation have been published by Minkowycz and Sparrow (1966) and Sparrow et al. (1967). These treatments include interfacial resistance, the process of mass transfer as a result of a temperature gradient (thermal diffusion) and energy transport as a result of

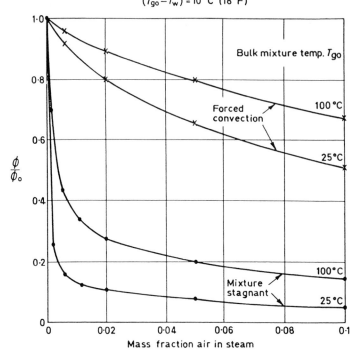

Fig. 10.4. The influence of non-condensable air in steam (Minkowycz and Sparrow 1966, Sparrow *et al.* 1967).

a concentration gradient. For the case where steam is the condensing vapour and air is the non-condensable gas these processes are unimportant.

Figure 10.4, prepared from the results of Minkowycz and Sparrow (1966) and Sparrow *et al.* (1967) shows the influence of non-condensable air concentrations on the heat transfer rate in the case when the steam–air mixture is either stagnant or flowing. The temperature difference between the bulk steam–air mixture and the cooling surface is held constant at 10°C (18°F). The ordinate represents the ratio of the heat flow with non-condensable air present to that which would have occurred at the same temperature difference with pure steam. The presence of air is most marked when the bulk mixture is stagnant. Reductions in heat transfer rate of more than 50 per cent are caused by air mass fractions as low as 0.5 per cent. The reductions become more marked as the total pressure is reduced. These findings are in qualitative and also reasonably quantitative agreement with the experimental results of Othmer (1929).

The heat transfer rate in forced convection condensation is much less sensitive to the presence of non-condensable air although the reduction in heat transfer become more serious as the pressure is reduced. The reductions in heat transfer rate for both stagnant and forced convection conditions increase slightly as the temperature difference $(T_{go} - T_W)$ is increased. For more information on these processes, refer to Section 10.8.1.

10.4 Film condensation on a planar surface

10.4.1 *The Nusselt equation for a laminar film*

The first attempt to analyse the film-wise condensation problem was that of Nusselt (1916) who made the following assumptions:

(a) The flow of condensate in the film is laminar
(b) the fluid properties are constant
(c) subcooling of the condensate may be neglected
(d) momentum changes through the film are negligible: there is essentially a static balance of forces
(e) the vapour is stationary and exerts no drag on the downward motion of the condensate
(f) heat transfer is by conduction only
(g) surface is isothermal.

The first step is to calculate the velocity distribution in a liquid film flowing down a plane surface of unit width inclined at an angle θ to the horizontal (Fig. 10.5). At a distance z from the top of the surface the thickness of the film is δ. Consider a force balance on an element of film lying between y and δ

$$(\delta - y)\,dz(\rho_f - \rho_g)g \sin \theta = \mu_f \left(\frac{du_y}{dy}\right) dz \qquad (10.49)$$

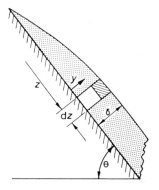

Fig. 10.5. Flow of a laminar film over an inclined surface (Nusselt).

Rearranging and integrating with the boundary condition $u_y = 0$ at $y = 0$, then

$$u_y = \frac{(\rho_f - \rho_g)g \sin \theta}{\mu_f}\left[y\delta - \frac{y^2}{2} \right] \tag{10.50}$$

The mass flow-rate per unit width Γ is given by

$$\Gamma = \rho_f \int_0^\delta u_y \, dy = \left[\frac{\rho_f(\rho_f - \rho_g)g \sin \theta \, \delta^3}{3\mu_f} \right] \tag{10.51}$$

Consider now the case where condensation occurs building up the film from zero thickness at the top of the plane surface. The rate of increase of film flow-rate with film thickness $(d\Gamma/d\delta)$ is therefore

$$\frac{d\Gamma}{d\delta} = \left[\frac{\rho_f(\rho_f - \rho_g)g \sin \theta \, \delta^2}{\mu_f} \right] \tag{10.52}$$

If the film surface temperature is T_{gi} and the wall temperature is T_W the heat transferred by conduction to an element of length dz is

$$d\phi = \frac{k_f}{\delta}(T_{gi} - T_W) \, dz \tag{10.53}$$

The mass rate of condensation on this area $(d\Gamma)$ is therefore

$$d\Gamma = \frac{k_f}{\delta i_{fg}}(T_{gi} - T_W) \, dz \tag{10.54}$$

Substituting eq. (10.54) into eq. (10.52), separating variables and integrating (noting $\delta = 0$ at $z = 0$)

$$\mu_f k_f(T_{gi} - T_W)z = \rho_f(\rho_f - \rho_g)g \sin \theta \, i_{fg}\left(\frac{\delta^4}{4} \right) \tag{10.55}$$

Thus the film thickness δ is given by

$$\delta = \left[\frac{4\mu_f k_f z(T_{gi} - T_W)}{g \sin \theta i_{fg}\rho_f(\rho_f - \rho_g)} \right]^{1/4} \tag{10.56}$$

The heat transfer coefficient $h_f(z)$ at any point z is given by

$$h_f(z) = \frac{k_f}{\delta} = \left[\frac{\rho_f(\rho_f - \rho_g)g \sin \theta i_{fg}k_f^3}{4\mu_f z(T_{gi} - T_W)} \right]^{1/4} \tag{10.57}$$

and the Nusselt number is given by

$$\mathrm{Nu}(z) = \left[\frac{h_f(z)z}{k_f} \right] = \left[\frac{\rho_f(\rho_f - \rho_g)g \sin \theta i_{fg}z^3}{4\mu_f k_f(T_{gi} - T_W)} \right]^{1/4} \tag{10.58}$$

Equations (10.57) and (10.58) give the local values of heat transfer coefficient and Nusselt number. The mean value of heat transfer coefficient over the whole surface, $\bar{h}_f$ is given by

$$\bar{h}_f = \frac{1}{z} \int_0^z h_f(z) \, dz = 0.943 \left[\frac{\rho_f(\rho_f - \rho_g)g \sin \theta i_{fg} k_f^3}{\mu_f z (T_{gi} - T_w)} \right]^{1/4} \tag{10.59}$$

The equation for $\bar{h}_f$ can be expressed in an alternative form

$$\bar{h}_f = \frac{\Gamma_z i_{fg}}{z(T_{gi} - T_w)} \tag{10.60}$$

where Γ_z is the condensate flow at a distance z from the top of the plane surface. Combining eq. (10.60) with eq. (10.54)

$$\delta = \frac{k_f \Gamma_z \, dz}{\bar{h}_f z \, d\Gamma} \tag{10.61}$$

Combining eq. (10.61) with eq. (10.51) yields

$$k_f \left[\frac{\rho_f(\rho_f - \rho_g)g \sin \theta}{3\mu_f} \right]^{1/3} \frac{dz}{z} = \frac{\bar{h}_f \Gamma^{1/3} \, d\Gamma}{\Gamma_z} \tag{10.62}$$

Integrating over z

$$\bar{h}_f = 0.925 \left[\frac{\rho_f(\rho_f - \rho_g)g \sin \theta k_f^3}{\mu_f \Gamma_z} \right]^{1/3} \tag{10.63}$$

The Reynolds number of a liquid film at a distance z below the top of the plane surface is given by

$$\mathrm{Re}_\Gamma = \frac{4\Gamma_z}{\mu_f} \tag{10.64}$$

Substituting into eq. (10.63) and rearranging for a vertical plane ($\sin \theta = 1$)

$$\frac{\bar{h}_f}{k_f} \left[\frac{\mu_f^2}{\rho_f(\rho_f - \rho_g)g} \right]^{1/3} = 1.47 \, \mathrm{Re}_\Gamma^{-1/3} \tag{10.65}$$

Since $h_f(z)$ varies as $z^{-1/4}$ (eq. (10.57)) the average coefficient $\bar{h}_f$ is 4/3 times the local coefficient $h_f(z)$. The validity of the above theory is limited to inclinations above which surface tension forces do not play a part.

10.4.2 *Improvements to the original Nusselt theory*

A number of papers have made significant improvements to the Nusselt theory. The original analysis was extended by Bromley (1952) who considered the effects of subcooling the condensate and by Rohsenow (1956) who also allowed for the non-linear distribution of temperature through the film due to energy convection. Rohsenow (1956) showed that the latent heat of vaporization, i_{fg}, in eq. (10.59) should be replaced by

$$i'_{fg} = i_{fg}\left[1 + 0.68\left(\frac{c_{pf}\,\Delta T_f}{i_{fg}}\right)\right]$$ (10.66)

where $\Delta T_f = (T_{gi} - T_w)$.

Sparrow and Gregg (1959) using a boundary layer treatment removed assumption (d) and considered momentum changes in the film. For common fluids with Prandtl numbers around unity the results obtained show that momentum effects are indeed negligible but for liquid metals with very low Prandtl numbers the heat transfer coefficient falls below the Nusselt prediction with increasing $(c_{pf}\,\Delta T_f/i_{fg})$.

More recently Chen (1961a), Koh et al. (1961) and others have considered the influence of the drag exerted by the vapour on the liquid film. There again the assumption made by Nusselt (assumption (e)) appears justified at Prandtl numbers around unity. For the condensation of liquid metals, however, the inclusion of the interfacial shear effect does cause a further substantial reduction of heat transfer. The ratio of the revised average heat transfer coefficient $(\bar{h}_f)$ to that given by eq. (10.59), $(\bar{h}_f)_{Nu}$ can be computed from the following approximate equation given by Chen (1961a) which includes both interfacial shear and momentum effects

$$\frac{\bar{h}_f}{(\bar{h}_f)_{Nu}} = \left[\frac{1 + 0.68A + 0.02AB}{1 + 0.85B - 0.15AB}\right]^{1/4}$$ (10.67)

where $A = [c_{pf}\,\Delta T_f/i_{fg}]$ and $B = [k_f\,\Delta T_f/\mu_f i_{fg}]$.

Equation (10.67) is valid for $A < 2$, $B < 20$, and for liquids with Prandtl numbers larger than unity or less than 0.05. A number of workers have considered the influence of variation in physical properties across the condensate film (assumption (b)). One method of introducing this influence is to use Nusselt's equation with the property values taken at some effective condensate film temperature, T_{film}. Thus,

$$T_{film} = T_w + F(T_{gi} - T_w)$$ (10.68)

The value of F given by Drew (1954) is 0.25 and by Minkowycz and Sparrow (1966) it is 0.31.

10.4.3 The influence of turbulence

Even at relatively low film Reynolds numbers, the assumption that the condensate layer is in viscous flow is open to some question. Experiments aimed at measuring the average thickness of liquid films flowing down vertical surfaces do confirm the Nusselt equations but examination of the surface structure of the flow indicates considerable waviness. This waviness may account for the observed differences (Drew 1954) between experimental and theoretical values. The results of most experimental studies give heat transfer coefficients up to 40–50 per cent higher than the theoretical values and it is recommended that coefficients evaluated from eq. (10.59) should be multiplied by a correction factor of 1.2 to allow for this discrepancy in design calculation.

For long vertical surfaces it is possible to obtain condensation rates such that the film Reynolds number exceeds the critical value at which turbulence begins. Under such circumstances Kirkbride (1933–34) found heat transfer coefficients much greater than given by eq. (10.65). By plotting the experimental data in terms of the dimensionless groupings of eq. (10.65) he was able to obtain an empirical relation for heat transfer in the case of combined viscous and turbulent flow on the outside surface of a vertical tube. Colburn (1933–34) attempted to predict the heat transfer coefficient for the case of turbulent flow by analogy with the flow of liquids through pipes under conditions where the liquid completely filled the pipe but where the pressure drop for flow was purely due to gravitational forces. The resulting relation for the local condensing heat transfer coefficient at any point, z, is

$$\frac{h_f(z)}{k_f}\left[\frac{\mu_f^2}{\rho_f(\rho_f - \rho_g)g}\right]^{1/3} = 0.056\left(\frac{4\Gamma_z}{\mu_f}\right)^{0.2}\left(\frac{c_{pf}\mu_f}{k_f}\right)^{1/3} \tag{10.69}$$

integration of the local coefficient over the tube length using eq. (10.57) up to a Reynolds number of 2000 and eq. (10.69) above 2000 resulted in the relationship shown in Fig. 10.6. Seban (1954) applied the Prandtl–Karman analogy to the problem of turbulent film condensation with no vapour shear. The results were presented graphically for a range of Prandtl numbers. In this work the critical value of the film Reynolds number was taken as 1600. Figure 10.6 shows the predictions. It will be noted that there is reasonable agreement with Colburn's analysis at high Prandtl numbers but that, at low Prandtl numbers, the heat transfer rate can fall below the Nusselt prediction at high Reynolds number, presumably due to the increased thickness of the turbulent film.

10.4.4 Condensation on horizontal tubes

Laminar film condensation on the outside of a single horizontal tube was first examined by Nusselt, who, using a derivation exactly as that described in Section 10.4.1, obtained the following relationship for the local condensing coefficient around the circumference of a horizontal tube:

$$h_f(\alpha) = 0.693\left[\frac{\rho_f(\rho_f - \rho_g)g \sin \alpha k_f^3}{\Gamma_\alpha' \mu_f}\right]^{1/3} \tag{10.70}$$

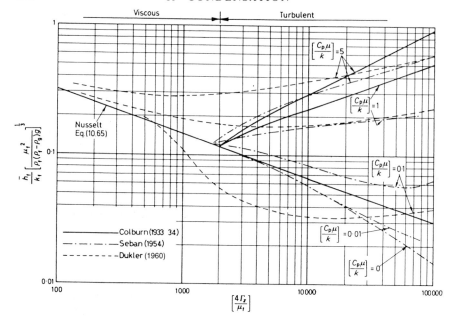

Fig. 10.6. Condensation on a vertical surface, no interfacial shear.

where α is the angle between any point on the tube surface and the vertical and Γ'_α is the local mass flow rate of condensate per unit length of tube. Integrating this expression around the tube gives the mean heat transfer coefficient as

$$\bar{h}_f = 0.725 \left[\frac{\rho_f(\rho_f - \rho_g)g\, i'_{fg} k_f^3}{D\mu_f(T_{gi} - T_W)} \right]^{1/4} \tag{10.71}$$

where D is the outside diameter of the tube, i'_{fg} is the modified latent heat of vaporization given by eq. (10.66) or in terms of film Reynolds number is

$$\frac{\bar{h}_f}{k_f} \left[\frac{\mu_f^2}{\rho_f(\rho_f - \rho_g)g} \right]^{1/3} = 1.51\, \mathrm{Re}_\Gamma^{-1/3} \tag{10.72}$$

where the Reynolds number is defined by eq. (10.64) with Γ_z equal to the mass flow-rate of condensate from the tube per unit length. This equation is valid up to values of Re_Γ of 3200.

In testing Nusselt's equations it is important to see that the experimental conditions comply with the requirements of the theory. The results of most experimental studies for horizontal tubes are within about 15 per cent of eq. (10.71). It is interesting to compare the performance of vertical versus horizontal tubes in stagnant vapour. From eqs. (10.59) and (10.71) it will be

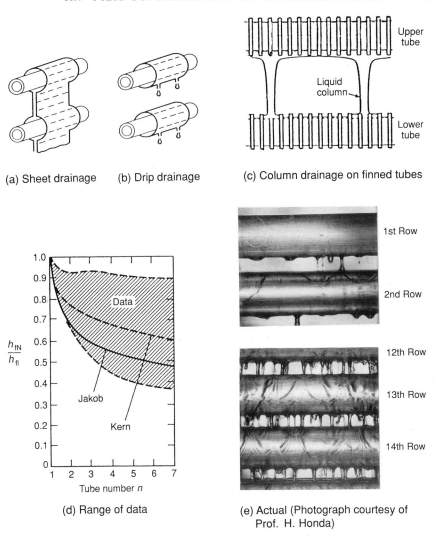

(a) Sheet drainage   (b) Drip drainage   (c) Column drainage on finned tubes

(d) Range of data

(e) Actual (Photograph courtesy of Prof. H. Honda)

Fig. 10.7. Film condensation on horizontal tubes.

seen that for the same temperature difference and fluid conditions the same average heat transfer coefficient will exist if

$$\frac{0.943}{z^{1/4}} = \frac{0.725}{D^{1/4}} \quad \text{or} \quad z = 2.87D \tag{10.73}$$

In a bank of horizontal tubes it is possible for the condensate to run off the bottom of the upper tube on to the next tube below (Fig. 10.7). Jakob (1936) has considered this problem and concluded that the mean heat transfer coefficient for a vertical column of n horizontal tubes with the same temperature difference is given by eq. (10.71) with D replaced by nD. Thus:

$$\frac{\bar{h}_{fN}}{h_{f1}} = n^{-1/4}$$

and (10.74)

$$\frac{\bar{h}_{fN}}{h_{f1}} = n^{3/4} - (n-1)^{3/4}$$

where $\bar{h}_{fN}$ is the mean coefficient for n tubes, h_{fN} is the coefficient for the nth tube in the column and h_{f1} is the coefficient for the first tube.

Experimental results (Fuks 1957; Grant and Osment 1968) suggest that the coefficients may well be higher than given by eq. (10.74) because the condensate does not pass from one tube to another as a continuous sheet but as drips or column flow (Fig. 10.7(b,c)).

Kern has suggested an alternative to eq. (10.74) such that:

$$\frac{\bar{h}_{fN}}{h_{f1}} = n^{-1/6} \quad \text{and} \quad \frac{\bar{h}_{fN}}{h_{f1}} = n^{5/6} - (n-1)^{5/6} \qquad (10.75)$$

Although Berman's (1969) data encompass both Kern and Jakob's predictions, the former (eq. (10.75)) is preferred as being the best recommendation.

For film condensation on horizontal rows of tubes as in Fig. 10.7(a), Chen (1961b) considered the additional effects of the momentum gain of the falling condensate and condensation on the falling, subcooled sheet of condensate between tubes. He developed the following approximate solution for these conditions:

$$\frac{\bar{h}_{fN}}{h_{f1}} = n^{-1/4}\left\{1 + 0.2\left[\frac{c_{pf}(T_{gi} - T_W)}{i_{fg}}\right](n-1)\right\} \qquad (10.76)$$

applicable when $[c_{pf}(T_{gi} - T_W)/i_{fg}]$ is less than or equal to 2.0.

The wide breadth of the data represented in Fig. 10.7(d) and also reported by Burman (1969, 1981) is primarily the result of inclusion of data on tube banks with vapour shear effects and also tests exhibiting drip or column drainage condensate as in Fig. 10.7(b) and (c), respectively, rather than the sheet mode shown in Fig. 10.7(a). Accordingly to Murata et $al.$ (1990), the condensate flow mode depends on the vertical tube spacing, the film flow breakdown thickness and the condensate flow-rate. At large spacings only the column and drip modes can occur. Figure 10.7(e) from Honda et $al.$ (1988) illustrates the drip and column drainage modes for R-113 at near atmospheric pressure condensing on an inline tube bundle at a vapour velocity of 2 m/s.

All three modes have been observed to occur, although finned tubes have a higher tendency to drain in columns and thus have less degradation due to condensate inundation on the lower tubes. For design, the expression of Kern, eq. (10.75), is normally recommended. For more on condensation on tube banks, refer to Section 10.5.4.

10.5   The influence of interfacial shear

10.5.1   *Laminar flow over inclined surface*

It has already been noted that the consideration of interfacial shear can considerably modify the laminar flow solution under certain conditions. Nusselt did derive equations for condensation in the presence of high vapour velocities on the assumption that the entire condensate layer remained in laminar flow. Thus, co-current downward flow over an inclined plane surface results in the following equation instead of eq. (10.49)

$$(\delta - y)\,dz(\rho_f g \sin\theta - (dp/dz)) + \tau_i\,dz = \mu_f \left(\frac{du_y}{dy}\right) dz \qquad (10.77)$$

Defining a fictitious vapour density ρ_g^* by the relationship

$$\left(\frac{dp}{dz}\right) = \rho_g^* g \sin\theta \qquad (10.78)$$

if the only pressure gradient is due to the vapour head, then $\rho_g^* = \rho_g$. Additional pressure gradient terms will occur when condensation takes place inside a tube. These will be discussed in Secton 10.5.2. Substituting eq. (10.78) into eq. (10.77) and integrating,

$$u_y = \frac{(\rho_f - \rho_g^*)g \sin\theta}{\mu_f}\left[y\delta - \frac{y^2}{2}\right] + \frac{\tau_i y}{\mu_f} \qquad (10.79)$$

$$\Gamma = \left[\frac{\rho_f(\rho_f - \rho_g^*)g \sin\theta\,\delta^3}{3\mu_f}\right] + \frac{\tau_i \rho_f\,\delta^2}{2\mu_f} \qquad (10.80)$$

$$\frac{d\Gamma}{d\delta} = \left[\frac{\rho_f(\rho_f - \rho_g^*)g \sin\theta\,\delta^2}{\mu_f}\right] + \left[\frac{\tau_i \rho_f\,\delta}{\mu_f}\right] \qquad (10.81)$$

Repeating the analysis outlined in Section 10.4.1 the following equation corresponding to (10.56) is derived:

$$\left[\frac{4\mu_f k_f z(T_{gi} - T_w)}{g \sin\theta\, i_{fg}' \rho_f(\rho_f - \rho_g^*)}\right] = \delta^4 + \frac{4}{3}\left[\frac{\tau_i\,\delta^3}{(\rho_f - \rho_g^*)g \sin\theta}\right] \qquad (10.82)$$

Rohsenow *et al.* (1956) have defined three dimensionless groups; a dimensionless film thickness, δ^*,

$$\delta^* = \delta \left[\frac{\rho_f(\rho_f - \rho_g^*)g \sin \theta}{\mu_f^2} \right]^{1/3} \tag{10.83}$$

a dimensionless distance, z^*,

$$z^* = \left(\frac{4k_f z(T_{gi} - T_w)}{i'_{fg}\mu_f \delta} \right) \delta^* \tag{10.84}$$

and a dimensionless interfacial shear stress, τ_i^*,

$$\tau_i^* = \left(\frac{\tau_i}{(\rho_f - \rho_g^*)\delta g \sin \theta} \right) \delta^* \tag{10.85}$$

Equation (10.82) becomes

$$z^* = \delta^{*4} + \tfrac{4}{3}\delta^{*3}\tau_i^* \tag{10.86}$$

also

$$\bar{h}_f^* = \frac{\bar{h}_f}{k_f} \left[\frac{\mu_f^2}{\rho_f(\rho_f - \rho_g^*)g \sin \theta} \right]^{1/3} = \frac{4}{3}\frac{(\delta_z^*)^3}{z^*} + 2\frac{\tau_i^*(\delta_z^*)^2}{z^*} \tag{10.87}$$

and

$$\frac{4\Gamma_z}{\mu_f} = \tfrac{4}{3}(\delta_z^*)^3 + 2\tau_i^*(\delta_z^*)^2 \tag{10.88}$$

The dimensionless group δ_z^* may be eliminated between eqs. (10.87) and (10.88). The results for downward vapour flow are plotted in Fig. 10.8. The dotted line shows the approximate boundary at which turbulence effects become important.

10.5.2   Condensation within a vertical tube

In order to apply the theory of Rohsenow *et al.* (1956) to the case of laminar condensation with interfacial shear within a tube it is necessary to evaluate ρ_g^* (eq. (10.78)) in terms of the pressure gradient in the vapour phase. For one-dimensional flow in the vapour phase

$$\frac{dp}{dz} = \rho_g g \sin \theta + \frac{2u_g}{(D - 2\delta)}\frac{d\Gamma}{dz} - \frac{2\tau_i}{(D - 2\delta)} \tag{10.89}$$

Therefore

$$\rho_g^* = \rho_g \left[1 + \frac{1}{\rho_g g \sin \theta} \left\{ \frac{2u_g}{(D - 2\delta)}\frac{d\Gamma}{dz} - \frac{2\tau_i}{(D - 2\delta)} \right\} \right] \tag{10.90}$$

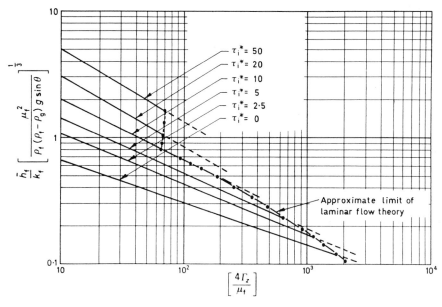

Fig. 10.8. The influence of vapour shear on the average laminar flow condensation heat transfer rate (Rohsenow *et al.*, 1956).

where

$$\frac{d\Gamma}{dz} = \frac{h_f(z)}{i_{fg}} (T_{gi} - T_W) \qquad (10.91)$$

This method proposed by Rohsenow *et al.* (1956) is in good agreement with a more rigorous treatment presented by Jacobs (1966), provided that the condensation surface is divided axially into a number of short sections over which the interfacial shear stress is assumed constant but of different magnitude to the neighbouring section.

Carpenter and Colburn (1951) also considered the influence of vapour velocity on film condensation inside tubes. In experiments where there was an appreciable co-current downward vapour velocity, heat transfer coefficients lay considerably above the static vapour case. The critical value of the film Reynolds number at which turbulence was initiated was greatly lowered (by a factor of up to ten or so) (see Fig. 10.8). Carpenter and Colburn (1951) reasoned, firstly, that the transition from laminar to turbulent flow in the film occurred at much lower Reynolds numbers than in the absence of vapour shear and, secondly, that the major resistance to heat transfer occurred across the laminar sublayer. They argued that, as the major force acting on the condensate film was not interfacial shear, the velocity profile in the film might be estimated from the analogous situation for flow of liquids in pipes.

The equation for the local heat transfer coefficient resulting from this treatment is

$$\left[\frac{h_f(z)\mu_f}{k_f \rho_f^{1/2}}\right] = 0.045\left[\frac{c_p\mu}{k}\right]_f^{1/2} \tau_W^{1/2} \tag{10.92}$$

where τ_W is the shear stress at the outer edge of the laminar sublayer (i.e., the wall shear stress). In calculating a value of τ_W account must be taken of the interfacial shear and the gravitational effects in the liquid film as discussed in Chapter 3 (viz., eq. (3.28)). In addition, the change of momentum of the condensing vapour and the influence of vapour mass transfer on the interfacial shear stress must be included (see Section 10.8).

With the help of further experimental data, Soliman *et al.* (1968) have improved the Carpenter and Colburn treatment and modified eq. (10.92):

$$\left[\frac{h_f(z)\mu_f}{k_f \rho_f^{1/2}}\right] = 0.036\left[\frac{c_p\mu}{k}\right]_f^{0.65} \tau_W^{1/2} \tag{10.93}$$

The shear stress at the outer edge of the laminar sublayer τ_W is made up of the sum of three components:

(a) the interfacial shear stress, τ_i due to vapour phase friction

$$\tau_i = \frac{D}{4}\left(-\frac{dp}{dz}F\right) \tag{10.94}$$

(b) the gravitational force τ_z in the liquid film

$$\tau_z = \frac{D}{4}(1-\alpha)(\rho_f - \rho_g)g \sin\theta \tag{10.95}$$

(c) the momentum change τ_a in the vapour phase

$$\tau_a = \frac{D}{4}\frac{G^2}{\rho_g}\left(\frac{dx}{dz}\right)\left[a_1\left(\frac{\rho_g}{\rho_f}\right)^{1/3} + a_2\left(\frac{\rho_g}{\rho_f}\right)^{2/3} + \cdots a_5\left(\frac{\rho_g}{\rho_f}\right)^{5/3}\right] \tag{10.96}$$

where the coefficients a_1 to a_5 are given by

$$\left.\begin{aligned}
a_1 &= 2x - 1 - \beta x \\
a_2 &= 2(1-x) \\
a_3 &= 2(1-x-\beta+\beta x) \\
a_4 &= (1/x) - 3 + 2x \\
a_5 &= \beta(2 - (1/x) - x)
\end{aligned}\right\} \tag{10.97}$$

x is the vapour mass quality and β is the ratio of the mean liquid film velocity to the interfacial velocity ($\beta = 2$ for laminar flow and 1.25 for turbulent flow).

The void fraction correlation due to Zivi (1964) is used in the derivation of eq. (10.96) and strictly should also be used in eq. (10.95). In the case of upward flow of a condensing vapour, the gravitational term must be subtracted rather than added. Equation (10.93) has been verified over the Prandtl number range 2 to 10.

For the average heat transfer coefficient, Carpenter and Colburn (1951) suggested a simplified relationship arrived at by neglecting the gravitational and momentum terms and assuming τ_w equal to the interfacial shear stress τ_i:

$$\left[\frac{\bar{h}_f \mu_f}{k_f \rho_f^{1/2}}\right] = 0.065 \left[\frac{c_p \mu}{k}\right]_f^{1/2} \tau_i^{1/2} \tag{10.98}$$

where

$$\tau_i = f_i^1 \left[\frac{\bar{G}_g^2}{2\rho_g}\right] \tag{10.99}$$

and f_i^1 is an 'apparent' interfacial friction factor evaluated as for single-phase pipe flow at a mean vapour mass velocity $\bar{G}_g$ given by

$$\bar{G}_g = \left[\frac{G_1^2 + G_1 G_2 + G_2^2}{3}\right]^{1/2} \tag{10.100}$$

G_1 and G_2 are the vapour mass velocities at the beginning and exit of the section considered. If all the vapour is condensed $G_2 = 0$ and $\bar{G}_g = 0.58 G_1$.

It is of interest to note that eq. (10.98) (also eqs. (10.93) and (10.92)) can be rearranged in terms of the dimensionless groups h_f^* and τ_i^* thus

$$\bar{h}_f^* = 0.065 \mathrm{Pr}_f^{1/2} (\tau_i^*)^{1/2} \tag{10.101}$$

Equation (10.101) is valid for Prandtl numbers between 1 and 5 and τ_i^* from 5 to 150.

Rohsenow et al. (1956) carried out a similar analysis to that of Carpenter and Colburn by extending Seban's treatment to the case of a positive interfacial shear stress. The Prandtl–Nikuradse 'universal velocity profile' was used and momentum effects within the liquid film were neglected. In addition it was necessary to make a similar assumption for the critical transition value of the film Reynolds number. The results of this analysis were presented graphically in dimensionless form as $\bar{h}_f^*$ against film Reynolds number ($4\Gamma_z/\mu_f$) with the interfacial shear stress τ_i^* as parameter. In the area where eq. (10.98) was applicable there was good agreement with the Carpenter and Colburn analysis.

Dukler (1960) presented a rather more refined method of predicting the hydrodynamics and heat transfer in vertical film-wise condensation. Working from the definition of eddy viscosity and using the Deissler equation for its variation near a solid boundary, Dukler obtained velocity distributions in the liquid film as a function of the interfacial shear and film thickness. The differential equations were solved digitally by means of a computer. While a single universal velocity distribution is usual in full pipe flow, Dukler found that the velocity distribution curve in the liquid film depends both on the interfacial shear and on the film thickness. Assuming that the ratio of the eddy thermal diffusivity to the eddy viscosity was unity and that the physical properties of the fluid do not change in the direction of heat transfer, equivalent temperature profiles were constructed.

From the integration of the velocity and temperature profiles, liquid film thickness and point heat transfer coefficients were computed. The evaluated results of these two properties were displayed as a function of the Reynolds and Prandtl numbers with the interfacial shear as a parameter. Average condensing heat transfer coefficients were evaluated from the point values and are shown in Fig. 10.6 for the case of zero interfacial shear. The results agree with the classical Nusselt values at very low Reynolds numbers and with the empirical equations of Colburn (1933–1934) in the turbulent region. There is also good agreement with Seban's analysis at high Reynolds and Prandtl numbers. Figures 10.9 and 10.10 show the local heat transfer coefficients obtained from Dukler's analysis for the case of positive interfacial shear for Prandtl numbers of 1 and 5 respectively.

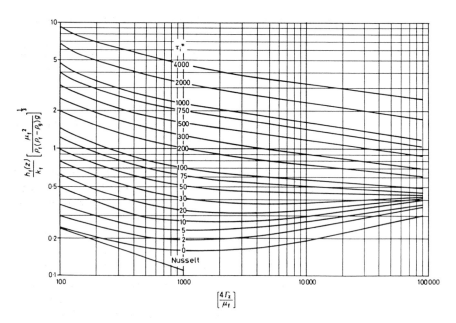

Fig. 10.9. Dukler analysis, Prandtl number $= 1.0$.

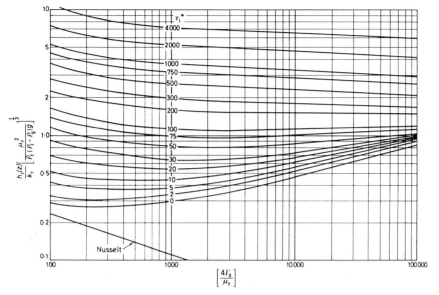

Fig. 10.10. Dukler analysis, Prandtl number = 5.0.

A number of objections can be raised to Dukler's analysis:

(a) the velocity distribution used was obtained from experiments in pipes with a different shear stress distribution;
(b) in defining the temperature profile at values of y^+ greater than 20, the molecular conductivity term is neglected with respect to the eddy conductivity.

Whilst this latter assumption is acceptable for fluids with values of Prandtl numbers around unity and higher, at low values of Prandtl number, the fall-off in heat transfer in the turbulent region is greatly over-estimated (Fig. 10.6). Lee (1964) has repeated Dukler's analysis for the case with no interfacial shear using an improved velocity profile and removing the above objection concerning the temperature profile. When this is done Lee obtains a solution which gives results very close to Dukler's (1960) at high Prandtl number and to Seban's (1954) at low Prandtl number.

There remains the question of correcting Dukler's analysis for positive values of interfacial shear stress. An analysis which does this has been given by Kunz and Yerazunis (1967). In addition to considering the points (a) and (b) raised above, these authors have also included interfacial resistance effects (Section 10.3) and the variation of the ratio of the eddy diffusivity of heat to the eddy diffusivity of momentum ($\varepsilon_H/\varepsilon$) with Prandtl and Reynolds number. Table 10.2 gives values of the local dimensionless heat transfer

Table 10.2   Values of $h_f^*(z)$ as a function of $[4\Gamma_z/\mu_f]$
and Prandtl number for the case of zero interfacial
shear (Kunz and Yerazunis, 1967)

Pr	$[4\Gamma_z/\mu_f]$			
	10^2	10^3	10^4	10^5
0.01	0.2661	0.1081	0.0357	0.0212
0.1	0.2661	0.1152	0.0852	0.1062
1.0	0.2700	0.2110	0.2993	0.495
5.0	0.3085	0.3778	0.6210	1.139

coefficient $h_f^*(z)$ (given by eq. (10.87)) as a function of the local Reynolds
number $[4\Gamma_z/\mu_f]$ and Prandtl number for the case of zero interfacial shear.
The usefulness of these theoretical predictions is confirmed by the remarkably
good agreement between values for local heat transfer coefficient along the
length of the tube calculated from Dukler's analysis and measured by
Carpenter and Colburn (1951).

10.5.3   *Condensation within a horizontal tube*

The problem of condensation within a horizontal tube is best examined by
reference to the flow patterns experienced in horizontal gas–liquid flow. Bell
et al. (1969) have reviewed the available correlations using such a framework.
The Baker flow pattern map (Fig. 1.6) forms a suitable starting point. A
useful suggestion made by Bell *et al.* (1969) concerns the preparation of an
overlay for use with Fig. 1.6. This overlay (Fig. 10.11), constructed to the
same scale, represents the path followed on the map during the course of
progressive condensation or evaporation at a constant mass velocity. If the
point marked $x = 0.5$ on the overlay is centred over the point on the Baker
map corresponding to the particular mass velocity G and a vapour mass
quality $x = 0.5$ then the line on the overlay can be used to determine the
range of quality over which each flow pattern may occur.

10.5.3.1   *Stratified flow*   The stratified flow pattern corresponds to a region
of small vapour velocity and low interfacial shear forces. In these circumstances the Nusselt equation for condensation on the outside of a horizontal
tube (eq. (10.71)) can be applied in a modified form. It is assumed that
laminar film of condensate runs down the inside upper surface of the tube and
collects as a stratified layer of liquid in the lower part of the tube (Fig. 10.12)

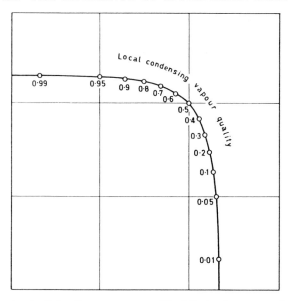

Fig. 10.11. Overlay for Baker flow pattern map (Fig. 1.6) showing course of condensation or evaporation (Bell *et al.* 1969).

$$\bar{h}_f = F\left[\frac{\rho_f(\rho_f - \rho_g)gi'_{fg}k_f^3}{D\mu_f(T_{gi} - T_w)}\right]^{1/4} \tag{10.102}$$

The factor F allows for the fact that the rate of condensation on the stratified layer of liquid running along the bottom of the tube is negligible. The value of F depends on the angle 2ϕ subtended from the tube centre to the chord forming the liquid level. Thus,

$$F = \left(1 - \frac{\phi}{\pi}\right)F^1 \tag{10.103}$$

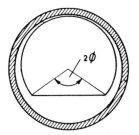

Fig. 10.12. Laminar condensation within a horizontal tube (after Nusselt).

Table 10.3 Values of the
factors F and F^1 for laminar
film stratified condensation
in a horizontal tube

ϕ^0	F^1	F
0	0.725	0.725
10	0.754	0.712
20	0.775	0.689
30	0.793	0.661
40	0.808	0.629
50	0.822	0.594
60	0.835	0.557
70	0.846	0.517
80	0.857	0.476
90	0.866	0.433
100	0.874	0.389
110	0.881	0.343
120	0.887	0.296
130	0.892	0.248
140	0.896	0.199
150	0.899	0.150
160	0.902	0.100
170	0.903	0.050
180	—	0

where values of F^1 and F are given as a function of ϕ in Table 10.3. In the limit as 2ϕ approaches zero F equals 0.725; cf. eq. (10.71).

Chato (1962) concluded that the heat transfer coefficient was not particularly sensitive to the angle 2ϕ and suggested a mean value of 120°. In the study by Chato and the earlier analysis of Chaddock (1957) it was assumed that the vapour is stagnant and that the condensate flows under a hydraulic gradient (Fig. 10.12a). In this case, the angle 2ϕ decreases with length along the tube. The alternative and perhaps more common situation (Rufer and Kezios 1966) is when there is a pressure gradient and the condensate at the tube outlet fills the tube cross-section (Fig. 10.13b). In this latter case, the angle 2ϕ increases with length along the tube and may be related to the void fraction α by the following relationship if the condensate film is of negligible thickness:

$$(1 - \alpha) = \frac{\phi - \frac{1}{2}\sin 2\phi}{\pi} \tag{10.104}$$

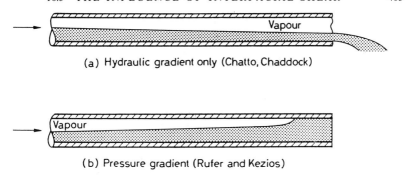

(a) Hydraulic gradient only (Chatto, Chaddock)

(b) Pressure gradient (Rufer and Kezios)

Fig. 10.13. Exit conditions in stratified condensation.

10.5.3.2   *Slug, plug and wavy flow*   These flow patterns, although still basically stratified, correspond to situations where the vapour shear forces are significant. Using the nomenclature of Lockhart and Martinelli the flow regime is viscous–turbulent (vt) ($G_gD/\mu_g > 2000$, $G_fD/\mu_f < 2000$). Point measurements of the condensing film coefficients for methanol and acetone have been reported by Rosson and Myers (1965) for condensation within a horizontal tube under such conditions. Coefficients were found to vary greatly from the top to the bottom of the tube. Near the top of the tube the film coefficient was independent of the amount of liquid in the tube and was correlated by including the effect of vapour shear on F in eq. (10.102). Thus

$$F = 0.31\left[\frac{G_gD}{\mu_g}\right]^{0.12}$$ (10.105)

Near the bottom of the tube the analogy between heat and momentum transfer was used to predict the coefficient. Thus

$$\left[\frac{h_f(z)D}{k_f}\right] = \frac{\phi_{fvt}\sqrt{[8(G_fD/\mu_f)]}}{5[1 + (1/Pr_f)\ln(1 + 5Pr_f)]}$$ (10.106)

where ϕ_{fvt} is the square root of the two-phase pressure drop multiplier (defined by eq. (2.48)) for the case of viscous–turbulent flow. The appropriate equation for evaluating ϕ_{fvt} is eq. (2.68) with C set to a value of 12.

An example of the variation of heat transfer coefficient around the tube circumference for methanol is given in Fig. 10.14. At some angle, θ_m, the local heat transfer coefficient is the arithmetic average of the upper and lower tube coefficients. Rosson and Myers found that the angle θ_m was a function of the liquid and vapour superficial Reynolds numbers and a grouping designated the Galileo number, Ga:

$$Ga = \left[\frac{D^3\rho_f(\rho_f - \rho_g)g}{\mu_f^2}\right]$$ (10.107)

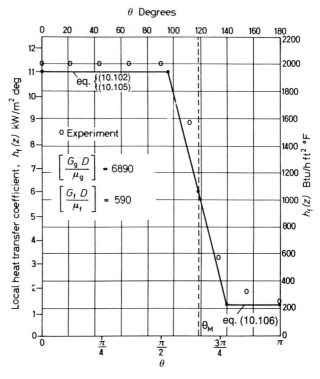

Fig. 10.14. Circumferential variation of local condensing heat transfer coefficient (Rosson and Myers 1965).

This relation is shown in Fig. 10.15 where (θ_m/π) is plotted against the superficial vapour Reynolds numbers with $[Ga/(G_fD/\mu_f)^{0.5}]$ as parameter. Marked on Fig. 10.15 is the value of (θ_m/π) corresponding to the conditions of the example given in Fig. 10.14. The slope $(d(h_f(z))/d\theta)$ at θ_m was found to be a function of the product of the superficial vapour Reynolds number and the mean temperature drop through the condensate film, $(T_{gi} - T_W)$. Thus

$$\left[\frac{d(h_f(z))}{d\theta}\right]_{at\ \theta_m} = 0.835\left[\frac{G_gD}{\mu_g}(T_{gi} - T_W)\right]^{-0.6} \tag{10.108}$$

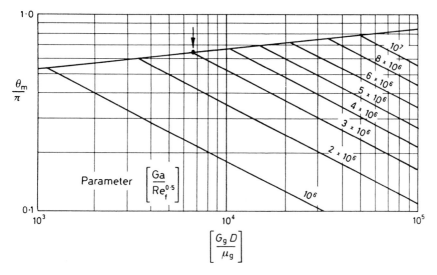

Fig. 10.15. Location of midpoint of heat transfer coefficient vs angle (Rosson and Myers 1965).

where $h_f(z)$ is measured in kW/m^2 °C, θ in radians, and $(T_{gi} - T_W)$ in °C, or

$$\left[\frac{d(h_f(z))}{d\theta}\right]_{\text{at } \theta_m} = 12{,}000\left[\frac{G_g D}{\mu_g}(T_{gi} - T_W)\right]^{-0.6} \qquad (10.109)$$

where $h_f(z)$ is measured in Btu/h ft^2 °F, θ in degrees, and $(T_{gi} - T_W)$ in °F.

10.5.3.3 *Annular flow* In this flow pattern, well away from the boundaries with wavy and slug flow, the correlations of Carpenter and Colburn (1951) (eq. (10.92)), of Soliman *et al.* (1968) (eq. (10.93), of Dukler (1960) and of Kunz and Yerazunis (1967) may be used. In the evaluation of the wall shear stress, τ_W, the gravitational force, τ_z (cf. eq. (10.95)), is not now included. Ananiev *et al.* (1961) propose a particularly simple relationship to correlate their data for the condensation of steam in a horizontal tube at elevated pressures. The analogy between liquid film flow and single-phase flow in a pipe is used (cf. Section 7.4) to correlate their results

$$h_f(z) = h_{fo}\sqrt{\frac{\rho_f}{\bar\rho}} \qquad (10.110)$$

where h_{fo} is the heat transfer coefficient to liquid flowing at the same total flow-rate in the tube and $\bar\rho$ is the homogeneous mean density of the vapour–liquid mixture given by eq. (2.23).

The average coefficient $\bar h_f$ for complete condensation of a vapour is approximately given by

$$\bar{h}_f = \frac{h_{fo}}{2}\left[1 + \left(\frac{\rho_f}{\rho_g}\right)^{1/2}\right] \tag{10.111}$$

These relationships are recommended for values of (ρ_f/ρ_g) less than 50.

10.5.3.4 *Design correlations*   For general design practice, completely empirical methods are most often used. Most of these are modifications of the Dittus–Boelter single-phase forced convection correlation, eq. (5.7). Akers *et al.* (1959) presented the following method based on data for several refrigerants and an organic, valid for the calculation of the local condensing coefficient $h_f(x)$ over the whole range of vapour quality:

$$\frac{h_f(x)D}{k_f} = C\,Re_e^n Pr_f^{1/3} \tag{10.112}$$

where the equivalent Reynolds number ($Re_e = DG_e/\mu_f$) for the two-phase flow is calculated for an equivalent mass flow-rate defined as

$$G_e = G\left[(1 - x) + x\left(\frac{\rho_f}{\rho_g}\right)^{1/2}\right] \tag{10.113}$$

where G is the total mass flow-rate. The values of parameters C and n are as follows:

$C = 0.0265$ and $n = 0.8$ for $Re_e > 50,000$
$C = 5.03$ and $n = 1/3$ for $Re_e < 50,000$

For annular flow regimes only, Cavallini and Zecchin (1974) developed a similar correlation.

Shah (1979) has developed an accurate correlation applicable to steam, refrigerants, and organics using the Dittus–Boelter correlation as a starting point and given as

$$\frac{h_f(x)D}{k_f} = 0.023Re_f^{0.8}\,Pr_f^{0.4}\left[(1 - x)^{0.8} + \frac{3.8x^{0.76}(1 - x)^{0.04}}{P_r^{0.38}}\right] \tag{10.114}$$

where P_r is the local reduced pressure. For an inlet vapour quality of 1.0, a value of 0.999 should be used to keep the method from breaking down. For complete condensation, this expression correctly goes to the single-phase value at $x = 0$. However, the flow regime may be transition or laminar flow at this point and an appropriate correlation should be used.

The Shah method is recommended for design when the mass velocity is greater than $200\ kg/m^2\,s$ while the Akers *et al.* method is better at lower mass velocities. Table 10.4 shows the Shah method evaluated for refrigerant R-134a condensing at 40°C (10.2 bar) inside a 3 m long, 8.72 mm internal

Table 10.4   Condensation of R134a in a horizontal
tube

Vapour quality	$h_f(x)$ (W/m² °C)	Vapour quality	$h_f(x)$ (W/m² °C)
1.0	3917	0.375	2704
0.875	4066	0.25	2212
0.75	3833	0.125	1628
0.625	3512	0.0	771
0.5	3135		

diameter tube at a mass velocity of 300 kg/m² s. As can be seen, the local value of $h_f(x)$ changes significantly along the tube with decreasing vapour quality.

10.5.4   Condensation outside a horizontal tube and tube bundles

The influence of vapour shear on the heat transfer coefficient for flow across a horizontal tube has been considered analytically by Fujii *et al.* (1972), by Lee and Rose (1982) and by Shekriladze and Gomelauri (1966). The latter assumed that the interfacial shear occurs solely as a result of the momentum lost by the condensing vapour. Assuming no separation of the laminar vapour boundary layer and laminar flow within the condensate film, the following equation is given for the case where the cross flow is downward over the tube:

$$\left[\frac{\bar{h}_f D}{k_f}\right] = 0.9 \left[\frac{\rho_f u_{g\infty} D}{\mu_f}\right]^{0.5} \tag{10.115}$$

where $u_{g\infty}$ is the vapour velocity approaching the tube. Equation (10.115) is a reasonable agreement with the results of Berman and Tumanov (1962). At high values of Reynolds number ($>10^6$) separation of the boundary layer occurs and the experimental results fall below eq. (10.115) and approach a parallel line corresponding to separation at an angle of $\theta = 82°$. The equation for this lower line is the same as eq. (10.115) but with the 0.9 replaced by a lower factor (0.59). Condensation, however, tends to inhibit boundary layer separation. For lower velocities where both gravitational and shear forces are important,

$$h_f = \left[\tfrac{1}{2}h_{SHEAR}^2 + (\tfrac{1}{4}h_{SHEAR}^4 + h_{GRAV}^4)^{1/2}\right]^{1/2} \tag{10.116}$$

where h_{SHEAR} is given by eq. (10.115) and h_{GRAV} is given by eq. (10.71).

Honda and Fujii (1974) extended this analysis to include vapour flows at various angles of incidence. Their results for horizontal flows are almost identical to the down flow result. However, at high vapour velocities and low temperature differences, Hawes (1976) has observed a marked reduction in the condensing coefficient by as much as a factor of 2. Butterworth (1977) has suggested that this is due to a dynamic effect whereby local reductions in static pressure occur in the regions of high velocity along the sides of the tube. This would effectively cause a reduction in the local saturation temperature and hence the temperature driving force.

Equations (10.115) and (10.116) can be applied to condensation within tube bundles providing a satisfactory method of estimating $u_{g\infty}$ is available. Butterworth proposes that this should be evaluated as (u_{gs}/α) where u_{gs} is the velocity calculated in the absence of any tubes and α is the void fraction for the bundle (i.e., free volume divided by total volume). This approach was tested using the data of Nobbs and Mayhew (1976) for downflow of atmospheric pressure steam in a tube bank (Fig. 10.16). The data tend to lie between the prediction with and without flow separation with the data at the higher temperature difference following the no flow separation line and at low temperature differences the theory allowing for flow separation.

Condensation on horizontal tube bundles is affected by the interactive effects of condensate inundation and vapour shear. The liquid film on the tubes is thinned by the downward cocurrent flow of high velocity vapour. The condensate drainage process from tube to tube is also affected by large vapour velocities. At low vapour velocities instead, the condensate flow is only governed by gravity. Recognizing the strong effect of vapour shear, McNaught (1982) proposed to model the process as two-phase forced convection similar to that occurring for in-tube and shell-side boiling. Thus, he arrived at the following expression for the shear-controlled heat-transfer coefficient

$$h_{\text{SHEAR}} = 1.26(1/X_{\text{tt}})^{0.78}h_{\text{f}} \tag{10.117}$$

where h_{f} is the liquid-phase convective heat-transfer coefficient for liquid flowing alone across the bundle and X_{tt} is the Lockhart–Martinelli parameter

$$X_{\text{tt}} = \left(\frac{1-x}{x}\right)^{0.9}\left(\frac{\rho_{\text{g}}}{\rho_{\text{f}}}\right)^{0.5}\left(\frac{\mu_{\text{f}}}{\mu_{\text{g}}}\right)^{0.1} \tag{10.118}$$

The gravity-controlled regime is calculated using the Nusselt equation for a single tube (eq. (10.71)) together with the inundation effect as

$$h_{\text{GRAV}} = h_{\text{f}1}[(W_{\text{fi}} + W_{\text{fN}})/W_{\text{fN}}]^{-\gamma} \tag{10.119}$$

where W_{fi} is the inundation rate from the above tubes and W_{fN} is the condensation rate on the Nth tube row. The exponent γ is 0.13 and 0.22 for triangular and square tube layouts, respectively. The local condensing heat transfer coefficient is then calculated from the superposition of these

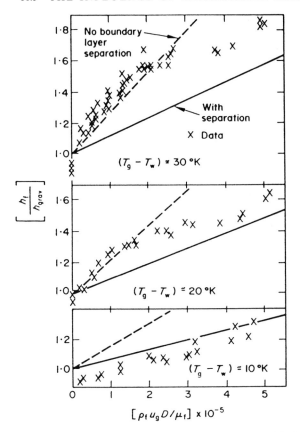

Fig. 10.16. Comparison of modified Shedriladze–Gomelauri model with data of Nobbs and Mayhew (1976) and Butterworth (1977).

contributions as

$$h = (h_{SHEAR}^2 + h_{GRAV}^2)^{1/2} \qquad (10.120)$$

The above method was developed with the data of Nobbs and Mayhew (1976) for steam condensing at atmospheric pressure.

More recently, a generalized method has been proposed by Honda et al. (1989). This method is more complex but accurately predicts R-11, R-12, R-21, R-113 and steam data from different sources. Its exactness also has been verified by Cavallini et al. (1990) for their recent R-11 and R-113 data, together with a suggested improvement for the first tube row effect in the shear flow regime.

For *staggered tube bundles* of plain tubes the Honda et al. (1989) general expression is

$$Nu = [Nu_{GRAV}^4 + (Nu_{GRAV}Nu_{SHEAR})^2 + Nu_{SHEAR}^4]^{1/4} \qquad (10.121)$$

where the Nusselt number in the gravity-controlled regime is

$$Nu_{GRAV} = Gr^{1/3}[(1.2Re_{GRAV}^{-0.3})^4 + (0.072Re_{GRAV}^{0.2})^4]^{1/4} \qquad (10.122)$$

and for the shear-controlled regime is

$$Nu_{SHEAR} = 0.165(s_1/s_2)^{0.7}\left[Re_g^{-0.4} + 1.83\left(\frac{\phi_n}{i_{fg}\rho_g u_n}\right)\right]^{1/2}(\rho_g/\rho_f)^{1/2}\,Re_f\,Pr_f^{0.4}$$

$$(10.123)$$

The Grashof number is given as

$$Gr = g\left(\frac{\rho_f - \rho_g}{\rho_f}\right)\left(\frac{D^3}{v_f^2}\right) \qquad (10.124)$$

The gravitational liquid Reynolds number is defined as

$$Re_{GRAV} = 2\pi D \sum_{j=1}^{(n+1)/2} [\phi_{2j-1}/(\mu_f i_{fg})] \qquad (10.125)$$

for odd number tubes n (1, 3, 5, etc.), and for the even number tubes n (2, 4, 6, etc.) is

$$Re_{GRAV} = 2\pi D \sum_{j=1}^{n/2} [\phi_{2j}/(\mu_f i_{fg})] \qquad (10.126)$$

The liquid shear Reynolds number is equal to Re_{GRAV} for the first tube row ($n = 1$) and for lower tube rows is obtained from

$$Re_{SHEAR} = 2\pi D\left[\sum_{j=1}^{n-1}(\phi_j D/s_1) + \phi_n\right]\bigg/(\mu_f i_{fg}) \qquad (10.127)$$

For the first tube row, the leading constant on the right-hand side of eq. (10.123) should be replaced by 0.13.

For *in-line tube bundles* of plain tubes the Honda et al. (1989) general expression is

$$Nu = [Nu_{GRAV}^4 + Nu_{SHEAR}^4]^{1/4} \qquad (10.128)$$

where the Nusselt number in the gravity-controlled regime is given by eq. (10.122) and for the shear-controlled regime is

$$Nu_{SHEAR} = 0.053\left[Re_g^{-0.2} + 18.0\left(\frac{\phi_n}{i_{fg}\rho_g u_n}\right)\right]^{1/2}\left(\frac{\rho_g}{\rho_f}\right)^{1/2}\frac{Re_f\,Pr_f^{0.4}}{Re_{SHEAR}^{0.2}} \qquad (10.129)$$

The Grashof number is given by eq. (10.124) and the gravitation Reynolds number is defined as

$$Re_{GRAV} = 2\pi D \sum_{j=1}^{n} [\phi_j/(\mu_f i_{fg})] \qquad (10.130)$$

The shear Reynolds number is equal to Re_{GRAV} in eq. (10.130) for the first tube row ($n = 1$) and for lower tube rows is obtained from

$$Re_{SHEAR} = 2\pi D \left[\sum_{j=1}^{n-1} (\phi_j D/s_1) + \phi_n \right] \Big/ (\mu_f i_{fg}) \qquad (10.131)$$

For the first tube row, the leading constant on the right-hand side of eq. (10.129) should be replaced by 0.042.

In the above equations, the vapour velocity at the nth row (based on minimum flow cross-sectional area) is

$$u_1 = \left(\frac{s_1}{s_1 - D} \right) \left(\frac{G}{\rho_g} \right) \qquad \text{for } n = 1 \qquad (10.132a)$$

$$u_n = \left[\frac{Gs_1 - \pi D \sum_{j=1}^{n-1} (\phi_j/i_{fg})}{(s_1 - D)\rho_g} \right] \qquad \text{for } n > 1 \qquad (10.132b)$$

where G is the vapour mass velocity at the tube bundle inlet and ϕ_j is the heat flux on the jth row tube. The liquid Reynolds number is defined as

$$Re_f = u_n D/\nu_f \qquad (10.133)$$

and the vapour Reynolds number is given as

$$Re_g = u_n D/\nu_g \qquad (10.134)$$

The Honda et al. (1989) correlation is for downward flowing vapour only.

10.6 Drop-wise condensation

Drop-wise condensation occurs on non-wettable surfaces. The necessary reduction in the interfacial tension between the condensate and the surface may be brought about as a result of contamination, accidental or deliberate, i.e., removal of wettable oxide films or by electroplating. The condensate appears in the form of droplets which grow and coalesce with adjacent droplets. Subsequently they roll or fall from the surface under the action of gravity or aerodynamic drag forces. New droplets then appear on the exposed clear surface.

The mechanism of drop-wise condensation is still a mystery and quite contradictory statements are common in the literature. The amount of published information is extensive since drop-wise condensation offers the attraction of increasing condensing heat transfer coefficients by an order of magnitude over film-wise condensation.

Two fundamentally different models have been proposed. In the first it is postulated that condensation initially occurs in a film-wise manner on a thin unstable liquid film covering all or part of the surface. On reaching a critical thickness the film ruptures and the liquid is drawn into droplets by surface tension forces. This process then repeats itself. This mechanism was first put forward by Jakob (1936) and has been reiterated since in a number of modified forms by Kast (1963) and Silver (1964) amongst others. The model has been supported by the findings of a number of studies including Baer and McKelvey (1958), Welch and Westwater (1961), and Sugawara and Katsuta (1966), which show condensation occurring entirely between drops in a very thin film. Welch and Westwater examined the process by taking highspeed photographs with a microscope. They concluded that droplets large enough to be visible (0.01 mm) grew mainly by coalescence leaving a 'lustrous bare area'. This lustre quickly faded and they explained this in terms of the build up of the thin film which fractured at a thickness of 0.5 to 1 μm.

Silver (1964) has attempted a quantitative treatment of drop-wise condensation based on radially inward drainage of condensate towards a droplet. This model gives the ratio between drop-wise and film-wise condensation rates as

$$\left[\frac{j_D}{j_F}\right] = \left[\frac{\rho_f^2 D^2 g}{24.2\mu_f j_F}\right]^{1/9} \tag{10.135}$$

where j_D is the drop-wise condensation mass flux and j_F is the film-wise condensation mass flux under identical conditions. For steam at atmospheric pressure for a value of j_F equal to 1.36×10^{-2} kg/m^2 s (10 lb/ft^2 h) this ratio is 6.5 and the mean thickness of the thin film is about 2 μm.

In the alternative model of drop-wise condensation it is assumed that droplet formation is fundamentally a heterogeneous nucleation process as discussed in Section 10.2.1. This concept was first proposed by Eucken (1937). Studies by Umur and Griffith (1965) and by Erb and Thelen (1965) support the nucleation mechanism. Using an optical method with polarized light to indicate changes in the thickness of liquid films of molecular dimensions, Umur and Griffith were able to establish that, for low temperature differences at least, the area between drops has no liquid film greater than a monolayer in thickness and that no net condensation takes place in that area.

McCormick and Baer (1963) have suggested that innumerable submicroscopic droplets are randomly nucleated at active sites on the condenser surface. These active sites are wetted pits and grooves in the surface which are continually being exposed by numerous drop coalescences and by large drops falling from the surface. Evidence to support this picture is being accumulated (McCormick and Baer 1963, Gose et al. 1967). Grigull has produced photographic evidence that preferred nucleation sites occur on the surface around the circumference of a previously departed condensate drop.

The drops grow by direct condensation and by coalescence. The rate of direct condensation on the larger droplets is less than for the smaller droplets because of the resistance to heat conduction through the drop. For the case of a hemispherical drop the growth rate is given by

$$\frac{dr^2}{dt} = 4.15 \left[\frac{k_f (T_{gi} - T_w)}{i_{fg} \rho_f} \right] \tag{10.136}$$

The larger drops therefore grow mainly by coalescence. Coalescence between smaller drops is negligible. The small drops grow mainly by condensation and are responsible for the major fraction of the heat transferred. During drop-wise condensation of steam at 1 bar pressure about 60 per cent of the surface is covered with drops greater than 50 μm, 10 per cent of the surface is bare and the remaining 30 per cent is occupied by drops of radius less than 50 μm. This latter fraction of the surface transfers 90 per cent of the heat.

Gose et al. (1967) have developed a model for drop-wise condensation which accounts for drop nucleation and growth, removal and re-nucleation on sites exposed by the removal and coalescence nucleation. Nucleation site densities are approximately 10^7 sites/cm^2.

Surface temperature fluctuations (Ohtani et al. 1973) occur as a result of the unsteady nature of the nucleation, growth and coalescence process and because of the large heat flows through the relatively much reduced surface area occupied by the smaller droplets. As a result the thermal diffusivity of the condenser surface has an important influence (Waas et al. 1982) through transient heat conduction processes near the surface. Typically the dropwise condensation heat-transfer coefficient on stainless steel ($k_w = 10 \, kW/m \, K$) is about half that on copper ($k_w = 400 \, kW/m \, K$).

A very considerable amount of work remains to be carried out to establish satisfactory explanations of all the published experimental results on drop-wise condensation. Whilst the droplet nucleation mechanism is certainly the more likely at low condensation rates (i.e., temperature differences up to 5°C) it is possible that at higher condensation rates there may be a film disruption mechanism as an intermediate stage before establishing fully developed film-wise condensation.

Figure 10.17 shows the results of Takeyama and Shimizu (1974) which indicate the transition to film-wise condensation at high temperature differences. At low temperature differences Le Fevre and Rose (1966) report drop-wise heat transfer coefficients which increase slightly with increasing temperature difference ($h_D = 0.2 \, MW/m^2 \, °C$ at 1°C and $h_D = 0.27 \, MW/m^2 \, °C$ at 4°C). At high temperature differences Welch and Westwater (1961) indicate a falling heat transfer coefficient with increasing temperature difference. This may be due to a change in mechanism or equally well to the presence of non-condensable gases in the vapour phase.

The influence of pressure on the drop-wise coefficient for steam at a fixed

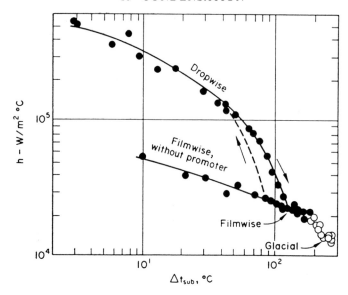

Fig. 10.17. Heat transfer coefficients for condensation of steam on a short vertical copper surface, from Takeyama and Shimizu (1974) (palmitic acid used as promoter for drop-wise condensation).

heat flux is shown in Fig. 10.18. For pressures above 3 bar, O'Bara *et al.* (1967) report that drop-wise condensation is gradually replaced by a combination of drop and film condensation and finally by film-wise condensation. This is probably brought about by the reduction in the surface tension of water at higher temperatures. These same workers also report on the influence of vapour velocity. For a given temperature difference, the heat transfer coefficient increased up to a maximum value with increased vapour velocity up to 2 m/s and then started to decrease. This was believed to be caused by the increased coalescence between droplets blanketing the surface with a film of condensate.

10.7   Pressure gradient in condensing systems

In condensing or evaporating systems the pressure gradient and the interfacial shear stress (in annular flow) may be conveniently estimated using the techniques developed in Chapters 2 and 3. In these conventional methods, the process of momentum transfer is considered to be independent of the process of vapour mass transfer to or from the liquid film. If the mass transfer rate is large, however, the momentum transfer will be influenced by the net vapour transport to or from the interface.

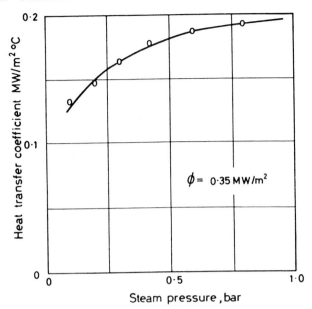

Fig. 10.18. Drop-wise heat transfer coefficients as a function of steam pressure (LeFevre and Rose 1966).

In two-phase heat transfer equipment, the two-phase pressure drop has a direct influence on the log-mean-temperature difference (LMTD) through the variation on the local saturation temperature. Typically, in practice the momentum recovery (from the accelerational pressure drop) in condensation is ignored since only part of it will actually be realized.

10.7.1   *The influence of condensation on the interfacial shear stress*

One simple way of considering this effect is in terms of the well-known Reynolds analogy. This approach has been discussed by Silver (1964) and elaborated by Wallis (1969). The Reynolds analogy, in its simplest form, considers the shear stress at a boundary to be due to some fraction of the flowing fluid hitting the surface and rebounding after giving up its momentum to the surface. If the mass flux hitting the interface is ε_{go} and mean gas and liquid velocities are u_g and u_f respectively (Fig. 10.19(a)) then the interfacial shear stress in the absence of phase change is

$$(\tau_i)_o = \varepsilon_{go}(u_g - u_f) \tag{10.137}$$

Consider what happens in the case of a condensation mass flux j towards

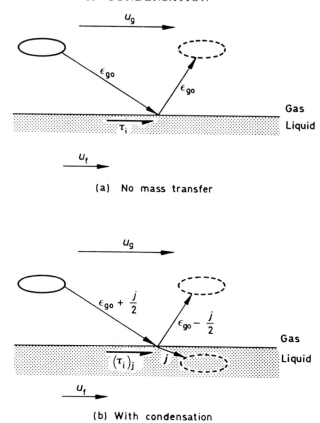

(a)   No mass transfer

(b) With condensation

Fig. 10.19. Reynolds analogy applied to interfacial shear stress in presence of condensation (after Silver 1964).

the interface from the gas phase. This can be brought about quite simply by superimposing a mass flux $j/2$ on the situation in the absence of phase change (Fig. 10.19(b)). The interfacial shear stress acting on the liquid in the presence of a condensation flux j is then

$$(\tau_i)_j = (\varepsilon_{go} - j/2)(u_g - u_f) + j(u_g - u_f) \qquad (10.138)$$

The first term in this expression can be related to the frictional pressure gradient and the second term to the momentum pressure gradient. Thus, although the interfacial shear stress increases as a result of the momentum lost by the condensing vapour, the frictional pressure gradient actually decreases. The interfacial shear stress acting on the vapour is therefore

$$(\tau_i)_g = (\varepsilon_{go} - j/2)(u_g - u_f) \tag{10.139}$$

$$= (\tau_i)_o \left[1 - \frac{1}{2} \left(\frac{j}{\varepsilon_{go}} \right) \right] \tag{10.140}$$

As the condensation flux j becomes comparable with ε_{go} eq. (10.140) becomes inaccurate, and a better expression has been derived by Spalding (1963)

$$(\tau_i)_g = (\tau_i)_o \left[\frac{a}{e^a - 1} \right] \tag{10.141}$$

where $a = j/\varepsilon_{go}$.

The value of ε_{go} may be estimated from the interfacial friction factor since

$$(\tau_i)_o = f_i \left[\frac{\rho_g(u_g - u_f)^2}{2} \right] \tag{10.142}$$

and

$$\varepsilon_{go} = f_i \left[\frac{\rho_g(u_g - u_f)}{2} \right] \tag{10.143}$$

If the interfacial friction factor is defined in terms of the superficial vapour velocity then eq. (10.143) becomes

$$\varepsilon_{go} = f_i' \left[\frac{\rho_g j_g}{2} \right] = f_i' \left[\frac{Gx}{2} \right] \tag{10.144}$$

The procedure for calculating pressure gradient and heat transfer coefficients in condensing systems with annular flow can be summarized in the following steps.

(a) From the known mass velocity, vapour mass quality, physical properties, and channel geometry calculate the frictional pressure gradient term using the methods given in Chapter 2.

(b) From this estimated pressure gradient compute consistent values for the local liquid film thickness and film flow-rate using the procedure outlined in Section 3.5.4.

(c) From a knowledge of the film thickness and the film Reynolds number estimate the local heat transfer coefficient across the liquid film using Table 7.1. Taking into account heat transfer resistances within the vapour core due to non-condensable gases, the tube wall and the cooling water, estimate a local overall heat transfer coefficient and condensing mass flux.

(d) Compute a value of $(\tau_i)_o$ from the previous estimated frictional pressure gradient and correct this for the influence of vapour mass transfer using eq. (10.140) or (10.141) to obtain $(\tau_i)_g$. From a knowledge of the condensing mass flux j calculate dx/dz $(= -4j/DG)$.

(e) Recompute the total pressure gradient in the vapour core taking into account the reduction in interfacial shear stress due to vapour mass transfer and the momentum lost by the condensing vapour. Return to

step (b) and repeat the calculation to obtain new estimates of the heat
transfer coefficient across the film and the local condensation rate.

As an alternative to the use of Table 7.1 in step (c) any of the methods
discussed in Section 10.5 may be used with the appropriate value of the
interfacial shear stress $(\tau_i)_g$ (for example in eq. (10.98)).

Sardesai *et al.* (1982) have examined the predictions from various interfacial
shear stress relationships for pressure drops in in-tube condensation in a
vertical tube in down flow. Starting with eq. (2.46) as a basis, the experimental
data were analysed using a simplified version of the annular flow model
discussed in Chapter 3. The data bank covered a wide range of fluids
including steam, methanol, R-113, R-11, R-21, and R-114. Tube diameters
varied from 7.5 to 25.4 mm and mass velocities up to 1000 kg/m² s. The
annular flow model was used to calculate the frictional component and the
void fraction was used for the gravitational and accelerational terms. The
correction for mass transfer effects as discussed above was included. For
some of the data reported the mass transfer effect increased the frictional
component by as much as 100 per cent. This effect was largest in regions of
high vapour shear and high condensation flux near the tube entrance. Two
interfacial roughness relationships were tested—that due to Wallis (eq. (3.35)
but with the coefficient multiplying δ/D as 360 rather than 300) and a
modification due to Whalley and Hewitt (1978):

$$\frac{f_i}{f_g} = 1 + 24 \left(\frac{\rho_f}{\rho_g}\right)^{1/3} \frac{\delta}{D} \qquad (3.35a)$$

It was found that the former tended to somewhat over-predict the pressure
drops, whereas the latter relationship (eq. 3.35a) led to an underprediction.

10.7.2   *Pressure drop in condenser tube banks*

The change in pressure when a condensing vapour passes over a tube bank
is particularly difficult to estimate, especially when the absolute pressures
are low and kinetic effects are important as in most large condensers. Little
published information has appeared in this area, primarily because of the
considerable difficulties in obtaining reliable and meaningful experimental
data. Many condensers are constructed as baffled shell-and-tube heat
exchangers and the study of two-phase flow in such geometries is at a very
early stage.

Comments in this section are therefore limited to the situation where the
condensing vapour passes in crossflow through a tube bundle. Pressure
changes occur due to both frictional (drag) effects and momentum changes.
Changes in velocity due to changes in cross-section and due to condensation

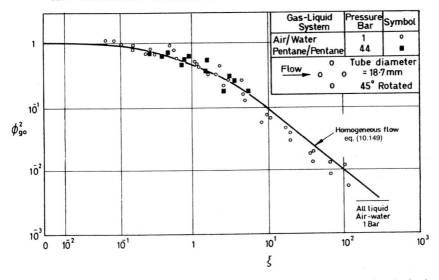

Fig. 10.20. Two-phase pressure drop for turbulent horizontal cross flow through tube bank with 45° layout (Diehl and Unruh 1958).

both contribute to the latter component. The frictional effects are composed of both skin friction and form drag components. Two-phase pressure drop tests on tube banks using various gas–liquid mixtures have been reported by Diehl (1957) and Diehl and Unruh (1958). Their experimental results were plotted as ϕ_{go}^2 versus a parameter ξ where

$$\xi = \frac{Q_f}{Q_f + Q_g} \frac{v_g}{v_f} = (1 - \beta)\frac{v_g}{v_f} \qquad (10.145)$$

Experimental data for horizontal cross flow through a tube bank with a rotated square pitch are shown in Fig. 10.20 plotted as ϕ_{go}^2 versus ξ. Wallis (1969) has compared this result with that expected from the homogeneous mode. In this case

$$\phi_{go}^2 = \frac{(dp/dzF)}{(dp/dzF)_{go}} = \frac{2f_{TP}G^2\bar{v}}{2f_{go}G^2v_g} = \frac{f_{TP}\bar{v}}{f_{go}v_g} \qquad (10.146)$$

Now

$$\frac{1}{\bar{v}} = \frac{(Q_f/v_f) + (Q_g/v_g)}{Q_f + Q_g} = \frac{1 - \beta}{v_f} + \frac{\beta}{v_g} \qquad (10.147)$$

Multiplying through by v_g and substituting into eq. (10.146) gives

$$\phi_{go}^2 = \frac{f_{TP}}{f_{go}}\left[\frac{1}{(1-\beta)(v_g/v_f)+\beta}\right] = \frac{f_{TP}}{f_{go}}\left[\frac{1}{\xi+\beta}\right] \qquad (10.148)$$

Assuming that the two-phase friction factor is equal to that for single-phase gas flow and substituting for β in terms of ξ from eq. (10.145) yields

$$\phi_{go}^2 = \left[\frac{1}{1+\xi(1-v_f/v_g)}\right] \approx \left[\frac{1}{1+\xi}\right] \qquad (10.149)$$

As can be seen from Fig. 10.20 this relationship is in satisfactory agreement with the experimental data.

These results were taken in the absence of mass transfer and it is to be expected that some reduction of form drag as well as skin friction may result when condensation occurs within a tube bundle. Wallis cites the experiments of Drummond (1966) which confirm this expectation. The reduction in form drag can be attributed to the retardation or perhaps a complete prevention of flow separation behind the tubes as a result of the boundary layer suction effect produced by condensation.

Ishihara et al. (1980) have reported on the results of their comprehensive and critical review of existing two-phase frictional pressure drop correlations for two-phase flows across tube bundles and compared them against available published data. They found no correlation that satisfactorily predicted all the data. Re-examining the Martinelli separated flow approach, they tried various mathematical forms of pressure drop correlations based on the Lockhart–Martinelli parameter X_{tt} given by eq. (10.118). Similar to eq. (2.68), they proposed the following general expression for $Re_f > 2000$:

$$\phi_{fo}^2 = 1 + \frac{8}{X_{tt}} + \frac{1}{X_{tt}^2} \qquad (10.150)$$

as that which best fits the data. This equation quite accurately predicted two-phase frictional pressure drops for values of X_{tt} from 0.002 to 20, i.e. high vapour fractions. For $Re_f < 2000$ they proposed the expression

$$\phi_{go}^2 = 1 + CX_{tt} + X_{tt}^2 \qquad (10.151)$$

which is similar to eq. (2.69). The new value proposed for C was not revealed.

References and further reading

Ackermann, G. (1937) 'Simultaneous heat and mass transfer with large temperature and partial pressure differences' (in German). *Ver. deutscher Ing. Forsch.*, **8** (382), 1–16.

Akers, W. W., Deans, H. A., and Crosser, O. K. (1959) 'Condensation heat transfer within horizontal tubes'. *Chem. Eng. Prog. Symp. Ser.*, **55**, 171–176.

Akers, W. W., Davies, S. H., and Crawford, J. E. (1960). 'Condensation of a vapour in the presence of a non-condensing gas'. *Chem. Engng. Prog. Symp. Series* No. 30, **56**, 139–144.

Aladyev, I. T., Kondratyev, N. S., Mukhin, V. A., Kpshidze, M. E., Porfentyeva, I., and Kisselev, V. V. (1966) 'Thermal resistance of phase transition with condensation of potassium vapour'. *Proc. 3rd Int. Heat Transfer Conf.*, **2**, 313–317.

Ananiev, E. P., Boyko, L. D., and Kruzhilin, G. N. (1961) 'Heat transfer in the presence of steam condensation in horizontal tube'. *Int. Develop. in Heat Transfer*, Part II, Paper 34, pp. 290–295. International Heat Transfer Conference, held Boulder, Colorado, USA, August 1961.

Baer, E., and McKelvey, J. M. (1958) 'Heat transfer in dropwise condensation'. *Proc. Symp. Heat Transfer in Dropwise Condensation*, 24, Newark, University of Delaware.

Bell, K. J., Taborek, J., and Fenoglio, F. (1969). 'Interpretation of horizontal in-tube condensation heat transfer correlations using a two-phase flow regime map'. AIChE preprint No. 18, presented in 11th Nat. Heat Transfer Conf., Minneapolis, Minn., August.

Berman, L. D. (1969) 'Heat transfer during film condensation of vapour on horizontal tubes in transverse flow', pp. 1–41 of *Convective heat transfer in two-phase and one-phase flows*, edited by Borishanskii and Paleev, English translation. TT 68-50328, AEC-tr-6877, UC-38.

Berman, L. D. (1981) 'Heat transfer with steam condensation on a bundle of horizontal tubes'. *Thermal Eng.*, **28**, 218–224.

Berman, L. D. and Tumarov, Yu. A. (1962) 'Condensation heat transfer in a vapour flow over a horizontal tube'. *Teploenergetika*, 10, 77–84.

Bonacci, J. C., Myers, A. L., Nongbri, G., and Eagleton, L. C. (1976) 'The evaporation and condensation coefficient of water, ice and carbon tetrachloride'. *Chem. Engng. Sci.*, **31**, 609–617.

Bromley, L. A. (1952) 'Effect of heat capacity of condensate'. *Ind. Eng. Chem.*, **44**, 2966.

Butterworth, D. (1977) 'Developments in the design of shell and tube condensers'. Paper 77-WA/HT-24 presented at ASME Winter Annual Meeting, Atlanta, Georgia, 27 November–2 December.

Butterworth, D., Sardesai, R. G., Griffith, P., and Bergles, A. E. (1983) 'Condensation'. In *Heat exchanger design handbook*, Hemisphere Publ. Corp., Washington, D.C., Chapter 2.6.

Carpenter, E. F. and Colburn, A. P. (1951) 'The effect of vapour velocity on condensation inside tubes'. *Proc. of General Discussion on Heat Transfer*, 20–26, IMechE/ASME.

Cavallini, A. and Zecchin, R. (1974) 'A dimensionless correlation for heat transfer in forced convection condensation'. Paper presented at the 5th Int. Heat Transfer Conference, Tokyo, 3, 309–313.

Cavallini, A., Longo, G. A., and Rossetto, L. (1990) 'Refrigerant vapour condensation on a horizontal tube bundle'. Paper presented at 9th Int. Heat Transfer Conference, Jerusalem, **3**, 63–68.

Chaddock, J. B. (1957) 'Film condensation of vapours in horizontal tubes'. *Refrig. Engng.* **65**, 36–41; 90–95.

Chato, J. C. (1962) 'Laminar condensation inside horizontal and inclined tubes'. *ASHRAE J.*, **4**, 52–60.

Chen, M. M. (1961) 'An analytical study of laminar film condensation. Part I—flat plates'. *J. Heat Transfer, Series C*, **83**, 48–55.

Chen, M. M. (1961) 'An analytical study of laminar film condensation: Part 2—Single and multiple horizontal tubes'. *J. Heat Transfer*, **83**, 55–60.

Colburn, A. P. (1933–34) 'The calculation of condensation where a portion of the condensate layer is in turbulent motion'. *Trans. AICHE*, **30**, 187.

Colburn, A. P. and Hougen, O. A. (1934) 'Design of cooler condensers for mixtures of vapours with non-condensing gases'. *Ind. Engng. Chem.*, **26** (11), 1178–1182.

Danon, F. (1962) 'Topics in statistical mechanics of fluids'. *USAEC Report* UCRL 10029.

Diehl, J. E. (1957) 'Calculate condenser pressure drop'. *Petroleum Refiner*, **36**, 147–153.

Diehl, J. E. and Unruh, C. H. (1958) 'Two-phase pressure drop for horizontal cross-flow through tube banks'. ASME Paper No. 58-HT-20.

Drew, T. B. (1954). See McAdams, W. H., *Heat Transmission*, 3rd edn, McGraw-Hill, New York.

Drummond, G. (1966) Heriot-Watt University, Edinburgh, Scotland—Personal communication to G. B. Wallis.

Dukler, A. E. (1960) 'Fluid mechanics and heat transfer in vertical falling film systems'. *Chem. Engng. Prog. Symp. Series* No. 30, **56**, 1, together with appendix supplied by author.

Erb, R. A. and Thelen, E. (1965) 'Dropwise Condensation'. First International Symposium on Water Desalination, Washington, D.C. See also Erb, R. A., Ph.D. Diss., Temple University.

Eucken, A. (1937) *Naturwissenschaften*, **25**, 209.

Fujii, T., Vehara, H., and Kurata, C. (1972) 'Laminar filmwise condensation of flowing vapour on a horizontal cylinder'. *Int. J. Heat. and Mass. Transfer*, **15**, 235–246.

Fuks, S. N. (1957) 'Heat transfer with condensation of steam flowing in a horizontal tube bundle'. *Teploenergetika*, **4** (1), 35–38. NEL Trans. 1041.

Gardner, G. C. (1963–64) 'Events leading to erosion in the steam turbine'. *Proc. Inst. Mech. Engrs.*, **178**, Pt. I, No. 23.

Gose, E., Mucciordi, A. N., and Baer, E. (1967) 'Model for dropwise condensation on randomly distributed sites'. *Int. J. Heat Mass Transfer*, **10**, 15–22.

Grant, I. D. R. and Osment, D. D. J. (1968) 'The effect of condensate drainage on condenser performance'. NEL Report 350.

Hawes, R. I. (1976) 'Effect of vapour cross-flow velocity on condensation'. NEL Report, 619, 55–58.

Hill, P. G., Witting, H., and Demetri, E. P. (1963) 'Condensation of metal vapours during rapid expansion'. *J. Heat Transfer*, 303–317, November.

Honda, H. and Fujii. (1974) 'Effect of the direction of an oncoming vapour on laminar filmwise condensation on a horizontal cylinder'. *Heat Transfer 1974*, **3**, 299–303, paper presented at 5th International Heat Transfer Conference, Japan.

Honda, H., Fujii, T., Uchima, B., Nozu, S., and Nakata, H. (1988) 'Condensation of downward flowing R-113 vapour on bundles of horizontal smooth tubes', *Trans.*

JSME, **54** (502B), 1453–1460, and also in *Heat Transfer-Jap. Res.*, **18**, (6), 31–52 (1989).

Ishihara, K., Palen, J. W., and Taborek, J. (1980) 'Critical review of correlations for predicting two-phase flow pressure drop across tube bundles'. *Heat Transfer Eng.*, **1** (3), 23–32.

Jacobs, H. R. (1966) 'An integral treatment of combined body force and forced convection in laminar film condensation'. *Int. J. Heat Mass Transfer*, **9**, 637–648.

Jakob, M. (1936) *Mech. Engng.*, **58**, 729.

Johnstone, R. K. M. and Smith, W. (1966). 'Rate of condensation or evaporation during short exposures of a quiescent liquid'. *Proc. 3rd Int. Heat Transfer Conf.*, **2** (paper 78), 348–353.

Kast, W. (1963) *Chemie-Ingr. Tech.*, **35**, 163.

Kirkbride, C. G. (1933–34) 'Heat transfer by condensing vapour on vertical tubes'. *Trans. AIChE*, **30**, 170.

Knacke, O. and Stranski, I. N. (1956) 'The mechanism of evaporation. In Chalmers, B. and King, R. (eds.), *Progress in Metal Physics*, No. 6, pp. 181–235, Pergamon Press.

Koh, J. C. Y., Sparrow, E. M., and Hartnett, J. P. (1961) 'The two-phase boundary layer in laminar film condensation'. *Int. J. Heat Mass. Transfer*, **2**, 69–82.

Kucherov, R. Ya. and Rikenglaz, L. E. (1960) 'The problem of measuring the condensation coefficient'. *Doklady Akad. Nauk. SSSR*, **133** (5), 1130–1131.

Kunz, H. R. and Yerazunis, S. (1967) 'An analysis of film condensation, film evaporation and single phase heat transfer'. Paper 67-HT-1 presented at 9th ASME–AIChE Heat Transfer Meeting, Seattle, August.

Labuntsov, D. A. and Smirnov, S. I. (1966) 'Heat transfer in condensation of liquid metal vapours'. *Proc. 3rd Int. Heat Transfer Conf.*, **2**, 329–336.

Lednovick, S. L. and Fenn, J. B. (1977) 'Absolute evaporation rates for some polar and non-polar liquids'. *AIChE J.*, **23** (4), 454–459.

Lee, J. (1964) 'Turbulent film condensation'. *AIChE J.*, **10**, 540–544.

Lee, W. C. and Rose, J. W. (1982) 'Film wise condensation on a horizontal tube-effect of vapour velocity'. *Heat Transfer*, **5**, 101–106. *Proc. Int. Heat Transfer Conf.*, Munich.

Le Fevre, E. J. and Rose, J. W. (1966). 'A theory of heat transfer by dropwise condensation'. *Proc. 3rd Int. Heat Transfer Conf.*, **2** (paper 80), 362–375.

Maa, J. R. (1970) 'Rates of evaporation and condensation between pure liquids and their own vapours'. *Int. Engng. Chem.*, **9**, (2), 283–287.

McCormick, J. L. and Baer, E. (1963) 'On the mechanism of heat transfer in dropwise condensation'. *J. Colloid Sci.*, 18, 208.

McNaught, J. M. (1982). 'Two-phase forced convection heat-transfer during condensation on horizontal tube bundles'. Paper presented at 7th Int. Heat Transfer Conf., Munich, **5**, 125–131.

Marto, P. J. (1984) 'Heat transfer and two-phase flow during shell-side condensation'. *Heat Transfer Engng.*, **5**, (1–2), 31–61.

Mason, B. J. (1951). *Proc. Phys. Soc.*, P. 64, 773.

Minkowycz, W. J. and Sparrow, E. M. (1966) 'Condensation heat transfer in the presence of non-condensables, interfacial resistance, superheating, variable properties and diffusion'. *Int. J. Heat Mass Transfer*, **9**, 1125–1144.

Misra, B. and Bonilla, C. F. (1956). 'Heat transfer in condensation of metal vapours: Mercury and sodium up to atmospheric pressure'. *Chem. Engng. Prog. Symp. Series* No. 18, **52**, 7–21.

Murata, K., Abe, N., and Hashizume, K. (1990) 'Condensation heat transfer in a bundle of horizontal integral-fin tubes'. Paper presented at the 9th Int. Heat Transfer Conf., Jerusalem, **4**, 259–264.

Nobbs, D. W. and Mayhew, Y. R. 'Effect of downward vapour flow and inundation on condensation rates on horizontal tube banks', NEL Report, 619, 39–54. Nat. Engng. Lab., Glasgow.

Nusselt, W. (1916) 'Die Oberflächenkondensation des Wasserdampfes'. *Zeitschr. Ver. deutsch. Ing.*, **60**, 541, 569.

O'Bara, J. T., Killion, E. S., and Roblee, L. H. S. (1967) 'Dropwise condensation of steam at atmospheric and above atmospheric pressure'. *Chem. Engng. Sci.*, **22**, 1305–1314.

Ohtani, S., Chiba, Y., and Ohwaki, M. (1973) 'Heat transfer in dropwise condensation of steam; correspondence of drop behavior and temperature fluctuation'. *Heat Transfer—Japan Res.* **1** (3), 83–90.

Oswatitsch, K. (1942) *Z. Angew. Math. u. Mech.*, **22**, 1–13.

Othmer, D. F. (1929) 'The condensation of steam'. *Ind. Chem. Engng.* **21**, 576–583.

Rohsenow, W. M. (1956) 'Heat transfer and temperature distribution in laminar film condensation'. *Trans. ASME*, **79**, 1645–1648.

Rohsenow, W. M., Webber, J. H., and Ling, A. T. (1956) 'Effect of vapour velocity on laminar and turbulent film condensation'. *Trans. ASME*, **78**, 1637–1643.

Rosson, H. F. and Myers, J. A. (1965) 'Point values of condensing film coefficients inside a horizontal tube'. 'Heat Transfer—Cleveland', *Chem. Engng. Prog. Symp. Series*, No. 59, **61**, 190–199.

Rufer, C. E. and Kezios, S. P. (1966) 'Analysis of two-phase one component stratified flow with condensation'. *J. Heat Transfer*, 265–275, August.

Ryley, D. J. (1961) 'Phase equilibrium in low pressure steam turbines'. *Int. J. Mech. Sci.*, **6**, 28–46.

Saha, P. (1977) 'Development of constitutive relations. Effect of pressure change'. Reactor Safety Research Programs QPR, 1 April–30 June 1977, BNL-NUREG-50683.

Sardesai, R. G., Owen, R. G., and Pulling, D. J. (1982) 'Pressure drop for condensation of a pure vapour in down flow in a vertical tube'. *Heat Transfer 1982*, **5**, 139–145. Proc. 7th Int. Heat Transfer Conf., Munich.

Schrage, R. W. (1953) *A theoretical study of interphase mass transfer*. Columbia University Press, New York.

Seban, R. A. (1954) 'Remarks on film condensation with turbulent flow'. *Trans. ASME*, **76**, 299.

Shah, M. M. (1979) 'A general correlation for heat transfer during film condensation inside pipes'. *Int. J. Heat Mass Transfer*, **22**, 547–446.

Shekriladze, I. G. and Gomelauri, V. I. (1966) 'Theoretical study of laminar film condensation of flowing vapour'. *Int. J. Heat Mass Transfer*, **9**, 581–591.

Silver, R. S. (1964) 'An approach to a general theory of surface condensers'. *Proc. Inst. Mech. Engrs.*, **179**, Part 1, No. 14, 339–376.

Silver, R. S. and Simpson, H. C. (1961) 'The condensation of superheated steam'. *Proc. Nat. Engng. Lab. Conf.*, Glasgow, Scotland.

Soliman, M., Schuster, J. R., and Berenson, P. J. (1968) 'A general heat transfer correlation for annular flow condensation'. *J. Heat Transfer*, May, 267–276.

Spalding, D. B. (1963) *Convective mass transfer*. McGraw-Hill, New York.

Sparrow, E. M. and Eckert, E. R. G. (1961) 'Effects of superheated vapour and non-condensable gases on laminar film condensation'. *AIChE J.*, **7** (3), 473–477.

EXAMPLE   485

Sparrow, E. M. and Gregg, J. L. (1959) 'A boundary-layer treatment of laminar film condensation'. *J. Heat. Transfer, Series C*, **81**, 13.

Sparrow, E. M., Minkowycz, W. J., and Saddy, M. (1967) 'Forced convection condensation in the presence of non-condensables and interfacial resistance'. *Int. J. Heat Mass Transfer*, **10**, 1829–1845.

Stever, H. G. (1958) 'Fundamentals of gas dynamics', edited by H. W. Eammons, Princeton.

Subbotin, V. I, Ivanoskii, M. N., Sorokin, V. P., and Chulkov, B. A. (1964) 'Heat exchange in the condensation of potassium vapour'. *Teplofizika Vyso. Temp.*, **2** (1), 612–622.

Sugawara, S. and Katsuta, K. (1966) 'Fundamental study on dropwise condensation'. *Proc. 3rd Int. Heat. Transfer Conference*, **2**, 354–361.

Sukhatme, S. P. and Rohsenow, W. M. (1964) 'Heat transfer during film condensation of a liquid metal vapour'. Report MIT-2995-1, April.

Takeyama, T. and Shimizu, S. (1974) 'On the transition of dropwise-film condensation', *Heat Transfer 1974*, **3**, 274. Paper CS 2.5 presented at 5th International Heat Transfer Conference, Tokyo, August.

Tolman, R. C. (1949) *J. Chem. Phys.*, **17**, 333–337.

Umur, A. and Griffith, P. (1965) *J. Heat Transfer*, 275–282, ASME paper 64-WA/HT-3, May.

Waas, P., Straub, J., and Grigull, U. (1982) 'The influence of the thermal diffusivity of the condenser material on the heat transfer coefficient in dropwise condensation'. Paper CS5, *Heat Transfer 1982*, **5**, 27–31. *Proc. 7th Int. Heat Transfer Conf.*, Munich.

Wallis, G. B. (1969) *One dimensional two-phase flow*. McGraw-Hill, New York.

Webb, R. L. (1984) 'The effects of vapour velocity and tube bundle geometry on condensation in shell-side refrigeration condensers'. *ASHRAE Trans.*, **90**, Part 1B, 39–59.

Webb, R. L. (1984) 'Shell-side condensation in refrigerant condensers'. *ASHRAE Trans.*, **90**, Part 1B, 5–25.

Welch, J. F. and Westwater, J. W. (1961) 'Microscopic study of dropwise condensation'. *International Developments in Heat Transfer*, ASME, Part II.

Whalley, P. B. and Hewitt, G. F. (1978) 'The correlation of liquid entrainment fraction and entrainment rate in annular two-phase flow'. AERE-R-9187.

Wilhelm, D. J. (1964) 'Condensation of metal vapours and kinetic theory of condensation'. ANL-6948.

Zivi, S. M. (1964) 'Estimation of steady state steam void fraction by means of the principle of minimum entropy production'. *J. Heat Transfer*, 247–257, May.

Example

Example 1
A 56 mm o.d. vertical tube condenser operates at a pressure of 1.52×10^4 N/m^2 and condenses steam free from non-condensable gas at a rate of 25 kg/h per tube. Determine the heat transfer coefficient and the length of tube required if the temperature drop across the condensate film is 5°C.

Solution
The saturation temperature of steam at $1.53 \times 10^4 \, \text{N/m}^2$ is 54.4°C and the latent heat of vaporization is 2.37 MJ/kg. The physical properties of the condensate are:

$\mu_f = 5.32 \times 10^{-4} \, \text{Ns/m}^2$
$k_f = 0.64 \, \text{W/m °C.}$
$\rho_f = 958 \, \text{kg/m}^3$

Calculate the film Reynolds number

$$\text{Re}_\Gamma = \frac{4W}{\pi D \mu_f} = \frac{4 \times 25}{\pi \times 3600 \times 56 \times 10^{-3} \times 5.32 \times 10^{-4}} = 297$$

The flow in the condensate film is laminar and eq. (10.65) is used.

$$\frac{\bar{h}_f}{k_f}\left[\frac{\mu_f^2}{\rho_f(\rho_f - \rho_g)g}\right]^{1/3} = 1.47 \, \text{Re}_\Gamma^{-1/3} \qquad (10.65)$$

In this case $\rho_f \gg \rho_g$ and rearranging

$$h_f = 1.47\left[\frac{\mu_f^2}{k_f^3 \rho_f^2 g}\right]^{-1/3} \text{Re}_\Gamma^{-1/3} = 1.47\left[\frac{k_f^3 \rho_f^2 g}{\text{Re}_\Gamma \mu_f^2}\right]^{1/3}$$

$$= 1.47\left[\frac{0.64^3 \times 985^2 \times 9.806}{207 \times (5.32 \times 10^{-4})^2}\right]^{1/3} = 1.47(2.97 \times 10^{10})^{1/3}$$

$$= \underline{4.54 \, \text{kW/m}^2 \, °C}$$

In line with the recommendation of Section 10.4.3 this coefficient will be multiplied by a correction factor 1.2 to allow for the discrepancy between the theoretical Nusselt value and the observed higher experimental values.

$$\bar{h}_f = 1.2(\text{say}) \times 4.54 = \underline{5.54 \, \text{kW/m}^2 \, °C}$$

The length of tube required can be determined by equating the heat given up by the condensing vapour to the heat transfer rate across the condensate film. Thus,

$$Wi_{fg} = \pi Dzh_f(T_{gi} - T_w)$$

$$\frac{25 \times 2.37 \times 10^6}{3600} = \pi \times 56 \times 10^3 \times z \times 5.45 \times 10^3 \times 5$$

$$z = \frac{25 \times 2.37 \times 10^6}{\pi \times 56 \times 5.45 \times 5 \times 3600} = \underline{3.43 \, \text{m}}$$

Problems

1. For the case of vapour condensation in the presence of non-condensable gas derive an approximate expression for the effective temperature difference at the interface in terms of the mass concentration of gas at the interface (c_{ai}). Assume that the interface is saturated and the concentration of gas in the bulk mixture is small

   Answer:
   $$\left[(T_{go} - T_{gi}) = -\frac{RT_{go}T_{gi}}{Mi_{fg}} \ln(1 - c_{ai}) \right]$$

2. Condensation of heptane takes place in a horizontal tube 20 mm i.d. The vapour enters the tube dry saturated at a pressure of 10 bar and a flow-rate of 100 kg/h. Compute the variation of heat transfer coefficient with vapour quality for a constant temperature difference between vapour and tube wall of 10°C. How long will the tube require to be for complete condensation?

3. If the vapour flow-rate in Problem 2 is reduced to 10 kg/h, what will be the flow pattern in the horizontal tube? If the condensate at the tube outlet fills the tube cross-section calculate the variation of heat transfer coefficient with vapour quality. How long will the tube require to be for complete condensation in this case?

4. Determine the mean heat transfer coefficient, condensate thickness and film Reynolds number for a vertical plate 500 mm high for refrigerant R-134a at a saturation temperature of 40°C and a uniform wall temperature of 35°C. (Properties: liquid density = 1147 kg/m³, vapour density = 50 kg/m³, latent heat = 163.1 kJ/kg, liquid thermal conductivity = 0.0746 W/m °C, liquid viscosity = 0.000163 N s/m², liquid specific heat = 1.514 kJ/kg °C.

5. For the same conditions as in Problem 4, determine the heat transfer coefficient for a 25.4 mm diameter horizontal tube at wall temperatures of 35 and 37.5°C.

6. For saturated steam condensing on a vertical row of 40 tubes, determine the heat transfer coefficient on the first tube, the 20th tube and the bottom tube and the mean coefficient for all the tubes using the Kern method. The tubes are 19.05 mm in diameter. The steam is at 30°C and all the tubes can be assumed to have a uniform temperature of 26°C.

7. A horizontal reboiler with 249 tubes in a single tube pass condenses steam at 180°C inside the tubes. The tubes are 19.05 mm external diameter with a wall thickness of 2.11 mm. Calculate the condensing heat transfer coefficient at the inlet, at a vapour quality of 50 per cent and at the exit for complete condensation. The total heat duty is 14 MW. Assume no subcooling of the condensate at the exit.

11

CONDITIONS INFLUENCING THE PERFORMANCE OF BOILING AND CONDENSING SYSTEMS

11.1 Introduction

In this chapter consideration will be given to methods of improving the performance of boiling and condensing heat transfer surfaces. Two aspects will be considered; firstly, methods of improving the heat transfer coefficient in boiling and condensing systems, and secondly, methods of increasing the critical heat flux for a boiling system. The latter part of the chapter is taken up with the problem of including the non-linear behaviour of the boiling side heat transfer characteristic in design methods and also a short comment on the related problems of corrosion and deposition in boiling systems, and their interaction with the hydraulic and thermal characteristics.

As part of a comprehensive study of methods of augmenting convective heat transfer, Bergles (1978) and Bergles and Morton (1965) have examined the various ways of enhancing heat transfer in boiling and condensing equipment. Heat transfer coefficients in such equipment are normally much higher than for single-phase heat exchange units. However, there are circumstances when, for thermodynamic reasons, it is essential to transfer a given heat load at the lowest possible temperature driving force. Examples of such equipment are multi-stage long tube evaporators for use in desalination plants and condenser–reboiler units in low temperature tonnage oxygen production. Considerable efforts have therefore been made to improve the heat transfer coefficients in boiling and condensation by various means.

Highly rated heat exchange equipment is often limited by the value of the critical heat flux on the evaporating side. Examples where such is the case include water-tube boilers, nuclear reactors cooled by high pressure water, and rocket motors cooled by cryogenic fluids. A considerable amount of attention has been and is being paid to seeking methods of improving the value of the critical heat flux. Much of this work is empirical; a systematic approach will only come with an understanding of the various mechanisms of the critical heat flux phenomenon in forced convective boiling.

For a comprehensive review of the fundamentals of enhanced boiling heat transfer, the reader is referred to the recent book of Thome (1990). As yet, a similar publication for enhanced condensation is not available.

11.2 Methods of improving the heat transfer coefficient in nucleate boiling

This section will deal with methods that have been suggested for the improvement of heat transfer coefficients in the nucleate boiling region. In the 'two-phase forced convection region' where evaporation occurs at the liquid film–vapour core interface, heat transfer coefficients are considerably higher than for nucleate boiling and are amongst the highest recorded.

Methods for the improvement of performance in the nucleate boiling region depend on achieving sustained nucleation at a lower superheat than with a normal smooth surface. In Chapter 4 it was noted that nucleation occurs at gas or vapour-filled cavities in the heating surface and that the superheat at which nucleation occurs depends critically on the size and distribution of these cavities and on the temperature gradient away from the heating surface. Not all surface cavities can retain vapour or gas and therefore be active. Some cavities may become deactivated as the surface is 'out-gassed' by prolonged boiling. Liquid rushing back into the cavity after the departure of the previous bubble is able to condense the vapour remaining in the cavity completely and thus deactivate the site. This mechanism, which will be termed 'nucleation instability' is of considerable importance in the boiling of liquid metals (Marto and Rohsenow 1965).

One route to improved performance, then, is the provision of more nucleation sites which are stable for prolonged periods. Ideally these cavities should be at or close to the optimum size for the given temperature gradient from the surface.

11.2.1 *Roughened surfaces*

Roughening the boiling surface is one obvious method of providing a potentially greater number of nucleation sites. Two methods of providing surface roughness have been used in boiling experiments. In some studies a systematic pattern of grooves or scratches (Jakob and Fritz 1931, Bonilla *et al.* 1963) has been applied to the surface; in others the surface has been roughened by using various grades of emery cloth or by sandblasting (Corty and Foust 1955, Berenson 1960).

The first study of roughened surfaces was that of Jakob and Fritz (1931) who boiled water at atmospheric pressure on a horizontal copper plate. Figure 11.1 shows some of the results obtained with different prepared surfaces. In all cases, however, after prolonged boiling much of the gain in heat transfer coefficient due to roughness disappeared, indicating that the new larger cavities were not stable.

Corty and Foust (1955) used surfaces roughened by emery paper. This method of preparation appeared to produce a large number of cavities of

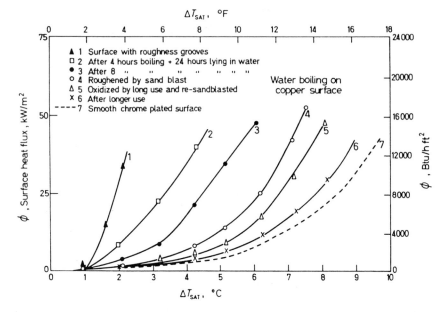

Fig. 11.1. Effect of surface roughness on initial and prolonged boiling heat transfer rates (Jakob and Fritz 1931).

uniform size since there was little if any increase in surface temperature with increasing heat flux. Under these conditions the roughest surfaces yielded heat transfer coefficients for various organic liquids up to 60 per cent higher than the smoothest (Fig. 11.2).

A study of boiling heat transfer from scored copper surfaces using water and mercury is reported by Bonilla *et al.* (1963). The highest heat transfer coefficients were produced on surfaces where the scratches were 2 to 2.5 bubble departure diameters apart, i.e., 6.35 mm spacing for water at atmospheric pressure. The maximum improvement in the heat transfer coefficient for boiling water varied from 80 per cent to 30 per cent greater than the smooth surface with the value decreasing as the heat flux increased.

It may be concluded that mechanical roughening of a surface is not altogether a satisfactory practical method of improving performance. The major difficulty is the long-term stability of the larger nucleation sites.

11.2.2   *Specially prepared surfaces*

Other surfaces which provide far more stable sites are potentially more attractive. Whilst the mouth of the cavity determines the superheat required to nucleate, the internal shape and volume together with the wetting

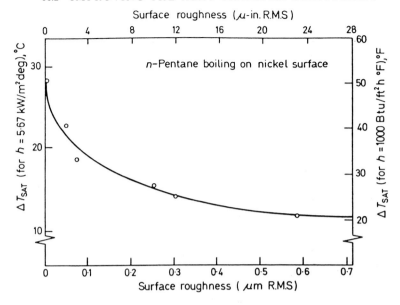

Fig. 11.2. Temperature difference needed to support a heat transfer coefficient of 5.67 K W m^2 °C (1000 Btu h ft °F^{-1}) as a function of surface roughness (Corty and Foust 1955).

characteristics of the cavity walls determine the stability of the nucleating cavity. The thermal diffusivity of the surface is also an important parameter. It has been shown (Marto and Rohsenow 1965) that re-entrant cavities of the shape shown in Fig. 11.3(a) are very stable and will continue to nucleate when conical cavities have been snuffed out. An artificial cavity of this type is shown in Fig. 11.3(b). It would be impractical to produce commercial heating surfaces with artificial cavities formed in this manner. However, certain types of surface finish are found to have a large number of naturally occurring re-entrant type cavities. Such surfaces include porous welds and sprayed stainless steel coatings (Marto and Rohsenow 1965). Boiling off such surfaces shows a significant improvement over the smooth surface case which does not deteriorate with time, under clean conditions.

The Linde Division of Union Carbide (Anon 1970) has developed surfaces (High Flux*) with a bonded porous metallic coat 0.1 to 1 mm thick. The pore size range is from 1 to 150 µm. The boiling side performance of this commercial tubing, compared with that of a fluted tube and a plain copper surface, is shown in Fig. 11.4. Corrugating the tube (in the case of a single-phase heating medium) or alternatively fluting the tube, either inside or outside (in the case of a condensing heating medium; see Section 11.3.1),

* 'High Flux' is now a Trade Name of UOP Inc.

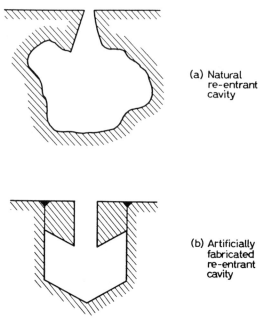

(a) Natural
    re-entrant
    cavity

(b) Artificially
    fabricated
    re-entrant
    cavity

Fig. 11.3. Re-entrant cavities.

allows some enhancement of the heating side surface coefficient which would otherwise limit the overall heat transfer rate and thus reduce the potential benefit from the porous surface. This surface has been used in refrigerant-flooded water chillers (Czikk *et al.* 1970) as well as light hydrocarbon and natural gas liquifiers (Milton and Gottzmann 1972). Comparative tests on tube bundles have been reported by Palen and Taborek (1971). This type of surface is probably only suitable for clean, non-corrosive conditions. The mechanism of boiling within porous media is such that salts will tend to concentrate in the pores resulting possibly in enhanced corrosion. This question will be returned to at the end of this chapter when the problem of deposition and corrosion in boiling systems is discussed.

Hitachi (Nakayama *et al.* 1975) have introduced a high performance heat transfer surface for both evaporating and condensing duties. Starting with a close-pitched low-fin surface, saw-toothed notches are cut in the fins. This surface is effective for condensation and is termed THERMOEXCEL-C.* For evaporation the notched fin is subsequently bent over to touch the base of the adjacent notched fin. This leaves a surface structure which comprises a series of tunnels connected to the evaporating liquid by regularly spaced

* THERMOEXCEL is a Trade Name of Hitachi Ltd.

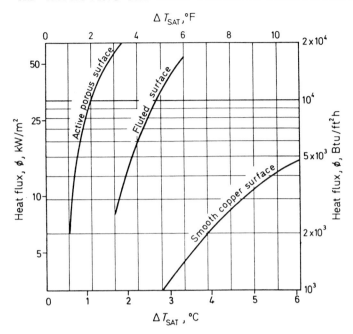

Fig. 11.4. The performance of commercial tubing with porous metallic coat in water at atmospheric pressure (Source: UOP Inc).

pores. This surface is termed THERMOEXCEL-E.* This method of construction provides regularly spaced re-entrant cavities and greatly enhanced nucleation properties.

An entirely different approach to the problem of stabilizing nucleation sites has been taken by Young and Hummel (1964). They have shown that if a heated surface which is normally wetted by the liquid is provided with a number of poorly wetted spots then very large improvements in performance are achieved. In the study reported by these workers, water was boiled on a stainless steel surface made non-wetting over certain small areas by Teflon spots.

Figure 11.5 shows the relative performance of four types of surface; acid cleaned, pitted, pitted with Teflon inclusions, and a smooth surface with sprayed Teflon spots. Marto and Rohsenow (1965) have shown that cavities with walls that are not wetted are stable under all conditions. This may provide the reason for the dramatic improvement noted with the Teflon prepared surfaces. Vapour formation was noted at superheats of 1 to 1.5°C (2 to 3°F) indicating that sites of the order of 250 μm (0.010 in.) in diameter are active. Preliminary studies by Young and Hummel suggest that 30 to 60 poorly wetted spots per cm², 250 μm or less in diameter is about the right

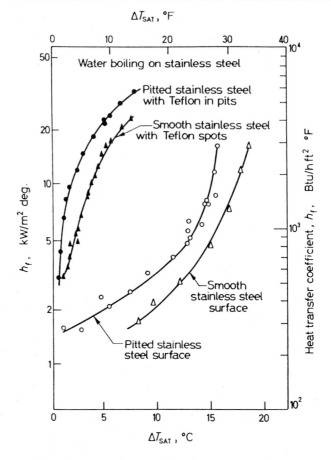

Fig. 11.5. Results of Young and Hummel (1964) for various specially prepared surfaces.

site density for stainless steel surfaces. There is, however, some evidence that after prolonged operation Teflon spots may tend to separate from the base metal (Anon, 1970).

Further studies of nucleate pool boiling on stainless steel surfaces coated with very thin layers of Teflon (7.5 μm to 35 μm thick) have been reported by Vachon *et al.* (1969). An improvement in performance was observed for the thinnest coatings (7.5 μm) only. With the thicker coatings the insulating effect of the coating outweighed any advantage from the improved nucleation characteristics.

Many other types of commercially successful surfaces are utilized to augment boiling heat transfer. For external boiling, low finned tubes, T-shaped low finned tubes named Gewa-T (Trade Name of Wieland-Werke

A.G.) and compressed, doubly cut knurled tubes known as Turbo-B (Trade Name of Wolverine Tube Inc.) are used in addition to those mentioned earlier. For in-tube boiling, twisted tapes and aluminium star inserts have been widely used but are giving way to the new internally microfinned tubes with a large number of helical fins only about 0.2 mm high. Figure 11.6 depicts a selection of commercially available enhanced boiling geometries.

11.2.3   *Mechanisms of enhanced nucleate boiling*

Besides providing more stable nucleation sites and wetted surface area as discussed earlier, enhanced boiling surfaces also modify the nucleation process itself and the underlying mechanisms of heat transfer. According to Thome (1992), the boiling process is improved as follows:

Nucleation. The re-entrant channels and cavities facilitate vapour trapping processes even for low wetting fluids and are adept at retaining large vapour nuclei. Because the nuclei reside and activate within the enhancement, they activate under uniform superheat conditions without the size-range limitation imposed on external nuclei by the thermal boundary layer. Thus, large nuclei can be activated at small superheats at multiple sites.

Evaporation mechanisms. Vapour bubbles grow in the external thermal boundary layer as they emerge from the re-entrant channels and pores in a similar fashion to bubbles growing on plain surfaces. In addition, re-entrant channels facilitate the formation of evaporating liquid films and menisci *inside* the enhancement itself. Figure 11.7 depicts the liquid films observed to form inside re-entrant channels of three different cross-sectional shapes (Thome 1990). Thin-film and capillary evaporation processes can be active over a large area and may be responsible for the majority of the vapour generated from these surfaces.

Liquid convection mechanisms. The increased number of active boiling sites for some enhanced surfaces also improves the external liquid-phase convection mechanisms of heat transfer, namely cyclic thermal boundary layer stripping and bubble agitation of the liquid. In addition, some liquid is 'pumped' through the narrow re-entrant channels and pores and out again into the bulk liquid pool. With hydraulic diameters of 1 mm and much less, very large 'entrance region', laminar flow heat transfer coefficients are attained.

The performance of a particular enhanced boiling surface thus depends on its specific geometry, its characteristic dimensions and its ability to promote enhanced nucleation and heat-transfer processes. Heat-transfer rates may also be increased by application of various force fields. These effects are discussed relative to their influence on the critical heat flux in Section 11.4.

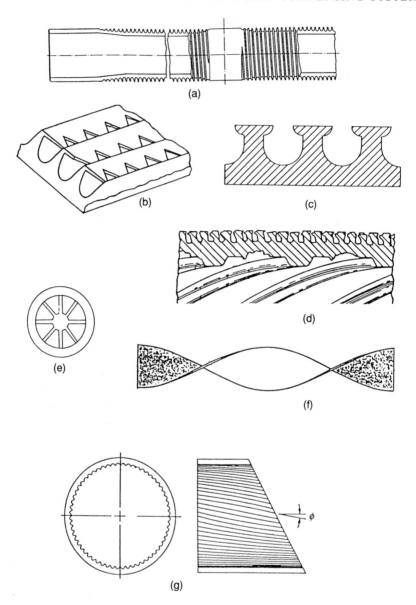

Fig. 11.6. Boiling augmentation geometries: (a) low finned tube; (b) Thermoexcel-E; (c) Gewa-T; (d) Turbo-B; (e) star insert; (f) twisted tape; (g) microfinned tube.

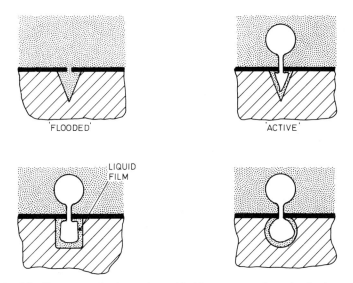

Fig. 11.7. Thin film evaporating layers formed inside re-entrant channels of enhanced boiling surfaces (Thome 1990).

11.3 Methods of improving the heat transfer coefficient in condensation

This section deals with methods that have been used to enhance heat-transfer coefficients during condensation. An early review of such methods was prepared by Williams *et al.* (1968) and a more recent one by Webb (1988). The methods used to improve heat transfer rates fall into three categories:

(a) changes to the geometry of the surface to increase the available area or to promote more rapid removal of condensate and thinner liquid films;
(b) treatment of the surface of promote drop-wise condensation;
(c) use of force fields.

11.3.1 *The influence of surface geometry*

The influence on film-wise condensation of artificially roughened surfaces has been examined by Spencer and Ibele (1966) and by Medwell and Nicol (1965). At low film Reynolds numbers the coefficients lie below the smooth tube values because the roughened surface apparently retains condensate due to increased surface tension forces. At high Reynolds numbers (>140) the heat-transfer coefficients are larger than for smooth surfaces.

One of the most promising methods of enhancing film-wise condensation coefficients is by fluting the tube. This improvement was first demonstrated by Gregorig (1954) who obtained condensation coefficients many times those for plain tubes. The enhancement comes about from the fact that the condensate is drawn into and runs down within the grooves in the surface leaving a very thin film on the ridge of each flute of the tube. The movement of condensate into the groove is brought about by surface tension forces resulting from changes in the radii of curvature of the surface and to a lesser extent by temperature gradient effects. The local heat transfer coefficient is thus high at the ridge and low at the groove with the average coefficient being much greater than that for a plain tube. When the condensate flow becomes too great to be carried entirely within the grooves, flooding occurs and a sharp drop in the heat transfer coefficient results. The increase in heat transfer coefficient and the onset of flooding are clearly functions of the design of a particular tube. Typical values for heat transfer coefficients during the condensation of steam are 50–70 kW/m^2 °C for the non-flooded condition (Lustenader *et al.* 1959). The drainage of the condensate into the groove means that the heat transfer coefficients are practically independent of length. A comprehensive series of experiments on the performance of fluted tubes investigating such features as vapour velocity, tube orientation and the presence of non-condensables has been reported by Carnovos (1965).

A similar enhancement to that caused by fluting can be achieved by placing wires along a vertical condenser tube. Again, surface tension forces cause the condensate to flow towards the wires and to drain as rivulets alongside the wire. Thomas (1967) made a theoretical and experimental study to optimize the number of wires around the tube circmference. He found that the highest heat transfer coefficient was achieved when the number of wires (*n*) was given by

$$n = \frac{0.18\pi D}{d} \tag{11.1}$$

where D is the tube diameter and d the wire diameter. Some further improvement may be achieved by spirally winding the wires around the tube.

Integral low finned tubing has been used for many years in horizontal-tube condensers. The fins are formed by passing plain tubes through a series of planetary rotating discs that press into the tube and raise the fins. Beatty and Katz (1948) and Beatty and Forbes (1950) made comprehensive measurements for six different fluids on a low finned tube and demonstrated that there is a substantial increase in condensation heat-transfer rates compared to a plain tube, significantly beyond that of the additional surface area.

The fin density of a low finned tube is the most important geometrical characteristic affecting performance. Since surface tension forces act to retain the condensate between the fins while gravity acts to drain the condensate from the tube, each fluid and operating condition has an optimal fin density (or spacing). For instance, for condensing steam under vacuum Wanniarachchi *et al.* (1985) found that a fin density of 667 fins per meter (17 fins/inch) gave the best performance as shown in Fig. 11.8. Instead, for the new

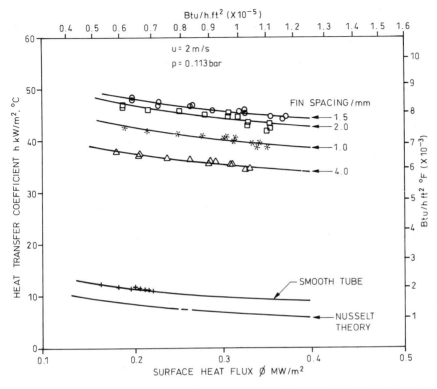

Fig. 11.8. Condensation performance of steam on horizontal low finned tubes (Wanniarachchi *et al.* 1985). Values are based on nominal area at root diameter.

refrigerant R-123 Murata *et al.* (1990) demonstrated the superior performance of 1184 fins/m (30 fins/in) tubing compared to 1026 fins/m (26 fins/in.).

The shape of the fins can also be important. Marto *et al.* (1986) found that a near-parabolic fin shape with a gradually decreasing curvature from the fin tip to the fin root resulted in a 10–15 per cent increase in performance over other fin shapes. Use of radiused fillets at the base of the fins reduces surface tension forces on the condensate and improves drainage and hence increases heat-transfer rates according to Briggs *et al.* (1992). Bella *et al.* (1992) ran extensive tests on the effect of vapour shear on condensation rates for a low finned tube using refrigerants R-11 and R-113. Heat-transfer enhancement due to vapour shear was less marked than for the plain tube tested and was nearly negligible for the finned tube at vapour Reynolds numbers below 100,000.

A number of commercial tube manufacturers (Yorkshire Imperial Metals Ltd, 1973, Withers and Young 1971) produce roped, spirally dented or deeply corrugated tubes for horizontal-tube condensing units. These tubes not only augment the condensation process on the outside but also the single-phase coefficient for the fluid flowing inside the tube. These tubes can also be effective for in-tube condensation. The microfinned tubes described in Section 11.2 are also very effective for increasing in-tube condensation rates with only a small adverse effect on the two-phase pressure drop. For more information on the augmentation of in-tube condensation and the effect of oil, refer to Schlager et al. (1989).

11.3.2   Predictive methods for condensation on low finned tubes
11.3.2.1   Beatty–Katz model

Beatty and Katz (1948) proposed a condensation model for horizontal low finned tubes based on the Nusselt equations for film condensation on plain surfaces described earlier in Section 10.4. In utilizing his theory they applied the same conditions as Nusselt to low finned tubes, principally that

(a) gravity forces drain the condensate film from the vertical fins and from the base area between adjacent fins;
(b) no surface tension forces act on the films;
(c) no vapour shear forces act on the films.

The condensation heat-transfer coefficient h (based on the nominal external area $\pi D_0 L$ at the fin tips) is obtained as the area weighed average of the condensation coefficients on the base area between fins h_h and that on the fins h_f as

$$h\eta = h_h \frac{A_r}{A} + \eta_f h_f \frac{A_f}{A} \tag{11.2}$$

where the total surface area is

$$A = A_r + A_f \tag{11.3}$$

A_r is the base area between the fins and A_f is the surface area of the fins. The parameter η is the surface efficiency and is equal to the ratio of the nominal to the effective surface area. The effective area A_{eff} is obtained using eq. (11.3) with the fin efficiency applied to A_f. The Nusselt expression for horizontal tubes is used for calculating h_h for the base area between the fins as

$$h_h = 0.725 \left[\frac{\rho_f(\rho_f - \rho_g)gk_f^3 i_{fg}}{\mu_f D_r(T_{gi} - T_w)} \right]^{1/4} \tag{11.4}$$

and the corresponding equation for vertical plates is utilized for the fins to obtain h_f as

$$h_f = 0.943 \left[\frac{\rho_f(\rho_f - \rho_g)gk_f^3 i_{fg}}{\mu_f L_f(T_{gi} - T_w)} \right]^{1/4} \tag{11.5}$$

The characteristic length L_f of the fin for film condensation was taken as the surface area of one side of the fin divided by the fin diameter D_0 such that

$$L_f = \frac{\pi(D_0^2 - D_r^2)}{4D_0} \tag{11.6}$$

The Beatty–Katz equations can be rearranged to give the following convenient form where the analytical coefficient 0.725 was replaced by 0.689 to better match their experimental data:

$$h\eta = 0.689 \left[\frac{\rho_f(\rho_f - \rho_g)gk_f^3 i_{fg}}{\mu_f(T_{gi} - T_w)} \right]^{1/4} \left[\frac{A_r}{A} D_r^{-0.25} + 1.3 \frac{\eta_f A_f}{A} L_f^{-0.25} \right] \tag{11.7}$$

11.3.2.2 Webb–Rudy–Kedzierski model

The Beatty–Katz model assumes no condensate is retained between the fins by surface tension forces. In contrast Rudy and Webb (1985) observed that a significant portion of the finned tube is flooded with condensate, rendering that portion of the tube less effective at transferring heat. They developed the following analytical expression for the lower half angle of the tube that retains condensate:

$$\alpha = \cos^{-1} \left\{ 1 - \left[\frac{2\sigma[(2e/\cos \varphi) + t_t - t_b]}{D_0 \rho_f g(p_f e - A_p)} \right] \right\} \tag{11.8}$$

where α is in radians, e is the fin height, t_t is the fin thickness at the tip, t_b is the fin thickness at the base, p_f is the fin pitch, A_p is the cross-sectional area of the fin and φ is half the apex angle of a trapezoidal or triangular shape fin ($\varphi = 0°$ for a rectangular fin). This expression accurately predicted measured retention angles and was used by Webb et al. (1985) to correct the Beatty–Katz model as follows:

$$h\eta = (1 - c_b) \left[h_h \frac{A_r}{A} + \eta_f h_f \frac{A_f}{A} \right] + c_b h_b \tag{11.9}$$

where the condensate-flooded fraction of the tube circumference is $c_B = \alpha/\pi$ and the unflooded fraction is $(1 - c_b)$. The heat transfer coefficient h_b in the lower flooded portion of the finned tube was estimated from a simple finite element analysis. An easier approach instead is to assume that the condensate flow is laminar and thus use an appropriate expression for fully developed channel flow. The heat-transfer coefficient on the fin h_f is predicted from a surface tension drainage model of Adamek (1981) given as

$$
h_f = 2.149 \frac{k_f}{S_m} \left\{ \left[\frac{i_{fg} \theta_m S_m}{v_f k_f (T_{gi} - T_w)} \right] \left[\frac{\zeta + 1}{(\zeta + 2)^3} \right] \right\}^{1/4}
\tag{11.10}
$$

rather than the original gravity controlled model. θ_m is the rotation angle in radians from the normal to the fin surface ($\theta_m = \pi/2$ for a horizontal rectangular finned tube and about 1.483 radians, i.e. 85°, for trapezoidal fins). S_m is the length of the surface profile of one side of the fin. The dimensionless parameter ζ characterizes the aspect ratio of the fin cross-section (height/thickness) and is obtained from fitting the following profile equation to the shape of the actual fin:

$$
\frac{1}{r} = \frac{\theta_m}{S_m} \frac{\zeta + 1}{\zeta} \left[1 - \left(\frac{s}{S_m} \right)^\zeta \right] \qquad \text{for } -1 \leqslant \zeta < \infty
\tag{11.11}
$$

The condensation coefficient h_h is calculated using the gravitational expression of Nusselt for a horizontal tube as described earlier with the additional condensate flow rate from the fins onto the base area between the fins.

11.3.2.3 *Newer models* Other improved models for condensation on low finned tubes have been developed by Honda and Nozu (1986) and Adamek and Webb (1990). These methods are complex and computer simulation is needed to apply them; the reader is recommended to refer to the original publications directly. For quick hand calculations, it is recommended that eq. (11.9) be used with Beatty–Katz expressions for h_h and h_f.

11.3.3   *The promotion of drop-wise condensation*

One obvious method of enhancing condensation heat transfer rates is inducing drop-wise rather than film-wise condensation upon the surface. Since water and organics wet most metal surfaces, it is necessary to treat the surface in some way so as to make it non-wetting (i.e., hydrophobic, in the case of water). Three methods have been tried. These are:

(a) chemically coated surfaces
(b) polymer coated surfaces
(c) electroplated surfaces.

Much of the early work on drop-wise condensation was concerned with the identification of various chemicals to act as efficient promoters. A suitable substance to act as a promoter consists of two parts; an active and inactive group. The active group attaches itself to the surface and usually consists of COOH, OCSS, SH, SS, or Se radicals whilst the inactive group consists of a long-chain hydrocarbon which is exposed to the steam. An excellent review of the work on chemical promoters has been given by Osment and Tanner (1962). Chemical promoters tend to be removed from the surface within a short time period and there is a gradual reversal from drop-wise to film-wise condensation. Periodic injection of the promoter into the condensing vapour has been attempted but, in general, the use of chemical coated surfaces has not been particularly successful.

In an attempt to find more permanent hydrophobic coatings attention has been directed to the use of polymer coated surfaces. In particular, consider-able attention has been directed at the use of Teflon (polytetrafluorethylene) coatings. As with the application in boiling, the coating thickness must be sufficiently thin to ensure that the enhancement of the condensation coeffici-ent is not offset by the insulating effect of the coating. Much attention has therefore been paid to the development of processes for coating surfaces with very thin (0.25 to 1 μm) uniform films which adhere to the metal base. A number of users (Anon 1970) of Teflon-covered surfaces have reported a tendency to peel as they age or become damaged. For many years it was thought that only steam could be made to condense in a drop-wise manner. However, the very low surface free energy of Teflon coatings renders them non-wetting to many liquids including a number of organics.

A further method of promoting drop-wise condensation is by the use of electroplated coatings of the noble metals. These coatings are ideal in that they are highly hydrophobic, have high thermal conductivity, and can be diffusion bonded to the base metal. However, the coatings are expensive and this is limiting the introduction of such surfaces. Erb and Thelen (1965) have reported the excellent behaviour of a 1.25 μm gold layer deposited on to a cupro-nickel alloy tube which had been precoated by a 7.5 μm film of nickel. Other suitable metals are platinum, palladium, and rhodium. The absence of surface oxides appears to be the prime reason for the hydrophobic nature of the noble metal surfaces.

11.3.4 The use of force fields to enhance condensation

In film-wise condensation the condensate film constitutes the major resist-ance to heat transfer and a variety of external forces, centrifugal, vibrational, and electrostatic, have been used to reduce its thickness and to promote drainage.

A number of studies have been carried out where condensation is made to occur on a horizontal rotating disc. For a plain disc Nandapurkar and Beatty (1960) derived the following equation on the basis of no interfacial shear:

$$\bar{h}_f = 0.904 \left[\frac{k_f^3 \rho_f^2 \omega^2 i_{fg}}{\mu_f (T_{gi} - T_w)} \right]^{0.25} \qquad (11.12)$$

where ω is the angular rotation (radians/s). Further information is available on grooved discs (Bromley *et al.* 1966) and on rotating horizontal tubes (Singer and Preckshot, 1963).

The influence of vibration on condensation has also been studied. Mathewson and Smith (1963) condensed organic vapours within a vertical tube under conditions where acoustic vibrations were induced in the vapour phase. The vibration caused an enhancement of about 50 per cent. Experiments in which the tube itself was oscillated either longitudinally (Haughey 1965) or transversely (Raben *et al.* 1961) have also been shown to enhance the film coefficient. For longitudinal vibrations Haughey (1965) gives the ratio of the coefficient with vibration to that without as

$$\frac{h_{\text{vib}}}{h_{\text{f}}} = 1 + 0.0018FA \qquad (11.13)$$

where F is the frequency (0–150 Hz) and A is the amplitude (mm). For transverse vibrations of a vertical tube, Raben *et al.* (1961) suggest

$$\frac{h_{\text{vib}}}{h_{\text{f}}} = 0.619[AF^{1.2}]^{0.205} \qquad (11.14)$$

where $AF^{1.2} > 5.2$.

Electrostatic forces can be used to enhance condensing heat transfer rates of organic vapours. Choi (1964) has reported a study in which Refrigerant 113 was condensed within a vertical tube. The condensate film was acted upon by an electrostatic field generated from an electrode placed at the centre of the tube. The motion induced in the film was similar to that on the outside of a vertical rotating cylinder and led to enhanced condensation heat transfer coefficients. More recent experimental studies have been conducted for pseudo-dropwise condensation (Yabe *et al.* 1986) and for a 12-tube condenser unit (Poulter and Allen 1986). A theoretical model for the influence of electric fields on the condensation of nonconducting fluids on a horizontal tube has been developed by Trommelmans and Berghmans (1986).

It is doubtful if any of the methods of improving condensation by means of force fields will find widespread use because of the considerable disadvantages of the additional equipment to cause rotary or vibrating motion and the likely reduced reliability compared with heat exchangers having no moving parts.

11.4   Conditions influencing the critical heat flux in forced convective boiling

A considerable number of studies have been devoted to methods for increasing the critical heat flux in forced convection boiling. The various proposed methods can be loosely divided into three categories depending on whether the improvement is brought about by (a) geometrical changes (b) force fields, or (c) trace additives.

11.4.1   *The influence of geometrical changes*

Changes to the tube geometry starting from a simple vertical uniformly heated tube can affect the critical heat flux both favourably and adversely. Bernstein *et al.* (1964) have suggested a variety of geometries which might offer improvement over an unmodified tube. These are:

(a) a serpentine tube where the duct is bent in one plane in the form of a hairpin,
(b) a straight tube with a twisted ribbon insert to induce swirl flow,
(c) a straight tube with internal integral helical finning,
(d) a straight tube with a separate helical wire insert,
(e) a straight tube with alternative short sections of larger then smaller internal diameter.

Tubes of each type were tested in a horizontal position using a low-pressure steam–water loop. The heating source was condensing steam and the 'apparent' heat transfer coefficients were evaluated at various points along the tube from the local film temperature difference and the overall average heat flux. All of the specially prepared tubes showed high heat transfer rates to considerably greater steam qualities than with the plain unmodified tube.

The tubular elements of waste heat boilers and of nuclear steam generators, particularly of the once-through type, may be arranged in a variety of configurations. For example, horizontal serpentine (hairpin), vertical U-tube, and helical coil elements have been used. The presence of gravitational and centrifugal forces have a pronounced effect on the various heat transfer processes including the critical heat flux.

11.4.1.1   *Horizontal and inclined tubes*   Evaporation and condensation within inclined and horizontal tubes is encountered frequently. Stratification of the flow occurs so that the majority of the liquid flows along the bottom of the tube. To a degree this is beneficial in the case of condensation but can cause problems during evaporation in that the liquid film at the top of the tube may dry out (Fig. 1.4).

Dryout in a horizontal tube differs from that in a vertical tube in two ways:

(a) *stratification* of the flow can occur at low velocities for both low quality and subcooled conditions. Such conditions can lead to overheating of steam boiler tubes at quite modest heat fluxes;

(b) *dryout* of the tube at high vapour qualities occurs over a relatively long tube length, starting at the top of the tube where the film thickness and flow rate are lowest and ending up with the final evaporation of the rivulet running along the bottom of the tube. Under these conditions the vapour flow in the upper part of the tube may become superheated before dryout occurs at the base of the tube.

Rounthwaite and Clouston (1961) and Rounthwaite (1968) have studied the problem of evaporation in horizontal hairpin elements representing sections of a full sized unit for a United Kingdom nuclear power station. The test section (Rounthwaite 1968) consisted of two 6.25 m lengths of straight horizontal mild steel tubing of 41.3 mm bore connected by a return bend 6.35 cm mean radius. The test conditions covered the entire quality range at pressures between 14 and 65 bar, mass velocities between 100 and 800 kg/m² s and heat fluxes between 25 and 80 kW/m². These tests showed that serious stratification with greatly reduced heat transfer coefficients (by a factor of twenty) over the upper section of the tube, occurred for values of the superficial liquid velocity (j_f) of 0.2 m/s and below.

In work by Lis and Strickland (1970) the influence of the 180° return bend has been studied. These tests have shown that the bend produces severe flow disturbances over a wide range of operating conditions. The disturbances cause large amplitude oscillations of wall temperature and the formation of dry patches along the top of the tube up to 70 diameters downstream of the bend. At low heat fluxes (≈ 100 kW/m²) these disturbances appear to reach a maximum in the wavy or the low velocity annular flow regions ($\alpha = 0.6$–0.65) and are confined to mass velocities below 1000–1200 kg/m² s.

Styrikovich and Miropolskii (1950) reported the effects of stratification of a high pressure steam–water mixture in a horizontal pipe. These caused wide temperature differences between the top and bottom of the boiler tube. Experiments were carried out on a single 7.5 m, 56 mm i.d. tube at pressures between 10 bar and 220 bar with heat fluxes in the range 22 to 135 kW/m² and inlet velocities between 0.24 and 1 m/s. It was found that there was a critical two-phase velocity, j, before which stratification occurred and above which it did not. This critical velocity, j, increased with heat flux up to 60 kW/m² and decreased with increase in pressure (Fig. 11.9). An investigation into the effect of the angle of slope showed that overheating only occurred if the angle of slope was less than 9.6° to the horizontal.

Using an alcohol–water analogue for steam–water flow Gardner and Kubie (1976) established the following expression for the critical velocity, j:

$$j^2 = 6.6 \frac{\{\sigma g \cos \alpha (\rho_f - \rho_g)\}^{0.5}}{\rho_f f K^2} \tag{11.15}$$

where α is the angle of inclination of the tube to the horizontal, f is the

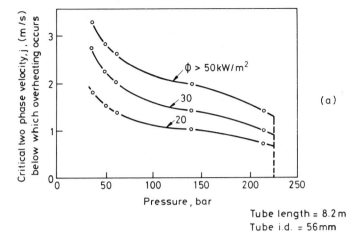

Tube length = 8.2 m
Tube i.d. = 56 mm

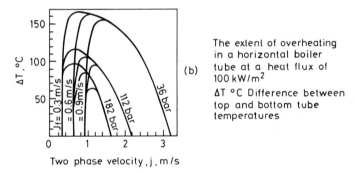

The extent of overheating in a horizontal boiler tube at a heat flux of 100 kW/m²

ΔT °C Difference between top and bottom tube temperatures

Fig. 11.9. Stratification and overheating of horizontal boiler tubes (Styrikovich and Mirpolskii 1950).

friction factor and K is an empirical constant found to be 2.17. Substituting for f and K leads to the following equation for j:

$$j^{1.8} = 30.43 \frac{D^{0.2}\{\sigma g(\rho_f - \rho_g)\}^{0.5}}{\rho_f^{0.8}\mu_f^{0.2}} \qquad (11.16)$$

Table 11.1 gives the values of j calculated from eq. (11.16) for steam–water flows over a range of pressures and tube diameters. Excellent agreement is seen with the data in Fig. 11.9 and eq. (11.16) is recommended as giving the minimum single- or two-phase velocity below which stratification will occur in a horizontal tube.

Overheating of the upper tube surfaces and local drying out of the liquid film after a bend when combined with deposition of particulate material may

Table 11.1 Values of j (m/s) to prevent stratification for
steam/water flow in horizontal tubes

	p (bar)					
D(mm)	33.5	64.2	112.9	146	165	187
20	2.71	2.47	2.11	1.82	1.63	1.37
40	2.92	2.67	2.28	1.97	1.77	1.48
60	3.06	2.80	2.38	2.06	1.85	1.55

contribute to tube failures in boilers by the process of 'on-load' corrosion.
This is discussed later in the chapter.

11.4.1.2   *Helical coils and bends*   A further method of influencing the critical
heat flux is to form the tube into a helical coil. Stratification can also occur
in helical coils and in vertical bends. For the case of a vertical bend, Bailey
(1977) showed that the gravitational force maintaining stratified flow is
supplemented by centrifugal forces so that $g \cos \alpha$ in equation (11.15) is
replaced by $(g + j^2/r)$. Incorporating this modification into eq. (11.16)

$$j^{1.8} = 30.43 \frac{D^{0.2}\{\sigma(g + j^2/r)(\rho_f - \rho_g)\}^{0.5}}{\rho_f^{0.8}\mu_f^{0.2}} \qquad (11.16a)$$

The derivation of eq. (11.16a) neglects the secondary flow which is known
to exist in both bends and coils. This secondary flow tends to enhance the
stratification and increase the value of j. For the example examined
experimentally by Bailey, a tube 19 mm i.d. with a bend radius (r) of 0.45 m,
the constant in eq. (11.16a) is increased to 37.4 and the critical velocity, j,
by 17–20 per cent. Experimental results confirmed that the secondary flow
effect should be taken into account at low void fractions ($\alpha < 0.5$) but was not
appropriate for high void fractions ($\alpha > 0.8$).
  For the case of a helical coil since the gravitational and centrifugal forces
do not act in the same direction, $g \cos \alpha$ in eq. (11.15) must be replaced by
$\{(g \cos \alpha)^2 + (j^2/r)^2\}^{0.5}$ the resultant force, where α is the angle of inclination
of the coiled tube.
  Helical coil evaporators have been used in nuclear steam generators and
cryogenic evaporators. Miropolskii *et al.* (1966) and Miropolskii and Pickus
(1969) in the USSR and Carver *et al.* (1964) in the USA have reported critical
heat flux data for coiled tubes. The former workers found that in the
subcooled region the critical heat flux was reduced from that in a straight
vertical tube due to stratification of the type discussed earlier. Both sets of
workers agree that the transition to the liquid deficient condition in the

evaporator occurs at a higher average steam quality than the corresponding condition in a straight pipe. In most cases the steam quality for the transition at all parts of the tube circumference is higher than the straight tube value. Dryout first occurs at locations on the tube wall between the point closest to the centre of curvature and the top of the tube. At low pressures it is possible for the centrifugal forces acting on the gas phase to be higher (Banerjee et al. 1967) than on the liquid phase due to the different phase velocities, i.e. $[\rho_g u_g^2/R] > [\rho_f u_f^2/R]$. The liquid can then flow on the inside rather than on the outside parts of the coil. Experiments at low pressure with boiling in a helical coil reported by Owhadi and Bell (1967) support this picture and dryout occurs first near 100 per cent quality at the top and bottom positions around the tube circumference rather than at the point nearest the centre of curvature. Nucleate boiling heat transfer coefficients are not greatly affected by the centrifugal forces or the secondary flow.

11.4.1.3 *Swirl flow* Swirl flow has attracted a considerable amount of attention as a method of increasing the critical heat flux in tubes. One of the earliest investigations in this field, by Gambill et al. (1961) at Oak Ridge, obtained dramatic improvements in heat transfer to subcooled water flowing in short, small diameter tubes. In the 1960s other teams started experiments in this area, first at low pressure (Rosuel and Sourioux 1963, Volterras and Tournier 1963) and then with geometries and pressures found in conventional and nuclear steam generators (Matzner et al. 1965, Moeck et al. 1964, Viskanta 1961, Rosuel et al. 1963, Bahr et al. 1968). The usual way of obtaining swirl flow is to insert a twisted ribbon, of width equal to the tube inside diameter, along the full length of the heated tube. The amount of ribbon twist is expressed in terms of the twist ratio, defined as the number of tube inside diameters per 180° twist. Table 11.2 summarizes the range of variables investigated for swirl flow of water in uniformly heated tubes.

Gambill and Bundy (1962) have reviewed the status of swirl flow heat transfer up to 1962. Even though they found some disagreement in the results presented by the various authors, they have shown a definite relationship between the radial acceleration of the coolant and the critical heat flux. For the case of zero exit subcooling, they found that the incremental critical heat flux above that without acceleration (axial flow) rises at an increasing rate for accelerations in excess of about 100 times earth gravity. For example, for water at a radial acceleration of 6.0×10^4 times earth gravity the incremental critical heat flux was estimated to be 60 MW/m² (1.9×10^7 Btu/ft² h) above that for axial flow.

However, this is not the highest steady state heat flux achieved. Liu and Lienhard (1992) have succeeded in removing heat fluxes as high as 400 MW/m² (1.3×10^8 Btu/ft² h) by ultra high speed impinging subcooled liquid jets.

Table 11.2 Range of variables for critical heat flux experiments in uniformly heated tubes with swirl flow

Source and reference	Twist ratio y	Tube dimensions				Inlet subcooling or quality	Range of experimental variables				Outlet quality % by wt.
		Inside diameter		Heated length			Pressure		Mass velocity		
		mm	in.	cm	in.		bar	psia	kg/m² s	lb/h ft² (×10⁻⁶)	
SNECMA, France Rosuel and Sourioux (1963)	6, 12, 18	10–20	0.394–0.788	10.65	4.19	subcooled	1	15	122–680	0.09–0.50	28–99
Volterras and Tournier (1963)	6	10	0.394	80	31.5	subcooled and quality 1%, 4%	1	15	54–95	0.04–0.07	
Rosuel et al. (1963)	6	10	0.394	80	31.5		68.9	1000	4500, 5850	3.3, 4.3	15, 19
ORNL, USA Gambill et al. (1961)	4.2–24	3.45–10.25	0.136–0.402	3.81–44.2	1.5–17.4	subcooled	1.72–50.2	25–730	4500–47.500	3.3–35.0	subcooled-17
Columbia University, USA Matzner et al. (1965)	30	10.15	0.400	488	192	subcooled	68.9	1000	1200–4600	0.9–3.4	35–88
AECL, Canada Moeck et al. (1964)	11.1, 69.0	11.4	0.450	101.5	40	38%–85%	70	1015	400–1150	0.29–0.85	74–95
ANL, USA Viskanta (1961)	5, 10	7.95	0.313	45.7	18	subcooled	138	2000	680–2700	0.5–2.0	subcooled-50
ISPRA, Italy Bahr et al. (1968)	6	20	0.788	250	98.5	subcooled-55%	140–185	2030–2700	1000–2250	0.737–1.66	>70

While swirl flow increases the pumping requirements, the gain in critical heat flux more than offsets the pumping loss. Gambill and Bundy (1962) showed that at constant pumping power, the ratio of the swirl flow to the axial flow critical heat fluxes can be as much as 2.5 at centrifugal accelerations of 6×10^4 times earth gravity.

Swirl flow also has an influence on the boiling heat transfer coefficient below the critical heat flux. Gambill and Bundy quote the following equation to relate the wall superheat, ΔT_{SAT} °F, for water with the twist ratio, y, and the heat flux, ϕ Btu/ft^2 h,

$$\Delta T_{SAT} = 2.94 y^{0.81} (\phi/10^6)^{1.35} \tag{11.17}$$

Equation (11.17) obviously becomes invalid at large values of y. Blatt and Adt (1963) report experiments on swirl flow heat transfer in nucleate boiling of Refrigerant 11 and also water. They also found that boiling Refrigerant 11 at a given temperature difference and mass velocity the heat transfer coefficient increased with radial acceleration. For boiling water near atmospheric pressure and at a mass velocity of about 122 kg/m^2 s (9×10^4 lb/h ft^2) twisted ribbons improved the heat transfer coefficient only at low accelerations and low heat fluxes. These findings are in agreement with those of Merte and Clark (1959) who studied the influence of increased acceleration fields on the pool boiling of water. It would appear that the radial acceleration increases only the convective or single-phase liquid component ϕ_{SPL} of the total heat flux whilst the nucleate boiling component remains unchanged.

The increase in surface temperature at dryout in tests with twisted tapes is often very gradual and the heat transfer coefficient in the liquid deficient and superheated vapour conditions appears to be enhanced as a result of the radial acceleration by up to 15 per cent. The approximate increases in the critical heat flux due to swirl flow in the saturated boiling region are indicated by the AECL (Moeck et al. 1964) and Columbia University (Matzner et al. 1965) data. At a pressure of 69 bar (1000 psi) and for a given tube diameter, length, and inlet enthalpy, the critical heat flux in swirl flow is 1.25, 1.33, and 1.66 times that for axial flow for twist ratios of 69.0, 30, and 11.1 respectively at a mass velocity of about 1360 kg/m^2 s (1.0×10^6 lb/h ft^2). Figure 11.10 shows the results at a twist radio of 11.1 plotted as a heat flux efficiency (η) against the mass velocity, G. The heat flux efficiency, η, is defined as the ratio of the actual critical heat flux to the heat flux required to evaporate all the water fed to the test section.

At exit qualities in excess of about 85 per cent and mass velocities below 680 kg/m^2 s (0.5×10^6 lb/h ft^2), the critical heat fluxes with twisted tapes may be less than those for plain tubes. At these low water flow-rates, the twisted ribbon captures some of the water in the channel as a film on its surface. This water is then not available to cool the heated surface as it would

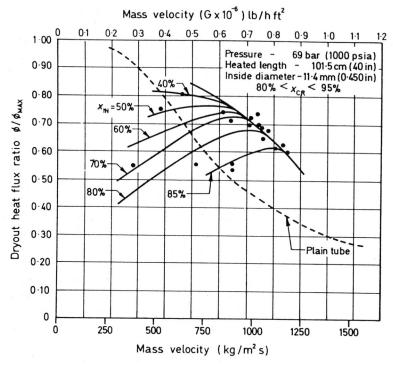

Fig. 11.10. Dryout heat flux ratio for a twist ratio of 11:1 (Moeck *et al.* 1964).

otherwise have been if the ribbon were not present. Therefore, in applying twisted ribbons to once-through boiler applications, the ribbons should terminate some distance before the end of the evaporation region.

In addition to the above work, other studies have been performed with twisted tape inserts over the total length of the tube (Gambill 1965, Cumo *et al.* 1974) or provision of short, twisted ribbons at strategic locations along the heated length (Matzner *et al.* 1965). Coiled wires, meshes and springs have also been successfully utilized (Megerlin *et al.* 1974, Celata *et al.* 1992). Tubes with internal helical fins (Bernstein *et al.* 1964) and internal ribs (Swensen *et al.* 1961, Kitto and Wiener 1982, Kohler and Kastner 1986, Kitto and Albrecht 1987, Matsuo *et al.* 1987), the latter, also referred to as rifled tubes, can provide a remarkable improvement in the CHF value (up to a factor of three) and also counteract the adverse effects of non-uniform heat fluxes and tube inclination on CHF (Kitto and Wiener 1982). Use of ribbed tubes at supercritical pressures to suppress pseudo-film boiling has also been successful (Ackerman 1970).

Integral internally ribbed tubes have been in commercial service for many years in conventional once-through boilers (Vallourec Industries 1988).

On the other hand, tube inserts have some practical problems associated with their application, such as oscillations caused by the two-phase flow. Tubes coated internally with porous metallic coatings have also been shown to be effective in experiments conducted in Russia (Kuzma-Kichta *et al.* 1989). Because of deposits that build up inside boiler tubes, porous coatings may have fouling problems in these services, however.

11.4.1.4 *Improvements in critical heat flux due to geometrical changes in non-circular ducts* Twisted ribbon tapes have also been used to increase the critical heat flux in both an internally heated annulus (Rosuel and Sourioux 1963, Volterras and Tournier 1963) and in a rod cluster (Rosuel *et al.* 1963) type fuel element for a nuclear reactor. Improvements of up to 30 per cent have been noted at 69 bar (1000 psia).

Roughening of the heated surface is another possible way of improving the performance of a boiler system. Burck *et al.* (1969) report the influence on the critical heat flux for subcooled conditions of various macro surface finishes including sand roughness, knurls, and threads. Only with a very coarse knurl (roughness > 0.3 mm) was the critical heat flux increased significantly above the smooth surface value and even in this case the improvement was small compared with the increase in single-phase convective heat transfer coefficients expected with this surface. These results are at variance with those presented by Durant *et al.* (1962) where, for conditions near atmospheric pressure, increases in the critical heat flux of up to 80 per cent at constant velocity and 60 per cent at constant pumping power were reported for an internally heated annulus when the heating surface was roughened by coarse knurling (roughness ~ 0.15–0.3 mm).

Janssen *et al.* (1963) reported tests carried out at higher steam qualities in an internally heated annulus with a heated rod whose surface had been sandblasted to a 7.5 μm (300 μin.) roughness. The critical heat flux was *reduced* below the smooth rod value by 35 per cent at the low flows and as much as 50 per cent at the higher flows. Roughening of the heated surface by sandblasting thus has an adverse effect on the critical heat flux for saturated boiling at medium and high exit steam qualities. This is understandable since the roughening would be expected to reduce the flow-rate of the liquid film on the surface and thus bring about dryout at a lower heat flux.

The use of thick massive fins, ribs, or studs on the heater surface has been investigated in France (Le Franc *et al.* 1964) and the USSR (Ornatskii and Sheherbakov 1963). The concept is known in France as the 'Vapotron' and is used to cool electronic power tubes dissipating heavy heat loads. The principle of the Vapotron is shown in diagrammatic form in Fig. 11.11. This diagram shows two fins with boiling taking place in interspace. The outer surface temperature of the fins varies with location according to the modes of heat transfer taking place. Under typical operation, the boiling heat

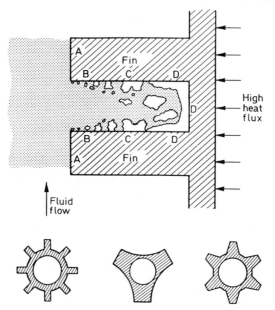

Typical shapes of "Vapotron" fins

Fig. 11.11. The 'Vapotron' concept.

Table 11.3

Location	Mode	ΔT between surface and liquid
A—tops of fins	Non-boiling convection	8°C
B—sides of fins	Stable nucleate boiling	8 to 25°C
C—sides of fins	Transition boiling (past critical heat flux)	25° to 125°C
D—base of fins and surface between fins	Film boiling	> 125°C

transfer modes shown in Table 11.3 might occur (Haley and Westwater 1965).

The temperature distribution in the fins results from the balance between the heat conduction ΔT within the fin and the boiling heat transfer coefficient at the surface. The detailed temperature distribution in a fin, the fin efficiency, and the optimum fin shape have been considered by Haley and Westwater (1966), by Lai and Hsu (1967), and by Cumo *et al.* (1965). Because the heat transfer coefficient is a strong function of surface temperature, the usual formulae for fins are seriously in error. For cooling of electronic tubes, copper

Table 11.4

Quality % by wt.	Mass velocity G kg/m² s (lb/h ft²)	Improvement in critical heat flux
15	680 (0.5 × 10⁶)	none
15	2300 (1.7 × 10⁶)	48%
25	680 (0.5 × 10⁶)	75%
25	2300 (1.7 × 10⁶)	> 100%

is used to obtain high conductivity in the fins. Portions of the surface can operate past the critical heat flux without destruction of the surface because the high thermal conductivity and the large thickness of the material allow high heat fluxes for the maximum permissible temperature gradient.

One further method which has been used very successfully to increase the critical heat flux in an internally heated annulus or rod cluster (Moeck *et al.* 1975, Hesson *et al.* 1965) operating at moderate steam qualities, is roughening of the *unheated* outer wall. This geometrical modification stems from the philosophy that the water in the form of a thick liquid film on the unheated wall does not normally contribute to the cooling process and hence should be directed to the heated surfaces. Experiments in this area have been performed at GEAP, San Jose (Janssen *et al.* 1963, Howard 1963) and at Chalk River (Moeck *et al.* 1964). Howard (1963) has noted that the provision of roughness rings on the inside of the flow tube of an annulus heated internally gave no improvement in the critical heat flux at outlet qualities below 5 per cent by wt. For higher qualities, the improvement in the critical heat flux increases with increasing quality and mass velocity (Table 11.4).

The AECL tests (Moeck *et al.* 1964) performed at exit qualities between 65 per cent and 95 per cent and mass velocities below 680 kg/m² s (0.5 × 10⁶ lb/h ft²) showed improvements in the critical heat flux up to a factor of 7 for a 0.75 mm (0.030 in.) roughness height at a 3.8 cm (1.5 in.) roughness ring spacing. At a given roughness ring spacing, mass velocity and inlet quality, the dryout heat flux increases by up to a factor of 5 when the roughness height is increased from 0 to 1.25 mm (0 to 0.050 in.). The actual ring spacing has only a relatively weak effect on the dryout heat flux value in the range investigated, viz., 3.8 to 11.4 cm (1.5 to 4.5 in.).

11.4.2 *Improvement in critical heat flux through the use of force fields*

Clearly, the considerable increases brought about by the use of swirl flow is one example of the application of a force field. Other methods which come under this category are electrostatic fields and ultrasonic vibrations.

11.4.2.1 *Electrical fields*   The use of electrical fields to increase boiling heat transfer rates has been successfully demonstrated in laboratory type equipment. Bonjour *et al.* (1962) report increases in critical heat flux by factors of 2 to 6 by application of a.c. fields to boiling organic liquids and pure water heated by 0.1 and 0.2 mm wires. Markels and Durfee (1962) report an increase in burnout heat flux of over a factor 4.5 by application of 7000 volts d.c. to a 9.5 mm (0.375 in.) tube acting as a heater causing pool boiling of isopropanol at one atmosphere. One effect of a voltage applied across a vapour–liquid interface is to produce a force on the liquid phase tending to accelerate the fluid towards the solid surface. This results in the vapour film in the film boiling region being destabilized and hence increased wetting of the heater surface. Heat transfer coefficients in the nucleate boiling region are also increased by the electrical field. Experiments under forced convection conditions have been reported by Nichols *et al.* (1965). Experiments conducted with water at 103 bar (1500 psia) with high inlet subcoolings showed that the critical heat flux can be increased by up to 25 per cent by the application of electrical potentials of 1000–2000 a.c. volts.

11.4.2.2 *Flow and ultrasonic vibrations*   Both Isakoff (1956) and Ornatskii and Sheherbakov (1960) report that ultrasonic vibrations considerably increase the critical heat flux in pool boiling. The paper from the USSR (Ornatskii and Sheherbakov 1960) gives results for atmospheric pressure subcooled pool boiling of distilled water from a horizontal Nichrome wire. The strength of the ultrasonic waves emitted from the transducer at 1 Mhz was 1.5 to 2.0 watts/cm^2. The ultrasonic field greatly increased the bubble separation frequency. The observed increases in the critical heat flux varied with the degree of subcooling from approximately 30 per cent at zero subcooling to 80 per cent at ΔT_{SUB} of 80°C. However, Romie and Aronson (1961), using a forced flow system, obtained a contrary result. These investigators obtained critical heat flux data for water flowing vertically upward in an internally heated annulus. An acoustic transducer was located at the inlet end of the flow channel and it propagated ultrasonic waves parallel to the surface of the heating element. No effect on the critical heat flux was detectable although a considerable increase in the frequency of bubble formation was noted. The conditions of this series of tests differed from those mentioned above for pool boiling in that (a) the direction of the field was parallel rather than normal to the heater surface and (b) the coolant convection was forced rather than free. Bergles (1964) reported the influence of flow vibration on forced convection boiling heat transfer of water. Using a similar arrangement of equipment to that of Romie and Aronson (except that the vibrator was placed at the downstream end of the test section), Bergles also found no influence of flow vibrations on the critical heat flux. Heat transfer in the region of the onset of boiling, however, was considerably enhanced as a result of an increase in the single-phase forced convection heat transfer coefficient.

11.4.3 *The influence of trace additives*

The addition of a small quantity of a second component to water has been reported as a means of increasing the critical heat flux in pool boiling. There is, however, little published information for forced convection boiling. In some cases the critical heat flux is higher than the value for either component alone. Further details are given in Section 12.6.2 in respect of pool boiling and Section 12.7.3 in respect of forced convective boiling.

Although it has been shown that considerable increases in the critical heat flux can be obtained from the addition of small quantities of a second component, the practical application of additives in commercial plant must await a considerably increased understanding of the phenomena involved.

11.5 The design of boiling systems where heat flux is a dependent variable

Convective boiling has been discussed in previous chapters for the situation where the heat flux is an independent imposed variable. Close approximations to this situation occur in a radiantly heated water-tube boiler and in a water-cooled nuclear reactor. However, in many instances heating of the evaporating liquid is carried out by heat transfer from a circulating primary fluid. One example is the waste-heat boiler. It is, therefore, necessary to have some method for transposing results taken under conditions of constant imposed heat flux to the case where heating is carried out by a primary fluid. This section discusses this problem and suggests a graphical method as an alternative to the current design methods.

The conventional method of designing heat exchangers usually involves the use of a length-mean (or point) overall heat transfer coefficient U evaluated from

$$\frac{1}{U} = \frac{1}{h_i} + \frac{t A_i}{k_w A_w} + \frac{1 A_i}{h_o A_o} \tag{11.18}$$

where A_i, A_o, and A_w are the internal, external, and log mean tube areas. Such a relationship is useful when the heat transfer coefficients, h_i and h_o, are relatively invariant with heat flux. In convective boiling the relationship between surface heat flux and boiling side temperature difference is not even approximately linear and the conventional method breaks down.

The alternative method proposed in this section considers the heat flow to and from the inside tube surface as separate entities. Under steady state conditions the heat supplied must be equal to the heat removed and the intersection of the heat input characteristic and the heat removal characteristics will provide the operating point.

The proposed method is best illustrated by way of an example. Consider a point on the inside tube surface. By assuming that the shell side heat transfer coefficient, h_o, is constant, it is possible to arrive at an expression giving the tube inside wall temperature, $(T_w)_i$, in terms of the thermal conductivity, k_w, and the thickness of the tube wall, the shell side heat transfer coefficient, h_o, and the local primary fluid temperature, T_o. Thus

$$(T_w)_i = T_o - \phi\left(\frac{D_i}{D_o}\right)\left[\frac{1}{h_o} + \frac{D_o \ln(D_o/D_i)}{2k_w}\right] \qquad (11.19)$$

Figure 11.12 shows a typical heat removal characteristic for a steam–water mixture. The solid lines depict the data of Herkenrath *et al.* for a series of local conditions at some point on the inner tube surface. Such measurements were made with a fixed pressure, tube diameter, and mass flow rate under conditions where the heat input characteristic was that of constant heat flux. Also shown are a series of dotted lines joining the solid curves in the liquid deficient region with the corresponding dryout heat flux for that particular steam quality. These lines were drawn quite arbitrarily and in practice this part of the heat removal characteristic would be estimated from one of the transition boiling correlations given in Chapter 7.

Also shown in Fig. 11.12 are straight lines corresponding to eq. (11.19) for a variety of steam qualities each intersecting the vertical axis at a value of T_{Wi} equal to the corresponding local primary fluid temperature T_o which is obtained from a heat balance. The intersection of one of these lines with the heat removal characteristic for that particular quality provides both the inside wall temperature and the local surface heat flux.

It can be seen from Fig. 11.12 that stable liquid deficient region operation occurs at point B (39 per cent quality) and at higher steam qualities. Stable saturated nucleate boiling occurs at point A (33 per cent quality) and at lower steam qualities. At mass qualities between 33 per cent and 39 per cent, the lines interact in a region which is unstable when the heat input characteristic is a constant heat flux situation. In the example shown a stable operating point will occur in the transition boiling region (point C) for say a steam quality of 35 per cent. Because of the nature of the transition boiling region this condition may well be accompanied by surface temperature fluctuations. If the heat removal characteristic has a more negative slope than the heat input characteristic then point C will not be a stable operating point.

This method of analysing the local conditions in a vapour generator heated by a circulating primary fluid depends on a premise, not yet discussed, which is that the heat removal characteristic is independent of the method used to supply the heat. This question is a vital one. There is very little experimental

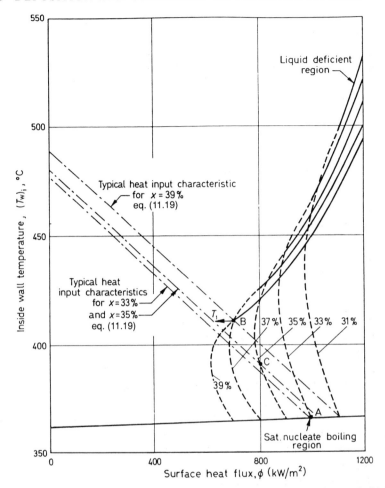

Fig. 11.12. The boiling characteristic for a high pressure water–steam mixture. Solid lines represent data of Herkenrath; $G = 2250 \text{ kg m}^2$, $D = 10 \text{ mm}$, $p = 195 \text{ bar}$. Broken lines show arbitrary transition region.

evidence on this point and in the absence of such evidence it is difficult to argue convincingly that the heat removal characteristic is independent of the heat source.

11.6 Deposition and corrosion on boiling surfaces

So far the discussion of convective boiling heat transfer has been limited to the condition where the heat transfer surface is clean. This, however, is rarely the case in operating plant (Styrikovich 1978) and it is necessary to discuss briefly the incidence of deposition and corrosion particularly on steam-generating surfaces and the influence this deposition and/or corrosion has on the heat transfer and pressure drop characteristics.

11.6.1   The characteristic of deposition in once-through evaporators

With a once-through evaporator all dissolved salts and suspended particulate matter entering with the liquid feed to the evaporator must either:

(a) remain in the evaporator forming a deposit,
(b) be mechanically entrained in the vapour outflow, or
(c) be actually dissolved in the vapour outflow.

A comprehensive review of the solubility of various salts in steam has been given by Styrikovich and Martynova (1963). For supercritical fluid it is possible to consider the solubility of the salt as a function of temperature and pressure. For subcritical systems information on the partitioning of the salt between the water and steam phases is required. The partition (or distribution) coefficient denoted by K_D is

$$K_D = \frac{C_g}{C_f} = \mathrm{fn}(\rho_g/\rho_f) = (\rho_g/\rho_f)^m \qquad (11.20)$$

where C_g and C_f are the concentrations of salt in the steam and water phases respectively. The partition coefficient is found to be a simple power law of the ratio of the phase densities, the value of m varying from salt to salt (Fig. 11.13).

If it it not possible to establish that all or most of the salts present in the feed are soluble in the effluent steam to a greater extent than the feed concentration then a continuous build-up of salt deposits will result. It is, of course, not possible to state the converse, i.e., that evaporators in which the saturation levels are not exceeded will be free from deposits.

Collier and Pulling (1961) have studied the deposition of silica from unsaturated solutions in a low-pressure steam-heated vertical tube evaporator. The pattern of deposition appeared to be closely related to the heat transfer regions encountered. In the saturated nucleate boiling region, deposition appeared in the form of annular rings around a large number of discrete sites on the surface. An explanation for the form of these deposits can be given in terms of an examination of the micro-structure of the boiling process. Vapour bubbles grow at active nucleation sites leaving, as they grow, a very thin 'microlayer' of liquid beneath the bubble. This thin layer is totally evaporated leaving the solid content of the micro-layer deposited on the surface. The bubble detaches and unsaturated liquid contacts the surface

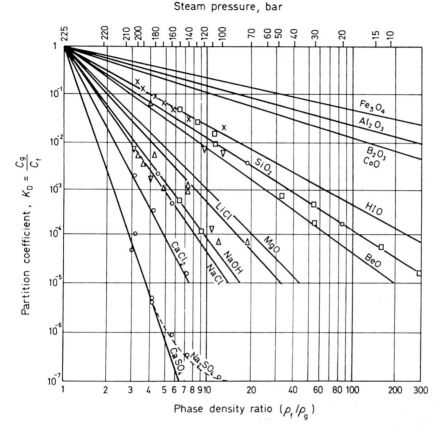

Fig. 11.13. The variation of partition coefficients with the ratio of densities of the two phases.

again, removing some of the deposit. A new bubble forms and the process is repeated. Hospeti and Mesler (1965) employing ^{35}S as a tracer in a saturated solution of calcium sulphate, have made use of this phenomenon to measure the thickness of the micro-layer. Since the number of active sites is proportional to the square of the heat flux, the results of Palen and Westwater (1966), which indicate a sharp increase in the rate of fouling of surfaces with calcium sulphate as the heat flux is increased under pool boiling conditions, are readily explained.

In the two-phase forced convection region where nucleation had been suppressed Collier and Pulling (1961) found the surface of the evaporator free from deposits. Further up the tube a region of continuous deposit was found and it was suggested that this corresponded to the 'liquid deficient' region. However, because of the use of steam heating, the tube wall temperature did not rise appreciably at dryout and, although there was no liquid film on the surface, water droplets were able to impinge on the tube wall and subsequently evaporate to dryness. In a constant heat flux situation, corresponding to a sharp rise in temperature at the dryout point, droplets would be unable to 'wet' the wall and the formation of a deposit by this means is unlikely. Deposition, however, may occur as a result of decreased solubility in the steam phase as its temperature is raised.

Careful studies have been carried out on the deposition of magnetite in once-through boilers. The magnetite exists in boiler feed water as a colloidal suspension of particles and agglomerates in the size range from below 1 μm to 50 μm. For the turbulent flow of water in unheated pipes, the rate of deposition increases with the Reynolds number of the flow. In the core region, away from the wall, turbulent eddies carry the magnetite particles in a radial direction. As the wall is approached the eddies become smaller until, in the immediate vicinity of the wall, turbulence is completely suppressed and there exists a laminar boundary layer. The thickness of this laminar boundary layer is inversely proportional to the Reynolds number. Particles entering the boundary layer will only reach the wall if their entry momentum is sufficient. Deposition is enhanced by increases in velocity because the particle radial velocities are increased in proportion and the boundary layer is also thinner. For the particles to remain at the wall the adhesion forces must exceed the separative forces. In practice, deposition of magnetite takes place in the natural valleys formed in commercially rough tubes. The subsequent infilling of these valleys may often cause the tube to become hydraulically smooth with a consequent reduction in the turbulence level at the same Reynolds number. This effect may account for a fall in the rate of deposition with increasing time.

The rate of deposition of magnetite is relatively insensitive to the presence of a heat flux through the tube wall as long as boiling does not occur. Once subcooled boiling and subsequently saturated boiling occur, the rate of deposition is dramatically increased. Mankina et al. (1960, 1961) have carried out extensive tests in experimental and operating boilers and have recommended the following relationship between the deposited mass, the heat flux, magnetite concentration, and time under boiling conditions:

$$D = K\phi^n C t \qquad (11.21)$$

where D is the mass of magnetite deposit per unit area of tube (kg/m^2); ϕ is the local heat flux (W/m^2); C is the iron concentration in the feed water (kg/m^3); t is the time (h); and K is a proportionality constant.

Mankina et al. (1960) found that the rate of deposition increased sharply with increases in heat flux and concluded that $n = 2$. The value of K varies with the water chemistry but typically for water at 350°C with trisodium phosphate dosing, K had a value 4.2–6.2 × 10^{-13}. Other workers have concluded that the value of n should be nearer unity and have shown a strong dependence of the deposition rate on the local mass quality (Klein and Rice 1965, Goldstein *et al.* 1967) the rate of deposition increasing as the quality is increased. It is generally observed that mass velocity has little effect on deposition under boiling conditions.

At the dryout point, under conditions of constant heat flux, the rate of deposition drops sharply. Other impurities may deposit with the magnetite. Mankina found that copper will deposit at about one-tenth the rate of iron at similar concentrations and will be distributed uniformly through magnetite deposit. The presence of copper within a magnetite deposit will make it more adherent and cause a serious resistance to heat transfer.

11.6.2 *Corrosion at boiling surfaces*

'On-load corrosion' is a particularly rapid and severe attack of boiler tubes in zones where steam is raised. As a result of intensive research into this problem over the past few years a considerable amount is known about the reasons for this type of corrosion. In particular, work carried out by Masterson *et al.* (1969) has elucidated three mechanisms by which corrosive salts may be concentrated in the boiler. The first mechanism is that of dryout, either complete as in a once-through boiler, or partial as may occur due to stratification in the horizontal tubes of waste-heat boilers.

Mann (1975) has developed a model for the behaviour of aggressive agents such as sodium chloride or sodium hydroxide in a once-through boiler. His results show that the equilibrium indicated by eq. (11.20) is not achieved at dryout and that concentrated solutions form as rivulets which penetrate a short distance beyond the dryout point. Mann postulated that the level of solute concentration in the film is determined primarily by the temperature of the tube wall and the pressure in the system. Finally, a point is reached where sodium chloride will exceed its solubility limit in the liquid phase and will be precipitated. However, the concentration of sodium hydroxide can continue to increase to very high levels.

A second concentration mechanism is that occurring in crevices in heated surfaces particularly where the heating is applied asymmetrically (Goldstein and Burton 1969). The third mechanism is closely linked with the build up of porous deposits on the heat transfer surfaces. Liquid is drawn into the deposit between the micro-size magnetite particles by a 'wicking' effect whilst steam is liberated into 'tunnels' in the deposit. Very high concentration factors in excess of 10^4 have been measured for the liquid held within the deposit. Picone *et al.* (1963) have measured the *in-situ* concentration of a ^{22}Na tracer within a 25 μm thick magnetite deposit having 70 per cent porosity. No concentration was found within the deposit under non-boiling conditions. Table 11.5 gives the results obtained under boiling conditions.

Table 11.5 Concentration of sodium salt in porous magnetite deposit (Picone *et al.* 1963)

Test	T_{SAT} °F	T_b °F	T_W °F	D_H-D_B* Counts/min	ϕ Btu/h ft²	W_g Steaming rate lb/h ft²	$\dfrac{\bar{C}}{C_b}$†
1	540	528	554	125	126,300	193	98
2	551	530	565	250	147,300	231	127
3	561	531	574	550	168,300	271	284
4	569	532	580	650	189,400	311	348
5	583	534	593	1025	210,700	343	545
6	588	535	597	1425	231,500	404	765
7	600	536	608	2250	252,000	460	1220
8	620	539	626	5250	294,600	586	2950

* Background $D_B = 650$ counts/min.
† Assumes deposit voids are filled only with liquid.

The concentration is very strongly dependent upon heat flux. After each test the heat flux was lowered and this showed a complete reversibility of the concentration effect. Clearly the maximum concentration possible in the deposit is the inverse of the partition coefficient.

It is clear that both corrosion and deposition must be considered together and where deposition occurs in a steam generator then conditions are also right for corrosion.

11.6.3   *The influence of deposition upon heat transfer and pressure drop*

Deposits can and do have an influence on the hydrodynamic and heat transfer performance of steam generators. As mentioned previously, magnetite particles reaching the wall fill in the micro-roughness of the boiler tube internal surface causing an initial decrease in the pressure drop. This may be accompanied by a corresponding decrease in the heat transfer rate. Reductions in the dryout heat flux and the liquid deficient heat transfer rates have been observed to result from deposition (Goldstein and Burton 1969). Further deposits can cause an increase in pressure drop and this increase can be many times that anticipated from the change in flow cross-section since the deposit is often rippled and acts as a surface roughness.

Cohen (1969) discusses the heat transfer behaviour of fouled surfaces. Under non-boiling conditions heat transfer through the deposit is by conduction. The effective conductivity of the deposit may be computed

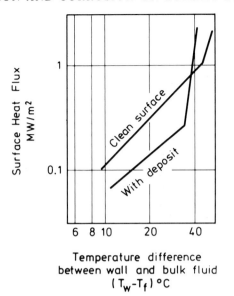

Fig. 11.14. Nucleate boiling on clean and fouled surfaces (Cohen and Taylor 1967).

from Maxwell's formula,

$$k_{eff} = k_f \left[\frac{1 - (1 - ak_p/k_f)b}{1 + (a - 1)b} \right] \qquad (11.22)$$

where k_f is the conductivity of the liquid (or vapour) phase; k_p is the conductivity of the particulate phase; $a = 3k_f/(2k_f + k_p)$; $b = V_p/(V_f + V_p)$; V_p is the total volume of the particulate phase; and V_f is the total volume of the fluid phase.

The equation is valid up to values of $b \approx 0.5$. For magnetite deposits $k_p \simeq 1.0$ Btu/h ft^2 °F/ft. For a deposit of 65 per cent porosity in water at 600°F, k_{eff} is calculated to be 0.46 Btu/h ft^2 °F/ft. If the continuous phase is steam then the value drops to 0.1 Btu/h ft^2 °F/ft. The temperature rise of a constant heat flux surface when dryout occurs within a magnetite deposit is therefore accentuated by the reduction in effective conductivity of the deposit.

Nucleate boiling from surfaces covered by porous magnetite deposits has been studied by Cohen and Taylor (1957). The number of nucleation sites in the case of the fouled surface was considerably greater than for a clean surface. Figure 11.14 shows the experimental boiling curves for both fouled and clean surfaces. In the single-phase forced convection region the fouled surface temperature exceeded that of the clean surface by up to 15°C. Nucleation occurred at a lower heat flux in the case of the fouled surface and for heat fluxes above 0.9 MW/m^2 the temperature of the fouled

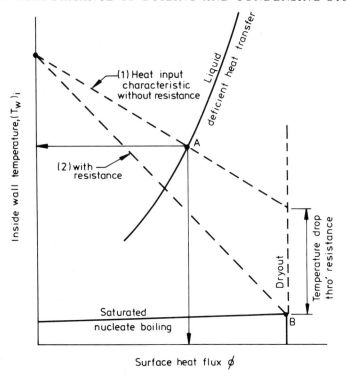

Fig. 11.15. The influence of a resistance in increasing the overall heat flux.

surface fell below that of the clean surface. The remarkable improvement of performance of the deposit-covered surface may be better understood when the behaviour of the tubing with a porous metallic coat discussed in Section 11.2.2 is recalled. Clearly the deposit acts to provide large stable nucleation sites.

In special cases the presence of an additional thermal resistance in a system heated by a primary fluid may significantly improve the overall thermal performance of the system. This is illustrated in Fig. 11.15. The heat input characteristic for the initial system is shown as curve (1). This intersects with the heat removal characteristic corresponding to the liquid deficient heat transfer region at point A. The inclusion of an additional resistance either on the primary fluid side or on the boiling surface side will produce a new characteristic (curve 2) which will now intersect the heat removal characteristic corresponding to the nucleate boiling heat transfer region at point B. This results in an appreciably higher heat flux than in the case without the resistance. The optimum value of the resistance is one which allows point B to be just below the dryout heat flux corresponding to the local boiling side conditions.

References and further reading

Ackerman, J. W. (1970) 'Pseudoboiling heat transfer to supercritical pressure water in smooth and ribbed tubes'. *J. Heat Transfer*, **92**, 490–497.

Adamek, T. A. (1981) 'Bestimmung der kondensationgroß auf feingewellten oberflachen zur ausle-gun aptimaler wandprofile, *Warme-und Stoffubertragung*', **15**, 255–270.

Adamek, T. A. and Webb, R. L. (1990) 'Prediction of film condensation on horizontal integral fin tubes'. *Int. J. Heat Mass Transfer*, **33**, 1721–1735.

Anon (1970) 'New heat exchanger tubes make their commercial debut', *Chem. Engng.*, 60–62, March.

Bähr, A., Herkenrath, H., and Mörk-Mörkenstein (1968) 'The development of the heat transfer crisis in tubes with and without swirl generators'. EUR/C-IS/785/68e.

Bailey, N. A. (1977) 'Dryout in the bend of a vertical U tube evaporator'. Personal communication.

Banerjee, S., Rhodes, E., and Scott, D. S. (1967) 'Film inversion of co-current two-phase flow in helical coils. *AIChE.*, 13(1), 189.

Beatty, K. O. and Forbes, A. V. (1950) *Chem. Engng. Prog.*, **46**, 531.

Beatty, K. O. and Katz, D. L. (1948) *Trans. Am. Inst. Chem. Engrs.*, **44**, 55.

Bella, B., Cavallini, A., Longo, G. A., and Rossetto, L. (1992) 'Pure vapour condensation of refrigerants 11 and 113 on a horizontal integral finned tube at high vapor velocity'. Paper presented at 10th National Heat Transfer Congress held at Genoa, Italy, UIT, 599–610.

Berenson, P. (1960) 'Transition boiling heat transfer from a horizontal surface'. Sc.D. Thesis, Mech. Eng. Dept., MIT Tech. Report No. 17.

Bergles, A. E. (1964) 'The influence of flow vibrations on forced convection heat transfer', *J. Heat Transfer*, 559–560, November.

Bergles, A. E. (1978) 'Survey of augmentation of two-phase heat transfer'. *Two-phase Transport and Reactor Safety*, **1**, 457–478, Hemisphere Publishing Corp.

Bergles, A. E. and Morton, H. L. (1965) 'Survey and evaluation of techniques to augment convective heat transfer'. MIT report 5382–34. See also notes by Bergles, A. E., 'Augmentation of boiling heat transfer', given at Two-phase Gas–Liquid Flow and Heat Transfer Special Summer Program, MIT, 10–21 July (1967).

Bernstein, E., Petrek, J. P., and Meregian, J. (1964) 'Evaluation and performance of once-through, zero gravity boiler tubes with two-phase water'. PWAC (Pratt & Whitney)-428.

Blatt, T. A. and Adt, R. R., Jr. (1963) 'The effects of twisted tape generators on the heat transfer rates and pressure drop of boiling Freon 11 and water'. ASME Paper 63-WA-42.

Bonilla, C. F., Grady, J. J., and Avery, G. W. (1963) 'Pool boiling heat transfer from scored surfaces'. AIChE preprint 32 presented at 6th National Heat Transfer Conference, Boston, August.

Bonjour, E., Verdier, J., and Weil, L. (1962) 'Improvements of heat exchanges in boiling liquids under the influence of an electric field'. AIChE paper 7 presented at 5th National Heat Transfer Conf., Houston, August.

Briggs, A., Wen, X. L., and Rose, J. W. (1992) 'Effect of radiused fin root fillets on condensation heat transfer of integral-fin tubes'. Paper presented at 10th National Heat Transfer Congress held at Genoa, Italy, UIT, 611–622.

Bromley, L. A., Humphreys, R. F., and Murray, W. (1966) *J. Heat Transfer*, **88**, 80.

Burck, E., Hufschmidt, W., and de Clereq, E. (1969) 'The influence of artificial roughness on the increase in the critical heat flux of water in annuli with forced convection'. EUR 4040d.

Carnovos, T. C. (1965) Paper SWD/17 presented at First International Symposium on Water Desalination, Washington.

Carver, J. R., Kakarala, C. R., and Slotnik, J. S. (1964) 'Heat transfer in coiled tubes with two-phase flow'. TID 20983.

Celata, G. P., Cumo, M., and Mariani, A. (1992) 'Water cooled CHF flow boiling in tubes with and without turbulence promoters'. Paper presented at 10th National Heat Transfer Congress held at Genoa, Italy, UIT, 157–169.

Choi, H. Y. (1965) Tufts Univ. Dept. of Mech. Engng. Report 64–1.

Cohen, P. (1969) *Water coolant technology of power reactors*. Gordon and Breach.

Cohen, P. and Taylor, G. R. (1967) Discussion of paper entitled 'Boiling heat transfer at low heat flux', by W. C. Elrod *et al.*, *Trans. ASME (C) J. of Heat Transfer*, **89**, 242.

Collier, J. G. and Pulling, D. J. (1961). 'The deposition of solids in a vertical tube evaporator'. *Industrial Chemist*, **39**, 129–133, 200–203 (1963); also AERE-R 3807.

Corty, C. and Foust, A. (1955) 'Surface variables in nucleate boiling'. *Chem. Eng. Prog. Symp. Series* No. 17, **51**, 1–12.

Cumo, M., Lopez, S., and Pinchera, G. C. (1965) *Chem. Eng. Prog. Symp. Series* No. 59, **61**, 225.

Cumo, M., Farello, G. E., and Ferrari, G. (1974)'Influence of twisted tapes in subcritical, once-through vapor generators in counter flow'. *J. Heat Transfer*, **96**, 365–370.

Czikk, A. M. *et al.* (1970) ASHRAE Paper No. 2132.

Durant, W. S., Towell, R. H., and Mirshak, S. (1962) 'Improvement of heat transfer to water flowing in an annulus by roughening the heated wall'. AIChE preprint 128 presented at 55th Ann. Meeting, Chicago, December.

Erb, R. B. and Thelen, E. (1965) *Ind. Engng. Chem*, **57**(10), 49.

Gambill, W. R. (1965) 'Subcooled swirl-flow boiling and burnout with electrically heated twisted tapes and zero wall flux'. *J. Heat Transfer*, **87**, 342–348.

Gambill, W. R. and Bundy, R. D. (1962) 'An evaluation of the present status of swirl flow heat transfer'. Paper 62-HT-42 presented at 5th National Heat Transfer Conference, Houston, August.

Gambill, W. R., Bundy, R. D., and Wansbrough, K. W. (1961) ORNL-2911 (1960); see also *Chem. Eng. Prog. Symp. Series* No. 32, **57**.

Gardner, G. C. and Kubie, J. (1976) 'Flow of two liquids in sloping tubes: an analogue of high pressure stem and water'. *Int. J. Multiphase Flow*, **2**, 435–451.

Goldstein, P. and Burton, C. L. (1969) 'A research study on internal corrosion of high pressure boilers—final report'. *J. Engng. for Power* (ASME), **91** (A), 75–101.

Goldstein, P., Dick, I. B., and Rice, J. K. (1967) 'Internal corrosion of high pressure boilers'. *J. Engng. for Power* (ASME), **89** (A), 378–394.

Gregorig, R. (1954) *Z. angew. Math. Phys.*, **5**, 39.

Haley, K. W. and Westwater, J. W. (1965) 'Heat transfer from a fin to a boiling liquid'. *Chem. Eng. Sci.*, **20**, 711.

Haley, K. W. and Westwater, J. W. (1966) 'Boiling heat transfer from single fins'. *Proc. 3rd Int. Heat Transfer Conf.*, **2**, 245–253.

Haughey, D. P. (1965) *Trans. Inst. Chem. Engrs.*, **43**, T40.

Hesson, G. M., Fitzsimmons, E. E., and Batch, J. M. (1965) 'Experimental boiling burnout heat fluxes with an electrically heated 19-rod bundle test section'. BNWL-206.

Honda, H. and Nozu, S. (1986) 'A prediction method for heat transfer during film condensation on horizontal low integral-fin tubes'. *J. Heat Transfer*, **108**, 218–225.

Hospeti, N. B. and Mesler, R. B. (1965) 'Deposits formed beneath bubbles during nucleate boiling of radioactive calcium sulphate solutions'. *AIChE J.*, **11**, 662.

Howard, C. L. (1963) 'Methods of improving the critical heat flux for BWRs'. GEAP 4203.

Isakoff, S. E. (1966) 'Effect of an ultrasonic field on boiling heat transfer—exploratory investigation'. Heat Transfer and Fluid Mech. Inst. Preprints of Papers, Stanford Univ., Stanford, CA.

Jakob, M.and Fritz, W. (1931) *Forsch. Gebiete Ingenieur*, **2**, 435.

Janssen, E., Levy, S., and Kervnen, J. A. (1963) 'Investigations of burnout in an internally heated annulus cooled by water at 600 to 1450 p.s.i.a.'. Paper 63-WA-149 presented to ASME Winter Annual Meeting, Philadelphia, November.

Kitto, J. B. and Albrecht, M. J. (1987) 'Elements of two-phase flow in fossil boilers'. *Two-Phase Flow Heat Exchangers Thermal Hydraulic Fundamentals and Design*, NATO ASI Series 3, **143**, 495–551.

Kitto, J. B. and Wiener, M. (1982) 'Effects of nonuniform circumferential heating and inclination on critical heat flux in smooth and ribbed bore tubes'. Presented at 7th Int. Heat Transfer Conf., Munich, **4**, 297–302.

Klein, H. A and Rice J. K. (1965) 'A research study on internal corrosion of high pressure boilers'. Paper presented as ASME Annual Meeting, Chicago, 7-11 November.

Kohler, W. and Kastner, W. (1986) 'Heat transfer and pressure loss in rifled tubes'. Paper presented at 8th Int. Heat Transfer Conf., San Francisco, **6**, 2861–2865.

Kuzma-Kichta, Y. A., Komendantov, A. S., Vasil'eva, L. T., and Hasanov, Y. G. (1989) 'The investigation of heat transfer intensification by porous coating in forced convection at the uniform and nonuniform heating'. *Advances in Phase Change Heat Transfer*, Chongqing, 20–23 May, 425–430.

Lai, F. S. and Hsu, Y. Y. (1967) 'Temperature distribution in a fin partially cooled by nucleate boiling'. *AIChE J.* **13**(4), 817–822.

Le Franc, J. D., Bruchner, H., Domenjoud, P., and Morin, R. (1964) 'Improvements made to the thermal transfer of fuel elements by using the Vapotron Process'. Paper A/CONF 28/P/96 presented at 3rd International Conference on Peaceful Uses of Atomic Energy.

Lis, J. and Strickland, J. A. (1970) 'Local variations of heat transfer in a horizontal steam evaporator tube'. Paper presented at 1970 International Heat Transfer Conference, Paris, August.

Liu, X. and Lienhard, J. H. (1992) 'Extremely high heat flux removal by subcooled liquid jet impingement'. Presented in *Fundamentals of Subcooled Flow Boiling*, HTD-Vol. 217, pp. 11–20, at ASME Winter Ann. Mtg, Anaheim, CA.

Lustenader, E. L., Richter, R., and Neugebauer, F. V. (1959) *J. Heat Transfer*, **81**(4), 297.

Mankina, N. N. (1960) 'Investigation of conditions of formation of iron oxide deposit. *Teploenergetika*, **7**,(3), 8–12.

Mankina, N. N. and Przhiyalkovskü, M. M. (1961) 'Formation of iron oxide deposits in recirculation steam boilers'. *Teploenergetika*, **6**(2), 79 (1959); see also *Brit. Power Engng.*, 60, March (1961).

Mann, G. M. W. (1975) 'Distribution of sodium chloride and sodium hydroxide between steam and water at dryout in an experimental once-through boiler'. *Chem. Engng. Sci.*, **30**(2), 249–260 February.

Markels, M., Jr. and Durfee, R. L. (1962) 'The effect of applied voltage on boiling heat transfer'. AIChE paper 157 presented at 55th Ann. Meeting, Chicago, December.

Marto, P. J. and Rohsenow, W. M. (1965) 'The effect of surface conditions on nucleate pool boiling heat transfer to sodium'. MIT-3357-1, Mech. Eng. Dept. MIT, January.

Marto, P. J., Mitrou, E., Wanniarachchi, A. S., and Rose, J. W. (1986) 'Film condensation of steam on horizontal finned tubes: effect of fins shape'. Paper presented at 8th Int. Heat Transfer Conf., **4**, 1695–1700.

Masterson, H. G., Castle, J. E., and Mann, G. M. W. (1969) 'Waterside corrosion of power station boiler tubes'. *Chemistry and Industry*, 1261–1266, September.

Mathewson, W. S. and Smith, J. C. (1963) *Chem. Eng. Prog. Symp. Series* No. 41, **59**, 173.

Matsuo, T., Iwabuchi, M., Kanzaka, M., Haneda, H., and Yamamoto, K. (1987) 'Heat transfer correlations of rifled tubing for boilers under sliding pressure operating condition'. *Heat Transfer—Jap. Res.*, **16**(5), 1–13.

Matzner B., Casterline, J. E., Moeck, E. O., and Wikhammer, G. A. (1965) 'Critical heat flux in long tubes at 1000 psia with and without swirl promoters'. Paper 65-WA/HT-30 presented at ASME Winter Annual Meeting.

Medwell, J. O. and Nicol, A. A. (1965) Paper 65-HT-43 presented at ASME–AIChE Heat Transfer Conference, Los Angeles, August.

Megerlin, F. E., Murphy, R. W., and Bergles, A. E. (1964) 'Augmentation of heat transfer by use of mesh and brush inserts'. *J. Heat Transfer*, **96**, 145–151.

Merte, J., Jr. and Clark, J. A. (1959) 'Study of pool boiling in an accelerating system'. Univ. of Michigan Report No. 2646-21-T, Tech. Report No. 3.

Milton, R. M. and Gottzmann, C. F. (1972) *Chem. Engng. Prog.*, **68**, 56. See also O'Neill, P. S., Gottzmann, C. F., and Terbot, J. W. *Adv. Cryo. Engng.*, **27**, 420.

Miropolskii, Z. L. and Pickus, V. Yu. (1969) 'Critical boiling heat fluxes in curved channels'. *Heat Transfer—Sov. Research*, (1), 74–79.

Miropolskii, Z. L., Picus, V. J., and Shitsman, M. E. (1966) 'Regime of deteriorated heat transfer at forced flow of fluids in curvilinear channels'. *Proc. 3rd Int. Heat Trans. Conf.*, Vol. 2, paper 50, 95–101.

Moeck, E. O., Wikhammer, G. A., Macdonald, I. P. I., and Collier, J. G. (1964) 'Two methods of improving the dryout heat flux for high pressure steam/water flow'. AECL 2109.

Moeck, E. O., Wikhammer, G. A., Stern, F., and Dempster, R. T. (1965) 'Dryout in a 19 rod bundle cooled by steam–water fog at 515 psia'. ASME paper 65-HT-50 presented at Heat Transfer Conference, Los Angeles, August.

Murata, K., Abe, N., and Hashizume, K. (1990) 'Condensation heat transfer in a bundle of horizontal integral-fin tubes'. Paper presented at 9th Int. Heat Transfer Conf., Jerusalem, **4**, 259–264.

Nakayama, W., Daikoku, T., Kuwahara, H., and Kakizaki, J. (1975) 'High performance heat transfer surface THERMOEXCEL'. *Hitachi Review*, **24**(8) 329–334.

Nandapurkar, S. S. and Beatty, K. O. (196·) *Chem. Eng. Prog. Symp.*, *Series*, No. 30, **56**, 129.

Nichols, C. R., Spurlock, J. M., and Markels, M. (1964, 1965) 'Effect of electrical fields on boiling heat transfer'. NYO-2404-5 and NYO-2404-70.

Ornatskii, A. P. and Sheherbakov, V. K. (1960) 'Intensification of heat transfer in the critical area with the aid of ultrasonics'. Quoted in report NYO-9500.

Ornatskii, A. P. and Sheherbakov, V. K. (1963) 'Determination of the critical heat flow of heat transfer surfaces with studs and transverse fins in the case of electrical heating'. *Tepolfizika Vyso. Tempr.* **1**(3), 425–430.

Osment, B. D. J. and Tanner, D. W. (1962) 'Promoters for the dropwise condensation of steam'. NEL Report No. 34.

Owhadi, A. and Bell, K. J. (1967) 'Forced convection boiling inside helically-coiled tubes'. *Int. J. Heat Mass. Transfer.* **10**, 397–401.

Palen, J. W. and Taborek, J. (1971) AIChE preprint 8 presented at 12th National Heat Transfer Conference.

Palen, J. and Westwater, J. W. (1966) *Chem. Eng. Prog. Symp. Series, No. 64,* **62**, 77–86.

Picone, L. F., Whyte, D. D., and Taylor, G. R. (1963) 'Radio-tracer studies of hide out at high temperature and pressure'. WCAP 3731 June.

Poulter, R. and Allen, P. H. G. (1986) 'Electrohydrodynamically augmented heat transfer in the shell/tube heat exchanger'. Paper presented at 8th Int. Heat Transfer Conf., San Francisco, **6**, 2963–2968.

Raben, I. A., Commerford, G., and Diertent, R. (1961) U.S. Dept. of Int. O.S.W. R and D Prog. Report No. 49.

Romie, F. E. and Aronson, C. A. (1961) 'Experimental investigation of the effects of ultrasonic vibrations on burnout heat flux with boiling water'. ATL-A-123.

Rosuel, A. and Sourioux, G. (1963) *Rapport Special Euratom* No. 6, SNECMA; also EURAEC 146.

Rosuel, A. *et al.* Rapport Special Euroatom No. 16, SNECMA (October 1963); also EURAEC.

Rounthwaite, C. (1968) 'Two-phase heat transfer in horizontal tubes'. *J. Inst. Fuel,* 66–76.

Rounthwaite, C. and Clouston, M. (1961) 'Heat transfer during evaporation of high quality water–steam mixtures flowing in horizontal tubes'. *Int. Heat Transfer Conf., Boulder, Colorado, Part 1,* 200–211.

Rudy, T. M. and Webb, R. L. (1985) 'An analytical model to predict condensate retention on horizontal integral-fin tubes'. *J. Heat Transfer,* **107**, 361–368.

Schlager, L. M., Pate, M. B., and Bergles, A. E. (1989) 'Performance of micro-fin tubes with refrigerant-22 and oil mixtures'. *ASHRAE J.,* November, 17–28.

Singer, R. M. and Preckshot, G. W. (1963) *Proc. Heat Transfer and Fluid Mechanics. Inst.,* **14**, 205.

Spencer, D. L. and Ibele, W. E. (1966) *Proc. 3rd Int. Heat Trans. Conf.,* **2**, 337.

Styrikovich, M. A. (1978) 'The role of two-phase flows in nuclear power plants'. *Int. Seminar for Momentum, Heat and Mass Transfer in Two-phase Energy and Chemical Systems,* Dubrovnik, Yugoslavia, 4–9 September.

Styrikovich, M. A. and Martynova, O. O. (1963) 'Contamination of the steam in boiling reactors from solution of water impurities'. *Sov. At. Energy,* **15**(3), 917.

Styrikovich, M. A. and Miropolskii, Z. L. (1950) *Dokl. Akad. Nauk, SSSR* **71**(2).

Swenson, H. S., Carver, J. R., and Szoeke, G. (1961) 'The effects of nucleate boiling versus film boiling on heat transfer in power boiler tubes'. ASME Paper 61-WA-201.

Thomas, D. G. (1967) *Ind. Eng. Chem. Fundamentals*, **6**(1), 97.

Thome, J. R. (1990) *Enhanced Boiling Heat Transfer*, Hemisphere Publ. Corp., New York.

Thome, J. R. (1992) 'Mechanisms of enhanced nucleate pool boiling'. Paper presented at Engineering Council Conference on Pool and External Flow Boiling held in Santa Barbara, 22–27 March.

Trommelmans, J. and Berghmans, J. (1986) 'Influence of electric fields on condensation heat transfer on nonconducting fluids on horizontal tubes'. Paper presented at 8th Int. Heat Transfer Conf., San Francisco, **6**, 2969—2974.

Vachon, R. I., Nix, G. H., Tanger, G. E., and Cobb, R. O. (1969) 'Pool boiling heat transfer from Teflon coated stainless steel'. *J. Heat Transfer*, 364–370 August.

Vallourec Industries (1988) 'Multi-rifled seamless boiler tubes'. Product Bulletin No. V 4900, May.

Viskanta, R. (1961) 'Critical heat flux for water in swirling flow'. Letter to the Editor of *Nuclear Science & Engng.*, **10**, 202.

Volterras, J. and Tournier, G. *Rapport Special Euratom* No. 9, SNECMA (1963); also EURAEC 639.

Wanniarachchi, A. S., Marto, P. J., and Rose, J.W. (1985) 'Film condensation of steam on horizontal finned tubes: effect of fin spacing, thickness and height'. *Multiphase Flow and Heat Transfer*, ASME HTD Vol. 47, 93–99.

Webb, R. L. (1988) 'Enhancement of film condensation', *Int. Comm. in Heat and Mass Transfer*, **15**(4), 475–508.

Webb, R. L., Rudy, T. M., and Kedzierski, M. A. (1985) 'Prediction of condensation on horizontal integral-fin tubes'. *J. Heat. Transfer*. **107**, 369–376.

Williams, A. G., Nandapurkar, S. S., and Holland, F. A. (1968) 'A review of methods for enhancing heat transfer rates in surface condensers'. *The Chemical Engineer*, CE 367–373, November.

Withers, J. G. and Young, E. H. (1971) 'Steam condensing on vertical rows of horizontal corrugated and plain tubes'. *Ind. Eng. Chem. Process Des. Dev.*, **10**, 19–20.

Yabe, A., Taketani, T., Kikuchi, K., Mori, Y., and Maki, H. (1986) 'Augmentation of condensation heat transfer by applying electro-hydro-dynamical pseudo-dropwise condensation'. Paper presented at 8th Int. Neat Transfer Conf., San Francisco, **6**, 2957–2962.

Yorkshire Imperial Metals Ltd. (1973), Technical Bulletin No. 17, England.

Young, R. K. and Hummel, R. L. (1964) 'Improved nucleate boiling heat transfer'. *Chem. Engng. Prog.* **60**(7), 53–58.

Problems

1. The operator of a tonnage oxygen plant plans to replace the plane tubes in the conventional condenser–reboiler by tubes having a porous coating giving an enhanced boiling side coefficient. The aluminium tubes are 0.375 in. o.d. × 0.040 in. wall thickness. The saturation temperatures of the evaporating oxygen and the condensing nitrogen (assumed pure) are 165°R and 170°R respectively. The condensing coefficient on the nitrogen side is constant at 1000 Btu/h ft^2 °F. The 'active' nucleation cavity radius for the plain tube surface on the boiling oxygen side is 5 μm. If the total heat load for the existing reboiler is 10^6 Btu/h calculate

the increased heat load if the reboiler is re-tubed by tubes with the enhanced heat transfer surface.

Use the method outlined in Section 11.5 to calculate the heat input and removal characteristics for each boiling surface. Assume that the heat removal characteristic for the boiling surface on a ϕ versus T_W plot is a vertical line intersecting the abscissa at the wall temperature necessary to cause the onset of nucleate boiling corresponding to the two active cavity sizes.

2. Figure 16.6 represents the model first proposed by Macbeth (AEEW-R 711) for the process of boiling heat transfer upon a heating surface overlaid by a porous deposit. It is assumed that liquid is drawn into the deposit (thickness h) through fine pores (d_1) by surface tension and steam is ejected via larger diameter pores (d_2). In the limit as the heat flux is increased all the liquid drawn into the deposit

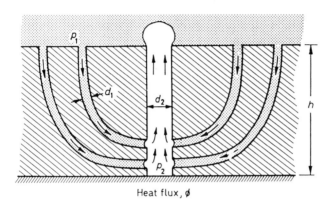

Heat flux, ϕ

Fig. 11.16.

is vaporized and dryout occurs. Construct a model of this process accounting for surface tension. frictional and inertial terms within the two sizes of pore. Assume that the frictional and inertial terms for the two-phase flow in the larger pores are approximated by the homogeneous model. Calculate the concentration of a salt within the deposit for which the partition coefficient between the liquid and vapour is K_D.

3. A waste-heat boiler is designed in which the tubes are formed into a serpentine composed of horizontal 56 mm i.d. tubes connected by $180°$ return bends. The unit is to be operated with an exit mass quality of 25 per cent and at a pressure of 55 bar. The initial choice of mass velocity is 600 kg/m² s. What will be the maximum tube wall temperature if the surface flux is 30 kW/m²? What will be the consequences of prolonged operation under these conditions? What is the minimum value of mass velocity required to prevent stratification and overheating of the upper tube surfaces?

4. Calculate the condensing heat transfer coefficient for steam condensing on a plain 19.05 mm diameter horizontal tube. The steam is saturated at 30°C and the tube outer wall is at 25°C.

5. Repeat problem 4 using the Beatty–Katz correlation for the following low finned tube: 1181 fins/meter (30 fins/inch), fin height = 0.813 mm, fin thickness = 0.279 mm and fin tip diameter = 19.05 mm. Assume the fin cross-section is rectangular and a fin efficiency of 0.7.

6. For the finned tube in Problem 5, determine the retention angle of the steam condensate at 30, 100 and 150°C.

12

MULTI-COMPONENT BOILING AND CONDENSATION

12.1 Introduction

Previous chapters have only considered the behaviour of pure single component fluids during evaporation and condensation.

However, in the petrochemical industries in particular, many processes involve the evaporation (and condensation) of binary ($n = 2$) and multi-component ($n > 2$) mixtures where n is the number of components. In the boiling of binary and multicomponent mixtures, the heat transfer and the mass transfer processes are closely linked, with the evaporation rate usually being limited by the mass transfer processes. This is significantly different from single-component systems where interfacial mass transfer rates are normally very high.

This chapter will first consider the differences to be expected in the basic physical processes when mixtures are evaporated and condensed. This will lead on to a review of the available information on pool boiling of mixtures. Finally, the problems of predicting heat transfer rates in forced convective evaporation and condensation of mixtures will be examined.

There is now a considerable body of published information on pool boiling of binary mixtures and a growing number of experimental studies relating to forced convective evaporation of binary mixtures. The evaporation of multi-component mixtures has been addressed in several studies but is still a largely unexplored area. For in-depth state-of-the-art reviews of mixture boiling, the reader is referred to Thome and Shock (1984) and Stephan (1982).

12.2 Elementary phase equilibria

12.2.1 Definitions

It may be useful to define some of the variables which will be used in this chapter:

The *molar concentration* $\tilde{c}_i$ of component i is given by

$$\tilde{c}_i = \frac{\rho_i}{M_i}$$

where ρ_i is the density of component i and M_i is its molecular weight.

The *total molar concentration* of the mixture is given by

$$\tilde{c} = \sum_{i=1}^{n} \tilde{c}_i$$

The *mol fraction* in the liquid $\tilde{x}_i$ or the vapour $\tilde{y}_i$ is defined by

$$\tilde{y}_i \ (\text{or } \tilde{x}_i) = \frac{\tilde{c}_i}{\tilde{c}}$$

The *molar flux* $\dot{n}_i$ is the number of moles of component i passing normally through unit area in unit time.

The *diffusive molar flux* J_i is the flux of component i, *relative* to the mixture total molar flux $\dot{n}$.

The *mixture total molar flux* $\dot{n}$ is defined as

$$\dot{n} = \sum_{i=1}^{n} \dot{n}_i$$

12.2.2   *Binary systems*

In a mixture of vapours the partial pressure, p_A, of component A is that pressure which would be exerted by A alone in a volume appropriate to the concentration of A in the mixture at the same temperature. Since $p = \sum p_A$, then for an ideal mixture the partial pressure, p_A, is proportional to the mole fraction of A in the vapour phase:

$$p_A = \tilde{y}_A p \qquad (12.1)$$

Two laws relate the partial pressure in the vapour phase to the concentration of component A in the liquid phase.

(a) *Henry's law* which states that the partial pressure p_A is directly proportional to the mol fraction $\tilde{x}_A$ of component A in the liquid phase

$$p_A = \text{constant } \tilde{x}_A \qquad (12.2)$$

This holds for dilute solutions only where $\tilde{x}_A$ is small.

(b) *Raoult's law* which states that the partial pressure p_A is related to the mol fraction $\tilde{x}_A$ and the vapour pressure of pure component A at the same temperature

$$p_A = p_A^0 \tilde{x}_A \tag{12.3}$$

where p_A^0 is the vapour pressure of pure component A at the same temperature. This holds for high values of $\tilde{x}_A$ only.

If a mixture follows Raoult's law it is said to be 'ideal' and values of $\tilde{y}$ can be calculated for various values of $\tilde{x}$ from a knowledge of the vapour pressures of the two pure components at various temperatures.

Few mixtures behave in an 'ideal' manner. If the ratio of the partial pressure p_A to the mol fraction in the liquid $\tilde{x}_A$ is defined as the 'volatility' then for a binary mixture of components A and B

$$\text{volatility of A} = \frac{p_A}{\tilde{x}_A}, \qquad \text{volatility of B} = \frac{p_B}{\tilde{x}_B}$$

The ratio of these two volatilities is known as the 'relative volatility' α, given by:

$$\alpha = \frac{p_A \tilde{x}_B}{p_B \tilde{x}_A} \tag{12.4}$$

Substituting eq. (12.1)

$$\alpha = \frac{\tilde{y}_A \tilde{x}_B}{\tilde{y}_B \tilde{x}_A} \quad \text{or} \quad \frac{\tilde{y}_A}{\tilde{y}_B} = \alpha \frac{\tilde{x}_A}{\tilde{x}_B} \tag{12.5}$$

Clearly, from eq. (12.3) for an 'ideal' mixture the relative volatility is just the ratio of the vapour pressures of the pure components A and B at the chosen temperature. Values of α vary somewhat with temperature and tables for different binary systems are available in chemical engineering handbooks.

Consider a container initially filled with a binary liquid mixture having a mol fraction $\tilde{x}_0$ of the more volatile component and held at constant pressure, p, and temperature, T_1 (Fig. 12.1). What happens when the liquid mixture is heated at constant pressure can be shown conveniently on a *temperature–composition diagram* (Fig. 12.1). In this diagram the mol fraction of the more volatile component in the liquid is plotted as the abscissa and the temperature at which the mixture boils as the ordinate.

The initial conditions in the container are denoted by point Q. The container is now heated at constant pressure. When the temperature reaches T_2 the liquid will start to boil as shown by point R. Some vapour is formed of composition $\tilde{y}_0$ as shown by point S. This initial vapour is rich in the more volatile component. If the experiment is repeated for the complete range of values of $\tilde{x}_0$ then a series of points such as R and S can be determined. The locus of points like R is a curve AXTRB which is known as the 'bubble point' or 'boiling point' curve. The locus of points like S is a curve AWVSB which is known as the 'dew point' curve.

If the container is heated further the composition of the liquid will change,

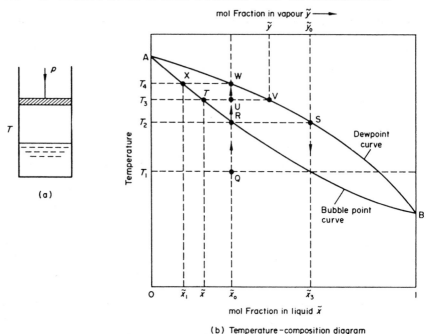

Fig. 12.1. Elementary phase equilibria for binary systems.

because of the loss of the more volatile component to the vapour. The boiling point will therefore rise to some temperature T_3. Both the liquid and vapour phases become richer in the less volatile component with the liquid having a composition represented by point T and the vapour having a composition represented by point V. Since no material is lost from the container, the proportion of liquid to vapour must be given by

$$\frac{L}{V} = \frac{UV}{TU} = \frac{\tilde{y} - \tilde{x}_0}{\tilde{x}_0 - \tilde{x}} \tag{12.6}$$

The ratio $\tilde{y}/\tilde{x}$ at constant temperature is referred to as the *equilibrium ratio, K*

$$K = \tilde{y}/\tilde{x} \tag{12.7}$$

Combining eqs. (12.6) and (12.7):

$$\tilde{x} = \frac{\tilde{x}_0 \left(\dfrac{L}{V} + 1\right)}{K + \dfrac{L}{V}} \tag{12.8}$$

On further heating to a temperature of T_4, all the liquid is vaporized to give a vapour of the same composition as the original liquid (point W). The last drop of liquid to disappear is very rich in the less volatile component ($\tilde{x}_1$) (point X). During this hypothetical experiment, the mass fraction vaporized, the specific volume and the heat added to the system to bring it to each temperature condition would also have been recorded. Note the analogy between this experiment and the once-through vaporization of a liquid mixture fed to the bottom of a tube.

In the design of evaporation equipment for liquid mixtures, thermodynamic equilibrium is usually assumed. However, clearly this must be an approximation since temperature and concentration differences must exist for evaporation to occur. Equilibrium is invariably assumed to exist at vapour–liquid interfaces, i.e., where the phases remain in intimate contact with one another. However, it is possible to conceive a situation where the vapour bubbles formed rise to the surface and into a vapour space such that there is no longer any close contact with the liquid phase. Further evaporation of the liquid must occur with a heavier (less volatile) liquid than originally designed for. There will be a corresponding rise in the boiling point, the effective temperature driving force for evaporation will be reduced and the available surface area may become insufficient to accomplish the required duty. Such a separation of the vapour from the liquid may occur to some extent when evaporating a mixture on the shellside of a kettle reboiler, especially at low circulation rates. Alternatively, it could also occur within tubes where there is stratification or poor distribution of the flows.

Temperature composition diagrams such as Fig. 12.1 are helpful in interpreting different physical situations. For example, consider the case where subcooled boiling of a binary liquid mixture of composition $\tilde{x}_0$ and temperature T_1 occurs (point Q). The vapour formed at the heated surface will have a composition $\tilde{y}_0$. If the vapour bubble detaches from the surface, passes though the liquid and condenses, then the condensed liquid will have a composition $\tilde{x}_3$ corresponding to $\tilde{y}_0$. Thus the liquid adjacent to the heating surface will tend towards a composition $\tilde{x}_1$, while the liquid some way away from the heating surface will tend towards $\tilde{x}_3$. These composition changes will be accompanied by changes in physical properties of the liquid.

Some temperature–composition curves look like those shown in Fig. 12.2. For these binary systems there are critical compositions $\tilde{x}_A$ where the vapour has the same composition as the liquid such that there is no change of composition upon evaporation. It will be seen later that for these special mixtures, called *azeotropes*, mass transfer limitations are absent and the mixture behaves when it is boiled or condensed like a pure component having mean mixture physical properties.

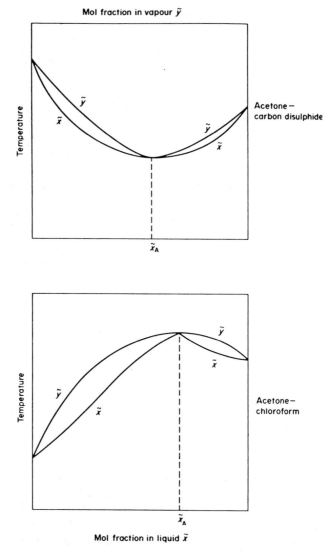

Fig. 12.2. Temperature–composition diagrams for binary mixtures forming azeotropes.

12.2.3   *Multi-component mixtures*

The above concepts can readily be extended to multi-component mixtures. If Raoult's law applies, then eq. (12.8) can be used with the appropriate $\tilde{x}_0$ and K values. Various values of L/V are assumed for a known liquid composition at a given pressure and temperature. The correct L/V ratio is

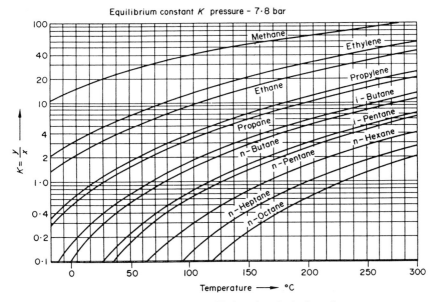

Fig. 12.3. Vapour–liquid equilibrium data for hydrocarbons.

that which satisfies the constraint

$$\sum \tilde{x} = 1 \qquad (12.9)$$

The application of eq. (12.8) to multi-component non-ideal mixtures is more difficult and involves a further constraint

$$\sum K\tilde{x} = 1 \qquad (12.10)$$

The calculations can be carried out in terms of the number of moles of each component in each phase instead of using mol fractions as above. K values have been measured for a wide range of hydrocarbons at various pressures and some values are shown in Fig. 12.3. The methods used to measure or predict phase equilibrium data are, however, outside the scope of this chapter. Generalized methods for predicting equilibrium states are described in Prausnitz (1969).

The evaporation of any binary or multi-component liquid can be expressed in terms of a curve of the equilibrium temperature (T) against an amount of heat (Q)—the conventional 'cooling' or 'condensation' curve. It is a unique curve provided the pressure (p) throughout the evaporator is constant. Another, and in some ways more convenient, way of expressing this curve is in terms of the temperature against the specific enthalpy ($\tilde{i}$) (Fig. 12.4). Transferring one curve to another is quite straightforward and

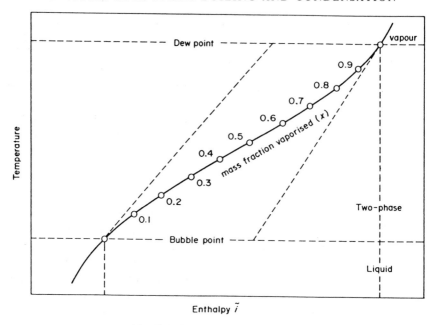

Fig. 12.4. Temperature–enthalpy curve.

can be done by

$$\tilde{\imath} = \frac{Q - Q_{\mathrm{r}}}{N} \tag{12.11}$$

where Q_{r} is a reference value of Q which could conveniently be *either* the value at the lowest temperature reached in the system *or* the 'bubble point' value at the lowest pressure, p; N is the total number of moles of fluid.

The *condensation* curve must always be provided to the heat exchanger designer. Typical information will be in a tabular form for a series of pressures spanning the range of interest for each mixture. Since the heat exchanger designer has to work with local conditions within the exchanger, the local pressure will always be known at any point in the calculation and it is, therefore, possible to allow for the effect of pressure on the equilibrium temperature by interpolation between tables at given pressures. For the sake of clarity, however, the effect of pressure is omitted from the rest of the discussion.

In addition to there being a unique T–$\tilde{\imath}$ relationship at each pressure level, so there is also a unique x–$\tilde{\imath}$ relationship where x is the quality of the vapour phase mass fraction. When the condensation curve (T–$\tilde{\imath}$) is specified the corresponding values of x must also be given (Fig. 12.4).

It is possible to consider the change in enthalpy of a binary or multi-component fluid with temperature as made up of three terms, *viz.*

$$d\tilde{\imath} = (1 - x)\, dT\tilde{c}_{pf} + x\, dT\tilde{c}_{pg} + \tilde{\imath}_{fg}\, dx \qquad (12.12)$$

These are (i) the sensible heat change in the liquid phase $(1 - x)$; (ii) the sensible heat change in the vapour phase (x); (iii) the heat of vaporization of an amount of liquid (dx).

This concept differs from that for a single component system since, in that case, enthalpy changes are *either* sensible heat changes *or* latent heat changes (at constant temperature).

12.2.4 *Mixture physical properties*

When one or more components are mixed together, the physical properties of the resulting mixture are required to describe the new fluid. In some cases the mixing is ideal, which means that the mixture properties can be obtained by a simple linear mixing law approach using the component concentrations and the pure fluid properties at the same temperature. For a mixture at its saturation point, the liquid component concentrations are used for predicting the liquid properties and the vapour concentrations are used for the vapour properties.

In most instances the mixing is *not* ideal and the heat of mixing, non-ideal variations in viscosity, surface tension, density, etc. must be estimated with more sophisticated methods. Some of the resulting properties may not lie between those of the pure components. Very large variations can occur, especially for liquid viscosity and latent heat. For further information on suitable predictive methods, refer to Reid *et al.* (1987).

A physical property important to boiling, nucleation, and drop-wise condensation is the contact angle. Figure 12.5 depicts data obtained by Shakir and Thome (1986) for several binary systems in the presence of nitrogen gas at 25°C. The contact angle varies linearly with liquid mol fraction for ethanol–benzene and methanol–water but is very non-linear for the n-propanol–water system.

12.3 Nucleation in a binary system

As with a single component system, vapour may be formed from a binary liquid mixture at either a planar (stable equilibrium) or curve interface (unstable equilibrium). The superheat $(T_g - T_{SAT})$ required to maintain a bubble embryo of radius r^* in equilibrium is given by

$$(T_g - T_{SAT}) = \frac{2\sigma}{(\partial p_{SAT}/\partial T)r^*} \qquad (12.13)$$

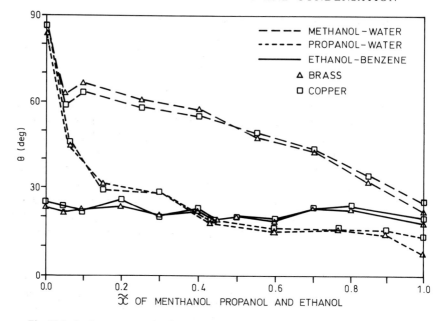

Fig. 12.5. Static contact angles for selected binary mixtures (Shakir and Thome 1986).

For mixtures, both $(\partial p_{SAT}/\partial T)$ and the surface tension σ depend on the composition as well as the temperature. The extended Clapeyron equation is given by Stein (1978) and Malesinsky (1965):

$$\left(\frac{\partial p_{SAT}}{\partial T}\right) = \left(\frac{\partial p}{\partial T}\right)_{\tilde{x}} + \frac{p}{\tilde{R}T}\left(\frac{\partial \tilde{x}}{\partial T}\right)_p (K-1)\tilde{x}\left(\frac{\partial^2 g}{\partial \tilde{x}^2}\right)_{T,p} \qquad (12.14)$$

where K is the equilibrium constant and g is the Gibbs free energy.

For ideal systems, $(\partial^2 g/\partial \tilde{x}^2)_{T,p}$ is given by

$$\left(\frac{\partial^2 g}{\partial \tilde{x}^2}\right)_{T,P} = \frac{\tilde{R}T}{\tilde{x}(1-\tilde{x})} > 0 \qquad (12.15)$$

The slope of the boiling point curve on an equilibrium diagram at constant pressure can be evaluated by rearranging eq. (12.14) and noting that $(\partial p_{SAT}/\partial T) = 0$ at constant pressure and $(\partial p/\partial T)_{\tilde{x}} = p\tilde{i}_{fg}/\tilde{R}T^2 = \rho_g\tilde{i}_{fg}/T$. Rearranging eq. (12.14) with these constraints:

$$\left(\frac{\partial T}{\partial \tilde{x}}\right)_p = -\left(\frac{T}{\tilde{i}_{fg}}\right)(K-1)\tilde{x}\left(\frac{\partial^2 g}{\partial \tilde{x}^2}\right)_{T,p} \qquad (12.16)$$

Alternatively, substituting for $(\partial^2 g/\partial \tilde{x}^2)_{T,p}$ from eq. (12.15):

$$\left(\frac{\partial T}{\partial \tilde{x}}\right)_p = -\left(\frac{pT}{\rho_g \tilde{l}_{fg}}\right)\left(\frac{K-1}{1-\tilde{x}}\right) \leqslant 0 \qquad (12.17)$$

This expression can be seen to be almost independent of pressure.

The second term on the right-hand side of eq. (12.14) is always negative. This means that the value of $(\partial p_{SAT}/\partial T)$ is always somewhat less than the value for an equivalent pure liquid. Nevertheless, Shock (1977) has shown that $(\partial p_{SAT}/\partial T)$ for a binary mixture can be *higher* than the value for the pure components alone. Considering eq. (12.13), the influence of this change in $(\partial p_{SAT}/\partial T)$ is usually small compared with changes in both the surface tension, σ, and the size of the active nucleation site on the heated surface, r^*.

The addition of relatively small quantities of a second component to water often greatly reduces the surface tension. Figure 12.6 shows the changes in

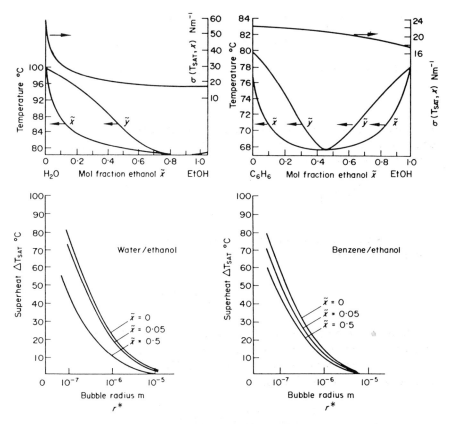

Fig. 12.6. Nucleation in binary mixtures.

bubble–dew point temperature and in surface tension (σ) for the two systems, ethanol–water and ethanol–benzene. This figure also shows the calculated superheat values for these same two systems. For the ethanol–water system, the reduction in surface tension as the mol fraction of ethanol is increased causes a sharp drop in the superheat required to maintain a given size embryo in equilibrium. Conversely, for the ethanol–benzene system the variation of surface tension and also of superheat with composition is small.

Homogeneous nucleation in a binary or multi-component fluid has been considered by Holden and Katz (1978). Equation (4.13) remains valid; however, the collision frequency λ is modified to

$$\lambda = \left(\frac{\tilde{y}_A}{\sqrt{M_A}} + \frac{\tilde{y}_B}{\sqrt{M_B}} \right) \left(\frac{2\sigma}{\pi} \right)^{1/2} \tag{12.18}$$

For multi-component fluids, further terms of the type ($\tilde{y}/\sqrt{M}$) are required to be added to the first term. The use of this equation requires both phase equilibria data and a knowledge of the variation of surface tension with composition and temperature. The theory agrees reasonably well with available experimental data. For multi-component hydrocarbon mixtures, the data of Porteous and Blander (1975) indicate a linear relationship between superheat and mol fraction.

Heterogeneous nucleation has been studied experimentally by Shakir and Thome (1986) for four different binary mixture systems on smooth plates and tubes and on porous coated tubes. Figure 12.7 shows the boiling activation and deactivation superheats they measured for ethanol–benzene on a smooth tube at 1.0 bar. The activation superheat for nucleation of the first active boiling site is consistently larger than the deactivation superheat for the last active boiling site upon decreasing the heat flux. These superheats are a strong function of composition and tend to be larger than those which would otherwise be expected from a linear molar interpolation between the pure fluid and azeotrope values.

Figure 12.7(b) depicts eq. (12.13) evaluated for several mixture systems for a cavity radius of 1 μm. Thus, neither eq. (12.13) nor the vapour-trapping process using the contact angles in Fig. 12.5 predicts the experimental variation of the activation and deactivation superheats for this system.

Thome and Shakir have proposed that the maxima in the activation and deactivation superheats is a result of a mass diffusion effect on the vapour nuclei as follows. When a boiling site deactivates, the surrounding liquid is partially depleted of the more volatile component due to mass diffusion effects on bubble growth (as will be discussed in Section 12.5) and the vapour in the cavity has a vapour composition corresponding to this local liquid composition. When the heated surface returns to the saturation temperature of the bulk liquid and the liquid composition to that of the bulk, the vapour

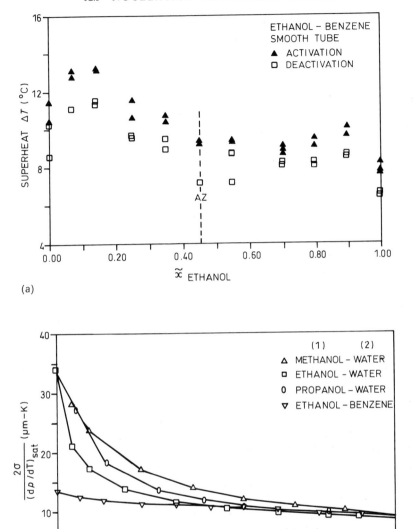

(a)

(b)

Fig. 12.7. Nucleation superheats in binary mixtures: (a) experimental results (Shakir and Thome 1986); (b) eq. (12.13).

in the nucleus will be supersaturated with respect to the less volatile component and a fraction of this component condenses at the phase interface to attain equilibrium conditions. Hence, the size of the nucleus and its radius are reduced and a larger nucleation superheat is required to activate the site. This process cannot occur for a pure fluid or a mixture at the azeotrope composition. Thus, maxima to both sides of the azeotrope point are expected and were observed to occur experimentally.

12.4   Size of nucleation sites

Before eq. (12.13) can be used to estimate the superheat required of a mixture upon a heated surface it is necessary to know how the maximum size of nucleation site on the surface varies with composition. This maximum size is determined by the penetration of the liquid into pre-existing cracks and flaws in the surface.

This penetration is influenced both by changes in surface tension and in contact angle. Experimental evidence from Shock (1977) suggest that these changes in the maximum size of nucleation site with composition can be profound. Table 12.1 shows the variation of the computed size of nucleation site on the onset of boiling on a nickel plated surface for the ethanol–water system. On this evidence it would appear that changes in mixture composition, particularly when one of the components is water, may produce significant changes in the superheat required to initiate and sustain nucleation primarily as a result of changes in the contact angle and hence the maximum active nucleation site.

Table 12.1   Variation of size of active nucleation site with composition for boiling of ethanol/water on a nickel plated surface (Shock, 1977)

Composition mol fraction EtOH $\tilde{x}$	Wall temperature at ONB °C	$(T_g - T_{SAT})$ at ONB °C	r_{MAX} µm
0	138.2	9.9	1.05
0.058	143.5	26.6	0.23
0.058	143.0	26.7	0.24
0.058	146.5	29.3	0.20
0.058	145.4	28.6	0.21
0.197	145.2	36.8	0.095
0.197	145.0	36.7	0.095

$p = 2.5$–2.6 bar

12.5 Bubble growth in a binary system

Bubble growth in a single component system is limited very rapidly by the rate at which heat can diffuse to the interface to provide the latent heat of vaporization. In a binary liquid mixture, however, a limitation also occurs as a result of the liquid close to the bubble interface becoming depleted in the more volatile component (Scriven, 1959). For vaporization and bubble growth to continue the more volatile component must now diffuse from the bulk liquid through the depleted region. This is shown diagrammatically in Fig. 12.8. The initial bulk liquid contains a mass fraction (x_0) of the more volatile liquid and it is superheated by an amount ΔT_{SAT} (to point E) above the boiling point corresponding to the initial liquid composition $T(x_0)$ (point D). At the bubble interface the mass concentration of the more volatile component in the liquid phase has fallen to x (point A) whilst the composition of the vapour within the bubble is y (point B). The corresponding rise in the saturation temperature *at the bubble wall* ($T(x) - T(x_0)$) is denoted by ΔT.

Van Stralen (1967) and Van Stralen and Zijl (1978) have given the theory for bubble growth in binary liquid mixtures. A mass balance at the interface for the more volatile component is

$$\rho_g(y - x)\dot{R} = \rho_f D \left(\frac{\partial x}{\partial r}\right)_{r=R} \tag{12.19}$$

If $(\partial x/\partial r)_{r=R}$ is approximated by assuming that the concentration falls from x_0 to x across a diffusion shell of thickness z_M the bubble growth rate can be given as:

$$\dot{R} = \left(\frac{\rho_f}{\rho_g}\right)\left(\frac{x_0 - x}{y - x}\right)\left(\frac{D}{z_M}\right) = \left(\frac{\rho_f}{\rho_g}\right)\left(\frac{x_0 - x}{y - x}\right)\frac{D}{(\pi/3)^{1/2}(Dt)^{1/2}} \tag{12.20}$$

Van Stralen has termed $(x_0 - x)/(y - x)$ the *vaporized mass diffusion fraction* (G_d).

It will be seen from Fig. 12.8 that the effective superheat conducting heat to the bubble wall is not ΔT_{SAT} but ($\Delta T_{SAT} - \Delta T$). Thus the bubble growth rate is also given by

$$\dot{R} = \left(\frac{\rho_f}{\rho_g}\right)\left[\frac{(\Delta T_{SAT} - \Delta T)c_{pf}}{i_{fg}}\right]\frac{\alpha_f}{(\pi/3)^{1/2}(\alpha_f t)^{1/2}} \tag{12.21}$$

Equating eqs. (12.20) and (12.21), substituting for G_d and rearranging,

$$\frac{\Delta T}{G_d} = \left(\frac{D}{\alpha_f}\right)^{1/2}\left(\frac{i_{fg}}{c_{pf}}\right)\left(\frac{\Delta T_{SAT}}{\Delta T} - 1\right)^{-1} \tag{12.22}$$

It may be assumed that $(\Delta T_{SAT}/\Delta T)$ and therefore $(\Delta T/G_d)$ are independent of the actual value of ΔT_{SAT}:

$$\frac{\Delta T}{G_d}(x_0) = -(K-1)x\left(\frac{\partial T}{\partial x}\right)_p$$

It may be evaluated graphically from equilibrium data as shown in Fig. 12.8.

The asymptotic bubble growth equation for binary fluid mixtures is obtained by substituting eq. (12.22) in eq. (12.21):

$$\dot{R} = \frac{\Delta T_{SAT}}{\left(\dfrac{\rho_g}{\rho_f}\right)\left[\dfrac{i_{fg}}{c_{pf}} + \left(\dfrac{\alpha_f}{D}\right)^{1/2}\dfrac{\Delta T}{G_d}\right]}\left(\frac{3}{\pi}\frac{\alpha_f}{t}\right)^{1/2} \tag{12.23}$$

It can be seen that a minimum value of $\dot{R}$ for a binary mixture corresponds to a maximum value of $(\Delta T/G_d)$. This was confirmed by Van Stralen using experimental observations. Substituting the approximation for $(\Delta T/G_d)$ into eq. (12.23) and integrating yields an equation which can be compared with eq. (4.27):

$$R = \left(\frac{12\alpha_f}{\pi}\right)^{1/2}\left(\frac{\rho_f c_{pf} \Delta T}{\rho_g i_{fg}}\right)\mathrm{Sn}\, t^{1/2} \tag{12.24}$$

where Sn is defined as

$$\mathrm{Sn} = \left[1 - (y-x)\left(\frac{\alpha_f}{D}\right)^{1/2}\left(\frac{c_{pf}}{i_{fg}}\right)\left(\frac{\partial T}{\partial x}\right)_p\right]^{-1} \tag{12.25}$$

Scriven (1959) previously arrived at a similar expression to eq. (12.24) from an analytical solution of the heat and mass diffusion process in a superheated liquid surrounding a growing vapour bubble. The Scriven number Sn was defined by Thome (1981) for this non-dimensional mass diffusion term to commemorate Scriven's original contribution. The value of Sn is always less than or equal to 1.0.

Thome and Davey (1981) measured bubble growth rates for bubbles growing from artificial cavities in a heated horizontal disc for nitrogen–argon mixtures at 1.3 bar. They observed a wide variation in the growth rates of sequential bubbles emanating from the same site and hence determined the statistically mean bubble growth rates for a complete sequence of 20–30 bubbles using a computerized image analysis system. A qualitative agreement with the Scriven and Van Stralen models was found, correctly accounting for the mass diffusion effect.

Equation (4.28) given by Mikic and Shock (1983) remains valid for binary

Mass fraction of more volatile component in vapour (y)

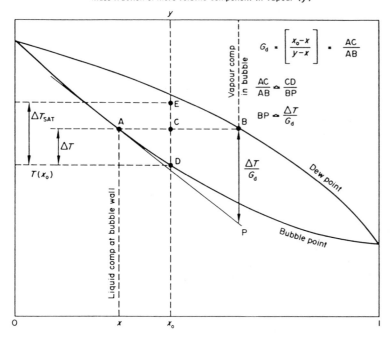

Mass fraction of more volatile component in liquid (x)

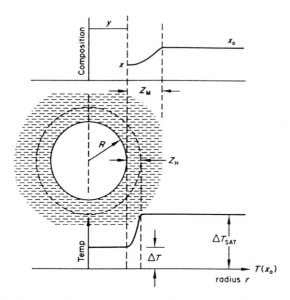

Fig. 12.8. The growth of vapour bubbles in binary systems (Van Stralen, 1967).

systems provided B is evaluated from

$$B = \left(\frac{12\alpha_f}{\pi}\right)^{1/2} \text{Ja Sn} \tag{12.26}$$

The bubble growth rate in a binary system is sharply reduced from that for a single component system since the liquid mass diffusivity, D, of the more volatile component is an order of magnitude smaller than the liquid thermal diffusivity (α_f).

The much lower growth rate also results in much smaller bubbles at the point of departure from the surface. Thome (1982) shows that the ratio of the actual departure diameter to that for the equivalent pure fluid can be approximated by

$$\frac{D_d}{D_{d_i}} = \text{Sn}^{4/5} \tag{12.27}$$

Similarly for bubble growth times in binary fluids, Thome (1982) shows for inertia-controlled bubble departure that the ratio of the growth time in a binary mixture to that for the equivalent pure fluid is

$$\frac{t_g}{t_{g_i}} = \text{Sn}^{2/5} \tag{12.28}$$

Thus, both the bubble departure diameters and the bubble growth times are predicted to decrease in mixtures, as was observed experimentally by Thome (1978) and Shock (1984) for nitrogen–argon mixtures.

A further important point is that the proportion of the heat transferred to the bubble while still attached to the heating surface is greatly reduced in the case of mixtures and, in the latter case, the bubbles continue to grow significantly after their departure.

Zeugin et al. (1975) have studied the behaviour of the microlayer underneath a bubble formed from a binary liquid mixture and growing at a heated surface. As the bubble grows initially the surface temperature under the bubble falls sharply from the initial superheat to just above the boiling point of the liquid mixture. The surface temperature then recovers more slowly than in the case of pure liquids with the temperature rise following the increase in boiling point of the microlayer as the more volatile component is depleted. Evidently, vapour produced from evaporation of the microlayer will be progressively less rich in the more volatile component than that produced from the rest of the bubble wall. If the microlayer evaporates completely, then the composition of the vapour produced will be that of the original liquid.

12.6 Pool boiling

The pool boiling curve is considerably altered when the fluid being evaporated is a binary mixture rather than a pure single component liquid. The principal changes are shown diagrammatically in Fig. 12.9. Firstly, the onset of boiling is delayed to higher wall superheats as a result of the temperature gradients set up in the pool to accommodate the corresponding gradients in liquid composition. Heat transfer coefficients in the nucleate boiling region are sharply reduced. The critical heat flux may be increased or reduced depending on the extent of the contribution from convection in the pool. The minimum heat flux and the corresponding wall superheat are increased. Finally, heat transfer rates in the film boiling region are also somewhat higher.

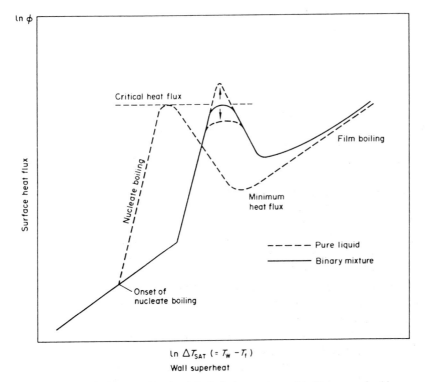

Fig. 12.9. Schematic diagram showing principal changes to pool boiling curve for binary mixtures.

12.6.1 Nucleate boiling

Provided the additive is not surface active even small amounts of a second component cause considerable reductions in the heat transfer rate under nucleate pool boiling conditions compared with that measured for the pure liquid. The reason for this reduction can be traced back to the influence of the second component upon the bubble growth rate. The minimum bubble growth rate, the minimum heat transfer coefficient, and the occurrence of a maximum critical heat flux all occur at the same liquid composition corresponding to a maximum value of $|\tilde{y} - \tilde{x}|$. In the case of surface active agents added to water, small increases ($\sim 25\%$) in boiling heat transfer coefficient have been observed.

A good deal of useful experimental data for nucleate pool boiling of binary liquids has been presented by Sternling and Tichacek (1961). They used fourteen binary systems with components ranging between water, light alcohols, and heavy oils. All the mixtures had a wide boiling range, at least $90°C$. For all the systems the heat transfer coefficient for a given heat flux was less than would be expected for an 'ideal' single component fluid with the same physical properties.

The first empirical mixture correlation for the nucleate pool boiling of mixtures was that of Palen and Small (1964). Their expression characterizes the mixture using its boiling range and is given as follows:

$$\frac{h}{h_\text{I}} = \exp(-0.027\,\Delta\theta_\text{bp}) \tag{12.29}$$

where the boiling range $\Delta\theta_\text{bp}$ is the temperature difference between the dew point and bubble point temperatures of the bulk mixture.

Stephan and Körner (1969) and Stephan and Preusser (1978) have suggested a simple method whereby heat transfer coefficients in binary systems may be computed from data for the pure components. The method is illustrated in Fig. 12.10. The upper part of this diagram represents an equilibrium diagram for the system A–B and shows the mol fraction of component B in the vapour phase ($\tilde{y}$) plotted against the mol fraction of component B in the liquid phase ($\tilde{x}$) for a constant pressure. The lower part of the diagram shows the difference in temperature between the heating surface T_W and the boiling point $T_\text{SAT}(\tilde{x})$ corresponding to a liquid composition ($\tilde{x}$). The curve of $[T_\text{W} - T_\text{SAT}(\tilde{x})]$ passes through a maximum corresponding to the maximum value of $|\tilde{y} - \tilde{x}|$. The value of $[T_\text{W} - T_\text{SAT}(\tilde{x})]$ may be expressed as

$$[T_\text{W} - T_\text{SAT}(\tilde{x})] = \Delta\theta_\text{I} + \Delta\theta_\text{E} = \Delta\theta_\text{I}(1 + \Theta) \tag{12.30}$$

$\Delta\theta_\text{I}$ is the 'ideal' value of the temperature difference and is evaluated from the values of $\Delta\theta_\text{SAT}$ for the two pure components A and B calculated from

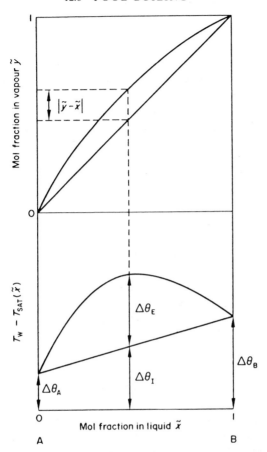

Fig. 12.10. The evaluation of nucleate boiling heat transfer coefficients for binary systems (Stephan and Körner 1969).

the same pressure and heat flux as the mixture. These wall superheat values will be denoted $\Delta\theta_A$ and $\Delta\theta_B$ respectively. Then $\Delta\theta_I$ is given by

$$\Delta\theta_I = (1 - \tilde{x})\,\Delta\theta_A + \tilde{x}\,\Delta\theta_B \qquad (12.31)$$

Stephen and Körner found that Θ in eq. (12.30) could be correlated against the value of $|\tilde{y} - \tilde{x}|$. Thus

$$\Theta = \Lambda|\tilde{y} - \tilde{x}| \qquad (12.32)$$

Λ was found to vary with pressure and over the pressure range 1–10 bar the following expression was used:

$$\Lambda = \Lambda_0(0.88 + 0.12p) \qquad (12.33)$$

where p is the system pressure in bar and Λ_0 is a constant dependent on the

Table 12.2 Value of Λ_0 for use in eqs. (12.32) and (12.33) for the estimation of nucleate boiling heat transfer coefficients for binary mixtures (Stephan and Körner 1969)

Sysrem	Λ_0
Acetone–ethanol	0.75
Acetone–butanol	1.18
Acetone–water	1.40
Ethanol–benzene	0.42
Ethanol–cyclohexane	1.31
Ethanol–water	1.21
Benzene–toluene	1.44
Heptane–methyl cyclohexane	1.95
Isopropanol–water	2.04
Methanol–benzene	1.08
Methanol–amylalcohol	0.8
Methyl–ethyl ketone–toluene	1.32
Methyl–ethyl ketone–water	1.21
Propanol–water	3.29
Water–glycol	1.47
Water–glycerine	1.50
Water-pyridine	3.56

particular binary system studied. Values of Λ_0 are given in Table 12.2 for seventeen common systems. A value of Λ_0 of 1.53 was recommended when no experimental data for the system under consideration were available. This method is not reliable if one of the components is strongly surface active.

Happel and Stephan (1974) have reported the results of pool boiling experiments with benzene–toluene, ethanol–benzene, and water–isobutanol mixtures over the entire range of compositions and at pressures 0.5, 1, and 2.0 bar. The results were presented in terms of heat transfer coefficient rather than temperature difference. If there is no influence of mass transfer on the boiling process the heat transfer coefficient ('ideal' value) at any liquid composition ($\tilde{x}$) can be related to the coefficients of the pure components at the same condition (h_A, h_B):

$$h_I = h_A(1 - \tilde{x}) + h_B\tilde{x} \tag{12.34}$$

This differs from eq. (12.31) in that now

$$\frac{1}{\Delta\theta_I} = \frac{(1 - \tilde{x})}{\Delta\theta_A} + \frac{\tilde{x}}{\Delta\theta_B} \tag{12.35}$$

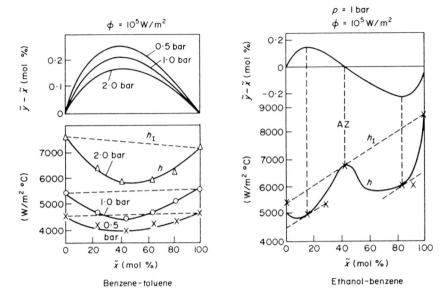

Fig. 12.11. Variation of heat transfer coefficient for binary mixtures with composition and pressure (Happel and Stephan 1974).

Figure 12.11 shows results from benzene–toluene mixtures and confirms that the minimum coefficient does occur at the maximum value of $|\tilde{y} - \tilde{x}|$. For mixtures which form azeotropes (where the liquid and the vapour phases in equilibrium have the same compositions; $|\tilde{y} - \tilde{x}| = 0$) such as ethanol–benzene, one might expect the heat transfer coefficient to approach the 'ideal' value at the azeotropic composition and this also is confirmed. The ratio of the actual to the ideal coefficient was correlated as a function of $|\tilde{y} - \tilde{x}|$.

$$\frac{h}{h_I} = \frac{\Delta\theta_I}{\Delta\theta} = 1 - B[\tilde{y} - \tilde{x}]^n \qquad (12.36)$$

where B depends on the mixture and the pressure. For benzene–toluene mixtures at 1 bar $B = 1.5$ and $n = 1.4$, for ethanol–benzene mixtures $B = 1.25$ and $n = 1.0$, and for water–isobutanol mixtures over their range of miscibility $B = 0.9$ and $n = 0.7$.

The first mechanistic type of nucleate pool boiling mixture correlation was that of Calus and Leonidopoulus (1974), who equated the decrease in the heat transfer coefficient to the reduction in the bubble growth rate and obtained

$$\frac{h}{h_I} = Sn \qquad (12.37)$$

Subsequently Thome (1981) analysed the effects of mass diffusion controlled bubble growth on the cyclic thermal boundary layer stripping mechanism of boiling heat transfer (Fig. 4.18b). He arrived at a similar expression:

$$\frac{h}{h_1} = Sn^{7/5} \tag{12.38}$$

Equation (12.38) predicts a larger reduction in the heat-transfer coefficient than eq. (12.37).

The Stephan method with an empirical constant for each binary mixture combination is not practical for designing petrochemical reboilers and there is no simple method for evaluating the liquid mass diffusivity D to calculate Sn in eqs. (12.37) and (12.38). A method by Thome (1981) overcomes this difficulty. He argues that in eq. (12.30) the term $\Delta\theta_E$ can be related to the elevation of the boiling point due to the depletion of the volatile component. The maximum value of this elevation, which might be approached at the critical heat flux, is the temperature difference between the dew point and the bubble point temperatures, $\Delta\theta_{bp}$, as shown in Fig. 12.12. Thus,

$$\frac{h}{h_1} = \frac{\Delta\theta_1}{\Delta\theta_1 + \Delta\theta_{bp}} \tag{12.39}$$

where $\Delta\theta_1$ is evaluated from eq. (12.31). Thome has compared eq. (12.39) with experimental data from 15 different sources. Figure 12.13 shows one such comparison for nitrogen–methane mixture data of Ackerman et al. (1975). Heat-transfer coefficients are predicted with an accuracy of about 30 per cent for these diverse mixture systems.

This method is directly applicable to multi-component mixtures with the temperature difference between the dew point and bubble point temperature evaluated at the bulk liquid composition. It is also interesting to note the similarity between this simple maximum degradation method and the earlier empirical correlation of Palen and Small (1964).

Most previous methods were compared to binary mixture boiling data, although these are the exception rather than the norm in industrial practice. Consequently, Bajorek et al. (1989) obtained boiling data for the complete composition range of numerous 2-, 3-, and 4-component mixtures for plain and low finned tubes. As an example, the triangular plots in Fig. 12.14 show measured plain tube boiling heat transfer coefficients (in $kW/m^2 \, °C$) at a heat flux of $100 \, kW/m^2$ and $\Delta\theta_{bp}$ (in $°C$) for the acetone–methanol–water system. Depending on the particular ternary mixture system, the minimum in h may either be for a ternary composition or lie along one of the binary lines. The binary and ternary minima in h correspond quite closely to those for the respective maxima in $\Delta\theta_{bp}$, substantiating the use of this parameter to correlate mixture data rather than $(\tilde{y} - \tilde{x})$.

The effect of heat flux on $\Delta\theta_E$ in eq. (12.30) is also important. At heat

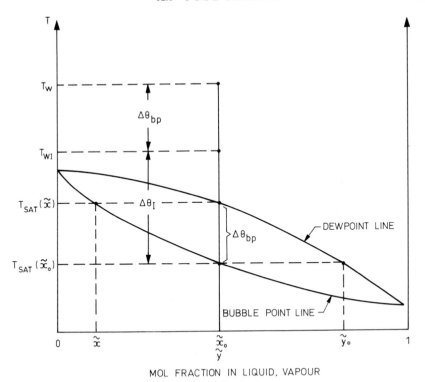

Fig. 12.12. Vapour–liquid phase equilibrium diagram showing the temperature differences used in the Thome (1983) model.

fluxes below that corresponding to the nucleation superheat, the process corresponds to the single-phase convection mode, such that $\Delta\theta_E$ is zero. At the critical heat flux ϕ_{CRIT}, $\Delta\theta_E$ approaches $\Delta\theta_{bp}$. Thus, Schlünder (1983) developed a new boiling model that includes the effect of heat flux using a film theory diffusion model on the mass diffusion 'shell' formed around a growing vapour bubble. For a mixture of n components, his equation is

$$\frac{h}{h_1} = \left\{ 1 + \frac{h_1}{\phi} \left[\sum_{j=1}^{n-1} (T_{SAT_n} - T_{SAT_j})(\tilde{y}_j - \tilde{x}_j)\left(1 - \exp\frac{-B_0\phi}{\rho_f i_{fg}\beta_f} \right) \right] \right\}^{-1} \quad (12.40)$$

where B_0 is a scaling factor relating the fraction of total heat flux converted to latent heat, which he assumed to be 1.0. T_{SAT_n} is the bubble point temperature of the highest boiling point component and temperatures T_{SAT_j} are those of the other components. Schlünder set the liquid mass transfer coefficient β_f to a fixed value of 0.0002 m/s based on a study showing typical values varying from 0.0001 to 0.0005 m/s in mass-transfer processes. He also assumed that the slope of the bubble point curve $(dT/d\tilde{x})$ is equal to the

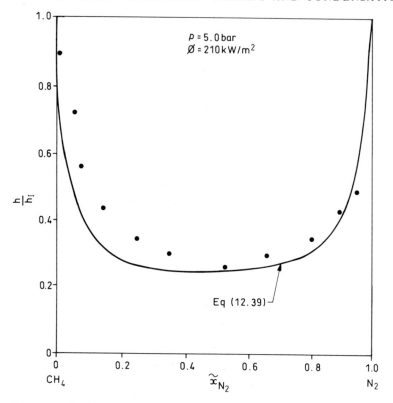

Fig. 12.13. Normalized boiling heat transfer coefficients for liquid nitrogen–methane mixtures (Ackermann *et al.*, 1975) compared to Thome (1983) model.

difference in the pure fluid saturation temperatures (i.e., the temperature difference between points A and B in Fig. 12.1 for a binary mixture) and thus is *not* a function of mixture composition. The weakness of this assumption is apparent from examination of the two-phase equilibrium diagrams shown in Fig. 12.5. Even so, the method is accurate over a wide range of reduced pressures. However, at high and low reduced pressures some of the components may be above their critical pressure or below their freezing point, respectively, which prevents their boiling point temperatures from being determined. Since many petrochemical mixtures evaporated in reboilers have 20 or more components, a method using a $(\tilde{y} - \tilde{x})$ term is cumbersome if not impossible to apply.

Because of the aforementioned problems, Thome and Shakir (1987) rederived Schlünder's model in such a way to include the parameter $\Delta\theta_{bp}$, whose value is readily available for multi-component mixtures from flash calculations and is a function of composition. The Thome and Shakir

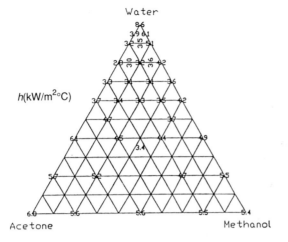

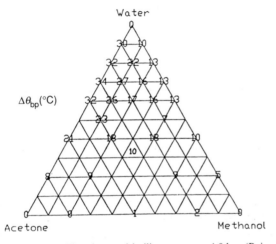

Fig. 12.14. Ternary mixture boiling data and boiling ranges at 1.0 bar (Bajorek *et al.* 1989).

correlation is

$$\frac{h}{h_{\mathrm{I}}} = \left\{1 + \frac{h_{\mathrm{I}}}{\phi}\Delta\theta_{\mathrm{bp}}\left(1 - \exp\frac{-B_0\phi}{\rho_{\mathrm{f}}i_{\mathrm{fg}}\beta_{\mathrm{f}}}\right)\right\}^{-1} \qquad (12.41)$$

B_0 remains equal to 1.0 but they recommend using $\beta_{\mathrm{f}} = 0.0003$ m/s. This method predicts plain tube (Thome and Shakir 1987) and low finned tube (Bajorek *et al.* 1991) data for 2-, 3-, and 4-component mixtures to a standard deviation of 15 per cent.

  In the foregoing methods, the value of h_{I} is for the equivalent 'ideal' pure fluid obtained by evaluating a single-component boiling correlation using the physical properties of the mixture. This is the recommended approach since it includes the effect of non-linear variations in the physical properties on h_l, which eqs. (12.31) and (12.34) do not.

12.6.2   Critical heat flux

Changes in the pool boiling critical flux occur when small amounts of a second component are added to a pure liquid. The critical heat flux may be higher or lower than the value for either component alone. As with the heat transfer coefficient in nucleate boiling, the change induced by the second component would appear to be large compared with the small changes in physical properties (surface tension; viscosity and density) attributed to the second component. It is, therefore, not possible to predict, a priori, the influence a second component might have on the critical heat flux simply by substituting the altered physical properties into an equation of the type proposed for pool boiling critical heat flux by Kutateladze (1961).

The experimental and theoretical understanding of the critical heat flux phenomenon for binary mixtures has been increased very considerably by the work of Van Stralen (1966–68). The early investigations suggested the action of several conflicting mechanisms regarding such variables as the volatility, molecular weight of the second component, and the geometry of the heating section. Some of this early work and the resulting explanations are briefly reviewed below.

Lowery and Westwater (1957) suggested that their observed increases in the critical heat flux for methanol were due to the high molecular weight additive providing particles which acted as nucleation sites within the superheated boundary layer. Van Stralen (1959) concluded that the increases he observed in the critical heat flux for water with the addition of a more volatile component were due to the formation of smaller bubbles. Dunskus and Westwater (1960) obtained data on the effects of various additives on the heat transfer characteristics of boiling isopropanol. They concluded that when the second component came from the same homologous series the magnitude of the increase was greater the higher the molecular weight. A decreased volatility of the second component also increased the critical heat flux. These authors suggested that the increases could be attributed to a depletion of the solvent with respect to the non-volatile second component at the bubble wall leaving the bubble surrounded by a fluid of high viscosity which tended to reduce coalescence between bubbles, promote foaming and also to reduce the motion of eddies.

An effect of geometry has observed by Kutateladze (1961) for boiling on a wire and a plate. In this instance, butanol was used to influence the critical heat flux for water at a pressure of 10 bar. For a wire heater the critical heat flux had a *maximum* at about 10 per cent butanol of 2.5 times the value for pure water. For a plate heater the critical heat flux has a *minimum* at about 6 per cent butanol of 0.56 times the value for pure water. A similar discrepancy was found by Owens (1964) when he compared his results for the boiling on a stainless steel tube of water containing various additives with the results obtained by Van Stralen (1959) for similar binary systems using a thin wire heater.

The more recent work of Van Stralen (1966–68) has presented an explanation for many of the phenomena reported above. Van Stralen has shown that the maximum often observed in the critical heat flux with binary mixtures corresponds to the minimum bubble growth rate and to the maximum value of $|\tilde{y} - \tilde{x}|$. The lower bubble growth rate considerably reduces the heat transfer coefficient from the heating surface to the boiling fluid with the result that the wall superheat is considerably increased. The critical heat flux can be considered as the sum of two terms; one resulting from direct vaporization at the heating surface; the second from convection of hot liquid away from the heating surface coupled with indirect vaporization into the bubble away from the heating surface. The work of Costello *et al.* (1965) and of Lienhard and Keeling (1970) suggests that even for pure liquids the second term is important.

Subcooled pool boiling of mixtures of water with 2 per cent (by weight) 1-pentanol has been studied by Scarola (1964) using a 1.7 mm diameter horizontal stainless steel tube. The addition of the second component produces a decrease in the critical heat flux except at very low or very high subcooling. If the explanation in the previous paragraph is accepted it will be seen that for this particular size of heater, the reduction in the heat flux resulting from direct vaporization outweighs any increase in the convective term. For smaller heaters the reverse might be the case. Refer to other reviews (Thome and Shock 1984, Shock 1982) for a more complete discussion of pool boiling critical heat flux for binary mixtures and the influence of heater geometry and orientation.

12.6.3 Transition boiling

Little has been published on transition boiling of binary mixtures. Happel and Stephan (1974) report that qualitatively the composition of the mixture has a similar influence as in the nucleate boiling region.

12.6.4 Minimum heat flux

The additional diffusional resistance in the liquid phase due to the mass transfer of the more volatile component into the vapour film at the heating surface results in the interface being at a higher temperature (T_i) than the rest of the liquid (T_∞). A significant proportion of the total heat flux passing through the vapour film is therefore conducted from the interface into the bulk rather than forming vapour directly. As a consequence the vapour film is thinner. Van Stralen *et al.* (1973) studying film boiling of water–2-butanone mixtures on a thin wire, observed that direct vapour production at the heating surface corresponded to only 53 per cent of the applied heat flux compared with a figure of 95 per cent for pure fluids. The remaining component of the heat flux is convected into the liquid phase and thence *indirectly* into the vapour bubbles. Thus the minimum heat flux may be expected to be greater for a binary mixture compared with an equivalent pure fluid.

Yue and Weber (1974) have examined the minimum heat flux for horizontal cylinders.

12.6.5   Film boiling

Yue and Weber (1973) have carried out a theoretical and experimental study of film boiling of binary mixtures on a vertical plate. The analysis assumes a two-phase boundary layer and involves vapour formation by evaporation at an approximately planar interface with no vapour removal by bubble formation. The heat flux, for a given wall superheat, is increased due to the heat convected from the interface into the bulk liquid. Contrast this with the situation in the nucleate boiling region where, for a given wall superheat, heat fluxes are reduced by the addition of a second component. A point is reached, however, when the flux is sufficiently high to reduce the concentration of the more volatile component at the interface to zero. The heat flux through the liquid phase is then at its maximum and as the overall heat flux is increased, this becomes a smaller fraction of the total. The effect of the second phase is predicted to disappear at values of wall superheat above that given by

$$(T_W - T_\infty) = \frac{3 i_{fg}}{c_{pg}} \cdot \mathrm{Pr}_g \qquad (12.42)$$

where T_∞ is the bulk liquid temperature remote from the vapour film. Experimental results generally confirm the analytical predictions. For acetone–cyclohexanol mixtures with a relative volatility of up to 80, the simple Bromley relationship yields heat fluxes as much as 30 per cent too low. The contribution from radiation must also be included.

12.7   Forced convective boiling

The published literature on forced convection vaporization of mixtures is still limited but growing. One of the earliest published studies is that of

McAdams *et al.* (1942) who carried out experiments using a four-pass horizontal tube evaporator heated by steam. Each pass had three separate steam jackets to allow the local heat flux to be measured. The fluid was a benzene–oil mixture. Bulk fluid temperatures were found to increase throughout the saturated boiling length as the liquid became richer in oil. Thus, some of the heat transferred to the liquid was retained in the form of sensible heat to maintain the fluid at saturation conditions and was not available for evaporation. Average boiling heat transfer coefficients were calculated for each pass where boiling occurred in all three jackets. At a given vapour mass quality the coefficient decreased as the oil content of the feed increased.

A number of workers (Bonnet and Gerster 1951, Shellene *et al.* 1968) have studied the performance of complete reboilers but such studies cannot provide information on the local conditions in the evaporating stream.

Recently, the beneficial thermodynamic and environmental aspects of replacing pure refrigerants and azeotropic mixture refrigerants (R500 and R502) with non-azeotropic (or zeotropic) binary and ternary refrigerant mixtures has spurred experimental interest in forced convective boiling of mixtures.

Singal *et al.* (1983) investigated pure R12 and R13–R12 mixtures evaporating in a horizontal, electrically heated stainless steel tube. They found that that the mean boiling coefficients for the test section were slightly higher for mixtures with 5, 10, 15, and 20 per cent R13 relative to those of pure R12. Radermacher *et al.* (1983) and Ross *et al.* (1987) ran experiments for R13B1–R152a mixtures. They observed local and mean heat transfer coefficients for the mixtures significantly lower than those of the two pure components, similar to nucleate pool boiling of mixtures on single tubes. Jung *et al.* (1989a,b) continued these tests with the R22–R114 and R12–R152a mixture systems. Figure 12.15 shows their local measurements as a function of inlet liquid composition at several vapour qualities at a mass velocity of $519 \text{ kg/m}^2 \text{ s}$, all at a reduced pressure of 0.08. Substantial heat transfer coefficient reduction for the mixtures was again observed.

Hihara *et al.* (1989) obtained local boiling coefficients for two binary systems in a 8 mm I.D. tube. Their mixture coefficients for R22–R12 were between those of the pure components, which can be explained by the maximum value of $\Delta\theta_{bp}$ of only 3°C. Instead, for R22–R114 mixtures the degradation in the mixture coefficients was as much as 55 per cent at low vapour qualities where nucleate boiling dominates. The level of degradation tended to be less with increasing vapour quality. Similar results were obtained by Murata and Hashizume (1990), who observed a nearly linear variation in the mixture heat transfer coefficients between the pure fluid values at a vapour quality of 0.8 while sizeable degradations occurred at vapour qualities of 0.2 and 0.5. Thus, the mixture mass diffusion effect should have a larger influence on h_{NcB} than h_c in convective boiling correlations.

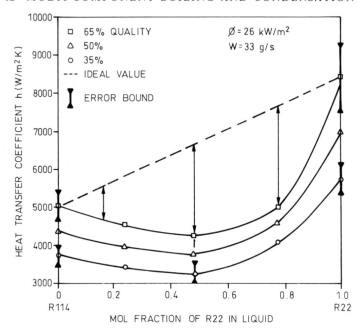

Fig. 12.15. Heat transfer coefficients of R22–R114 mixtures at $P_r = 0.08$ (Jung *et al.* 1989).

12.7.1   *Saturated nucleate boiling*

The nucleate boiling contribution to flow boiling will be influenced by the addition of one or more additional components in the same qualitative manner as nucleate pool boiling. To account for these effects Bennett and Chen (1980) have modified the original Chen correlation for single-component fluids described in Chapter 7 to include an Sn term influencing the value of h_{NcB} plus several additional terms affecting the value of h_c. However, nucleate *pool* boiling data of aqueous mixtures are not well predicted by the Sn type of correlations and this may have influenced the corrections made to h_c.

To determine if nucleate pool boiling mixture correlations are applicable to flow boiling, Thome (1989) compared a selection of methods against the Toral *et al.* (1982) data for ethanol–cyclohexane mixtures evaporating in a vertical tube. The study showed that the newer nucleate pool boiling mixture correlations, such as the Thome–Shakir (1987) correlation (eq. (12.41)), can accurately model the mixture effect on the nucleate boiling contribution to flow boiling. Similarly, Jung *et al.* (1989b) have correlated their flow boiling data for R22–R114 and R12–R152a mixtures using a binary mixture correlation of Unal (1986), Murata and Hashizume (1990) correlated their R114–R11 data using the approach of Stephan and Körner (1969), and Hihara and Saito (1990) correlated their data satisfactorily with the $\Delta\theta_{bp}$ method of Thome (1983).

So far, however, the methods available (Bennett and Chen 1980, Jung *et al.* 1989b, Mirata and Hashizume 1990, Hihara and Saito 1990) have not been compared to a comprehensive database and they also involve new single-component flow boiling correlations that may not be applicable for general use. Thus, at present it is recommended to use eq. (12.41) to modify the *nucleate boiling term* in the single-component flow boiling correlations described in Chapter 7 while using the local equivalent pure fluid properties of the mixture to evaluate the values of h_{NcB} and h_c.

In the bubble flow regime, the stripping of the more volatile component surrounding the bubble by the mass diffusion effect leaves each bubble in a shell of more viscous non-volatile liquid. This tends to inhibit bubble coalescence because of the increased time needed for drainage of the film between colliding bubbles. The consequence usually is the formation of a foam which will be stable to higher void fractions than is normal for pure liquids. In bubbly flow any increase in the convective heat transfer results primarily from the increased liquid velocity as a consequence of the increased liquid–vapour volumetric flow:

$$\frac{h_{TP}}{h_f} = \left(\frac{1}{1 - \alpha}\right)^{0.8} \tag{12.43}$$

The presence of vapour may also induce turbulence in an otherwise laminar liquid flow but, in any case, the increase over the single-phase heat transfer coefficient predicted by eq. (12.43) is relatively small (<2). In some instances, the two-phase convective heat transfer coefficient (h_{TP}) may turn out to be *lower* than the single-phase value (h_f). This is because the correct fluid properties for the calculation of h_f in eq. (12.43) are those of the more viscous non-volatile liquid surrounding the bubble cavity. The composition of this liquid may be estimated using the concept G_d, *the vaporized mass diffusion fraction* introduced by Van Stralen and discussed in Section 12.5.

12.7.2 Two-phase forced convection region

Further up the evaporator tube, annular flow is formed and this flow pattern will occupy most of the channel length. The liquid is displaced to the heating wall and it is necessary to evaluate mass transfer effects, firstly within the liquid film and, subsequently, within the vapour core. Shock (1976) has considered these effects in some detail.

Consider first the liquid film, Fig. 12.16. If the molar concentration of component A is represented by $\tilde{c}$ then the conservation equation in the liquid

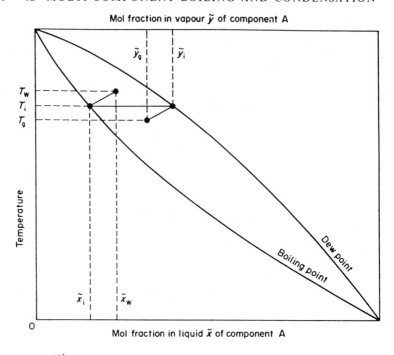

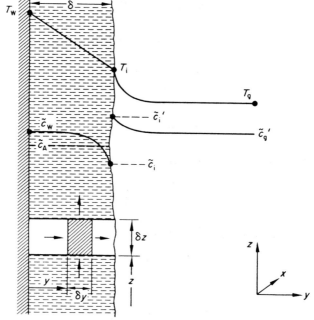

Fig. 12.16. Mass transfer from a thin liquid film (after Shock 1976).

film given by

$$\left[u_x \frac{\partial \tilde{c}}{\partial x} + u_y \frac{\partial \tilde{c}}{\partial y} + u_z \frac{\partial \tilde{c}}{\partial z}\right] = D_{EFF} \left(\frac{\partial^2 \tilde{c}}{\partial x^2} + \frac{\partial^2 \tilde{c}}{\partial y^2} + \frac{\partial^2 \tilde{c}}{\partial z^2}\right) \tag{12.44}$$

There is no net convection in the x and y direction and assuming that the concentration profile in the y direction is 'fully developed'

$$u_z \frac{\partial \tilde{c}}{\partial z} = D_{EFF} \left(\frac{\partial^2 \tilde{c}}{\partial y^2}\right) \tag{12.45}$$

This equation must be integrated using an expression for the mass diffusivity. D_{EFF}

$$J = -(D + \varepsilon_M) \frac{\partial \tilde{c}}{\partial y} = -D_{EFF} \frac{\partial \tilde{c}}{\partial y} \tag{12.46}$$

For a laminar film $D_{EFF} = D$ and

$$\int_0^\delta u_z \frac{\partial \tilde{c}}{\partial z} \, \mathrm{d}y = D \int_0^\delta \frac{\partial^2 \tilde{c}}{\partial y^2} \, \mathrm{d}y \tag{12.47}$$

Define a mean bulk molar concentration $\tilde{c}_A$ at any location z such that

$$\tilde{c}_A = \frac{1}{\Gamma} \int_0^\delta u_z \tilde{c} \, \mathrm{d}y \tag{12.48}$$

where Γ is the volumetric flow-rate per unit wetted perimeter

$$\Gamma = \int_0^\delta u_z \, \mathrm{d}y \tag{12.49}$$

also

$$\Gamma \frac{\partial \tilde{c}_A}{\partial z} = -J_i \tag{12.50}$$

where J_i is the molar flux at the interface.

$$\frac{\partial \tilde{c}}{\partial z} = \frac{\partial \tilde{c}_A}{\partial z} = -\frac{J_i}{\Gamma} \tag{12.51}$$

Substituting for $\partial \tilde{c}/\partial z$ in eq. (12.45) leads to

$$\frac{\partial^2 \tilde{c}}{\partial y^2} = -\frac{J_i u_z}{\Gamma D} \tag{12.52}$$

or

$$\frac{\partial \tilde{c}}{\partial y} = -\frac{J_i}{\Gamma D} \int_0^\delta u_z \, \mathrm{d}y \tag{12.53}$$

Define a molar mass transfer coefficient K_f:

$$K_f = \frac{J_i}{(\tilde{c}_A - \tilde{c}_i)} \tag{12.54}$$

(a) *For a laminar falling film* the Sherwood number is given by

$$Sh_f = \left[\frac{K_f \delta}{D}\right] = \frac{70}{17} \tag{12.55}$$

or, combining it with the relevant expression for film thickness to give an explicit relationship for the mass transfer coefficient (K_f),

$$\delta = \left[\frac{3\Gamma\mu_f}{(\rho_f - \rho_g)g}\right]^{1/3} \qquad K_f = \frac{70}{17}\left[\frac{(\rho_f - \rho_g)gD^3}{3\Gamma\mu_f}\right]^{1/3} \tag{12.56}$$

By analogy with the situation for heat transfer it seems probable that the actual mass transfer coefficient will be higher than that given by eq. (12.56) due to the presence of waves.

(b) *For a laminar film with constant shear stress across the film* (as in upwards annular flow at high interfacial shear) the Sherwood number is given by

$$Sh_f = \left[\frac{K_f \delta}{D}\right] = 5 \tag{12.57}$$

or, combining it with the relevant expression for film thickness to give an explicit relationship for the mass transfer coefficient (K_f),

$$\delta = \left[\frac{2\Gamma\mu_f}{\tau_i}\right]^{1/2}, \qquad K_f = 5\left[\frac{\tau_i D^2}{2\Gamma\mu_f}\right]^{1/2} \tag{12.58}$$

where τ_i is the interfacial shear stress. If the analogy with the heat transfer situation applies, mass transfer coefficients in practice may be lower than estimated from eq. (12.58) by about 30 per cent.

(c) *For a turbulent film.* Turbulence within the liquid film will enhance the mass transfer coefficient for a given film thickness. Equation (12.46) applies where ε_m is the 'eddy mass diffusivity'. Expressions for turbulent mass transfer in liquid films can be developed by an analogous procedure to that used to establish the hydrodynamic and heat transfer characteristics of the film. Shock (1976) assumed that ε_m for use in eq. (12.46) was identical to the 'eddy viscosity' (ε) and that the expression given by Deissler for ε could be used to establish the relationship between a dimensionless concentration, C^+, and a dimensionless distance, y^+. The concentration C^+ is defined as

$$C^+ = \frac{u^*}{J_i}(\tilde{c} - \tilde{c}_i) \tag{12.59}$$

where u^* is the friction velocity ($= \sqrt{(\tau/\rho_f)}$).

Mass-transfer coefficients were deduced from the dimensionless concentration profile evaluated at the appropriate value of the dimensionless film thickness (y_i^+). The molar mass transfer coefficient K_f is defined as

$$K_f = \frac{J_i}{(\tilde{c}_A - \tilde{c}_i)} = \frac{u^*}{C_b^+} \tag{12.60}$$

where C_b^+ is the bulk value of C^+ evaluated from the profiles of C^+ and y^+. A liquid film Sherwood number can also be defined as

$$Sh_f = \frac{K_f \delta}{D} = \frac{Sc\, y_i^+}{C_b^+} \tag{12.61}$$

For the case of constant shear stress across the film Shock (1976) has produced tables of Sh_f against the film Reynolds number Re_f with Schmidt number (Sc) as parameter. The following equation fits the results of Shock acceptably well:

$$Sh_f = 5 - 0.11\, Sc + 0.01\, Re_f Sc \tag{12.62}$$

12.7.3 Critical heat flux in forced convection boiling

The critical heat flux (CHF) for binary and multi-component mixtures in forced convective boiling is best considered in terms of the separate mechanisms of *departure from nucleate boiling* (DNB) which occurs under subcooled and low quality conditions, and *dryout* which occurs at higher vapour qualities.

12.7.3.1 Departure from nucleate boiling (DNB) Data for the critical heat flux of binary organic mixtures for subcooled and saturated forced convective boiling have been given by a number of workers (Table 12.3). Maxima in the CHF versus composition curves, corresponding to maximum values of $|\tilde{y} - \tilde{x}|$, are seen in these results also (Fig. 12.17) although not so marked as for the pool boiling case. The CHF for the binary mixture increases with both increased velocity and increased subcooling in a similar manner to that for pure liquids. The value of the critical heat flux for a binary mixture, $\phi_{cr,m}$ may be expressed, by analogy with the treatment by Stephan and Körner for pool boiling heat transfer, as the sum of two terms; $\phi_{cr,I}$, the 'ideal' value of the critical heat flux evaluated from the values of ϕ_{cr} for the two pure components A and B calculated at the same pressure, velocity and subcooling as the mixture, and an additional heat flux $\phi_{cr,E}$ which allows for the enhancement in CHF brought about by mass transfer effects, etc.:

$$\phi_{cr,m} = \phi_{cr,I} + \phi_{cr,E} = \phi_{cr,I}(1 + \chi) \tag{12.63}$$

where

$$\phi_{cr,I} = [\phi_{cr,A}(1 - \tilde{x}) + \phi_{cr,B}\tilde{x}] \tag{12.64}$$

Table 12.3 Experimental studies of DNB for binary mixtures

Reference	Fluids	Geometry	Experimental ranges				
			Pressure (bar)	Subcooling (°C)	Velocity (m/s)	Critical heat flux (MW/m^2)	
Tolubinsky and Matorin (1973)	ethanol–water acetone–water ethanol–benzene ethylene-glycol–water	tube 4 mm i.d. 60 mm long	3.3–13.2	10–110	2.5–10	2–13	
Andrews et al. (1968)	acetone–toluene benzene–toluene	annulus 6.35 mm i.d. 20.9 mm o.d. 76 mm long	1	20–50	1.09–4.36	1–4.5	
Sterman et al. (1968)	mono-iso-propyl di-phenyl–benzene	annulus 10 mm i.d. 16 mm o.d. 110 mm long	2	25–70	4–12	1–4.5	
Naboichenko, Kiryutin, and Gribov (1965)	mono-iso-propyl di-phenyl–benzene	annulus 6 mm i.d. 10 mm o.d. 80 mm long	2.94–16.37	25–125	4–8	1–6	
Carne (1963)	benzene–toluene acetone–toluene	internally heated annulus 6.35 mm i.d. 19.05 mm o.d. 76.2 mm long	2.05	20–80	2.18–4.36	0.5–3.0	
Bergles and Scarola (1966)	water–1-pentanol (2.2% by wt.)	tube 6.25 mm i.d. 170 mm long	2	0–50	3–6	0–10	

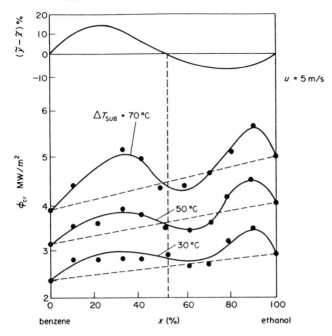

Fig. 12.17. Critical heat flux for forced convective boiling of benzene–ethanol mixtures (Tolubinsky and Matorin 1973).

Tolubinsky and Matorin (1973) and Sterman *et al.* (1968) found that χ in eq. (12.63) could be expressed as

$$\chi = \text{fn}\{|\tilde{y} - \tilde{x}|, [GD/\mu_f]_A, [T_{SAT,B}/(T_{SAT,m} - T_{SAT,B})]\} \qquad (12.65)$$

The Reynolds number $[GD/\mu_f]_A$ is based on the properties of the less volatile component since this component becomes concentrated at the heated surface.

Tolubinsky and Matorin (1973) and Sterman *et al.* (1968) found that χ in eq. (12.63) could be expressed as

$$\chi = 1.5(|\tilde{y} - \tilde{x}|)^{1.8} + 6.8(|\tilde{y} - \tilde{x}|)[(T_{SAT,m} - T_{SAT,B})/T_{SAT,B}] \qquad (12.66)$$

This expression correlated CHF data for ethanol–water, acetone–water, ethanol–benzene and water–ethylene glycol mixtures within ± 20 per cent. For mixtures forming an azeotrope, $(\tilde{y} - \tilde{x})$ and χ are both zero, so that $\phi_{cr,m} = \phi_{cr,I}$. This situation is confirmed by examination of the results for benzene–ethanol mixtures shown in Fig. 12.17.

In the experiments of Sterman *et al.* (1968) the values of χ varied from 0 to 0.8 and an alternative expression for χ was given:

$$\chi = 3.2 \times 10^5 \frac{(|\tilde{y} - \tilde{x}|)^3}{[GD/\mu_f]_A} + 6.9 \frac{(|\tilde{y} - \tilde{x}|)^{1.5}}{[GD/\mu_f]_A^{0.4}} \left[\frac{T_{SAT,B}}{T_{SAT,m} - T_{SAT,B}}\right] \qquad (12.67)$$

A somewhat different pattern of behaviour has been observed Bergles and Scarola (1966). They investigated the addition of 1-pentanol to water and its effect on forced convective subcooled boiling CHF. A distinct *reduction* in ϕ_{cr} occurred at low subcooling when 2.2 per cent by weight of 1-pentanol was added. Their explanation for the phenomenon was that the smaller bubbles formed in the mixture are a disadvantage under conditions of low pressure and subcooling. The large subcooled voids formed when pure water is boiled, tend to increase the velocity and thus prevent vapour blanketing of the surface. When the voids are reduced, this beneficial effect is less and ϕ_{cr} is lower. At higher subcooling, however, there is a consistent indication that ϕ_{cr} may be slightly improved by the addition of the volatile additive.

Celata *et al.* (1992) have recently measured CHF data for binary mixtures of R-12 and R-114. Deviations in CHF from ideal behaviour were observed but were not substantial. The Sterman *et al.* (1968) method accurately predicted their data.

12.7.3.2 *Dryout*

12.7.3.2 *Dryout*   For higher vapour qualities, associated with the annular flow pattern, dryout is likely to be determined by the point at which the flow-rate of the liquid film on the heated surface approaches zero. No experimental data or theoretical model have been published relating to dryout in binary or multi-component mixtures. However, it is possible to recommend the method proposed by Hewitt and his co-workers. On the evidence of Shock (1976) it is apparent that the flow-rates within the liquid film and of entrained liquid within the vapour core are dictated primarily by hydrodynamic effects and that mass transfer effects on the distribution of the phases will be small. In carrying out the step-by-step integration of the equations given by Hewitt it is recommended that equilibrium be assumed between the liquid and vapour at each axial location. If it is considered important the small departures from the equilibrium situation can be introduced using the equations given by Shock (1976). For positive mixtures, i.e. $(\tilde{y} - \tilde{x})_{\text{positive}}$, breakdown of the liquid film into rivulets may occur somewhat earlier than for a pure liquid due to surface tension and temperature gradient effects. Breakdown may be delayed for negative mixtures.

12.8 Simplified treatment for binary and multi-component evaporation and condensation

Consider a simple single-pass evaporator heated by a fluid stream in counter-current flow. The subscripts H and e stand for the heating and evaporating streams respectively

$$\frac{dQ}{dz} = \dot{M}_e \frac{di_e}{dz} = \dot{M}_H \frac{di_H}{dz} = UA(T_H - T_e) \qquad (12.68)$$

In eq. (12.68) A is the 'effective' heat exchange area per unit length (perimeter) allowing for any secondary surface which may be present. U is the overall heat transfer coefficient and T_H and T_e are respectively the local temperature of the heating stream and the equilibrium (*bubble point*) temperature of the evaporating stream.

If the local values of the enthalpies i_H and i_e are known, then the temperatures T_H and T_e are also known, together with the mass fraction vaporized (x_e). This, together with the flow-rates, provides sufficient information to calculate the overall coefficient U. Thus eq. (12.68) can be progressively integrated along the evaporator starting from known conditions at the evaporator inlet ($z = 0$).

Define Γ as the flow-rate per unit perimeter

$$\Gamma_e = \frac{\dot{M}_e}{A} \qquad (12.69)$$

Likewise, the product $U(T_H - T_e)$ represents the heat flux (ϕ_W) passing through the wall separating the two streams:

$$\phi_W = U(T_H - T_e) \qquad (12.70)$$

Substituting eq. (12.69) and eq. (12.70) into eq. (12.68)

$$\phi_W \, dz = \Gamma_e \, di_e \qquad (12.71)$$

Consider Fig. 12.18 which shows a section of an evaporator heated by a fluid stream in counter-current flow separated from the vaporizing stream by a wall. The heating stream is flowing downwards and the evaporating stream upwards. It is assumed that the flow pattern on the evaporating side is *annular flow* and the evaporating liquid is displaced to the wall as a thin liquid film.

Consider the situation at a position z from the evaporator inlet and over a length dz. There is a change of evaporating fluid bubble point temperature over this length so that

$$di_e = (1 - x_e) \, dT_e c_{pf} + x_e \, dT_e c_{pg} + dx_e i_{fg} \qquad (12.12)$$

Now the heat flux through the vapour–liquid interface, ϕ_i, is less than ϕ_W by the amount of the flux absorbed in heating the liquid flow $\Gamma_e(1 - x_e)$ through dT_e. Therefore

$$\phi_i = \phi_W - \Gamma_e(1 - x_e) \left(\frac{dT_e}{dz}\right) c_{pf} \qquad (12.72)$$

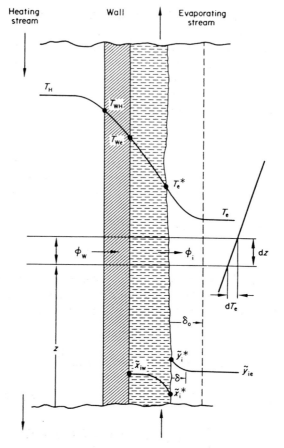

Fig. 12.18. Evaporation of a multi-component fluid in annular flow.

Of course, in pure single component vaporization (or condensation) $\phi_w \simeq \phi_i$ because there is no axial temperature gradient of any consequence.

The interfacial heat flux ϕ_i is, in turn, made up of two terms; an evaporation flux (ϕ_v) at the interface and a sensible heat flux (ϕ_g) required to heat the vapour phase through dT_e.

12.8.1   *Interfacial mass transfer during evaporation and condensation of binary and multi-component mixtures*

At the interface mass transfer and heat transfer processes occur in parallel and the resistances to mass and heat transfer must be considered. To induce these parallel mass and heat transfer processes there must be parallel concentration and temperature driving forces from the interface to the bulk vapour phase. Shock (1976) has shown that mass transfer resistances in the liquid phase are negligible by comparison and therefore it is assumed that the resistance to heat and mass transfer occurs in the vapour phase across a boundary layer of thickness δ_0.

From the definitions given earlier we see that the basic transport equation for the molar flux of component i from the interface is

$$\dot{n}_i = \dot{n}\tilde{y}_{ie} + J_i \tag{12.73}$$

where $\tilde{y}_{ie}$ is the mol fraction of component i in the bulk vapour.

These equations were first set down by Colburn and Drew (1937) for condensation in a binary system and have been reiterated by Price and Bell (1974).

The *diffusive molar flux* J_i can be evaluated from

$$J_i = K_{i,\text{eff}}(\tilde{y}_i^* - \tilde{y}_{ie}) \tag{12.74}$$

where $K_{i,\text{eff}}$ is an effective mass transfer coefficient. The sensible heat flux (ϕ_g) from the interface to the vapour core is the sum of the sensible heat flux due to the bulk flow of the vapour from the interface and the heat conducted through the boundary layer:

$$\phi_g = (T_e^* - T) \sum_{i=1}^{n} \dot{n}_i \tilde{c}_{pi} + h_g \frac{dT}{d\eta} \tag{12.75}$$

Integrating this equation from $\eta = 0$, $T = T_e^*$ to $\eta = 1$, $T = T_e$ gives

$$\phi_g = \frac{a}{1 - e^{-a}} [h_g(T_e^* - T_e)] = h_g'(T_e^* - T_e) \tag{12.76}$$

where

$$a = \frac{1}{h_g} \sum_{i=1}^{n} \dot{n}_i \tilde{c}_{pi}$$

The evaporative heat flux required to form the molar vapour flux $\dot{n}$ is

$$\phi_V = \dot{n}\tilde{\imath}_{fg} \tag{12.77}$$

where $\tilde{\imath}_{fg}$ is the molar latent heat of vaporization.

Thus, the interfacial heat flux is given by

$$\phi_i = \phi_g + \phi_V = \frac{a}{1 - e^{-a}} [h_g(T_e^* - T_e)] + \dot{n}\tilde{\imath}_{fg} \tag{12.78}$$

In general, the solution of such sets of equations is complicated by the fact that the conditions at the interface are not known and must be guessed.

An iteration is then necessary to achieve a solution. A *simplifying assumption* is to say that for situations where the thickness of the mass transfer and heat transfer boundary layers in the vapour phase are equal:

$$Le = \frac{Sc}{Pr} = \frac{\alpha}{D} \simeq 1 \tag{12.79}$$

then the resistance to heat transfer and the resistance to mass transfer will also be about equal. Thus, if we evaluate the resistance to the transfer of the sensible heat component ϕ_G in eq. (12.76) then it is argued that there will be sufficient mass transfer driving force to provide ϕ_v as well. Thus, the overall heat transfer coefficient on the evaporating side will be described by the equation

$$(T_{We} - T_e) = (T_{We} - T^*) + (T^* - T_e)$$

Now, if $\phi_i \simeq \phi_W$ (but see eq. (12.72)) then the overall local evaporating side coefficient. h_e, can be given by

$$\frac{\phi_W}{h_e} = \frac{\phi_W}{h_{fe}} + \frac{\phi_g}{h'_g} \tag{12.80}$$

where h_{fe} is the coefficient for the liquid film and h'_g is the 'modified' single-phase vapour heat transfer coefficient (defined by eq. (12.76)). Dividing both sides of eq. (12.80) by ϕ_W gives

$$\frac{1}{h_e} = \frac{1}{h_{fe}} + \frac{(\phi_g/\phi_W)}{h'_g} \tag{12.81}$$

The term (ϕ_g/ϕ_W) was denoted z by Bell and Ghaly (1972) and is given by

$$\left(\frac{\phi_g}{\phi_W}\right) = z = \left[x_e \left(\frac{dT_e}{di_e}\right) c_{pg} \right] \tag{12.82}$$

This term is easily computed from the known phase equilibrium information and the approximation can be used for both binary and multi-component systems. Although the liquid–vapour interface is rough, it is recommended that the smooth tube value of h_g be used in eq. (12.76) to ensure adequate mass transfer driving force.

The more rigorous approach given earlier, eq. (12.78), is strictly only correct for binary mixtures and can be solved using the methods described Price and Bell (1974).

The extension of these methods to multi-component systems, primarily in relation to condensation has been pursued by Standart *et al.* (1975), Krishna and Standart (1976a,b), and Krishna *et al.* (1976), although the interactions between the diffusive fluxes of the various components makes the solution complex. Krishna and Standart (1976a) have pointed out that eq. (12.74) is incorrect for multi-component evaporation and condensation. The correct relationship is

$$J_i = \sum_{j=1}^{n} K_{ij}(\tilde{y}_j^* - \tilde{y}_{je}); \qquad i = 1, \ldots, n \qquad (12.83)$$

Methods of calculating the matrix K_{ij} of mass transfer coefficients have been discussed by Krishna and Standart (1976a,b). The energy balance becomes

$$\phi_i = \frac{a}{1 - e^{-a}} [h_g(T_e^* - T_e)] + \sum_{i=1}^{n} \dot{n}_i(\tilde{l}_{fg})_i \qquad (12.84)$$

Using vapour–liquid equilibrium data eq. (12.73) and eq. (12.83) can be solved to provide T_e^* and $\dot{n}_i$. The heat balance relationships eq. (12.76) and eq. (12.84) are then used to calculate the bubble point at some point further down the evaporator. The vapour phase composition is calculated from

$$\frac{1}{A} \frac{d\dot{N}_i}{dz} = \dot{n}_i \qquad (12.85)$$

from which the vapour phase mol fractions $\tilde{y}_{ie}$ are readily calculated. No comparisons have been published between this method and the more simplified method given earlier for the case of multi-component evaporation. Some comparisons are, however, available for the analogous condensation situation and have been given by Price and Bell (1974), by Krishna and Standart (1976a), and by Webb (1982).

Generally, overall condensation rates are reasonably well predicted by the simplified methods discussed earlier. However, local condensation rates and liquid–phase compositions require the application of a more complex analytical method such as that of Krishna and Standart. Turning to boiling, Sardesai et al. (1982) have shown that the method of Silver (1947) and Bell and Ghaly (1972) for convective multi-component condensation can be adapted to apply to convective boiling, arriving at the expression

$$h_{TP} = \frac{\phi}{T_W - T_g} = \frac{1 + (h_{NcB}/h_c)}{1/h_c + (z/h_g)} \qquad (12.86)$$

where T_g is the local vapour temperature assuming equilibrium flash evaporation, i.e. $T_g = T_{SAT}$, T_W is the local wall temperature, ϕ is the total heat flux, h_g is the vapour-phase heat transfer coefficient, h_{NcB} is the nucleate boiling contribution and h_c is the two-phase convective contribution to the convective boiling process and z is given by eq. (12.82). The value of h_g is calculated for the vapour flowing alone in the tube without correcting for the presence of the liquid or waves on the liquid film and h_{NcB} includes a suitable nucleate boiling mixture correction factor, such as eq. (12.41). This appears to be the appropriate starting point for developing a new general convective mixture boiling correlation, as suggested also by Palen *et al.* (1964), who arrived at a similar expression.

References and further reading

Ackermann, H., Bewilogua, L., Knöner, R., Kretzschmar, B., Usyugui, I. P., and Vinzelberg, H. (1975) 'Heat transfer in liquid nitrogen–methane mixtures under pressure'. *Cryogenics*, **15**, 657–659.

Andrews, D. G., Hooper, F. C., and Butt, P. (1968) 'Velocity, subcooling and surface effects in the departure from nucleate boiling of organic binaries'. *Can. J. Chem. Engng.*, **46**, 194–199.

Bajorek, S. M., Lloyd, J. R., and Thome, J. R. (1989) 'Experimental study of multicomponent pool boiling on smooth and finned surfaces'. *AIChE Symp. Ser.*, **85**(269), 54–59.

Bajorek, S. M., Lloyd, J. R., and Thome, J. R. (1991) 'Prediction methods of boiling of pure and multicomponent fluids on a low finned tube'. *Fouling and Enhancement Interactions*, ASME HTD **164**, 101–108.

Bell, K. J. and Ghaly, M. A. (1972). 'An approximate generalized design method for multi-component partial condensers'. *Chem. Eng. Prog. Symp. Series*, No. 131, **69**, 72–79.

Bennett, D. L. and Chen, J. C. (1980) 'Forced convective boiling in vertical tubes for saturated pure components and binary mixtures'. *AIChE J.*, 26, 454–461.

Bergles, A. E. and Scarola, L. S. (1966) 'Effect of a volative additive on the critical heat flux for surface boiling of water in tubes'. *Chem. Engng Sci.*, **21**, 721–723.

Bonnet, W. E. and Gerster, J. A. (1951) 'Boiling coefficients of heat transfer—C_4 hydrocarbon/furfural mixtures inside vertical tubes'. *Chem. Engng. Prog.*, **77**(3), 151–158.

Calus, W. F. and Leonidopolous, D. J. (1974) 'Pool boiling—binary liquid mixtures'. *Int. J. Heat Mass Transfer*, **17**, 249–256.

Carne, M. (1963) 'Studies of the critical heat flux from some binary mixtures and their components. *Can. J. Chem. Enging.*, December, 235–240.

Celata, G. P., Cumo, M., and Setaro, T. (1992) 'Forced convective boiling in binary mixtures'. Paper E4 presented at the European Two-Phase Flow Group Meeting, Stockholm, 1–3 June.

Colburn, A. P. and Drew, T. B. (1937) 'The condensation of mixed vapours'. *Trans. Am. Inst. Chem. Engrs.*, **33**, 197–215.

Cooper, M. G. and Stone, C. R. (1981) 'Boiling of binary mixtures—study of individual bubbles'. *Int. J. Heat Mass Transfer*, **24**, 1937–1950.

Costello, C. P., Bock, C. O., and Nichols, C. C. (1965) 'A study of induced convective effects on pool boiling burnout'. *Chem. Engng. Prog. Symp. Series*, **61**, 271–280.

Dunskus, T. and Westwater, J. W. (1960) 'The effect of trace additions on the heat transfer to boiling isoproponal'. AIChE preprint 30, presented at AIChE–ASME Heat Transfer Conference, Buffalo, N.Y., August.

Happel, O. and Stephan, K. (1974) 'Heat transfer from nucleate boiling to the beginning of film boiling in binary mixtures'. Paper B7.8 presented at 5th Internatonal Heat Transfer Conference, Tokyo, September.

Hihara, E. and Saito, T. (1990) 'Forced convective boiling heat transfer of binary mixtures in a horizontal tube'. Paper Presented at 9th Int. Heat Transfer Conf., Jerusalem, **2**, 123–128.

Hihara, E., Tanida, K., and Saito, T. (1989) 'Forced convective boiling experiments of binary mixture'. *JSME Int. J.*, Series II, 32(1), 98–106.

Holden, B. S. and Katz, J. L. The homogeneous nucleation of bubbles in superheated binary mixtures'. *AIChE J.*, **24**, 260–267.

Jung, D. S., McLinden, M., Radermacher, R., and Didion, D. (1989) 'Horizontal flow boiling heat transfer experiments with a mixture of R22/R114'. *Int. J. Heat Mass. Transfer*, **32**, 131–145.

Jung, D. S., McLinden, M., Radermacher, R., and Didion, D. (1989) 'A study of flow boiling heat transfer with refrigerant mixtures'. *Int. J. Heat Mass. Transfer*, **32**, 1751–1764.

Krishna, R. and Standart, G. L. (1976) 'Determination of interfacial mass and energy transfer rates for multi-component vapour–liquid systems'. *Lett. Heat. Mass Transfer*, **3**(2), 173–182.

Krishna, R. and Standart, G. L. (1976) 'A multi-component film model incorporating a general matrix method of solution to the Maxwell–Stefan equation'. *AIChE J.*, **22**, 383.

Krishna, R., Panchal, C. B., Webb, D. R., and Coward, I. (1976) 'An Ackerman–Colburn and Drew type analysis for condensation of multi-component mixtures'. *Lett. Heat. Mass Transfer*, **3**, 163–172.

Kutateladze, S. S. (1961) 'Boiling heat transfer'. *Int. J. Heat. Mass Transfer*, **1**, 31–45.

Lienhard, J. H. and Keeling, K. B. (1970) 'An induced-convection effect upon the peak boiling heat flux'. *J. Heat Transfer*, **92**, 1–5.

Lowery, A. J., Jr. and Westwater, J. W. (1957) 'Heat transfer to boiling methanol—effect of added agents'. *Int. Engng. Chem*, **49**, 1445–1448.

McAdams, W. H., Woods, W. K., and Bryon, R. L. (1942) 'Vaporization inside horizontal tubes—II. Benzene–oil mixtures'. *Trans. ASME*, 193–200.

Malesinsky, W. (1965) *Azeotrophy and other theoretical problems of vapour-liquid equilibrium.* Interscience, Wiley, London.

Mikic, B. B. and Shock, R. A. W. (1983) 'Phenomena in pool boiling of binary mixtures'. AERE-R 10856, April.

Murata, K. and Hashizume, K. (1990) 'An investigation on forced convection boiling of nonazeotropic refrigerant mixtures'. *Heat Transfer-Jap. Res.*, **19**(2), 95–109.

Naboichenko, K. V., Kiryutin, A. A., and Gribov, B. S. (1965) 'A study of critical heat flux with forced flow of monoisopropyldiphenyl–benzene mixture'. *Teploenergetica*, **12**(11), 81–86. See also *Thermal Engng*, **12**(11), 107–114.

Owens, W. L. Jr. (1964) 'An analytical and experimental study of pool boiling with particular reference to additions'. UKAEA Report, Winfrith, AEEW–180.

Palen, J. W. and Small, W. M. (1964) 'A new way to design kettle and internal reboilers'. *Hydrocarbon Processing*, **43**(11), 199–208.

Palen, J. W., Yang, C. C., and Taborek, J. (1980) 'Application of the resistance proration method to boiling in tubes in the presence of inert gas'. *AIChE Symp. Ser.*, **76**(199), 282–288.

Porteous, W. and Blander, M. (1975) 'Limits of superheat and explosive boiling of light hydrocarbons, halocarbons and hydrocarbon mixtures'. *AIChE J.*, **21**, 560–566.

Prausnitz, J. M. (1969) *Molecular thermodynamics of fluid-phase equilibria*. Prentice-Hall, Englewood Cliffs, NJ.

Price, B. C. and Bell, K. J. (1974) 'Design of binary vapour condensers using the Colburn–Drew equations'. *AIChE Symp. Series*, **70**(138), 163–171.

Radermacher, R., Ross, H., and Didion, D. (1983) 'Experimental determination of forced convective evaporative heat transfer coefficients for nonazeotropic refrigerant mixtures'. ASME Paer 83-WA/HT54.

Reid, R. C., Prausnitz, J. M., and Poling, B. E. (1987) *The properties of gases and liquids* (4th edn), McGraw-Hill, New York.

Ross, H., Radermacher, R., di Marzo, M., and Didion, D. (1987) 'Horizontal flow boiling of pure and mixed refrigerants'. *Int. J. Heat Mass Transfer*, **30**, 979–992.

Sardesai, R. G., Shock, R. A. W., and Butterworth, D. (1982) 'Heat and mass transfer in multicomponent condensation and boiling'. *Heat Transfer Engng.*, **3**(3–4), 104–114.

Scarola, L. S. (1964) 'Effect of additive on the critical heat flux in nucleate boiling'. S.M. Thesis in Mech. Engng., MIT.

Schlünder, E. U. (1983) 'Heat transfer in nucleate boiling of mixtures'. *Int. Chem. Eng.*, **23**(4), 589–599.

Scriven, L. E. (1959) 'On dynamics of phase growth'. *Chem. Engng. Science*, **10**(1/2), 1–13.

Shakir, S. and Thome, J. R. (1986) 'Boiling nucleation of mixtures on smooth and enhanced boiling surfaces'. Paper presented at 8th Int. Heat. Transfer Conf., San Francisco, **4**, 2081–2086.

Shellene, K. R., Sternling, C. V., Church, D. M., and Snyder, N. H. (1968) 'Experimental study of a vertical thermosyphon reboiler'. *Chem. Eng. Prog. Symp. Series*, No. 82, **64**, 102–113.

Shock, R. A. W. (1976) 'Evaporation of binary mixtures in upward annular flow'. *Int. J. Multiphase Flow*, **2**, 411–433.

Shock, R. A. W. (1977) 'Nucleate boiling in binary mixtures'. *Int. J. Heat. Mass Transfer*, **20**, 701–709.

Shock, R. A. W. (1982) 'Boiling in multicomponent fluids'. *Multiphase Science and Technology*, Vol. 1, Chapter 3, Hemisphere, New York.

Silver, L. (1947) 'Gas cooling with aqueous condensation'. *Trans. Inst. Chem. Eng.*, **25**, 30–42.

Singal, L. C., Sharma, C. P., and Varma, H. K. (1983) 'Experimental heat transfer coefficient for binary refrigerant mixtures of R13 and R12'. *ASHRAE Trans.*, No. 2747, 165–188.

Standart, G. L., Krishna, R., and Cullinan, H. T. (1975) 'Multi-component isothermal mass transfer'. Paper presented at AIChE Conference, Tulsa, Oklahoma, March (1974). See also Krishna, R., 'Interphase transport of mass and energy in multi-component system'. PhD Thesis, UMIST.

Stein, H. N. (1978) Chapter 17 in Van Stralen, S. J. D. and Cole, R. (eds.), *Boiling phenomena*. Hemisphere Publishing Corp., Washington.

Stephan, K. (1982) 'Heat transfer in boiling mixtures'. Paper presented at 7th Int. Heat Transfer Conf., **1**, 59–81.

Stephen, K. and Körner, M. (1969) 'Calculation of heat transfer in evaporating binary liquid mixtures'. *Chemie-Ingenieur Technik*, **41**(7), 409–417.

Stephan, K. and Preusser, P. (1978) 'Heat transfer in natural convection boiling of polynary mixtures'. *Proc. 6th Int. Heat Transfer Conference*, **I**, 187–192. Toronto, 7–11 August.

Sterman, L., Abramov, A., and Cheketa, G. (1968) 'Investigation of boiling crisis at forced motion of high temperature organic heat carriers and mixtures'. Paper E2 presented at International Symposium on Research into Co-Current Gas–Liquid Flow, University of Waterloo, Waterloo, Ontario, Canada, September.

Sternling, C. V. and Tickacek, L. J. (1961) 'Heat transfer coefficients for boiling mixtures'. *Chem. Engng. Science*, **16**(4), 297–337.

Thome, J. R. (1978) 'Bubble growth and nucleate pool boiling in liquid nitrogen, argon and their mixtures'. Doctoral Thesis, Dept. of Engng. Sci., Oxford University.

Thome, J. R. (1981) 'Nucleate pool boiling in binary liquids—an analytical equation'. *AIChE Symp. Ser.*, **77**(208), 238–250.

Thome, J. R. (1982) 'Boiling heat transfer in binary liquid mixtures'. *NATO Advanced Research Workshop on Advances in Two-Phase Flow and Heat Transfer*, Spitzingsee, FRG, August 31–September 3.

Thome, J. R. (1983) 'Prediction of binary mixture boiling heat transfer coefficients using only phase equilibrium data'. *Int. J. Heat Mass Transfer*, **26**, 965–974.

Thome, J. R. (1989) 'Prediction of the mixture effect on boiling in vertical thermosyphon reboilers'. *Heat Transfer Eng.*, **10**(2), 29–38.

Thome, J. R. and Davey, G. (1981) 'Bubble growth rates in liquid nitrogen, argon and their mixtures'. *Int. J. Heat Mass Transfer*, **24**, 89–97.

Thome, J. R. and Shakir, S. (1987) 'A new correlation for nucleate pool boiling of aqueous mixtures'. *AIChe Symp. Ser.*, **83**(257), 46–51.

Thome, J. R. and Shock, R. A. W. (1984) 'Boiling of multicomponent liquid mixtures'. *Advances in Heat Transfer*, Academic Press, Princeton, **16**, 59–156.

Tolubinsky, V. I. and Matorin, P. S. (1973) 'Forced convection boiling heat transfer crisis with binary mixtures'. *Heat Transfer—Soviet Research*, **5**(2), 98–101.

Toral, H., Kenning, D. B. R., and Shock, R. A. W. (1982) 'Flow boiling of ethanol/cyclohexane mixtures'. Paper presented at 7th Int. Heat Transfer Conf., Munich, **4**, 255–260.

Unal, H. C. (1986) 'Prediction of nucleate boiling heat transfer coefficients for binary mixtures'. *Int. J. Heat Mass Transfer*, **29**, 637–640.

Van Stralen, S. J. D. (1959) 'Heat transfer to boiling binary liquid mixtures Pt. I and Pt. II'. *Brit. Chem. Engng.*, **4**(1), 8–17; **4**(2), 78–82.

Van Stralen, S. J. D. (1966–68) 'The mechanism of nucleate boiling in pure liquids and in binary mixtures Pt. I and Pt. II'. *Int. J. Heat. Mass Transfer*, **9**, 995–1046 (1966); **10**, 1469 (1967); **11**, 1467 (1968).

Van Stralen, S. T. D. (1967) 'Bubble growth rates in boiling binary mixtures'. *Brit. Chem. Engng.*, **12**(3), 390–395.

Van Stralen, S. J. D. and Zijl, W. (1978) 'Fundamental developments in bubble dynamics'. Paper presented at 6th International Heat Transfer Conference, Toronto, 7–11 August.

Van Stralen, S. J. D., Joosen, C. J. J., and Sluyter, W. M. (1972) 'Film boiling of water and an aqueous binary mixture'. *Int. J. Heat Mass Transfer*, **15**, 2427–2446.

Webb, D. R. (1982) 'Heat and mass transfer in condensation of multicomponent vapors'. Paper presented at 7th Int. Heat Transfer Conf., Munich, **5**, 167–174.

Yue, P. L. and Weber, M. E. (1973) 'Film boiling of saturated binary mixtures'. *Int. J. Heat Mass Transfer*, **16**, 1877–1887.

Yue, P. L. and Weber, M. E. (1974) 'Minimum film boiling flux of binary mixtures'. *Trans. Inst. Chem. Engng.*, **52**, 217–221.

Zeugin, L., Donovan, J., and Mesler, R. (1975) 'A study of microlayer evaporation for three binary mixtures during nucleate boiling'. *Chem. Engng. Science*, **30**, 679–683.

Problems

1. Examine the basis of the Chen correlation in relation to the prediction of heat transfer coefficients for vaporizing binary mixtures. The Forster–Zuber correlation for the nucleate boiling component uses a bubble radius and bubble growth velocity in its formulation. How is this to be modified for a binary system?

2. Calculate the mixture nucleate pool boiling heat transfer coefficient for the following conditions:

   heat flux $= 20 \, kW/m^2$
   pressure $= 5$ bar
   critical pressure $= 49$ bar
   surface roughness $= 1 \, \mu m$
   molecular weight $= 76$
   latent heat $= 350 \, kJ/kg$
   liquid density $= 1250 \, kg/m^3$
   dew point/bubble point temperature difference $= 30°C$.

3. Using Raoult's law, prepare a pressure–composition phase equilibrium diagram for the binary mixture of R22 and R114 at a temperature of 10°C. For these fluids, the vapour pressure curve can be represented by the simplified equation of state

$$\ln P = \frac{A}{T_{SAT}} + B$$

   where P is in bar and T_{SAT} is in K. For R22, $A = -2421.41$ and $B = 10.4507$; for R114, $A = -2845.26$ and $B = 10.2868$.

APPENDIX

Table A.1 Conversion factors

(a) Thermal conductivity

$\dfrac{W}{m^2\,°C/m}$	$\dfrac{Btu}{h\,ft^2\,°F/ft}$	$\dfrac{g\,cal}{s\,cm^2\,°C/cm}$	$\dfrac{W}{cm^2\,°C/cm}$	$\dfrac{kg\,cal}{h\,m^2\,°C/m}$
1	0.5779	0.002388	0.01	0.86
1.731	1	0.004132	0.01731	1.488
418.7	241.9	1	4.187	360
100	57.79	0.2388	1	86
1.163	0.672	0.002778	0.01163	1

(b) Coefficient of heat transfer

$\dfrac{W}{m^2\,°C}$	$\dfrac{Btu}{h\,ft^2\,°F}$	$\dfrac{g\,cal}{s\,cm^2\,°C}$	$\dfrac{W}{cm^2\,°C}$	$\dfrac{kg\,cal}{h\,m^2\,°C}$
1	0.1761	0.00002388	0.0001	0.86
5.678	1	0.0001355	0.0005678	4.882
41,870	7,373	1	4.187	36,000
10,000	1,761	0.2388	1	8,600
1.163	0.2048	0.00002778	0.0001163	1

(c) Heat flux

$\dfrac{W}{m^2}$	$\dfrac{Btu}{h\,ft^2}$	$\dfrac{g\,cal}{s\,cm^2}$	$\dfrac{W}{cm^2}$	$\dfrac{kg\,cal}{h\,m^2}$
1	0.3170	0.00002388	0.0001	0.86
3.154	1	0.00007535	0.0003154	2.712
41,870	13,272	1	4.187	36,000
10,000	3,170	0.2388	1	8,600
1.163	0.3687	0.00002778	0.0001163	1

(d) Viscosity

$\dfrac{\text{N s}}{\text{m}^2}$ (Poiseuille)*	Centipoises	$\dfrac{\text{lb}}{\text{s ft}}$	$\dfrac{\text{lbf s}}{\text{ft}^2}$	$\dfrac{\text{lb}}{\text{h ft}}$	$\dfrac{\text{kg}}{\text{h m}}$
1	1,000	0.672	0.0209	2,420	3,600
0.001	1	0.000672	0.0000209	2.42	3.60
1.49	1,490	1	0.0311	3,600	5,350
47.88	47,880	32.2	1	116,000	172,000
0.0004134	0.4134	0.000278	0.00000864	1	1.49
0.000278	0.278	0.000187	0.00000581	0.672	1

* 1 Ns/m² (poiseuille) = 10 poise.

(e) Physical quantities

Physical quantity	British units	SI	Conversion factor*	Reciprocal conversion factor*
Basic engineering units:				
Mass	lb	kg	0.4536	2.2045
Length	ft	m	0.3048	3.2808
Force	lbf	N	4.4482	0.2248
Energy	Btu	J	1055.06	9.4781×10^{-4}
	ft lbf	J	1.3558	0.7375
Power	550 ft lbf/s = 1 h.p.	W	745.69	1.3410×10^{-3}
Heat transfer units:				
ϕ	Btu/ft²h	W/m²	3.155	0.3169
h	Btu/ft²h °F	W/m² °C	5.678	0.1761
k	Btu/ft h °F	W/m °C	1.731	0.5777
c_p	Btu/lb °F	J/kg °C	4186.8	2.388×10^{-4}
ρ	lb/ft³	kg/m³	16.0185	0.06243
μ	lb/ft h	kg/m s, or N s/m²	4.134×10^{-4}	2.4189×10^3
α, ε, D	ft²/h	m²/s	2.581×10^{-5}	3.8744×10^4
τ, p	lbf/ft²	N/m²	47.880	0.02089
τ, p	lbf/in²	N/m²	6.8948×10^3	1.4053×10^{-4}

* Multiply the numerical value in British units by the conversion factor to obtain the equivalent in SI; multiply the numerical value in SI by the reciprocal conversion factor to obtain the equivalent in British units.

Table A.2 Viscosity of steam* Micropoiseuille (N s/m^2 × 10^{-6})

Absolute pressure bar	Saturated vapour	Temperature °C							
		0	100	200	300	400	500	800	1000
1	12.06	8.10	12.1	16.2	20.3	24.5	28.7	40.5	47.5
20	16.2	—	—	—	20.2	24.7	28.9	40.6	47.6
40	17.5	—	—	—	20.1	25.0	29.1	40.7	47.7
60	18.5	—	—	—	20.0	25.2	29.3	40.9	47.8
80	19.4	—	—	—	19.8	25.5	29.5	41.0	47.9
100	20.3	—	—	—	—	25.8	29.7	41.1	48.0
120	21.3	—	—	—	—	26.2	30.0	41.3	48.2
140	22.4	—	—	—	—	26.6	30.2	41.5	48.3
160	23.7	—	—	—	—	27.1	30.5	41.6	48.4
180	25.6	—	—	—	—	27.7	30.8	41.8	48.5
200	28.5	—	—	—	—	28.5	31.2	41.9	48.6
220	35.5	—	—	—	—	29.5	31.5	42.1	48.8
240	—	—	—	—	—	30.9	31.9	42.3	49.0

* From *U.K. Steam Tables in SI Units 1970*, Edward Arnold (Publishers) Ltd., London.

Table A.3 Steam tables

International Skeleton Table, 1963. Saturation line; specific heat capacity, surface tension and transport properties

Temp. °C	Pressure bar	Spec. vol. Water m³/kg	Spec. vol. Steam m³/kg	Spec. enthalpy Water kJ/kg	Spec. enthalpy Steam kJ/kg	c_{pf} kJ/kg K	$\sigma \times 10^3$ N/m	p_s bar	$\mu_f \times 10^6$ N s/m²	$\nu_f \times 10^6$ m²/s	k_f W/mK	$(Pr)_f$	c_{pg} kJ/kg K	$\mu_g \times 10^6$ N s/m²	$\nu_g \times 10^6$ m²/s	$k_g \times 10^3$ W/mK	$(Pr)_g$	Temp. °C
0.01	0.006112	1.0002×10^{-3}	206.146	0.000611	2501	4.218	75.60	0.00611	1786	1.786	0.569	13.2	1.863	8.105	1672	17.6	0.858	0.01
10	0.012271	1.0004×10^{-3}	106.422	41.99	2519	4.194	74.24	0.01227	1304	1.305	0.587	9.32	1.870	8.504	905	18.2	0.873	10
20	0.023368	1.0018×10^{-3}	57.836	83.86	2538	4.182	72.78	0.02337	1002	1.004	0.603	6.95	1.880	8.903	515	18.8	0.888	20
30	0.042418	1.0044×10^{-3}	32.929	125.66	2556	4.179	71.23	0.04241	798.3	0.802	0.618	5.40	1.890	9.305	306	19.5	0.901	30
40	0.073750	1.0079×10^{-3}	19.546	167.57	2574	4.179	69.61	0.07375	653.9	0.659	0.631	4.33	1.900	9.701	190	20.2	0.912	40
50	0.12335	1.0121×10^{-3}	12.045	209.3	2592	4.181	67.93	0.12335	547.8	0.554	0.643	3.56	1.912	10.10	121	20.9	0.924	50
60	0.19919	1.0171×10^{-3}	7.6776	251.1	2609	4.185	66.19	0.19920	467.3	0.473	0.653	2.99	1.924	10.50	80.6	21.6	0.934	60
70	0.31161	1.0228×10^{-3}	5.0453	293.0	2626	4.191	64.40	0.31162	404.8	0.414	0.662	2.56	1.946	10.89	54.9	22.4	0.946	70
80	0.47358	1.0290×10^{-3}	3.4083	334.9	2643	4.198	62.57	0.47360	355.4	0.366	0.670	2.23	1.970	11.29	38.5	23.2	0.959	80
90	0.70109	1.0359×10^{-3}	2.3609	376.9	2660	4.207	60.69	0.70109	315.6	0.327	0.676	1.96	1.999	11.67	27.6	24.0	0.973	90
100	1.01325	1.0435×10^{-3}	1.6730	419.1	2676	4.218	58.78	1.01330	283.1	0.295	0.681	1.75	2.034	12.06	20.2	24.9	0.987	100
110	1.4327	1.0515×10^{-3}	1.2101	461.3	2691	4.230	56.83	1.4327	254.8	0.268	0.684	1.58	2.076	12.45	15.1	25.8	1.00	110
120	1.9854	1.0603×10^{-3}	0.89171	503.7	2706	4.244	54.85	1.9854	231.0	0.245	0.687	1.43	2.125	12.83	11.4	26.7	1.02	120
130	2.7011	1.0697×10^{-3}	0.68832	546.3	2720	4.262	52.83	2.7011	210.9	0.226	0.688	1.31	2.180	13.20	8.82	27.8	1.03	130
140	3.6136	1.0798×10^{-3}	0.50866	589.1	2734	4.282	50.79	3.6138	194.1	0.210	0.688	1.21	2.245	13.57	6.90	28.9	1.05	140
150	4.7597	1.0906×10^{-3}	0.39257	632.2	2747	4.306	48.70	4.7600	179.8	0.196	0.687	1.13	2.320	13.94	5.47	30.0	1.08	150
160	6.1804	1.1021×10^{-3}	0.30685	675.5	2758	4.334	46.59	6.1806	167.7	0.185	0.684	1.06	2.406	14.30	4.39	31.3	1.10	160
170	7.9202	1.1144×10^{-3}	0.24262	719.1	2769	4.366	44.44	7.9202	157.4	0.175	0.681	1.01	2.504	14.66	3.55	32.6	1.13	170
180	10.027	1.1275×10^{-3}	0.19385	763.1	2778	4.403	42.26	10.027	148.5	0.167	0.677	0.967	2.615	15.02	2.91	34.1	1.15	180
190	12.533	1.1415×10^{-3}	0.15635	807.5	2786	4.446	40.05	12.551	140.7	0.161	0.671	0.932	2.741	15.37	2.40	35.7	1.18	190
200	15.550	1.1565×10^{-3}	0.12719	852.4	2793	4.494	37.81	15.549	133.9	0.155	0.664	0.906	2.883	15.72	2.00	37.4	1.21	200
210	19.080	1.1726×10^{-3}	0.104265	897.7	2798	4.550	35.53	19.077	127.9	0.150	0.657	0.886	3.043	16.07	1.68	39.4	1.24	210
220	23.202	1.1900×10^{-3}	0.086062	943.7	2802	4.613	33.23	23.198	122.4	0.146	0.648	0.871	3.223	16.42	1.41	41.5	1.28	220
230	27.979	1.2087×10^{-3}	0.071472	990.3	2803	4.685	30.90	27.976	117.5	0.142	0.639	0.861	3.426	16.78	1.20	43.9	1.31	230
240	33.480	1.2291×10^{-3}	0.059674	1037.6	2803	4.769	28.56	33.478	112.9	0.139	0.628	0.850	3.656	17.14	1.02	46.5	1.35	240
250	39.776	1.2512×10^{-3}	0.050056	1085.8	2801	4.866	26.19	39.736	108.7	0.136	0.616	0.859	3.918	17.51	0.876	49.5	1.39	250
260	46.941	1.2755×10^{-3}	0.042149	1135.0	2796	4.985	23.82	46.943	104.8	0.134	0.603	0.866	4.221	17.90	0.755	52.8	1.43	260
270	55.052	1.3023×10^{-3}	0.035599	1185.2	2790	5.134	21.44	55.058	101.1	0.132	0.589	0.882	4.575	18.31	0.652	56.6	1.48	270
280	64.191	1.3321×10^{-3}	0.030133	1236.8	2780	5.307	19.07	64.202	97.5	0.130	0.574	0.902	4.996	18.74	0.565	60.9	1.54	280
290	74.449	1.3655×10^{-3}	0.025537	1290	2766	5.520	16.71	74.461	94.1	0.128	0.558	0.932	5.509	19.21	0.491	66.0	1.61	290
300	85.917	1.4036×10^{-3}	0.021643	1345	2749	5.794	14.39	85.927	90.7	0.127	0.541	0.970	6.148	19.73	0.427	71.9	1.69	300
310	98.694	1.4475×10^{-3}	0.018316	1402	2727	6.143	12.11	98.700	87.2	0.126	0.523	1.024	6.969	20.30	0.372	79.1	1.79	310
320	112.89	1.4992×10^{-3}	0.015451	1462	2700	6.604	9.89	112.89	83.5	0.125	0.503	1.11	8.060	20.95	0.324	87.8	1.92	320
330	128.64	1.562×10^{-3}	0.012967	1526	2666	7.241	7.75	128.63	79.5	0.124	0.482	1.20	9.580	21.70	0.281	99.0	2.10	330
340	146.08	1.639×10^{-3}	0.010779	1596	2623	8.225	5.71	146.05	75.4	0.123	0.460	1.35	11.87	22.70	0.245	114	2.36	340
350	165.37	1.741×10^{-3}	0.008805	1672	2565	10.07	3.79	165.35	69.4	0.121	0.434	1.61	15.8	24.15	0.213	134	2.84	350
360	186.74	1.894×10^{-3}	0.006943	1762	2481	15.0	2.03	186.75	62.1	0.118	0.397	2.34	27.0	26.45	0.184	162	4.40	360
370	210.53	2.22×10^{-3}	0.00493	1892	2331	55	0.47	210.41	51.8	0.116	0.340	8.37	107	30.6	0.150	199	16.4	370
374.15	221.2	3.17×10^{-3}	0.00317	2095	2095	∞	0	221.2	41.4	0.131	0.240	∞	∞	41.4	0.131	240	∞	374.15

From U.K. Steam Tables in S.I. Units 1970, Edward Arnold (Publishers) Ltd, London

Table A.4 Physical properties of refrigerants

(a) Refrigerant R-134a (CF_3–CH_2F; 1,1,1,2-Tetrafluoroethane)

Temperature °C	Pressure kPa	ρ_f kg/m³	ρ_g kg/m³	i_{fg} kJ/kg	c_{pf} kJ/kg K	c_{pg} kJ/kg K	μ_f mNs/m²	μ_g mNs/m²	k_f W/m K	k_g W/m K	σ N/m
−10	200.71	1324.1	10.06	207.7	1.316	0.854	0.312	0.0103	0.0981	0.0111	0.0131
−5	243.45	1308.3	12.10	203.5	1.328	0.870	0.294	0.0106	0.0961	0.0115	0.0124
0	292.93	1292.2	14.46	199.2	1.341	0.889	0.277	0.0109	0.0940	0.0120	0.0116
5	349.83	1275.7	17.16	194.9	1.355	0.909	0.262	0.0111	0.0919	0.0125	0.0109
10	414.85	1258.9	20.26	190.6	1.371	0.931	0.248	0.0114	0.0899	0.0130	0.0102
15	488.69	1241.6	23.79	186.3	1.389	0.956	0.235	0.0117	0.0878	0.0135	0.0095
20	572.11	1223.8	27.81	181.8	1.408	0.984	0.223	0.0120	0.0858	0.0141	0.0088
25	665.85	1205.6	32.37	177.3	1.430	1.014	0.212	0.0123	0.0837	0.0146	0.0081
30	770.70	1186.7	37.53	172.7	1.453	1.047	0.201	0.0126	0.0817	0.0152	0.0075
35	887.47	1167.2	43.39	167.9	1.479	1.084	0.192	0.0129	0.0796	0.0159	0.0068
40	1016.99	1146.9	50.03	162.9	1.508	1.124	0.183	0.0132	0.0776	0.0165	0.0061
45	1060.14	1125.7	57.57	157.7	1.540	1.169	0.175	0.0135	0.0755	0.0172	0.0055
50	1317.83	1103.4	66.16	152.2	1.575	1.219	0.167	0.0138	0.0735	0.0180	0.0049
55	1491.04	1080.0	75.98	146.2	1.615	1.275	0.160	0.0140	0.0714	0.0188	0.0043
60	1680.81	1055.1	87.25	139.7	1.659	1.338	0.153	0.0143	0.0693	0.0197	0.0037

(T_{CRIT} = 374.205 K, P_{CRIT} = 40.46 bar, M = 102.03)

(b) Refrigerant R-123 ($CHCl_2$–CF_3; 1,1-Dichloro-2,2,2-Trifluoroethane)

(b) *Refrigerant R-123 (CHCl₂–CF₃; 1,1-Dichloro-2,2,2-Trifluoroethane)*

Temperature °C	Pressure kPa	ρ_f kg/m³	ρ_g kg/m³	i_{fg} kJ/kg	c_{pf} kJ/kg K	c_{pg} kJ/kg K	μ_f mNs/m²	μ_g mNs/m²	k_f W/m K	k_g W/m K	σ N/m
−10	20.67	1552.0	1.468	181.7	0.963	0.633	0.653	0.0092	0.0837	0.0071	0.0195
−5	26.30	1540.1	1.839	179.9	0.967	0.642	0.611	0.0094	0.0825	0.0073	0.0189
0	33.12	1529.0	2.280	178.1	0.972	0.651	0.574	0.0096	0.0813	0.0075	0.0183
5	41.30	1517.0	2.803	176.3	0.977	0.660	0.540	0.0098	0.0800	0.0078	0.0176
10	51.03	1505.0	3.417	174.4	0.983	0.669	0.508	0.0100	0.0787	0.0080	0.0170
15	62.53	1493.0	4.134	172.6	0.989	0.678	0.478	0.0102	0.0774	0.0083	0.0164
20	76.01	1481.0	4.964	170.7	0.995	0.688	0.451	0.0104	0.0760	0.0085	0.0158
25	91.69	1468.0	5.921	168.7	1.002	0.698	0.424	0.0106	0.0746	0.0087	0.0152
30	109.8	1456.0	7.018	166.7	1.009	0.708	0.399	0.0107	0.0733	0.0090	0.0146
35	130.7	1443.0	8.269	164.6	1.016	0.719	0.375	0.0109	0.0719	0.0092	0.0140
40	154.5	1430.0	9.690	162.6	1.024	0.729	0.352	0.0111	0.0705	0.0095	0.0134
45	181.6	1417.0	11.30	160.4	1.032	0.741	0.329	0.0113	0.0691	0.0097	0.0128
50	212.2	1404.0	13.11	158.2	1.040	0.752	0.318	0.0115	0.0677	0.0100	0.0122
55	246.7	1390.0	15.14	155.9	1.049	0.764	0.303	0.0117	0.0663	0.0102	0.0117
60	285.3	1376.0	17.42	153.5	1.058	0.777	0.288	0.0119	0.0649	0.0105	0.0111

($T_{CRIT} = 456.94$ K, $P_{CRIT} = 36.74$ bar, $M = 152.93$)

Index